Quantitative Methods in Soil Mineralogy

Quantitative Methods in Soil Mineralogy

Proceedings of a symposium sponsored by Division S-9 of the Soil Science Society of America. The symposium was held in San Antonio, Texas on October 23-24, 1990.

Organizing Committee
James E. Amonette, Chair
Joseph W. Stucki

Editorial Committee
James E. Amonette, Chair
Lucian W. Zelazny

Editor-in-Chief SSSA
R. J. Luxmoore

Managing Editor
Jon M. Bartels

SSSA Miscellaneous Publication

Soil Science Society of America, Inc.
Madison, Wisconsin, USA

1994

Cover Design: James E. Amonette

Soil Science Society of America, Inc.
677 South Segoe Road, Madison, WI 53711 USA

Library of Congress Cataloging-in-Publication Data

Quantitative methods in soil mineralogy : proceedings of a symposium
  sponsored by Division S-9 of the Soil Science Society of America,
  San Antonio, Texas on October 23–24, 1990 / organizing
  committee, James E. Amonette, chair ; editorial committee, Lucian
  W. Zelazny ; editor-in-chief, SSSA, R.J. Luxmoore.
        p.   cm. — (SSSA miscellaneous publication).
  Includes bibliographical references and index.
  ISBN 0-89118-806-1
    1. Soil mineralogy—Methodology—Congresses. 2. Soils—
  Analysis—Congresses. I. Amonette, James E. II. Zelazny, Lucian
  W. III. Soil Science Society of America, Division S-9. IV. Series.
  S592.55.Q36   1994
  631.4/16—dc20   93-48098
                                                              CIP

Printed in the United States of America

CONTENTS

FOREWORD

Knowledge of the minerals comprising the inorganic component is essential to understanding the behavior of soils. Methods for characterizing soil inorganic constituents are changing rapidly as soil scientists apply new techniques to the study of minerals. Many of these techniques were developed by chemists and physicists to study "pure" systems but have proven highly useful in the characterization of soil minerals.

Given the rapid changes underway in adoption of new analytical techniques, there is a need for periodic review and assessment of the strengths and weaknesses of alternative methods for characterizing soil minerals. This review requires not only a technological assessment of the methods but a realistic look at the costs and availability of very costly and sophisticated instruments.

Division S-9 (Soil Mineralogy) of the Soil Science Society of America held a symposium on "Quantitative Methods in Soil Mineralogy" at the 1990 annual meeting of the Tri-Societies. This publication presents the papers presented at the symposium. The papers cover all of the methods currently used for characterization of soil minerals and provide some insights on ways that developing technologies may be used in the future. The information contained in the publication will be of value to any scientist interested in soil mineralogy.

DARREL W. NELSON, *president*
Soil Science Society of America

PREFACE

At some point, most soil scientists will have a need to characterize a soil before they can move forward to solve a particular problem. The size of this characterization effort depends on the problem to be solved and the resources available. Thus, characterization may involve one or two simple analyses or it may require highly precise and accurate determinations of many soil properties. Regardless of the extent of the characterization effort, careful judgment must be exercised by the responsible scientist to ensure that: (i) samples to be analyzed are collected appropriately and they adequately represent the soil problem, (ii) the appropriate specimen preparation and analytical techniques are selected, (iii) the specimen preparation and analyses are performed correctly, and (iv) the limitations of the analyses are fully recognized when their results are interpreted.

Because of their impact on the physical and chemical properties of a soil, a determination of the types and relative amounts of the minerals present forms an essential component of most soil characterization efforts. In some instances, mineralogical characterization may be restricted to a particular class of minerals, say Fe oxides or layer silicates, whereas in others a modal analysis of all minerals present is needed. In many instances, a measurement of the reactivity of the minerals present is more important than their absolute quantity. In every instance, however, a thorough knowledge on the part of the scientist of the physical and chemical principles that underly the techniques used will add to the probability of success in solving the problem at hand.

The intent of this special publication is to provide the reader with a solid understanding of current and emerging analytical techniques and how they are best applied to the quantitative characterization of soil minerals. In so doing, we hope it will serve as a reference for soil chemists, pedologists, and mineralogists, as well as an introduction for those persons who are new to the field of soil mineralogy. Application of these analytical techniques is expected to help solve problems in such diverse areas as environmental soil chemistry, soil genesis and classification, soil fertility, civil engineering, agricultural engineering, soil physics, and hydrology.

The chapters in this special publication are based on papers presented as part of a symposium sponsored by Division S-9 (Soil Mineralogy) and held 23 to 24 Oct. 1990 at the annual meetings of the Soil Science Society of America in San Antonio, Texas. Chapters 1 and 2 discuss the suite of elemental analysis techniques including x-ray emission and atomic absorption spectrometry, particle-beam techniques, and inductively coupled plasma instrumentation. Chapter 3 reviews oxidation-state analyses with a focus on methods for the determination of Fe oxidation states. Chapter 4 presents different methods for calculating the structural formulae of layer silicates from elemental and oxidation-state data. Chapters 5 and 6 are concerned with light and transmission-electron microscopic techniques including a discussion of computer-based image analysis. Chapter 7 presents an introduction to x-ray photoelectron spectroscopy, whereas Chapter 8 summarizes the different techniques for preconcentrating soil minerals to enhance identification and quantification. Chapters 9 through 11 are concerned with applications of x-ray diffraction to soils and include treatments of the Rietveld method, curve-fitting techniques, and approaches to the analysis of oxide-rich soils and phyllosilicates. The application of thermal techniques to quantification of soil minerals is reviewed in Chapter 12, whereas Chapters 13 and 14 discuss the combination of selective dissolution and other analytical techniques such as differential x-ray diffraction or infrared spectroscopy to quantify allophane, imogolite, and several oxyhydroxide and carbonate phases in soils. Each chapter presents enough theory to give a working understanding of the techniques described, with emphasis placed on recent advances. The advantages, limitations, and future development of the techniques are critically assessed, and specific examples of the application of the techniques to soils are presented.

The idea for the symposium originally came from Dr. Joseph W. Stucki, Division S-9 Chair in 1990, who was instrumental in finding the financial and organizational support needed. In addition to the hard work provided by the authors and the forbearance shown by their families, this publication benefited from the technical reviews provided by our scientific colleagues, many of whom remain anonymous, and from the expertise of the editorial staff of the Soil Science Society of America.

We hope the reader finds our presentation interesting, informative, and useful.

JAMES E. AMONETTE, *lead editor*
Earth and Environmental Sciences Center
Battelle, Pacific Northwest Laboratories, Richland, Washington

LUCIAN W. ZELAZNY, *editor*
Department of Crop and Soil Environmental Sciences
Virginia Polytechnic Institute and State University, Blacksburg, Virginia

CONTRIBUTORS

James E. Amonette   Senior Research Scientist, Earth and Environmental Sciences Center, Pacific Northwest Laboratory, P.O. Box 999, MSIN K6-82, Richland, WA 99352

David L. Bish   Technical Staff Member, EES-1, Geology and Geochemistry, MS D469, Los Alamos National Laboratory, Los Alamos, NM 87545

Blahoslav Číčel   Director, Institute of Inorganic Chemistry, Slovak Academy of Sciences, 842 36 Bratislava, Skovakia

David L. Cocke   Gill Professor of Analytical Chemistry, Lamar University, P.O. Box 10022, Beaumont, TX 77710

Randy A. Dahlgren   Assistant Professor of Soil Science, Department of Land, Air, and Water Resources, 151 Hoagland Hall, University of California at Davis, Davis, CA 95616-8627

Robert H. Dowdy   Research Leader, Soil and Water Management Research Unit, USDA-ARS-MWA, 439 Borlaug Hall, University of Minnesota, St. Paul, MN 55108

L. Richard Drees   Associate Research Scientist, Department of Soil and Crop Sciences, Texas A&M University, College Station, TX 77843-2474

Huamin Gan   Postdoctoral Research Associate, Department of Agronomy, University of Illinois at Urbana-Champaign, 1102 South Goodwin Avenue, Urbana, IL 61801

Robert J. Gilkes   Associate Professor, Department of Soil Science and Plant Nutrition, The University of Western Australia, Nedlands, Western Australia, Australia 6009

Herbert D. Glass   Geologist Emeritus, Clay Minerals Unit, Illinois State Geological Survey, 615 East Peabody Drive, Champaign, IL 61820

Willie G. Harris   Associate Professor of Soil Mineralogy, Soil and Water Science Department, University of Florida, P.O. Box 110290, Gainesville, FL 32611-0290

Randall E. Hughes   Geologist, Clay Minerals Unit, Illinois State Geological Survey, 615 East Peabody Drive, Champaign, IL 61820

Rolly C. Jones   Professor of Soil Mineralogy, Department of Agronomy and Soil Science, University of Hawaii at Manoa, 1910 East-West Road, Honolulu, HI 96822

Anastasios D. Karathanasis   Associate Professor, Department of Agronomy, N-122 Agricultural Science Building-North, University of Kentucky, Lexington, KY 40546-0091

Faruque A. Khan   Postdoctoral Research Associate, Department of Agronomy, Iowa State University, Ames, IA 50011

Peter Komadel   Head, Department of Hydrosilicates, Institute of Inorganic Chemistry, Slovak Academy of Sciences, 842 36 Bratislava, Slovakia

David A. Laird   Soil Scientist, National Soil Tilth Laboratory, USDA-ARS-MWA, 2150 Pammel Drive, Ames, IA 50011

Richard H. Loeppert   Professor of Soil Chemistry, Department of Soil and Crop Sciences, Texas A&M University, College Station, TX 77843-2474

Hameed U. Malik   Staff Engineer, Masa Fujioka and Associates, Honolulu, HI 96822

Dewey M. Moore   Geologist, Clay Minerals Unit, Illinois State Geological Survey, 615 East Peabody Drive, Champaign, IL 61820

Michel D. Ransom   Associate Professor of Soil Genesis, Classification and Mineralogy, Department of Agronomy, Kansas State University, Manhattan, KS 66506-5501

Ronald W. Sanders   Senior Technical Specialist, Materials and Chemical Sciences Center, Pacific Northwest Laboratory, P.O. Box 999, MSIN P8-08, Richland, WA 99352

Brij L. Sawhney   Soil Chemist, The Connecticut Agricultural Experiment Station, P.O. Box 1106, New Haven, CT 06504

Darrell G. Schulze   Associate Professor of Agronomy, Agronomy Department, Purdue University, 1150 Lilly Hall, West Lafayette, IN 47907-1150

A. Duncan Scott   Professor Emeritus of Soils, Department of Agronomy, Iowa State University, Ames, IA 50011

David E. Stilwell   Assistant Analytical Chemist, The Connecticut Agricultural Experiment Station, P.O. Box 1106, New Haven, CT 06504

Joseph W. Stucki   Professor of Soil Physical Chemistry, Department of Agronomy, University of Illinois at Urbana-Champaign, 1102 South Goodwin Avenue, Urbana, IL 61801

Rajan K. Vempati   Visiting Professor, Gill Chair of Analytical Chemistry, Lamar University, P.O. Box 10022, Beaumont, TX 77710

Conversion Factors for Si and non-SI Units

Conversion Factors for SI and non-SI Units

To convert Column 1 into Column 2, multiply by	Column 1 SI Unit	Column 2 non-SI Unit	To convert Column 2 into Column 1, multiply by
Length			
0.621	kilometer, km (10^3 m)	mile, mi	1.609
1.094	meter, m	yard, yd	0.914
3.28	meter, m	foot, ft	0.304
1.0	micrometer, μm (10^{-6} m)	micron, μ	1.0
3.94×10^{-2}	millimeter, mm (10^{-3} m)	inch, in	25.4
10	nanometer, nm (10^{-9} m)	Angstrom, Å	0.1
Area			
2.47	hectare, ha	acre	0.405
247	square kilometer, km² (10^3 m)²	acre	4.05×10^{-3}
0.386	square kilometer, km² (10^3 m)²	square mile, mi²	2.590
2.47×10^{-4}	square meter, m²	acre	4.05×10^3
10.76	square meter, m²	square foot, ft²	9.29×10^{-2}
1.55×10^{-3}	square millimeter, mm² (10^{-3} m)²	square inch, in²	645
Volume			
9.73×10^{-3}	cubic meter, m³	acre-inch	102.8
35.3	cubic meter, m³	cubic foot, ft³	2.83×10^{-2}
6.10×10^4	cubic meter, m³	cubic inch, in³	1.64×10^{-5}
2.84×10^{-2}	liter, L (10^{-3} m³)	bushel, bu	35.24
1.057	liter, L (10^{-3} m³)	quart (liquid), qt	0.946
3.53×10^{-2}	liter, L (10^{-3} m³)	cubic foot, ft³	28.3
0.265	liter, L (10^{-3} m³)	gallon	3.78
33.78	liter, L (10^{-3} m³)	ounce (fluid), oz	2.96×10^{-2}
2.11	liter, L (10^{-3} m³)	pint (fluid), pt	0.473

Mass

To convert Column 1 into Column 2, multiply by	Column 1 SI Unit	Column 2 non-SI Unit	To convert Column 2 into Column 1, multiply by
2.20×10^{-3}	gram, g (10^{-3} kg)	pound, lb	454
3.52×10^{-2}	gram, g (10^{-3} kg)	ounce (avdp), oz	28.4
2.205	kilogram, kg	pound, lb	0.454
0.01	kilogram, kg	quintal (metric), q	100
1.10×10^{-3}	kilogram, kg	ton (2000 lb), ton	907
1.102	megagram, Mg (tonne)	ton (U.S.), ton	0.907
1.102	tonne, t	ton (U.S.), ton	0.907

Yield and Rate

To convert Column 1 into Column 2, multiply by	Column 1 SI Unit	Column 2 non-SI Unit	To convert Column 2 into Column 1, multiply by
0.893	kilogram per hectare, kg ha^{-1}	pound per acre, lb acre^{-1}	1.12
7.77×10^{-2}	kilogram per cubic meter, kg m^{-3}	pound per bushel, lb bu^{-1}	12.87
1.49×10^{-2}	kilogram per hectare, kg ha^{-1}	bushel per acre, 60 lb	67.19
1.59×10^{-2}	kilogram per hectare, kg ha^{-1}	bushel per acre, 56 lb	62.71
1.86×10^{-2}	kilogram per hectare, kg ha^{-1}	bushel per acre, 48 lb	53.75
0.107	liter per hectare, L ha^{-1}	gallon per acre	9.35
893	tonnes per hectare, t ha^{-1}	pound per acre, lb acre^{-1}	1.12×10^{-3}
893	megagram per hectare, Mg ha^{-1}	pound per acre, lb acre^{-1}	1.12×10^{-3}
0.446	megagram per hectare, Mg ha^{-1}	ton (2000 lb) per acre, ton acre^{-1}	2.24
2.24	meter per second, m s^{-1}	mile per hour	0.447

Specific Surface

To convert Column 1 into Column 2, multiply by	Column 1 SI Unit	Column 2 non-SI Unit	To convert Column 2 into Column 1, multiply by
10	square meter per kilogram, m^2 kg^{-1}	square centimeter per gram, cm^2 g^{-1}	0.1
1000	square meter per kilogram, m^2 kg^{-1}	square millimeter per gram, mm^2 g^{-1}	0.001

Pressure

To convert Column 1 into Column 2, multiply by	Column 1 SI Unit	Column 2 non-SI Unit	To convert Column 2 into Column 1, multiply by
9.90	megapascal, MPa (10^6 Pa)	atmosphere	0.101
10	megapascal, MPa (10^6 Pa)	bar	0.1
1.00	megagram per cubic meter, Mg m^{-3}	gram per cubic centimeter, g cm^{-3}	1.00
2.09×10^{-2}	pascal, Pa	pound per square foot, lb ft^{-2}	47.9
1.45×10^{-4}	pascal, Pa	pound per square inch, lb in^{-2}	6.90×10^3

(continued on next page)

Conversion Factors for SI and non-SI Units

To convert Column 1 into Column 2, multiply by	Column 1 SI Unit	Column 2 non-SI Unit	To convert Column 2 into Column 1, multiply by
Temperature			
$1.00\ (K - 273)$	Kelvin, K	Celsius, °C	$1.00\ (°C + 273)$
$(9/5\ °C) + 32$	Celsius, °C	Fahrenheit, °F	$5/9\ (°F - 32)$
Energy, Work, Quantity of Heat			
9.52×10^{-4}	joule, J	British thermal unit, Btu	1.05×10^{3}
0.239	joule, J	calorie, cal	4.19
10^{7}	joule, J	erg	10^{-7}
0.735	joule, J	foot-pound	1.36
2.387×10^{-5}	joule per square meter, J m^{-2}	calorie per square centimeter (langley)	4.19×10^{4}
10^{5}	newton, N	dyne	10^{-5}
1.43×10^{-3}	watt per square meter, W m^{-2}	calorie per square centimeter minute (irradiance), cal cm^{-2} min^{-1}	698
Transpiration and Photosynthesis			
3.60×10^{-2}	milligram per square meter second, mg m^{-2} s^{-1}	gram per square decimeter hour, g dm^{-2} h^{-1}	27.8
5.56×10^{-3}	milligram (H$_2$O) per square meter second, mg m^{-2} s^{-1}	micromole (H$_2$O) per square centimeter meter second, µmol cm^{-2} s^{-1}	180
10^{-4}	milligram per square meter second, mg m^{-2} s^{-1}	milligram per square centimeter second, mg cm^{-2} s^{-1}	10^{4}
35.97	milligram per square meter second, mg m^{-2} s^{-1}	milligram per square decimeter hour, mg dm^{-2} h^{-1}	2.78×10^{-2}
Plane Angle			
57.3	radian, rad	degrees (angle), °	1.75×10^{-2}

Electrical Conductivity, Electricity, and Magnetism

Column 1 SI Unit	Column 2 non-SI Unit	To convert Col. 1 into Col. 2, multiply by	To convert Col. 2 into Col. 1, multiply by
siemen per meter, S m^{-1}	millimho per centimeter, mmho cm^{-1}	10	0.1
tesla, T	gauss, G	10^4	10^{-4}

Water Measurement

Column 1 SI Unit	Column 2 non-SI Unit	To convert Col. 1 into Col. 2, multiply by	To convert Col. 2 into Col. 1, multiply by
cubic meter, m^3	acre-inches, acre-in	9.73×10^{-3}	102.8
cubic meter per hour, m^3 h^{-1}	cubic feet per second, ft^3 s^{-1}	9.81×10^{-3}	101.9
cubic meter per hour, m^3 h^{-1}	U.S. gallons per minute, gal min^{-1}	4.40	0.227
hectare-meters, ha-m	acre-feet, acre-ft	8.11	0.123
hectare-meters, ha-m	acre-inches, acre-in	97.28	1.03×10^{-2}
hectare-centimeters, ha-cm	acre-feet, acre-ft	8.1×10^{-2}	12.33

Concentrations

Column 1 SI Unit	Column 2 non-SI Unit	To convert Col. 1 into Col. 2, multiply by	To convert Col. 2 into Col. 1, multiply by
centimole per kilogram, cmol kg^{-1} (ion exchange capacity)	milliequivalents per 100 grams, meq 100 g^{-1}	1	1
gram per kilogram, g kg^{-1}	percent, %	0.1	10
milligram per kilogram, mg kg^{-1}	parts per million, ppm	1	1

Radioactivity

Column 1 SI Unit	Column 2 non-SI Unit	To convert Col. 1 into Col. 2, multiply by	To convert Col. 2 into Col. 1, multiply by
becquerel, Bq	curie, Ci	2.7×10^{-11}	3.7×10^{10}
becquerel per kilogram, Bq kg^{-1}	picocurie per gram, pCi g^{-1}	2.7×10^{-2}	37
gray, Gy (absorbed dose)	rad, rd	100	0.01
sievert, Sv (equivalent dose)	rem (roentgen equivalent man)	100	0.01

Plant Nutrient Conversion

Elemental	Oxide	To convert Col. 1 into Col. 2, multiply by	To convert Col. 2 into Col. 1, multiply by
P	P$_2$O$_5$	2.29	0.437
K	K$_2$O	1.20	0.830
Ca	CaO	1.39	0.715
Mg	MgO	1.66	0.602

1   Nondestructive Techniques for Bulk Elemental Analysis

J. E. Amonette and R. W. Sanders
Pacific Northwest Laboratory
Richland, Washington

Determination of the elemental composition of a soil sample is one of the steps usually taken in the process of quantifying the mineral phases present. Although knowledge of the amounts of each element present does not necessarily lead directly to quantification of the mineral phases present, such knowledge can eliminate some phases from consideration, help identify possible solid-solution and trace phases, and add confidence to the identification and quantification of phases by other techniques. The intensities of x-ray reflections, for example, depend on the types of atoms present as well as the relative locations of these atoms in crystallite structures. Similarly, knowledge of the elemental composition can help in the assignment of bands in the infrared spectrum of a specimen. By itself, however, bulk elemental analysis cannot directly distinguish among oxidation states [e.g., Fe(II) and Fe(III); Mn(II), Mn(III), and Mn(IV)] nor among polymorphs, i.e., minerals such as goethite, lepidocrocite, and akaganeite that have different structures but share the same elemental stoichiometry. Thus, although determination of the elemental composition may be an important step in the mineralogical characterization of a soil, the characterization process generally requires information from several complementary types of analysis (i.e., elemental, structural, physical) before it can be considered complete.

Techniques for elemental analysis may be categorized on the basis of whether or not the technique requires the complete decomposition of the specimen into ionic and/or elemental constituents before analysis. Thus, techniques that involve the decomposition of specimens in strong acids (e.g., HF-H_2SO_4) or in alkaline fluxes (e.g., $NaOH$, $Na_2O_2 \cdot LiBO_2$) or that involve the nebulization and injection of solutions or suspensions into flames, plasmas, or other sampling devices [e.g., atomic absorption spectrometry (AAS), inductively coupled plasma atomic emission spectrometry (ICP-AES), mass spectrometry] are considered to be sample destructive. On the other hand, nondestructive techniques generally obtain elemental information by direct analysis of the specimen, require a minimum of specimen preparation, and, in principle, allow retention of the same unaltered specimen for other analyses. Nondestructive elemental techniques typically involve exposing the specimen to photon- or particle beams and measuring the energies, intensities, and directions of the photons or particles emanating from the specimen as a result of exposure. In many instances, the exciting photons and particles or the emitted photons or particles will have low enough transmissivities that only the atoms very near the specimen surface (i.e., within 1 μm) contribute to the observed signal. Thus, methods for the analysis of surfaces [e.g., x-ray photoelectron spectroscopy (XPS), Auger emission spectroscopy (AES), Rutherford backscattering (RBS), and electron-induced x-ray emission (EDX)] are, almost by definition, nondestructive techniques. The ability to analyze much larger volumes (i.e., greater depths) of the specimen is required for bulk analysis, and nondestructive bulk techniques rely on high-transmissivity photons and particles for both excitation and detection. Because the depth analyzed will depend on the nature of the material as well as the energy of the photon or particle involved, nondestructive techniques for bulk analysis generally require the use of elaborate matrix correction algorithms for accurate estimates of elemental composition.

This chapter focuses on nondestructive techniques that allow rapid multielemental analysis of bulk samples for the purposes of soil mineralogical characterization. The techniques considered include x-ray fluorescence (XRF) spectroscopy, which is at a relatively mature stage of development; two particle-beam methods that have seen extensive development over the last 20 yr [particle-induced x-ray emission (PIXE) spectroscopy and particle-induced gamma emission (PIGE) spectroscopy]; and synchrotron radiation-induced x-ray emission (SRIXE) spectroscopy, a promising technique whose application to the characterization of soils and sediments is still in its infancy. Three of these techniques (XRF, PIXE, and SRIXE) rely on the emission of x-rays by elements in the specimen and differ mainly in the manner by which these x-rays are induced. The fourth technique, PIGE spectroscopy, is based on a different spectroscopic process (interactions with the nucleus rather than with inner-shell electrons) but uses particle-beam energies and experimental set-ups that are essentially the same as in PIXE spectroscopy. The chapter begins with a general review of the fundamental spectroscopic processes, excitation sources, detector technologies, and specimen preparation methods that form the basis for the analytical techniques. Then, specific attention is given to each of the nondestructive techniques, followed by a general comparison of their analytical strengths and weaknesses when they are used to characterize soils and sediments. The chapter concludes with a discussion of future trends in nondestructive bulk elemental analysis.

SPECTROSCOPIC FUNDAMENTALS

Physics

In general, all the nondestructive analytical techniques rely on excitation of atoms in the specimen to higher-energy electronic or nuclear states. The relaxation of an atom from its excited state to a lower-energy state carries a certain probability that photons or particles having energies characteristic to the atom will be emitted. The number of characteristic photons or particles detected is directly proportional to the number of atoms present in the specimen, after correction for the intensity of the excitation source, the excitation and emission probabilities for the atom, the transmissivities in the specimen of the exciting photon or particle and the emitted photon or particle, and the detection efficiency. Because the transmissivities of photons are generally much greater than those of particles and thus can be used to sample larger volumes (i.e., greater depths below the specimen surface), photon detection is used almost exclusively for these nondestructive bulk analytical techniques.

The relaxation of atoms from higher-energy *electronic* states results in a wide range of emitted photon or particle energies, which are typically expressed in terms of electron volts (eV). One electron volt is equal to the amount of energy acquired by an electron falling through a potential difference of 1 V (1.602×10^{-19} J). Electronic transitions between energy levels characteristic of outer-shell electrons are generally of low energy (< 10 eV) and give rise to photons in the visible and ultraviolet regions of the electromagnetic spectrum. These photon energies and transition probabilities depend on the oxidation state and bonding environment of the atom, due to the role of outer-shell electrons in electron-transfer processes and molecular orbital formation. Because of these complexities and the overlapping energy levels for outer-shell electrons in different atoms, outer-shell electronic transitions are not generally used for analyses in which only the elemental composition of a sample is desired. The energies of inner-shell electronic transitions, on the other hand, are much larger (~ 1 to 115 keV) and, except for atoms having few electrons, are essentially independent of the bonding environment and oxidation state of the atom. These transitions, therefore, are very useful for elemental analysis techniques in which bonding and oxidation state information are not needed. The photons generated by inner-shell electronic transitions are termed x-rays and form the basis for elemental analysis by XRF, PIXE, and SRIXE spectroscopies.

The "relaxation" of atoms from higher-energy *nuclear* states may also result in the emission of particles or photons having characteristic energies. These emission processes generally proceed as a result of a nuclear reaction in which a high-energy particle or photon strikes the nucleus of an atom and induces the higher-energy nuclear state. Because the nucleus of the atom is involved rather than the electrons, the probabilities and energetics of photon emission are dependent on the composition of the nucleus (i.e., the number of protons and neutrons) and thus are isotope specific. For example, ^{10}B reacts with protons having more than 2.6 MeV of energy to yield an alpha particle, ^{7}Be, and a 429-keV photon, whereas ^{11}B reacts with protons of the same energy to yield a backscattered proton, ^{11}B, and a 2125-keV photon. Although there is some overlap in energy, the photons produced by nuclear reactions generally have energies much higher than atomic x-rays, ranging up to several megavolts. These photons are distinguished from x-rays by their mode of origin and are termed gamma rays. Production of "prompt" gamma rays (i.e., those gamma rays emitted by relaxation processes occurring immediately after excitation of the nucleus) by bombardment of the specimen with high-energy protons forms the basis for PIGE spectroscopy.

X-Ray Production

Electronic energies in atoms are quantized at discrete levels referred to as shells or orbitals. The reference level for electron energies is usually taken as the energy of an unbound or free electron. The electrons having the lowest energy relative to a free electron are those bound most tightly to the nucleus and occupying the innermost shell. This shell is designated the K shell using Siegbahn notation and contains two 1s electrons. Higher-energy shells are designated L, M, N, etc. and may be split into subshells to distinguish the s, p, d, and f electronic energy levels. For example, the L shell may contain up to eight electrons, two 2s electrons in the L_I subshell, two $2p^{1/2}$ electrons in the L_{II} subshell, and four $2p^{3/2}$ electrons in the L_{III} subshell.

The most stable electronic configuration for an atom (i.e., the ground state) is one in which the lowest-energy shells are filled preferentially to higher levels. If an electron vacancy occurs in one of the shells that is normally filled in the ground state and electrons in higher-energy shells are present, the atom is in an excited state and will relax to a lower-energy state by an electronic transition, i.e., by moving electrons from higher-energy shells to lower-energy shells until vacancies exist only in the highest-energy shells. Each electronic transition results in the release of a photon that is characteristic to the atom and has an energy equal to the difference between the energies of the initial and final shells occupied by the electron.

The production of characteristic x-rays, therefore, is a two-step process involving (i) the creation of an inner-shell electron vacancy (ionization), and (ii) the filling of that vacancy by a higher-energy electron coincident with the release of an x-ray photon. To create a vacancy, the atom must receive a parcel of energy, in the form of either a photon or a particle, that is sufficiently large to overcome the binding energy of the inner-shell electron (Fig. 1-1 top left and right). The inner-shell electron that is removed by this process is emitted from the atom as a secondary electron, whereas the primary ionizing photon or particle loses energy and is scattered. Filling of the vacancy by an electron from a higher-energy shell proceeds via certain selection rules and may result in the emission of an x-ray (Fig. 1-1 bottom left). In some instances, however, this x-ray in turn creates a vacancy in a higher-energy shell instead of escaping from the atom (Fig. 1-1 bottom right). The secondary electron produced by this process has a distinct energy (because the energy of the ionizing photon and the initial energy of the electron are fixed) and is termed an Auger (pronounced oh-zhay') electron. The measurement of Auger-electron energies and yields is a standard analytical technique (Auger spectroscopy) for the elemental characterization of surfaces (Carlson, 1975; Briggs, 1983). The poor transmissivity of electrons in solid matrices, however, precludes the use of Auger spectroscopy for bulk analysis.

The probability that an inner-shell vacancy will be created when an atom interacts with a high-energy photon or particle is referred to as the ionization cross section (σ_i) and has units of barns (1 barn is equal to a cross-sectional area of 10^{-24} cm^2 and corresponds to a probability of 1 in 10^{24} that ionization will occur when a 1-cm^2 cross section of the specimen is irradiated by a single photon or particle). Cross sections for the ionization of inner-shell electrons by photons or particles of various energies are known to within 5% and have been tabulated for essentially all the elements (Veigele, 1973; Scofield, 1973; Chen & Craseman, 1985; Cohen & Harrigan, 1985; Berger & Hubbell, 1987; Paul & Sacher, 1989).

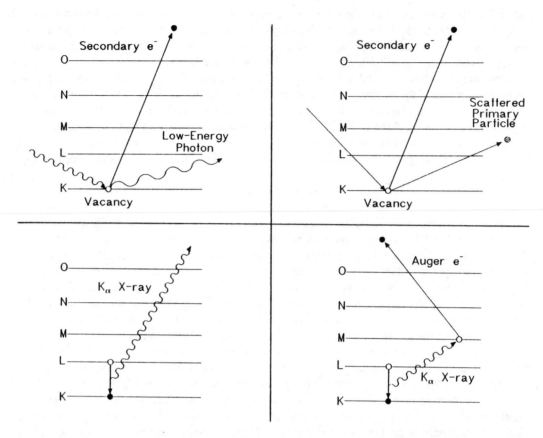

Fig. 1-1. Processes associated with the production of x-rays: creation of K-shell vacancy by a high-energy photon (top left); creation of K-shell vacancy by a high-energy particle (top right); emission of a Kα x-ray by an electronic transition from the L-shell to the K-shell (bottom left); ejection of an Auger electron by the Kα x-ray (bottom right).

Once a vacancy has been created, the probability that the electronic transition that fills the vacancy will result in the production of an x-ray is known as the fluorescence yield ($\omega_{K, L, or M}$, where the subscript refers to the atomic shell in which the vacancy occurs). A similar term, Auger yield, refers to the probability that an Auger electron will be emitted during the transition. Because the only two possible outcomes to the electronic transition are the emission of a photon or an Auger electron, the sum of the fluorescence and Auger yields for a single vacancy in an atomic shell is always equal to one, regardless of the manner by which the vacancy is produced. The fluorescence yields for the filling of K and L shell vacancies generally increase as the atomic number increases (Krause, 1979; Bambynek, 1984; Cohen, 1987), although L-shell fluorescence yields are typically much lower than K-shell yields (Fig. 1-2). For example, K-shell fluorescence yields range from about 2% for Na and 35% for Fe to near 98% for W, whereas the corresponding L-shell fluorescence yields are about 0.02, 0.5, and 30%.

As mentioned earlier, certain selection rules govern the electronic transitions that are responsible for filling vacancies. Transitions observed for the filling of K-shell vacancies in Zr atoms are shown in Fig. 1-3, along with the selection rules and principal, angular-momentum, and inner quantum numbers for each electronic energy level. The majority of the transitions follow the selection rules based on these quantum numbers, but a couple of very weak "forbidden" transitions (β_4 and β_5) are observed. Figure 1-3 also shows the relationship between the Siegbahn notation for electronic energy levels that was developed by the early x-ray spectroscopists and the modern spectroscopic notation developed by quantum chemists. Siegbahn assignments of subscripts to describe transitions to fill K-shell vacancies (e.g., $K\alpha_1$, $K\beta_1$) were initially based on the intensity of the x-ray peaks observed. Later, as better techniques and quantum mechanics revealed

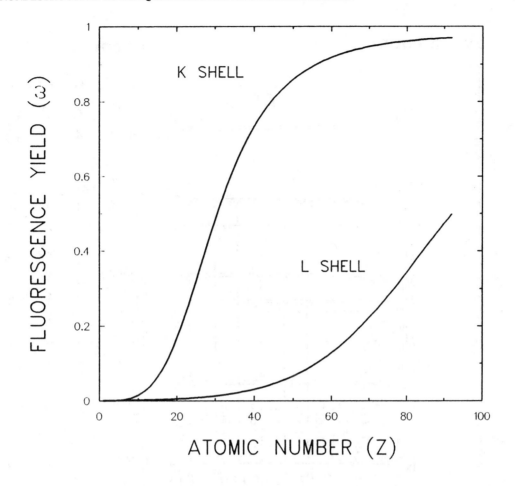

Fig. 1-2. Fluorescence yield values for K-shell and L-shell ionization processes of the naturally occurring elements (Bambynek, 1984).

previously hidden peaks, additional subscripts and superscripts were added to the Siegbahn notation in the order of discovery. Thus, in comparison with the modern spectroscopic notation, Siegbahn notation is neither as logical nor as descriptive of the actual energy levels involved in the transition. However, Siegbahn notation is certainly less cumbersome (e.g., it is much easier to say $K\alpha_1$ than $2p^{3/2} - 1s^{1/2}$), and this probably accounts for its continued use.

The observed energies and relative intensities of the x-rays arising from the Zr K-shell transitions are shown in Fig. 1-4. Clearly, the transitions of electrons from the L-shell (i.e., the $K\alpha$ transitions) are highly favored, whereas those from higher-energy shells ($K\beta$ transitions) are not nearly as intense. This result stems from differences in the degree of overlap between the K-shell electronic wave function and the wave functions for the outer-shell electrons. As the energy of the outer-shell electrons increase relative to the K-shell electrons, the wave-function overlap becomes smaller and lowers the transition probability. Intuitively, the relative transition probabilities can be described in terms of chemical reaction kinetics and thermodynamics. That is, a significant "activation energy" exists for transitions between shells (Fig. 1-5 top) and the total activation energy required for a transition from an N to a K shell is enough to limit the probability of such a transition to only 7% of that for an L to K transition (Fig. 1-5 bottom), even though the total energy change is 15% greater. Thus, placed in the context of chemical kinetics and thermodynamics, the reaction probability depends on both the total free-energy change *and* on the height of the activation-energy barrier.

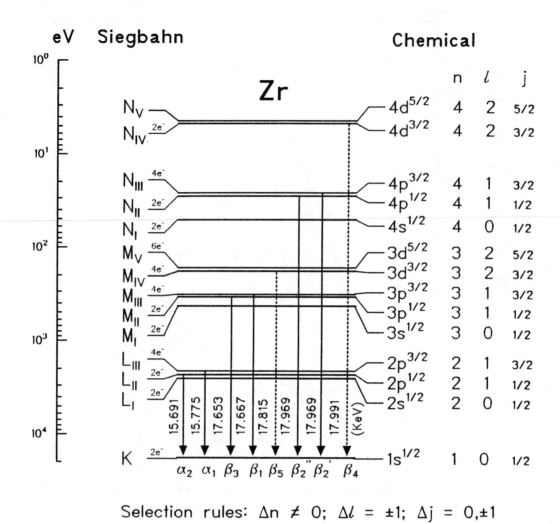

Fig. 1-3. Electronic structure of the Zr atom showing the atomic energy levels (Siegbahn and chemical notation), principle (n), angular-momentum (*l*), and inner (j) quantum numbers for each orbital, and the observed electronic transitions and predicted selection rules for filling of a K-shell vacancy and production of K x-rays.

The relative probabilities of the different K-shell transitions are known as the K-shell branching ratio. For Zr, the branching ratio would be 100:30:7 for $K\alpha$, $K\beta_{1,3}$, and $K\beta_2$ transitions (Fig. 1-5 bottom). Branching ratios for K-shell, L-shell, and M-shell transitions in the pure elements are reasonably well known (Salem et al., 1974; Scofield, 1974; Chen et al., 1981; Cohen & Harrigan, 1986) and, as for the fluorescence yields, are independent of the manner in which the vacancy was created. These branching ratios may vary slightly with the oxidation state and bonding environment of the atom.

From knowledge of the ionization cross section, the fluorescence yield, and the branching ratios, it is possible to calculate an x-ray production cross section for each x-ray line observed, by

$$\sigma_x(E) = k\ \omega_{K, L, \text{or } M}\ \sigma_i(E) \qquad [1]$$

where k is the branching ratio for the x-ray in transitions to a particular shell, $\omega_{K, L, \text{or } M}$ the overall fluorescence yield for the transitions to a particular shell, and $\sigma_i(E)$ the ionization cross section or probability of a vacancy being created by photons or particles of energy E. The units for $\sigma_x(E)$ are barns, the same as for $\sigma_i(E)$. Thus, for a single atom, it is possible to calculate with very

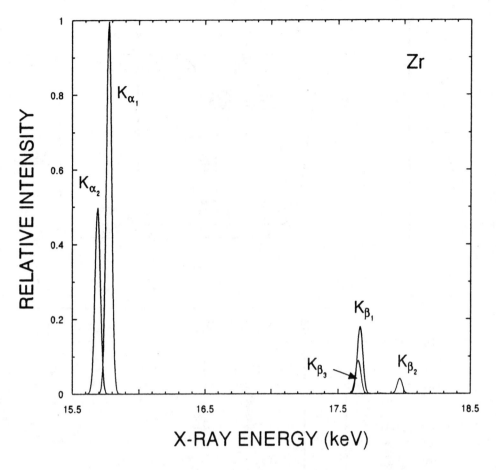

Fig. 1-4. Relative intensities and energies of Zr K x-rays.

good accuracy the absolute number of x-rays produced by each possible transition if the intensity and energy of the exciting photon or particle beam are known.

Interaction of X-Rays with Matter

When x-rays of a given energy pass through matter, they lose intensity as a result of scattering and of other interactions with electrons (Fig. 1-6). They may be scattered coherently (i.e., with no loss in energy) or incoherently (some energy is imparted to the atoms by which they are scattered). Coherent scattering forms the basis for structural studies of matter by x-ray diffraction. Because x-rays are also sufficiently energetic to induce ionization of the atoms present, they lose energy and cause the emission of secondary electrons, Auger electrons, and photons with energies characteristic of the atoms ionized. This latter set of emission processes is termed the photoelectric effect and is the dominant contributor to losses in x-ray intensity.

The total loss in intensity caused by scattering and the photoelectric effect is summarized in a factor known as the mass attenuation coefficient, $\mu(E)$, which can be calculated using an expression that is analogous to the Beer-Lambert law for absorption of light by a solution,

$$I(E) = I_0(E)\exp[-\mu(E)\rho x_i] \qquad [2]$$

where $I_0(E)$ is the incident intensity of the x-rays having energy E, and $I(E)$ is the intensity after passage through matter having a thickness of x_i, a density of ρ, and a mass attenuation coefficient equal to $\mu(E)$. Values of $\mu(E)$ for different elements and x-ray energies have been measured and

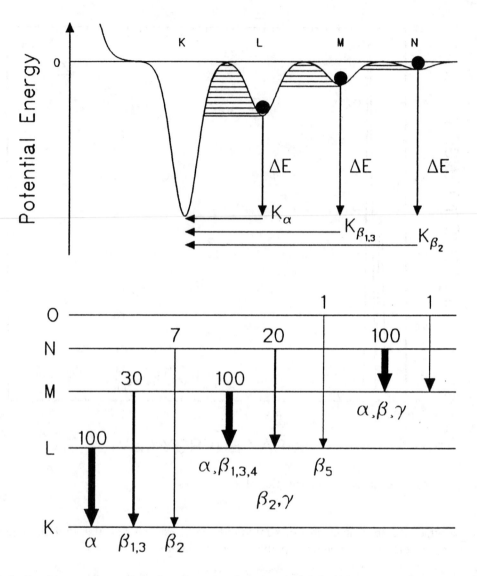

Fig. 1-5. Intuitive representation of total energy change and potential energy barriers encountered by Zr L-, M-, and N-shell electrons in transitions to fill a K-shell vacancy (top) and the relative probabilities of transitions to fill K-, L-, and M-shell vacancies giving rise to Zr K, L, and M x-rays (bottom).

tabulated (Thiesen & Vollath, 1967; Leroux & Thinh, 1977; Mitchell & Ziegler, 1977; Thinh & Leroux, 1979; Berger & Hubbell, 1987). Thus, if the elemental composition of a homogeneous substance is known, one can calculate directly (by weighted addition of the $\mu\rho$ products) the attenuation of the x-ray as it passes through a known thickness of the substance. In practice, it is the elemental composition that we seek to quantify; estimates of the attenuation of the x-ray as it passes through the specimen must be made and then refined with elemental composition data in an iterative fashion until the calculated and observed attenuation coefficients agree.

As shown by an example for a Zr specimen (Fig. 1-7), low-energy x-rays are attenuated much more readily than high-energy x-rays. At x-ray energies just above the absorption edge for a given subshell transition (i.e., the minimum energy needed to create an electron vacancy in the subshell), the mass attenuation coefficient is 10 to 20 times greater than at energies just below the absorption edge. Along with absorption of the exciting x-ray comes emission by the absorbing element of a characteristic x-ray having a lower energy than the exciting x-ray (Fig. 1-6). If the specimen

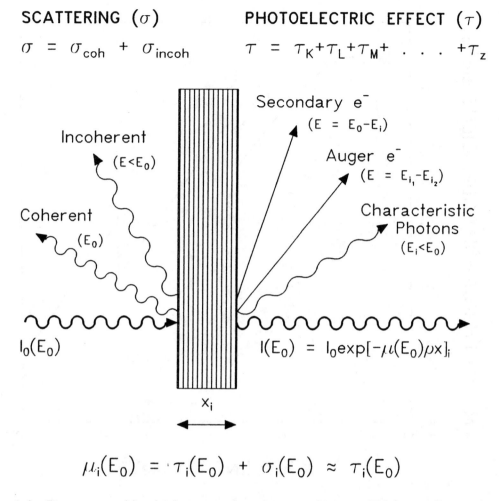

Fig. 1-6. Processes resulting in the attenuation of x-rays of energy (E_0) by a substance i, of thickness x_i and density ρ, and incorporated into the mass attenuation coefficient, $\mu_i(E_0)$.

contains a second element whose absorption edge is just below the energy of the x-ray emitted by the first element, reabsorption of the emitted x-ray and emission of a second x-ray of even lower energy may occur. This phenomenon is termed secondary fluorescence or enhancement and, if not considered in the quantification procedure, can lead to erroneously low values for the concentration of the first element and erroneously high values for the concentration of the second. Although secondary fluorescence is most pronounced in samples dominated by transition metals (e.g., Ni-Fe-Cr alloys, where corrections on the order of 20 to 50% are needed to give the true composition), it can also be seen in geological specimens dominated by Ca and S, or Si and Al, where the corrections are on the order of a few percentage points at most (Campbell et al., 1989; Reuter et al., 1975).

Absolute Yield Equations

If the elemental composition of a specimen is known, it is possible to calculate, absolutely, the x-ray or gamma-ray yield for each element in the specimen when it is placed in an x-ray or particle beam of known energy and intensity. Likewise, it is possible to work backwards from an observed spectrum and calculate the elemental composition of the specimen. These calculations are possible because the key processes affecting x-ray and gamma-ray production and transmission in matter

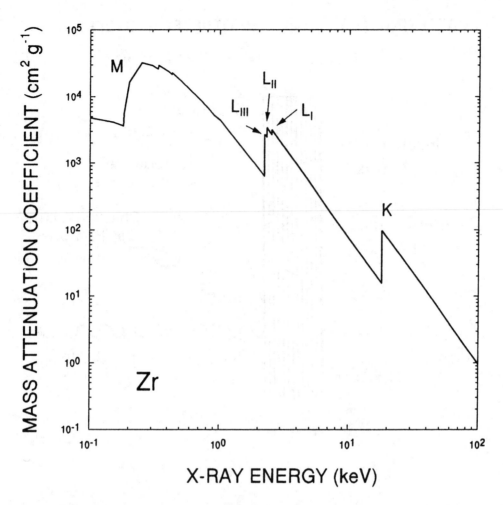

Fig. 1-7. Variation in the mass attenuation coefficient for Zr with incident x-ray energy showing the sharp increases for x-rays having energies just above the binding energies (i.e., the absorption edges) for the M-, L-, and K-shell electrons.

are well understood and the databases to support the calculations are available. The parameters and processes to be considered in an absolute calculation of yield include:

1. The incident energy and intensity of the exciting x-ray or particle beam.
2. The process by which the incident beam loses energy or intensity as it penetrates the specimen, i.e., the attenuating or stopping power of the specimen with respect to the incident x-ray or particle beam.
3. The x-ray or gamma-ray excitation efficiency of the incident beam as it encounters the analyte at different depths in the specimen, i.e., the attenuation-corrected x-ray or gamma-ray cross section for the analyte.
4. The concentration of the analyte in the specimen.
5. The attenuation of the x-ray or gamma-ray generated by the analyte in the specimen before it escapes the specimen.
6. The fraction of the total x-rays or gamma rays generated by the analyte that enter the detector, i.e., the detector solid angle.
7. The absolute efficiency of the detector and any filters between the detector and the specimen.

These fundamental parameters can be combined in a yield equation that has the general form

$$Y_z = Q_{exc} C_z D_{eff} \int_0^x \frac{\sigma(E_{exc}) T(E_z, x)}{S(E_{exc}, x)} \, \delta x \qquad [3]$$

where Y_Z is the x-ray or gamma-ray yield for a specific transition Z, Q_{exc} is the total number of photons or particles incident on the specimen, C_Z is the concentration of the analyte, D_{eff} is the detection efficiency (including the absolute efficiency of the detector, filters, and the solid angle), $\sigma(E_{exc})$ is the x-ray or gamma-ray production cross section for the analyte by the incident beam of energy E_{exc}, $T(E_Z, x)$ is the transmissivity of the analytical x-ray or gamma-ray of energy E_Z produced in the specimen as a function of depth x, and $S(E_{exc}, x)$ is the stopping power of the specimen for the incident beam. Equation [3] assumes that the specimen can be divided into a series of layers parallel to the surface, each having a thickness of δx. The terms within the integral estimate: (i) the attenuation of the incoming photon intensity or particle energy by all the layers between a given layer and the specimen surface, (ii) the probability of analytical photons being produced within the layer, and (iii) the attenuation of the analytical photon intensity by the specimen. The net result of the integral, therefore, is the probability that an analytical photon will leave the specimen. This result is then scaled by the detection efficiency, the concentration of analyte atoms present in the specimen, and the number of photons or particles incident on the specimen to give an analytical photon yield. The equation can be rearranged to solve for analyte concentration once the yield has been measured. However, the two attenuation parameters, $T(E_Z, x)$ and $S(E_{exc}, x)$, depend on the concentrations of the major constituents in the specimen (i.e., the matrix, which may or may not include the analyte). Final solution of the equation therefore must follow an iterative process, whereby an initial guess is made of the specimen matrix composition; this guess is used in the initial calculation of analyte and matrix composition based on the observed photon yields, and then the calculation is repeated using the results of the previous calculation to estimate the specimen matrix composition until the estimated and calculated matrix compositions agree within a certain tolerance.

Excitation Sources

The generation of x-rays or gamma-rays in the specimen for analytical purposes relies on a stable source of high-energy particles or photons that can be focused on the specimen. All the sources described below (with the exception of the radioactive source) begin with the acceleration of charged particles in a vacuum to create the necessary energy for x-ray or gamma-ray production.

Conventional X-Ray Tubes

For conventional XRF spectroscopy, the source of high-energy x-rays is a Coolidge tube focused either directly on the specimen or on a secondary target that produces x-rays to excite the specimen (Fig. 1-8). Primary radiation is generated in the tube by a stream of high-energy electrons that strike a metal anode, creating vacancies in the inner shells of the atoms that make up the anode. As discussed above, the filling of these vacancies by electrons from higher-energy shells causes the emission of x-rays having energies characteristic of the anode atoms. Most of the radiation produced, however, is in the form of bremsstrahlung, i.e., photons emitted by the electrons as they decelerate on striking the anode. The high-energy electrons are produced by passing a small current (up to 125 mA) through a tungsten filament and applying a large potential difference (up to 75 kV) between the filament and the anode. Thus, electrons emitted by the filament are accelerated by the electric field to energies as high as 75 keV before striking the anode. Because of the large amount of energy absorbed by the anode, it heats up rapidly and must be cooled by flowing chilled water through channels bored into the metal to maintain efficient x-ray production.

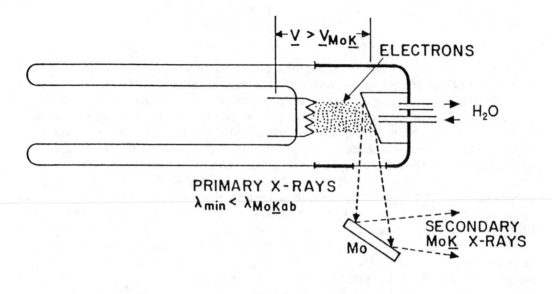

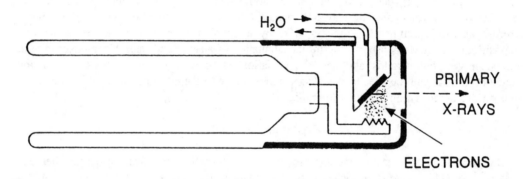

Fig. 1-8. Schematic of side-window (top) and end-window (bottom) Coolidge-type x-ray tubes.
Radiation from the side-window tube is being used to excite a secondary target to produce Mo
K-shell x-rays (after Bertin, 1975, 1981).

Primary radiation generated at the anode then passes through a thin (typically 75-125 μm) Be
window either directly onto the specimen or onto a secondary target outside the tube whose
absorption edge is below the energy of the primary x-rays. In the latter instance, x-rays
characteristic of the secondary target material are generated and, as a consequence of the
photoelectric effect, it is these x-rays that are focused on the specimen for analytical purposes. In
some instances, the same primary x-ray tube can be used with several different secondary targets
(e.g., a W x-ray tube can be used to excite Ti, Zr, Ag, and Gd secondary targets) for efficient and
cost-effective generation of x-rays across a wide range in energy.

Radioactive Sources

The decay of radioactive elements can result in the release of gamma rays having energies in
the x-ray region. If the rate of decay is sufficiently high, a planchet of radioactive material can
be used directly as a source of high-energy x-rays. One radioactive element that has been used
with success is [241]Am, which emits gamma rays of energy 59.6 keV and has a half-life of about
458 yr. A 100-m Ci source of [241]Am is quite adequate for XRF analysis of the K-shell transitions
of heavy elements (e.g., Cs and Ba) and most of the lanthanides. Other radioisotopes that have

been used include ^{55}Fe (5.9 keV), ^{109}Cl (19 and 87.8 keV), ^{57}Co (122-137 keV), and ^{155}Eu (71-77 keV) (Jenkins, 1974, p. 60).

Particle Beams

Several types of particle accelerators can be used to generate beams of high-energy protons, deuterons, and alpha particles for PIXE and PIGE spectroscopy. Of these, the Van de Graaff accelerator is probably the most widely used source, because it yields particles having energies up to 8 MeV (higher in some instances) in a stable, well-defined beam and is easily reconfigured for different particles or energies (Fig. 1-9). The high energies imparted to the particles are created by their acceleration across a potential difference created by forcing charge onto a high-voltage terminal using a rubber-belt pulley or an insulator/conductor charging chain. Charged particles (protons, deuterons, alpha particles) initially at the terminal voltage are formed by ionization of a source gas (H_2O, D_2O, or He) in a vacuum and then accelerated to ground and focused by passage through a tube consisting of a series of ring-shaped resistors (~ 50 MΩ each). The resistors create a voltage divider between the high-voltage terminal and ground and help focus the beam by providing equiplanar potentials perpendicular to the path of the particles. The high-energy particles then pass into a manifold where they are steered by a bending magnet through a narrow slit into the appropriate beam line. To prevent unplanned discharges of the high-voltage terminal, it is surrounded by an insulating gas (typically a mixture of compressed N_2, CO_2, and SF_6) and encased in a large pressurized tank. The ion source, the inside of the acceleration tube, and the inside of the beamlines, on the other hand are maintained at very high vacuum ($\sim 10^{-6}$ torr) to maintain the energy of the particles until they strike the specimen.

Protons are the most common particle used for x-ray emission analysis because they penetrate relatively deeply into the specimen, do not heat or damage the specimen as much as heavier particles, yield relatively low amounts of bremsstrahlung (background radiation caused by deceleration of charged particles), and have reasonable cross sections for x-ray and gamma-ray production. In addition to being heavier, deuterons have the side effect of generating neutrons when they collide with light elements and thus require additional shielding.

Synchrotron X-Rays

When accelerated at speeds that begin to approach the speed of light, charged particles will emit light, the energy of which is proportional to the acceleration. This property, originally perceived as a nuisance by physicists operating high-energy particle accelerators, soon was recognized as a valuable source of high-intensity x-rays for experimental and analytical purposes.

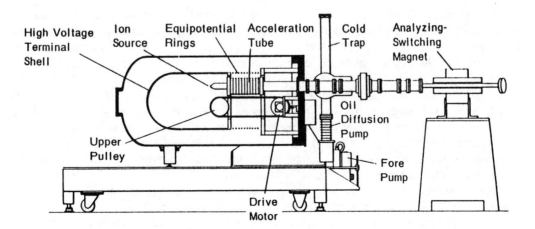

Fig. 1-9. Schematic of a Van de Graaff particle accelerator (Bygrave et al., 1970).

Although expensive to build and to operate, several synchrotron sources are available in the USA, Europe, and Japan, and more are in the planning or construction stages.

Synchrotrons operate by accelerating charged particles to very high energies (several GeV) using a cyclotron or a linear accelerator (Fig. 1-10). These high-energy particles are then transferred to a large storage ring equipped with bending magnets and magnetic insertion devices (wigglers and undulators) that force the particles to travel in a roughly circular path and, as a result of the centripetal acceleration, emit light. Each of the bending magnets and insertion devices has a port that allows the radiation created to pass into a beamline and thence to an experimental station at the end of the beamline. The energy of the particles and the strength of the magnets are the main factors controlling the x-ray energy at which maximum intensity is obtained.

The advantages of a synchrotron x-ray source are that: (i) intensities several orders of magnitude greater than those achievable by x-ray tubes are obtainable, (ii) x-rays of any energy within the range of the synchrotron can be selected rather than selections being restricted to particular absorption edges, and (iii) truly monochromatic x-rays are available without large losses in intensity. Some of the potential applications of synchrotron radiation to the characterization of soils and mineral surfaces have been summarized in workshop reports edited by Schulze and Smith (1990) and Smith and Manghnani (1988).

Detection

Detectors for quantification of photon intensities and energies operate in a proportional mode. That is, the height of the voltage pulse produced by the detector is directly proportional to the energy of the photon---the number of such pulses is proportional to the number or intensity of photons of that energy. The initial detector pulse is generally an increase in electrical charge caused by the ionization of a sensitive substance by the incoming photon. This charge burst is amplified and converted to a voltage by the detector (which basically functions as a capacitor), and the information stored as a count. Proportional detectors thus differ principally by the nature of the original sensitive material that is ionized by the incoming photon.

For energy-dispersive detection, the counts are separated by energy using an analog-to-digital convertor and stored in the channels of a multichannel analyzer set to cover the voltage (photon energy) range of interest. With wavelength-dispersive detection, the x-rays reaching the detector have already been separated according to wavelength (or energy) by diffraction with an analyzing crystal. A pulse-height discriminator, which eliminates counts having voltages less than a preset threshold, is used to improve the signal-to-noise ratio of the output before it is sent to a ratemeter, chart recorder, or single-channel analyzer.

X-Ray Detectors

For detection of x-rays, three types of energy-proportional detectors are in common usage: an Ar-filled gas-flow proportional counter, a Tl-doped NaI [NaI(Tl)] scintillation counter, and a Li-doped Si [Si(Li)] semiconductor detector (Fig. 1-11). In the gas-flow proportional counter, a steady flow of Ar gas is ionized by the incoming x-ray to produce Ar^+ and electrons. The electrons migrate to a positively biased wire in the middle of the detector cylinder and the Ar^+ ions to the cylinder wall, which is grounded. The separation of charge created by this migration is then registered as a voltage pulse. As the x-ray ionizes Ar atoms, it gives up energy in approximately 30-eV packets. The total number of such packets produced is proportional to the original energy of the photon and determines the size of the charge induced on the detector wire and cylinder wall. Other inert gases, such as Kr and Xe, can be used in sealed gas proportional counters for situations where very high counting rates ($\sim 2 \times 10^6$ s^{-1}) are needed.

The scintillation counter works by the delocalization of electrons in a NaI(Tl) crystal when struck by an x-ray. Each electron recombines with the atoms in the crystal and in so doing generates a photon of about 3 eV in energy, which is then detected, amplified, and converted to a voltage pulse by a photomultiplier. However, inefficiencies in the conversion of delocalized

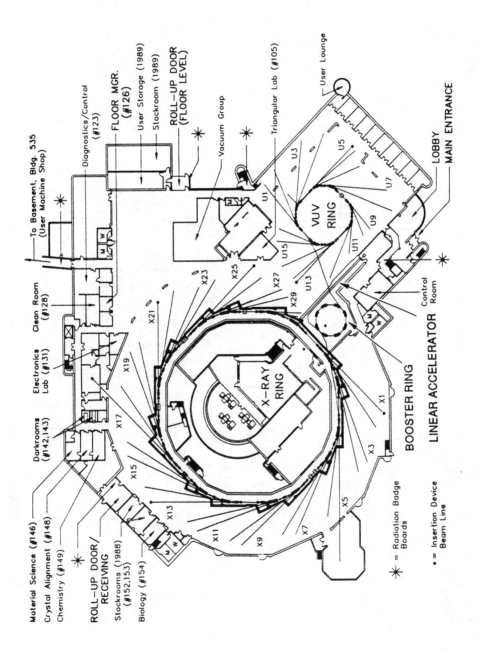

Fig. 1-10. Floor plan for the National Synchrotron Light Source showing the linear accelerator, booster ring, and the x-ray and vacuum ultraviolet (VUV) storage rings with insertion devices and beam lines (adapted from White-DePace et al., 1988).

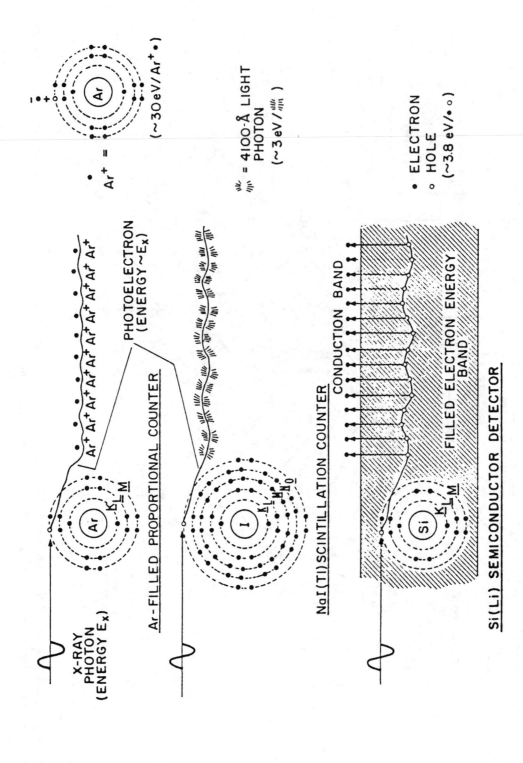

Fig. 1-11. Processes involved in the detection of x-rays by three types of energy-proportional detectors (Bertin, 1981).

electrons to light pulses and in the conversion of light pulses to electrons in the photomultiplier result in an actual detector resolution of about 300 eV (i.e., the incoming photons are measured in packets containing about 300 eV of energy). Because the scintillation counter is most efficient for x-rays having energies above 6 keV and the gas-flow proportional counter for x-rays below 8 keV, the two are often mounted side-by-side or in tandem to simultaneously cover the range from 0.6 to 60 keV.

The semiconductor detector works by the promotion of Si electrons from nonconducting to conducting bands in a Li-doped Si crystal. Like the gas-flow proportional detector, a high bias is maintained across the crystal so that any positive holes and conducting band electrons produced are quickly captured. Approximately 3.8 eV of energy are needed to promote an electron into the conducting band and, because this process produces nearly eight times as many primary electrons as in the gas-flow proportional counter (1 electron for 30 eV) and the conversion efficiency of the process is much better than for the scintillation counter, the inherent energy resolution of the Si(Li) detector (~80 to 120 eV) is superior to that of the other detectors. For this reason, Si(Li) detectors are typically used to measure x-rays directly rather than after diffraction by an analyzing crystal. The Si(Li) detector is most sensitive to x-rays in the region between 3 and 20 keV. Below 3 keV, attenuation of x-rays by the Be window on the front of the detector becomes increasingly important. Detection of elements as light as B can be achieved, however, by the use of ultrathin windows or windowless Si(Li) detectors. Above 20 keV, the detector efficiency drops off rapidly as a result of the decrease in the mass attenuation coefficient of Si for high-energy photons, and the detector is of little value above 40 or 50 keV.

A major problem with all the energy-proportional detectors is that they have poor resolution (and poor sensitivity in the instance of the semiconductor detector) for low-energy x-rays (Fig. 1-12). For these x-rays, the best results have been obtained using crystal diffraction (i.e., wavelength dispersion) to separate the x-ray peaks and then a gas-proportional or scintillation counter to measure the intensity of the diffracted peaks. Even though only one x-ray peak can be scanned at a time, the count-rates achievable with the gas-flow proportional or scintillation counters ($\sim 10^4$ to 10^5 s^{-1}) are so much higher than for a Si(Li) detector (10^3 to 10^4 s^{-1}) that the total analysis time for a limited number of elements may be much shorter. As the list of elements to be quantified grows longer, however, the full-spectrum approach involving energy-dispersive detection becomes more competitive.

The recent development of position-sensitive x-ray detectors (Fraser, 1984; Smith, 1984; Arndt, 1986; Osborn & Welberry, 1990; Zahorowski et al., 1991) may make full-spectrum analysis by wavelength dispersion competitive with the energy-dispersive systems. These detectors essentially consist of a gas-proportional counter with a long wire and the electronics to identify the exact point along the wire where the x-ray photon was detected. Although early versions have had relatively poor resolution, position-sensitive detectors have been used in synchrotron experiments for rapid collection of diffraction patterns (5-10 °2Θ at a time), and with continued improvement could be used to collect crystal-diffracted x-rays for several elements simultaneously, thus improving the quality and speed of low-energy x-ray data.

High-Energy X-Ray and Gamma-Ray Detectors

Because they generally are of higher energy, gamma-rays are attenuated only slightly in the small sensitive volume of Si-based semiconductor detectors. Consequently, a higher density semiconducting material such as Ge, which has a greater ability to attenuate (and hence be ionized by the gamma rays), is needed for adequate detection. The Ge semiconductor detectors operate on the same principles as the Si(Li) detectors and are generally most efficient for photons having energies between 3 keV and 1 MeV.

Particle Detectors

When charged particles are used to induce x-ray or gamma-ray emission, particle detectors are needed to measure the intensity of the particle beam striking the specimen. Two types of particle detectors are in wide use. The simplest detector is a Faraday cup, which consists of a deep metal

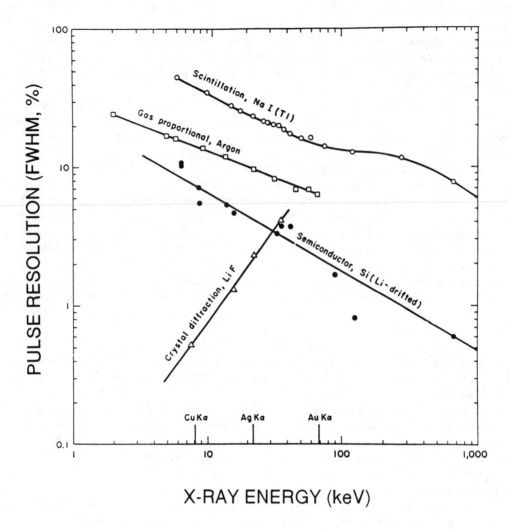

Fig. 1-12. Comparison of the resolution obtainable with three energy-proportional detectors alone, and with the additional wavelength dispersion provided by LiF crystal diffraction, over a wide range in x-ray energy (Burkhalter & Campbell, 1967).

cup, the bottom of which is biased with a charge opposite to the charge of the particle. The cup is placed into the path of the particle beam and collects the charged particles by electrostatic attraction. A current integrator placed between the cup and a suitable ground measures the total charge collected, which is directly proportional to the intensity of the particle beam. The Faraday cup can be used only to measure charge and is indifferent to the energy carried by the particle.

The second type of detector for charged particles is a surface barrier detector. This is a semiconductor detector (usually made from Si) in which a charge-depleted "sensitive" volume is maintained by a high bias between the front and back sides of the crystal. Charged particles that enter this volume will slow down and stop as a result of collisions with the Si atoms. With each collision, some of the electrons present in the Si will be promoted to conducting bands (i.e., electron-hole pairs will be created for each 3.8 eV of energy). Because the electron-hole pairs produced are in the "sensitive" volume of the detector they are subject to the electric field produced by the bias on the detector and will migrate to the oppositely charged contacts at the front and back of the crystal. Thus, the total charge produced by the particle in the sensitive volume will be collected as a pulse, which is converted to a voltage and amplified just as in the proportional photon detectors. This type of detector therefore can be used to establish both the energy and the intensity of a charged particle beam, although it can measure only relatively low-intensity particle beams.

Filters

Filters are often placed between the sample and the detector to eliminate unwanted radiation and particles from interfering with the detection of analytical x-rays. For example, most Si(Li) detectors have a very thin Be window in front of the sensitive semiconductor portion of the detector to prevent backscattered protons and secondary electrons from damaging the surface of the detector and interfering with the x-ray spectra. This filter also attenuates x-rays to a slight extent and its presence must be accounted for in analyzing the observed spectrum. Other filters may be deliberately placed between the specimen and the detector to eliminate very intense low-energy x-rays and allow a larger portion of the x-rays detected to come from elements at low concentration that emit more energetic x-rays.

This approach is used particularly effectively in PIXE spectroscopy to avoid saturating the Si(Li) detectors with low-energy x-rays and thus improve the detection limits for high-energy x-rays. With geological specimens at our laboratory, for example, three separate spectra are obtained: an unfiltered spectrum at low proton beam current (~ 1 nA), a second spectrum at a higher beam current (~ 8 nA) with a 14-μm Al filter placed between the specimen and the detector, and a third spectrum at very high beam current (~ 200 nA) with a compound filter containing O, Al, Si, K, and Cr. The second filter effectively removes the x-rays produced by Si, Na, and Mg, thus allowing a greater portion of the x-rays detected to be from other elements, such as K, Ca, and Fe, that are in the specimen. The third filter effectively removes x-rays emitted by all elements lighter than Ca as well as specifically those emitted by Fe (the Cr absorption edge is just below the main Fe-Kα peak energies), and thus allows much better detection limits for elements heavier than Fe.

Specimen Preparation

The ultimate goal of specimen preparation is to present a specimen to the analyzing beam that: (i) represents the sample as a whole, and (ii) closely fits within the constraints of the theoretical model used to estimate analyte concentrations. Equation [3], for example, assumes that the specimen is smooth at the surface and compositionally homogeneous at the molecular level. Soils, on the other hand, are rarely homogeneous, and generally consist of a mixture of particles that vary in composition, morphology, and size. In addition, most soil particles are electrical insulators and, when bombarded with charged particles, will build up an electrical charge that interferes with the analysis. Specimen preparation techniques for soils, therefore, are chiefly concerned with homogenizing the sample and presenting a smooth surface to the analyzing beam. When charged particle beams are used, however, some consideration must also be given to the elimination of sample charging.

Homogenization of the sample may be achieved by two techniques: (i) decomposition by fusion or (ii) grinding to very fine particle size. Fusion is the only technique that truly homogenizes the sample on an atomic scale, but the procedure is sample destructive, significantly dilutes the sample, and may result in the loss of trace constituents (e.g., S, Cl, Se, Br) because of the high temperatures needed for the decomposition. Various fluxes can be used (e.g., NaOH, $LiCO_3$, $Li_2B_4O_7$) in combination with an x-ray-absorbing diluent to minimize differences in mass absorption coefficients among specimens. Norrish and Hutton (1969), for example, describe a method in which a mixture of $LiBO_2$, Li_2O, $LiNO_3$, and La_2O_3 is used to fuse the specimens at about 1000 °C. Quenching and then annealing the resulting flux in a graphite mold yields a malleable smooth glass with a very high mass absorption coefficient that is ideally suited for analysis of the major elements. Eastell and Willis (1990) describe a low-dilution fusion method with an 80% $LiBO_2$/20% $Li_2B_4O_7$ mixture (containing $LiNO_3$ as an oxidant) that seems well suited for analysis of trace constituents. Some of the finer points in the annealing procedure required to produce robust glass specimens are described by Alvarez (1990).

Reduction of particle size by grinding also homogenizes the sample, but success with this technique relies on a uniform particle size being obtained, ideally on the order of 1 to 2 microns or smaller. If the sample contains minerals that respond differently to grinding, preferential concentration of the softer mineral in the smaller particle sizes may occur and result in inaccurately high concentrations being estimated for the constituents of the harder mineral (Jenkins, 1974, p. 115-117; Bertin, 1975, p. 524-527 and 747-751). Contamination of the sample from the grinder or from previous samples in the grinder can also pose a problem and, if wet grinding techniques are used, some loss of soluble constituents might be expected. The advantages of grinding, however, are that sample dilution is essentially nil and the procedure is simple and relatively rapid.

Once the sample has been homogenized, a specimen having a smooth planar surface must be prepared. Smoothness of the surface is important to ensure a constant x-ray path length for elements at a particular depth in the specimen, since the specimen surface is oriented at an angle to either the detector or the source beam (and in some instances, both). A smooth surface is particularly important for the analysis of low-energy "soft" x-rays because they are easily attenuated by the specimen. If a suitable mold is used for fused samples, the glassy surface of the cooled specimen may be suitable for analysis by x-ray fluorescence, although some polishing of the surface is usually required. Otherwise, the fused samples will have to be ground to a fine powder. Typically, with samples that have been homogenized by grinding (or by fusion and then grinding), a pelleted specimen is prepared using a die and a hydraulic press. To ensure the integrity of the specimen, the sample may be diluted with a binding agent such as cellulose, boric acid, polyvinylpyrrolidone, or briquetting graphite before the pellet is pressed. Dilution factors ranging from 0.01 to 5 or more have been used, depending on the diluent, the nature of the sample, and whether analysis for major or trace elements is desired. Graphite is the preferred diluent for analyses with charged particle beams because it is electrically conductive and thus can minimize the interferences due to sample charging. However, pellets may be pressed with no binding agent in some instances, or specimens may consist of loose powders supported by a thin plastic film if the sample chamber geometry and spectrum analysis algorithm are configured appropriately.

SURVEY OF ANALYTICAL TECHNIQUES

X-Ray Fluorescence

X-ray fluorescence (XRF) spectroscopy is the most mature of the nondestructive analytical techniques, with commercial instruments having been available since the late 1940s. The technique relies on a beam of high-energy photons to create inner-shell vacancies in the atoms that make up the specimen. The filling of these vacancies by electrons from higher atomic energy levels results in the emission of photons with energies characteristic of the atoms present (i.e., fluorescence). The intensities and quantities of the x-rays emitted by the specimen are measured with a detector and, by using a suitable algorithm based on an XRF yield equation, the amounts of each atom in the specimen are then determined. Only the distinguishing features of the technique are summarized here. For a more detailed discussion, the interested reader is referred to several excellent texts on the subject (Azaroff, 1974; Jenkins, 1974; Bertin, 1975; Bertin, 1978; Russ, 1984) as well as recent reviews (Jones, 1982; Jones, 1991).

Experimental Setup

For many years, the standard source for XRF spectrometers has been the side-window Coolidge x-ray tube, typically with a W anode. In recent years, modifications to this standard source have included (i) the development of end-window tubes to facilitate the use of multiple wavelength-dispersive detectors for a single specimen, and (ii) the placement of various secondary targets in the primary beam outside the tube for selective excitation of the sample (Fig. 1-8). In our laboratory, for example, a W-anode side-window Coolidge tube is used with Gd, Ag, Zr, and

Ti secondary targets for efficient excitation of the specimen over a wide range of x-ray energy (Fig. 1-13).

Although most XRF spectrometers still rely on crystal diffraction to separate the peaks in combination with gas-flow proportional counters and scintillation counters to measure their intensity, considerable inroads have been made in recent years by semiconductor detectors that perform both these functions (Fig. 1-14). The advantages of the semiconductor detectors are that simultaneous collection of data for a large number of elements is possible and spectral interferences (e.g., higher-order reflections from the analyzing crystal with wavelength dispersion) are fewer. On the other hand, the resolution of the peaks is better with crystal diffraction, particularly for the lighter elements (Fig. 1-12), and, because of the Be windows generally used to maintain a vacuum inside the Si(Li) detector (although new ultrathin windows and windowless detectors are also available), the sensitivity of the gas-flow proportional counter with an ultrathin Mylar window is superior for x-rays below about 2 keV in energy. Selection of the appropriate detector for XRF spectrometry, therefore, depends on the application. For rapid multielemental analysis of elements heavier than Ar, the Si(Li) semiconductor is appropriate. Where accurate measurements of light elements are needed, or where a small number of elements is being measured, the wavelength dispersive (i.e., crystal diffraction) technique using the gas-flow proportional and scintillation counters is the better choice. And, as mentioned earlier, the advent of position-sensitive detectors may make wavelength dispersion even more competitive for multielemental analysis in the future.

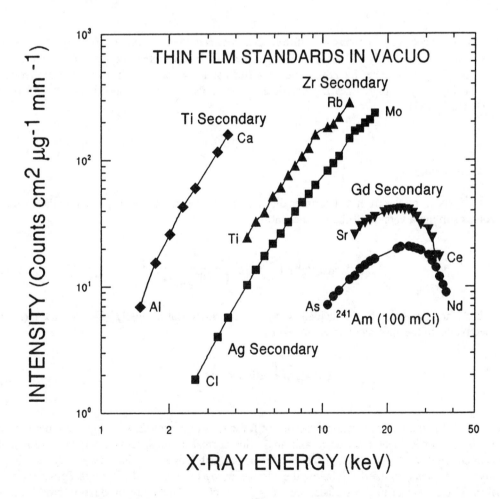

Fig. 1-13. Normalized intensities of Kα x-rays observed with an energy-dispersive Si(Li) detector for fluorescence of thin-film elemental standards during excitation with Kα x-rays from Ti, Zr, Ag, or Gd secondary targets or with 59.7-keV gamma rays from a 100-mCi [241] Am radioactive source.

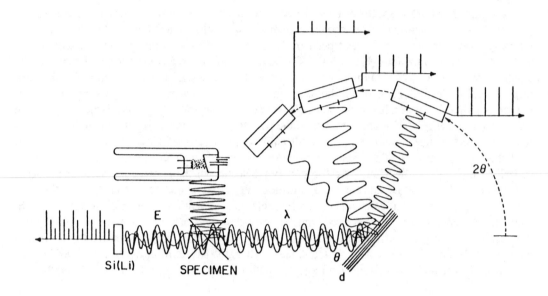

Fig. 1-14. Schematic showing the differences between the two types of x-ray fluorescence spectrometers. Analytical x-rays are fluoresced by the specimen and detected in an energy-dispersive spectrometer by a Si(Li) detector (left) and in a wavelength-dispersive spectrometer by using a crystal to diffract the analytical x-rays into a gas-filled proportional detector (right) (Bertin, 1981).

Spectrum Analysis

Analysis of the spectrum to determine the concentration of the elements present in the specimen involves solving a version of the yield equation

$$Y_z = Q_{exc} C_z D_{eff} \; \sigma(E_{exc}) \int_0^x \frac{T(E_z, x)}{T(E_{exc}, x)} \; \delta x \qquad [4]$$

in which the attenuation of the incident x-ray and the analytical x-rays [$T(E_{exc}, x)$ and $T(E_z, x)$, respectively] in the specimen are estimated by

$$T(E, x) = exp \left\{ -\mu(E) sec\theta \; \rho x \right\} \qquad [5]$$

where $\mu(E)$ is the mass attenuation coefficient for x-rays of energy E, Θ is the angle between the detector axis or incident x-ray beam axis and a line normal to the specimen surface, and ρ is the specimen density. For a planar specimen, the actual path length traveled in the specimen by the x-ray, ℓ, is related to the depth below the surface, x, by the secant of the angle Θ that the x-ray makes with a normal to the surface, i.e., $\ell = (x)(sec\Theta)$. The x-ray production cross section, $\sigma(E_{exc})$, depends solely on the *energy* of the incident photon, not the intensity, and thus, can be removed from the integral in Eq. [4]. Because $\mu(E)$ is a function of the composition of the sample, an initial estimate of the sample composition is made from standard yield data (Fig. 1-13) and photon-scattering intensities (see below) or from prior knowledge of the specimen [e.g., all specimens prepared by fusion using the technique of Norrish and Hutton (1969) have essentially the same $\mu(E)$]. This estimate is used to calculate an initial value for $\mu(E)$, and then the yield equation is solved iteratively until the estimated and calculated yields for the major elements agree.

One advantage in XRF spectroscopy deriving from the use of high-energy photons to excite the characteristic x-rays in the specimen is that the photon-scattering properties of the specimen can be used to estimate the matrix composition. When x-rays interact with matter they may be scattered with no loss in energy (coherent scattering), scattered with a small loss in energy (incoherent scattering), or absorbed (photoelectric effect). When scattered, the proportion of the x-rays that are scattered coherently depends on the number of electrons in the atoms causing the scattering, i.e., on the atomic number. The larger the number of electrons in the atom, the more likely it is that the x-ray will be scattered coherently. The cross sections for coherent and incoherent scattering have been tabulated for each element:photon combination (Veigele, 1973; Hubbell et al., 1975; Berger & Hubbell, 1987) and the ratio of these cross sections differs by a factor of approximately 40 in going from C to Ba (Fig. 1-15). Direct measurement of this ratio is a simple procedure for any specimen (Fig. 1-16) and the result can be used to estimate the average atomic number of the specimen. Because the ratio can be calculated from an estimate of the elemental composition of the specimen obtained from the x-ray spectrum as well as measured directly, a comparison of the calculated and measured values allows one to estimate the proportion of the specimen that is made up of elements that were not detected. For example, although elements lighter than Na (e.g., O and C) would not be detected with a Si(Li) detector having a Be window, an average light element composition can be estimated from a comparison of the

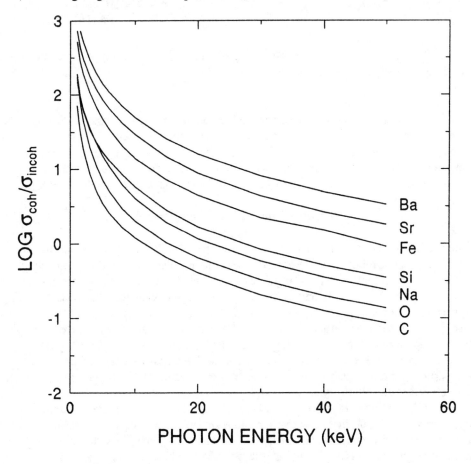

Fig. 1-15. The coherent:incoherent scattering-intensity ratios for x-rays between 1 and 50 keV incident on several elements covering a wide range in atomic mass (data from Veigele, 1973).

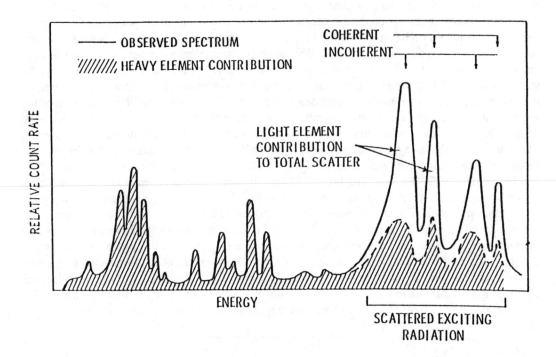

Fig. 1-16. Schematic of spectrum demonstrating the direct determination of coherent:incoherent scattering-intensity ratio for an unknown specimen.

calculated and measured scattering ratios for the high-energy incident x-rays that can be detected. This average composition of the undetected elements can then be used in matrix calculations to yield the appropriate mass attenuation coefficient for the specimen and, hence, the correct elemental concentrations. This technique forms the basis for the public-domain Back-Scatter Fundamental Parameter spectral analysis program used in our laboratory (Nielson & Sanders, 1982).

Particle-Induced X-Ray Emission

Over the last two decades, particle-induced x-ray emission (PIXE) spectroscopy has developed from a novelty into a robust, mature analytical technique. The technique was pioneered by the accelerator physics group at the Lund Institute of Technology in Sweden (Johansson et al., 1970; Johansson & Johansson, 1976) and initially applied to the elemental analysis of very thin atmospheric dust samples on filter membranes. Subsequent development of the technique by the Swedish group, as well as by groups in the USA, Canada, Australia, and elsewhere in Europe, has extended its applicability to infinitely thick specimens including those of geological origin. A recent development has been the proton microprobe, which takes advantage of the ability to focus charged particle beams for analysis of micron-sized areas (Cahill, 1980; Rogers et al., 1987; Cabri, 1988; Campbell et al., 1990; Ryan et al., 1990).

Briefly, PIXE relies on excitation of x-rays in the specimen by exposure to a collimated beam of high-energy charged particles, typically protons or alpha particles. As a result of collisions between the particles and inner-shell electrons, vacancies are created in the inner shells of the atoms that make up the specimen. Filling of these vacancies by outer-shell electrons results in the emission of x-rays, just as in XRF spectroscopy. An excellent and thorough treatment of the PIXE technique is given in a recent book by Johansson and Campbell (1988) and, in short form, as a review by Pineda (1992).

Experimental Setup

To maintain a monoenergetic collimated particle beam, the charged particles (typically 2- to 3-MeV protons) are delivered from the source (a Van de Graaff or tandem accelerator) to the specimen in an evacuated beamline. In our laboratory (a typical setup), the specimen and x-ray detector are also mounted inside an evacuated chamber attached to the end of the beamline (Fig. 1-17 top). The detector is mounted at an angle of 135° to the particle beam axis, is collimated to prevent detection of stray x-rays or particles, and has a filter wheel between it and the specimen to allow selective screening of portions of the x-ray spectrum. The incident intensity of the particle beam is monitored during the run by measuring the number of protons backscattered into a surface-barrier detector by a Au-coated paddle that rotates through the beam. The backscattered proton counts are calibrated to the proton current collected by the Faraday cup when the specimen target wheel is moved out of the way.

In some instances (e.g., soil monoliths, soil cores, large rocks), the specimen may not be amenable to placement inside an evacuated chamber, and an experimental setup like that shown at the bottom of Fig. 1-17 is needed. The quality of the data obtained with this setup will be lower than that obtained in the vacuum chamber as a result of (i) no direct measurement of proton beam intensity, (ii) absorption of x-rays by He gas between the specimen and the detector, (iii) charge build-up on insulating specimens, and (iv) a rough specimen surface, but the quality is still sufficient for qualitative analysis and sample screening purposes.

Spectrum Analysis

Because the incident particles change in energy (and hence in their ability to excite the atoms in the specimen) as they penetrate the specimen, the integral in the yield equation for PIXE analysis is summed over layers corresponding to changes in particle energy rather than specimen thickness. Thus,

$$Y_z = Q_{exc} C_z D_{eff} \int_{E_{exc}}^{0} \frac{\sigma(E_{exc}) T(E_z, E_{exc})}{S(E_{exc})} \delta E_{exc} \qquad [6]$$

and the attenuation of the x-rays produced in the specimen [$T(E_z, E_{exc})$] is now expressed in terms of changes in incident particle energy rather than distance traveled in the specimen. The relationship between incident particle energy and sample depth is summarized in the quantity referred to as the stopping power,

$$S(E_{exc}) = \frac{\delta E_{exc}}{\delta(\rho x)} \qquad [7]$$

which may be rearranged and integrated to yield

$$\rho x = \int_{E_{exc}}^{0} \frac{\delta E_{exc}}{S(E_{exc})} \qquad [8]$$

The right side of this expression may then be substituted in Eq. [5] to yield

$$T(E_z, E_{exc}) = \exp \left\{ -\mu(E_z) \sec\theta \int_{E_{exc}}^{0} \frac{\delta E_{exc}}{S(E_{exc})} \right\} \qquad [9]$$

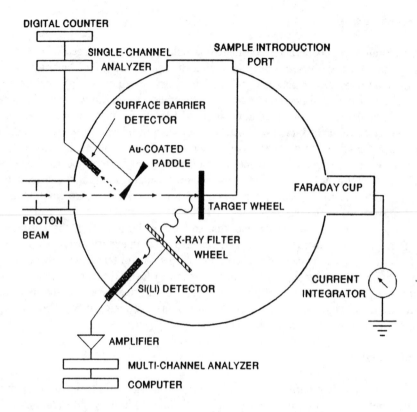

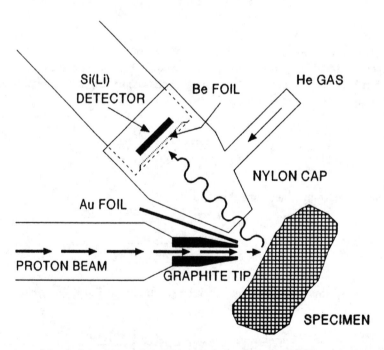

Fig. 1-17. Schematics of (top) the evacuated experimental chamber for PIXE analysis used in the authors' laboratory, and (bottom) a typical experimental set-up for PIXE analysis of samples outside an evacuated chamber (after Fleming & Swann, 1987).

for the attenuation of x-rays in the specimen. As with XRF, the values for $S(E_{exc})$ and $T(E_Z, E_{exc})$ are matrix dependent and, consequently, the yield equation is solved iteratively after initial estimates of the specimen composition are made.

The x-ray production cross sections for bombardment by high-energy particles are greatest for the light (low-Z) elements and drop off rapidly with increasing atomic number. With 2-MeV protons, for example, $\sigma(E_{exc})$ is about 100 times larger for the production of Al-Kα than for Fe-Kα x-rays. Consequently, for a given 2-MeV proton beam current, it takes 100 times as many protons to generate an Fe-Kα x-ray as it does for an Al-Kα x-ray, assuming the same concentrations of Fe and Al in the specimen. The obvious way to increase the sensitivity for high-Z elements is to bombard the specimen with a more intense proton beam. This approach has the drawback that so many x-rays are generated by the light elements that the detector becomes saturated and the data quality and collection efficiency decrease. The problem is resolved by placing x-ray filters between the specimen and the detector when higher beam currents are used. These filters preferentially attenuate the low-energy x-rays and thus allow efficient collection of high-energy x-rays generated by proton bombardment.

The effectiveness of x-ray filters in enhancing the detection of high-energy x-rays for a soil analyzed by PIXE are shown in Fig. 1-18. The unfiltered spectrum shows well-defined K-shell peaks for Na, Mg, Al, and Si (1-2 keV), and Kα and Kβ peaks for Ca (3-4 keV), Ti (4.5-5 keV), and Fe (6.4-7 keV). A less well-defined Kα peak for Mn appears at about 5.9 Kev (the Mn Kβ peak falls under the Fe Kα peak), along with a hint of Cu and Zn Kα peaks at 8 and 8.6 keV. The placement of a 14-μm Al filter in front of the detector and a tenfold increase in the number of protons striking the specimen eliminates the Si peak at 1.7 keV, yields a strong Al peak at 1.5 keV caused by secondary fluorescence of the filter, and yields much stronger and more clearly defined K, Ca, Ti, Mn, and Fe peaks. Use of a compound filter containing O, Al, Si, K, and Cr, together with an additional 45-fold increase in the number of protons, completely eliminates the Na, Mg, Al, and Si peaks, decreases the intensities of the K, Ca, and Ti peaks, and yields very well-defined Kα and Kβ peaks for Mn, Fe, Cu, Zn, and even Ga (9.2 keV, 15 ppm in specimen), which could not be seen in the other spectra. A strong Cr peak from secondary fluorescence of the filter caused by Fe K-shell x-rays from the specimen is also seen at 5.4 keV.

The spectral analysis for this specimen would involve separately calculating the yields for Na, Mg, Al, Si, S, and Cl from the unfiltered spectrum, those for K through Fe for the Al-filtered spectrum, and those for elements heavier than Fe from the compound-filtered spectrum using an initial assumption about the specimen matrix. These results would then be put together to calculate a new specimen matrix (assuming all elements in their common oxidation states as oxides unless otherwise known) and, if this matrix was sufficiently different from the initial matrix, repeating the yield calculation. The process would be repeated until the new and previous matrices agreed to within 1% relative for each element, normally three or four iterations. Fortunately, software packages to perform these calculations are readily available (Clayton, 1986; Wätjen, 1987; Johansson & Campbell, 1988, p. 144-159).

Particle-Induced Gamma Emission

As a companion technique to PIXE spectroscopy, particle-induced gamma emission (PIGE) spectroscopy has not received as much attention, perhaps because it is perceived as being suitable for determination of a relatively small number of mostly light isotopes. The fundamental difference between PIGE and PIXE spectroscopy is that the photons produced arise from reactions between high-energy charged particles and specific nuclei rather than inner-shell electrons. Two consequences stem from this reaction mechanism. First, there is no simple relationship between photon energy and atomic number analogous to that seen for the production of x-rays. Second, the energies of the photons produced cover a much wider range (e.g., from 50 keV-10 MeV). The technique may also be distinguished from other nuclear-reaction based techniques in that the photons measured are those that are released instantaneously at the time of the proton-nucleus collision (i.e., "prompt" gamma rays) rather than those released over a longer period of time

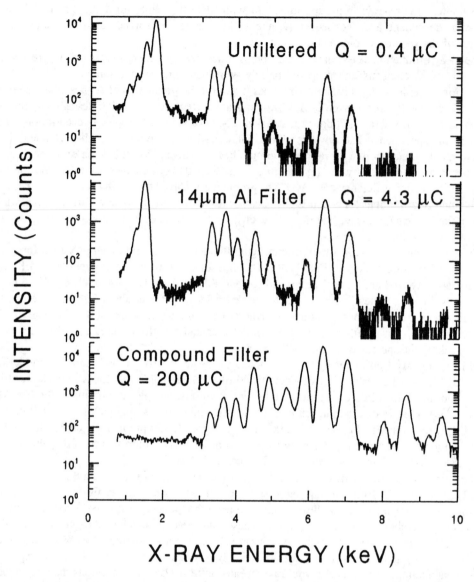

Fig. 1-18. The effects of selective x-ray filters placed between the specimen and the Si(Li)
detector and total charge collected by the specimen (Q) on the analytical x-ray spectrum for
a typical soil specimen (IAEA Soil 5) analyzed by PIXE spectrometry.

through an isotopic decay process as in neutron activation analysis. The interested reader is
referred to the recent review by Gihwala and Peisach (1992) for a more detailed discussion of
PIGE spectroscopy.

Selected nuclear reactions, analytical gamma-ray energies, and representative yields for analysis
by 2.4-MeV protons are shown in Table 1-1. From these data it can be seen that PIGE
spectroscopy complements the x-ray techniques by measuring some of the light elements that are
difficult to detect under routine conditions and that it is particularly sensitive to Li, B, F, Na, and
Al with somewhat less sensitivity to Be and Mg. In some instances (e.g., N), PIGE spectroscopy
is limited by low natural abundance of the reactive isotope. Unlike PIXE spectroscopy, however,
the use of heavier or more energetic ions as projectiles generally results in an increase in yields.
For example, with 9-MeV protons the log of the observed yield for ^{28}Si is 8.7; similarly, the use
of 55-MeV ^{35}Cl ions gives log yields on the order of four to five for the rare earth elements
(Gihwala & Peisach, 1992). Generation of ions with these energies, however, requires larger
accelerators than the typical 2.5- or 3.0-MeV Van de Graaff used for most PIXE analyses.

Table 1-1. Selected nuclear reactions, analytical gamma-ray energies, and representative photon yields obtained for thick-target specimens of several pure elements using a 2.4-MeV proton beam (Anttila et al., 1981).

Element	Reaction	E_γ	Yield
		keV	log Counts μC^{-1} sr^{-1}
Li	$^7Li(p,p'\gamma)^7Li$	478	7.4
Be	$^9Be(p,\alpha\gamma)^6Li$	3562	4.4
B	$^{10}B(p,\alpha\gamma)^7Be$	429	6.5
C	$^{12}C(p,\gamma)^{13}N$	3511	2.3
N	$^{15}N(p,\alpha\gamma)^{12}C$	4439	3.7
O	$^{16}O(p,\gamma)^{17}F$	495	3.0
F	$^{19}F(p,p'\gamma)^{19}F$	110	5.5
Na	$^{23}Na(p,p'\gamma)^{23}Na$	440	6.5
Mg	$^{25}Mg(p,p'\gamma)^{25}Mg$	585	4.8
Al	$^{27}Al(p,p'\gamma)^{27}Al$	1014	5.5
Si	$^{28}Si(p,p'\gamma)^{28}Si$	1779	2.4
P	$^{31}P(p,\gamma)^{32}S$	2230	3.5
S	$^{32}S(p,\gamma)^{33}Cl$	811	1.7

Because of their high energies, gamma rays are attenuated only slightly by the specimen matrix. Consequently, a greater volume of the specimen is sampled (i.e., to the full depth at which protons have the necessary energy to react with the nucleus) for the light elements than by the x-ray techniques, which are limited by absorption of the low-energy analytical photons by the specimen. Furthermore, because it is based on reactions with nuclei rather than electrons, PIGE spectroscopy avoids some of the other problems inherent to x-ray spectroscopy of light elements, namely increases in the tendency for multiple electron vacancies (yielding satellite peaks) and for incorporation of 2p and 3p electrons in molecular orbitals rather than atomic orbitals (yielding x-ray emission bands rather than peaks and altering the fundamental parameters on which quantification relies). For careful determinations of the amounts of certain light elements present in soil minerals (e.g., B, F, Na, Mg, Al, and Si) PIGE spectroscopy is probably the best nondestructive technique.

As shown by studies with thin targets (Boni et al., 1990a), the functions describing the production of gamma rays as a result of reactions with energetic particles do not vary smoothly with the energy of the particles (Fig. 1-19). The various peaks observed in the functions are referred to as resonances and can be very useful for depth-profiling studies of surfaces in which the gamma yield is monitored as a function of beam energy (and hence depth reached by particles of the resonance energy). Thus, although PIGE spectroscopy is generally perceived as a

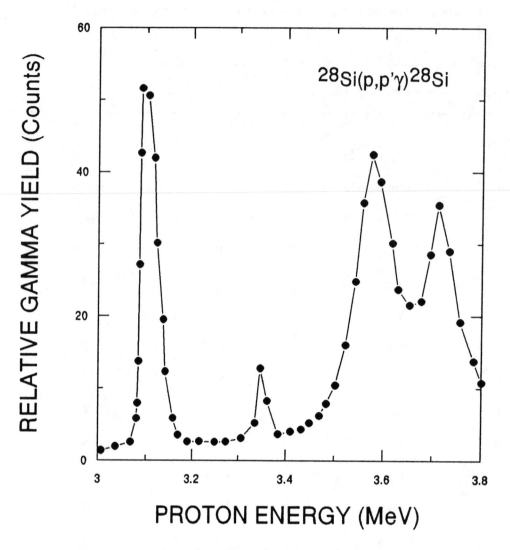

Fig. 1-19. Prompt gamma-ray yields observed for a thin-target Si specimen bombarded by protons
    ranging in energy from 3.0 to 3.8 MeV (after Boni et al., 1990a).

complementary technique nestled in the shadow of PIXE spectroscopy, it has unique capabilities
that are gaining appreciation as the analytical applications of particle accelerators are explored.

Experimental Setup

The experimental requirements for PIGE spectroscopy are similar to those for PIXE
spectroscopy (Fig. 1-17) except that a photon detector having a wider range in energy (e.g., 50
keV-10 MeV) is needed. Semiconductor detectors constructed from high-purity germanium crystals
typically provide this sensitivity with reasonable energy resolution. For optimal gamma yields,
particles having higher energies than those normally used in PIXE spectroscopy may be needed.
However, if simultaneous collection of PIXE and PIGE spectra is desired, a proton energy of about
3.2 MeV must not be exceeded to avoid the production of neutrons by the $^{13}C(p,n)^{13}N$ reaction
and a decrease in the x-ray detector signal-to-noise ratio (Raisanen, 1990).

Spectrum Analysis

A thin-target PIGE spectrum for "urban particulate matter" (which undoubtedly contains some soil minerals) is shown in Fig. 1-20. The spectrum is characterized by two or three sharp lines for each nuclide, some of which overlap with those for other nuclides. These spectral interferences need to be considered in selecting the lines used for analytical purposes.

The general yield equation for PIGE spectroscopy is the same as that given for PIXE (Eq. [6]) except that $\sigma (E_{exc})$ now refers to the gamma-ray production cross section and $T(E_Z, E_{exc})$ is essentially equal to one. Even though attenuation of the gamma rays produced by the specimen is minimal for most geological specimens, knowledge of the specimen matrix is still critical to the analysis because of the matrix effect on the particle stopping-power calculation $[S(E_{exc})]$ and, hence, on the estimated gamma yield. A complete description of the specimen matrix cannot be provided by PIGE spectroscopy and so the spectrum analysis must incorporate data from a PIXE or XRF analysis of the same specimen. Indeed, given the advantages of PIGE spectroscopy for the light elements (which are major components in most soil minerals), the data from a PIGE analysis of the specimen should be incorporated into the PIXE or XRF spectrum analysis procedures as well. Once the appropriate matrix data are available, the solution of the yield equation proceeds in an iterative fashion, as previously described for PIXE and XRF.

Synchrotron Radiation-Induced X-Ray Emission

The newest of the nondestructive analytical techniques, synchrotron radiation-induced x-ray emission (SRIXE) spectroscopy can be thought of as a specialized form of XRF spectroscopy. It differs from conventional XRF, however, in the nature and properties of the source radiation that is used to excite the atoms in the specimen. Synchrotron radiation is several orders of magnitude

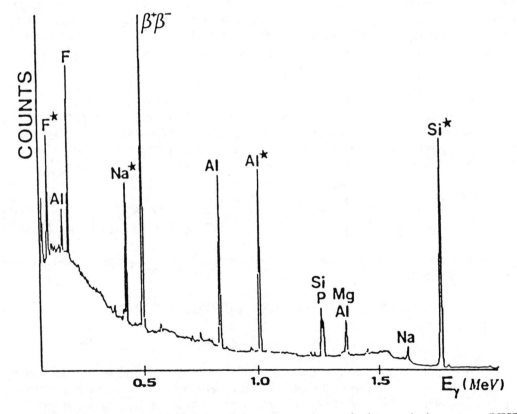

Fig. 1-20. PIGE spectrum obtained for a thin-target specimen of urban particulate matter (NIST SRM 1648) in which the analytical peaks selected are marked with asterisks and the $\beta^+\beta^-$ peak is an artifact of the detector (Boni et al., 1990b).

more intense than the radiation produced by an x-ray tube, is polarized and highly collimated, and is not restricted to certain atomic transition energies. With a suitable monochromator, therefore, it is possible to select essentially *any* x-ray energy of interest and produce a monochromatic beam of sufficient intensity and collimation for diffraction, absorption, or fluorescence studies. Alternatively, portions of the continuous spectrum above a certain cutoff energy (fixed by filter thickness) can be used to excite trace levels of high-Z elements in the specimen. The tremendous intensity offered by synchrotron sources has resulted in the development of x-ray microprobes with spot sizes on the order of 10 microns and detection limits for trace elements below 1 ppm (Jones & Gordon, 1989; Chen et al., 1990; Jaklevic et al., 1990). New, more powerful synchrotron sources and advances in x-ray focusing techniques promise to extend these limits to even smaller dimensions, perhaps to the point where the trace elemental analysis of individual clay-sized particles becomes possible. Two recent workshops (Smith & Manghnani, 1988; Schulze & Smith, 1990) as well as a book by Augustithis (1988) describe in more detail some of the potential applications of SRIXE and other synchrotron-based techniques to soil minerals. For the reader interested in synchrotron radiation in general, books by Winick and Doniach (1980), Margaritondo (1988), and Catlow and Greaves (1990) are recommended.

Experimental Setup

The experimental stations at synchrotron beamlines typically include an ionization chamber to monitor incident beam intensity, a monochromator, a collimator, an x-ray detector, and, where microbeam capabilities are needed, one of several possible arrangements for focusing the x-ray beam on the specimen (Fig. 1-21). X-ray bandwidths of a few electron volts are obtained with the crystal monochromators. With multilayer-coated mirrors arranged in the Kirkpatrick-Baez configuration (Fig. 1-21b), wider bandwidths, approaching 1 keV, are obtained. Of course, in the absence of a monochromator, radiation having bandwidths of tens of keV (i.e., "white radiation" or "white light") strikes the specimen. The use of wider bandwidths increases the intensity of the x-ray beam striking the specimen, thus improving detection limits but, at the same time, complicating the data reduction process needed to obtain the elemental concentrations.

Spectrum Analysis

When a monochromatic x-ray beam is used, the yield equation (Eq. [4]) and general approach already described for conventional XRF spectroscopy apply. With a wide-band x-ray source, however, Q_{exc} becomes a function of x-ray energy (E_{exc}) and, together with $\sigma(E_{exc})$, must be integrated over the energy range covered by the x-ray source to give

$$Y_z = C_z D_{eff} \int_{E_1}^{E_2} Q(E_{exc})\, \sigma(E_{exc}) \int_0^x \frac{T(E_z, x)}{T(E_{exc}, x)}\, \delta x\, \delta E_{exc} \qquad [10]$$

for the general form of the yield equation. As before, an iterative approach involving matrix estimates is followed.

COMPARISONS AMONG THE ANALYTICAL TECHNIQUES

Theoretical Comparisons

Of the four techniques discussed, three (XRF, PIXE, and SRIXE) involve the excitation and emission of characteristic x-rays by the specimen and differ chiefly in the manner and intensity of the excitation process. XRF and SRIXE use high-energy photons to ionize atoms in the specimen, whereas PIXE relies on high-energy particles (usually 2- to 3-MeV protons) for this purpose. Three major factors affect the x-ray yields resulting from the excitation process--the ionization cross section (i.e., the probability of ionization occurring for a single photon or particle interacting

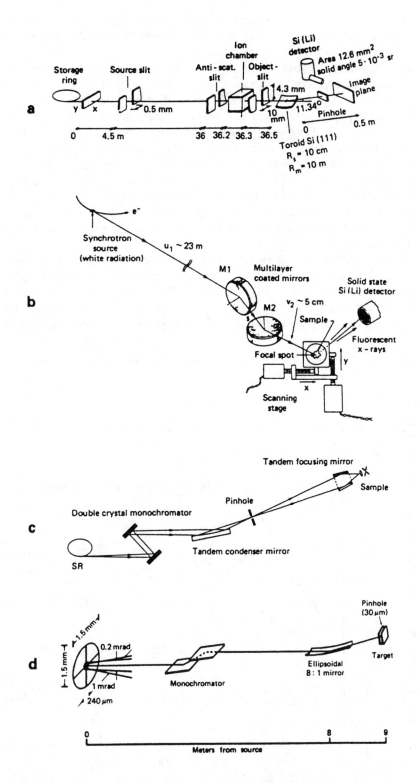

Fig. 1-21. Schematics of typical experimental set-ups used for microbeam analysis of specimens by synchrotron x-ray sources: *a*) a toroidal Si(111) crystal focuses and monochromatizes the beam which is then collimated by a pinhole; *b*) two multilayer-coated spherical mirrors arranged in the Kirkpatrick-Baez configuration focus and monochromatize the beam; *c*) a double-crystal monochromator yields a very narrow bandwidth beam which is then focused and collimated by Wolter type 1 optics; *d*) after monochromatization by a double-crystal channel-cut monochromator the beam is focused by an ellipsoidal mirror (Chen et al., 1990).

with an atom), the intensity of the ionizing agent, and the depth to which the ionizing agent penetrates into the specimen. A careful look at these three factors will lend some insight to the distinguishing features of the x-ray techniques.

The ionization cross sections for atoms interacting with high-energy photons or protons are known to within a few percentage points error, in most instances (Chen & Crasemann, 1985; Cohen & Harrigan, 1985; Berger & Hubbell, 1987). For the production of a vacancy in the K-shell, absorption-edge photon excitation is clearly the more efficient process when compared to 2-MeV protons (Fig. 1-22). For low-Z elements, there is relatively little difference in the ionization probabilities. However, with increasing atomic number, the ionization cross sections for proton excitation decrease much more rapidly than for photon excitation, so much so that photon excitation of the Ba K-shell (atomic number 56) is 100 000 times more efficient than proton excitation. Conventional XRF has a slight disadvantage with respect to SRIXE in that the available excitation energies are limited to a finite number of K x-rays (we have selected Ti-Kα, Zr-Kα, Ag-Kα, and Gd-Kα for illustration), whereas with SRIXE the precise K-edge photon may be selected with a suitable monochromator.

When these ionization cross sections are multiplied by the intensity of the ionizing agent (i.e., the number of photons or protons striking the specimen per second) an estimate of the ionization intensity can be made (Fig. 1-23). Here the initial efficiency of photon excitation is bolstered by the very high intensities offered by synchrotron radiation (10^{12}-10^{15} photons s^{-1}) to give SRIXE a clear advantage over the other techniques. The high proton beam currents (1-200 nA which is

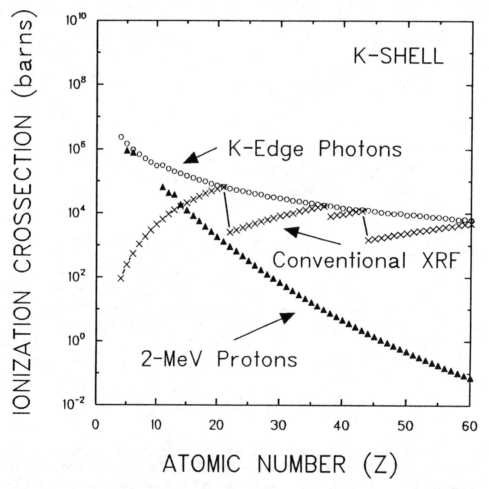

Fig. 1-22. Comparison of the cross sections obtained for K-shell ionization by photons having energies at the K-shell absorption edge, by photons having energies fixed by typical secondary targets used in conventional x-ray fluorescence spectrometry, and by 2-MeV protons (data from Veigele, 1973; and Cohen & Harrigan, 1985).

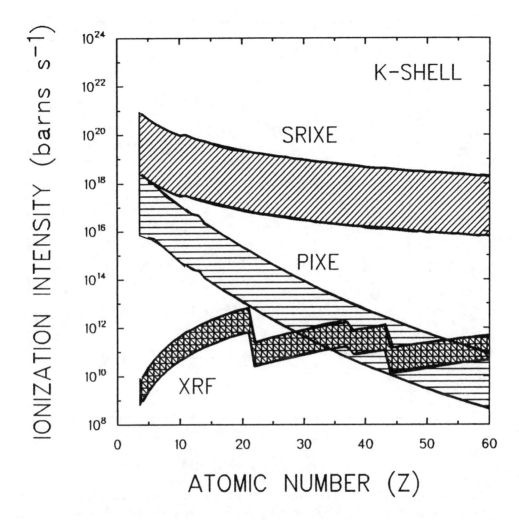

Fig. 1-23. Comparison of the intensities of K-shell ionization by synchrotron x-rays (SRIXE), typical 2-MeV proton beams (PIXE), and x-rays produced by conventional x-ray tubes exciting secondary targets (XRF).

about 10^{10} -10^{12} protons s^{-1}) that are typically used in PIXE partly offset the poor ionization efficiency for the high-Z elements and give it an advantage over conventional XRF for the low-Z elements. Even with a rotating-anode tube ($\sim 10^8$ photons s^{-1}), which is not typically used for XRF because of variability in the primary beam intensity, and a K-edge target, an ionization intensity gap of one to two orders of magnitude exists between PIXE and XRF for the light elements important in soil minerals. For the analysis of high-Z elements (Z > 40), however, the inherent efficiency of photon excitation makes conventional XRF much more competitive.

The depth of excitation can have a significant effect on the observed intensities of analytical x-rays because it helps determine the volume of the specimen sampled (i.e., the number of atoms that emit x-rays). An example calculation using cellulose as the matrix and Sm-Kα x-rays or 2.5-MeV protons as the ionizing agents was performed by Ahlberg (1977). Figure 1-24 shows the results of this calculation for analytical x-rays between 2 and 32 keV. The y-axis represents the depth in the specimen above which 75% of all the x-rays of a particular energy that are detected are produced. Thus, the greater the depth, the larger the number of atoms that are ionized and that contribute to the observed intensity for a particular element. For low-energy photons (<3 keV), there is little difference between the photon-induced and proton-induced excitation, because it is the attenuation of the analytical x-rays in the specimen that limits the number of x-rays that reach

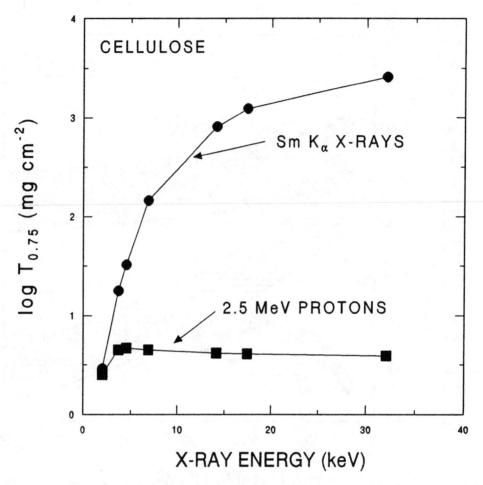

Fig. 1-24.   Comparison of the calculated depths of analytical x-ray excitation in a cellulose specimen bombarded by 2.5-MeV protons or Sm-Kα x-rays (data from Ahlberg, 1977).

the detector. As the analytical x-ray energy increases, however, excitation depth becomes more important and the photon-excitation method becomes more favorable. Even 2-MeV protons do not penetrate into a specimen more than about 10 to 30 μm (which, nevertheless, is a tremendous improvement over the 1- to 2-μm penetration depth observed with electrons) and thus cannot excite atoms below this depth. For Ba-Kα x-rays at 32 keV, on the other hand, approximately 1.8 cm of the cellulose specimen ($\sim$ 500 μm for an apatite specimen) is sampled to yield a significant volume advantage over proton excitation.

In terms of total x-ray yield, therefore, SRIXE is clearly the most sensitive technique of the three because of its high intensity coupled with the inherent efficiencies of photon ionization and photon transmittance in the specimen.   PIXE spectroscopy has a decided advantage over conventional XRF spectroscopy for the light elements because of its high ionization intensity. However, increasingly poor ionization cross sections and a small proton penetration depth cause this advantage to evaporate with increasing atomic number. By the middle of the first transition row (atomic number 25), PIXE and conventional XRF have about equal sensitivities and, by the end of the row (atomic number 35), conventional XRF is the more sensitive of the two. In some instances, however, the relatively constant depth of excitation offered by PIXE might be an advantage, especially in analyzing a heterogeneous specimen where composition changes significantly with depth.

We have not spoken about PIGE spectroscopy in this context, largely because it is usually limited to analysis of light elements that are difficult to quantify by conventional x-ray techniques and, consequently, is viewed as a complementary technique. It shares the excitation process used by PIXE, which limits the sample penetration depth to about 10 to 30 μm. The nuclear-reaction cross sections are generally lower than the analogous ionization cross sections for PIXE or XRF and vary widely among nuclides and particle energies, but the transmissivity of the analytical photons is generally much higher than for most x-rays. Because it does not involve electronic energy levels, PIGE spectroscopy is entirely independent of the bonding environment of the nuclides sampled and, thus, avoids some of the uncertainties in fluorescent yields and other fundamental parameters that tend to interfere in the x-ray analysis of light elements. For this reason alone, PIGE spectroscopy should be viewed as the best technique for these elements (Li through Si), even though absolute sensitivity might be less.

Experimental Comparison of X-Ray Fluorescence- and Particle-Induced X-Ray Emission Spectroscopy Results

A variety of samples representative of soils, sediments, and fly ash were analyzed in our laboratories in order to provide a limited comparison of the accuracy and precision of the XRF and PIXE techniques. Three of these samples were standard reference materials provided by various research organizations: SRM 1633a Fly Ash from the National Institute of Standards and Technology (NIST), Gaithersburg, Maryland; Soil 5 from the International Atomic Energy Agency (IAEA), Vienna, Austria; and Marine Harbor Sediment PACS-1 from the National Research Council of Canada (NRCC), Ottawa, Canada. The remaining two samples were a calcareous sediment from the Upper Ringold Formation on the Hanford Site near Richland, Washington, and a high-organic muck soil from Pierce County, Washington. These two samples were selected because (i) they contained high amounts of inorganic and organic C, respectively, and, as a consequence of neither technique being able to analyze for C, posed an analytical challenge, and (ii) their elemental compositions were completely unknown, thus freeing our results from any potential bias.

Methods

For analysis by PIXE, duplicate specimens were prepared as 13-mm-diam. pellets after grinding to pass a 53-μm sieve and diluting with one part of briquetting graphite to three parts sample. These pellets were mounted on the target wheel in our experimental chamber (Fig. 1-17 top) and bombarded with 2-MeV protons. Three separate runs, differing in proton beam intensity ($\sim$1, 8, and 200 nA, respectively), and total proton charge collected ($\sim$0.4, 4.3, and 200 μC, respectively), were made for each specimen. During the first run (1 nA and 0.4 μC), no filter was placed between the specimen and the x-ray detector. A 14-μm Al foil was used to filter the analytical x-rays for the second run, and a compound filter consisting of aluminized Mylar (2.5 μm Al and 38.1 μm Mylar), muscovite mica ($\sim$65 μm), and Cr foil (10 μm) was used for the third run. Typical spectra for these three runs are shown in Fig. 1-18. Total proton charge collected on each specimen was determined from the number of protons backscattered into a collimated surface barrier detector by the gold-coated paddlewheel (Fig. 1-17 top). The analytical x-rays were measured with a collimated Si(Li) detector, and the output was amplified, digitized, and stored as counts in a multichannel analyzer. The total data collection time was approximately 35 min for each specimen. Data reduction was performed on a personal computer using PNLPIXAN, a modified version of the PIXAN code developed by Clayton (1986). Spectra for a standard specimen (in this instance, BCR-1, a basalt rock standard from the U. S. Geological Survey, Reston, VA) that had been analyzed in each run were fit initially. The calibration parameters obtained from fitting the spectra for the standard specimen were then used to perform calculations on the unknown specimens.

Analysis by XRF proceeded with duplicate undiluted specimens that had been pressed into self-supporting wafers approximately 25 cm in diameter. These specimens were successively irradiated with Gd-, Ag-, Zr-, or Ti-K$\alpha\beta$ x-rays (fluoresced from a secondary target irradiated with W x-rays) for about 30 min to give a total irradiation time of about 2 h per specimen. The analytical x-rays were detected with a Si(Li) detector connected to an amplifier, digitizer, and multichannel analyzer system similar to the PIXE arrangement. The coherent and incoherent scattering intensities of the incident x-rays were also measured for each specimen in order to estimate the light element composition (e.g., C and O) of the specimen. Data reduction was performed with the SAP-BFP back-scatter fundamental parameters code developed at the Pacific Northwest Laboratory (Nielson & Sanders, 1982).

The concentrations of Na and Mg in these samples were determined by AAS after decomposition of subsamples in an $HF-H_2SO_4$ acid matrix. Subsamples from each of the five samples were also ignited for 1 h at 900 °C and the weight loss (LOI) determined for eventual mass balance reconciliation with the oxide sums obtained by the spectroscopic techniques.

Results and Discussion

The mean concentrations obtained by PIXE and XRF for five major and five trace elements, along with values given in reference compilations, are shown in Table 1-2 for the three standard reference materials analyzed. All elements shown were quantified from the intensities of their $K\alpha$ x-ray lines except Pb, for which the $L\alpha$ x-ray lines at 10.55 keV were used. These elements were selected to cover a wide range in analytical x-ray energy (1.47-32.07 keV) and, except for Pb, are ranked from lowest to highest in order of their x-ray energies. Using a similar format, the mean concentrations obtained for the two unknown samples are shown in Table 1-3. The results of the mass-balance calculations, in which the elemental concentrations are converted to oxide percents and summed along with the loss-on-ignition (LOI) values to (hopefully) total 1000 g kg^{-1}, are given in Table 1-4.

An initial inspection of the results in Table 1-2 shows reasonable agreement between the values obtained by the two techniques and the reference values. Notable exceptions include the values for Si and Ba in the NIST Fly Ash and the value for Si in IAEA Soil 5. Two of these reference values are not certified and it is quite possible that they are incorrect.

The PIXE and XRF data for the unknown samples (Table 1-3) also agree well. Comparison of the PIXE and XRF data in both tables shows that the standard deviations of the PIXE determinations of the low-Z elements (i.e., Al and Si) are generally much smaller than those of the XRF determinations. On the other hand, the XRF determinations generally have slightly lower standard deviations for the high-Z elements (i.e., Zn through Ba). In general, the results of the mass-balance calculations showed reasonably good agreement with 1000 g kg^{-1} for both techniques, even for the two unknown samples with large C contents (Table 1-4).

To help quantify the trends in accuracy and precision, two sets of calculations were performed using the data in Tables 1-3 and 1-4. First, the accuracy of each method was estimated by taking the absolute value of the difference between the experimentally determined concentration and the *certified* reference value, dividing this number by the reference value, and then multiplying by 100 to yield a percentage. The calculation was not performed if a certified value was not available (e.g., Si in IAEA Soil 5, or any of the results in Table 1-3). The result of this calculation was termed the "relative accuracy" because the deviations from the "true", i.e., certified, values are expressed as percentages of the true values rather than as absolute numbers. Second, the precision for each technique was estimated by taking the standard deviation for the experimentally determined concentration, dividing it by the mean for that determination, and then multiplying by 100 to yield a percentage point. This result was termed the "relative precision" and is identical to the relative standard error or relative standard deviation used in statistics. The use of these relative values allows comparisons to be made among different elements within a method, for the same element across methods, and, when averaged, for all elements across the two methods.

The results of the relative accuracy and relative precision calculations are shown in Fig. 1-25. On average, the PIXE results were less accurate than the XRF data (6.5 and 5.0% deviations, respectively), particularly for the elements present in the highest concentrations (Al, Si, and Fe). One reason for this might be the use of three databases to model the PIXE yields (one for the proton attenuation in the specimen, one for ionization cross sections, and one for the x-ray attenuation before emerging from the specimen), whereas only two databases are needed for the analogous models of XRF yields (attenuation of excitational X-rays and analytical X-rays are

Table 1-2. Concentrations of selected major and trace elements in three standard reference materials as determined by proton-induced x-ray emission (PIXE) and x-ray fluorescence (XRF) and as given in reference compilations. Values shown are means from analysis of two replicate specimens; standard deviations are enclosed in parentheses.

Element	NIST 1633a Fly Ash			IAEA Soil 5			NRCC Harbor Sediment PACS-1		
	PIXE	XRF	Reference[†]	PIXE	XRF	Reference[§]	PIXE	XRF	Reference[§]
						--- g kg^{-1} ---			
Al	156.1 (3.5)	154.9 (19.5)	143.	71.1 (1.0)	90.5 (7.9)	81.9	58.4 (1.6)	64.3 (2.7)	64.7
Si	257.0 (1.9)	241.5 (12.0)	228.	265.8 (0.1)	276.5 (7.9)	330.[¶]	258.0 (3.3)	257.8 (4.9)	260.4
K	19.2 (0.4)	19.0 (1.4)	18.8	20.3 (0.1)	19.2 (0.6)	18.6[¶]	12.4 (0.2)	11.6 (0.6)	12.5
Ca	10.6 (0.4)	12.6 (1.3)	11.1	22.7 (0.1)	24.2 (0.7)	22.[¶]	19.3 (0.0)	20.7 (0.4)	20.9
Fe	83.8 (6.7)	92.1 (0.0)	94.	49.1 (0.6)	46.8 (1.1)	44.5	51.5 (0.3)	47.3 (0.9)	48.7
						--- mg kg^{-1} ---			
Zn	225 (0)	225 (5)	220	418 (18)	382 (5)	368	834 (36)	765 (0)	824
Rb	136 (7)	134 (2)	131	130 (6)	129 (2)	138	40 (1)	42 (0.4)	--
Sr	844 (21)	819 (22)	830	343 (6)	347 (10)	330[¶]	278 (11)	270 (10)	277
Ba	1179 (28)	1284 (33)	1500[¶]	580 (29)	523 (12)	562	557 (28)	687 (2)	--
Pb	74 (3)	69 (4)	72	157 (13)	150 (2)	129	390 (8)	384 (6)	404

[†] National Institute of Standards and Technology SRM 1633a Certificate of Analysis.
[§] Potts et al. (1992).
[¶] Value not certified.

Table 1-3. Concentrations of selected major and trace elements in a calcareous sediment from the Upper Ringold Formation (Hanford Site, Washington) and an organic soil (Pierce Muck, Pierce County, Washington) as determined by proton-induced x-ray emission (PIXE) and x-ray fluorescence (XRF). Values shown are means from analysis of two replicate specimens; standard deviations are enclosed in parentheses.

Element	Upper Ringold sediment		Pierce Muck soil	
	PIXE	XRF	PIXE	XRF
	-- g kg^{-1} --			
Al	60.5 (1.5)	71.7 (7.7)	47.0 (0.1)	49.0 (4.7)
Si	237.0 (4.0)	228.3 (11.4)	171.2 (0.6)	161.2 (5.0)
K	16.7 (0.2)	14.0 (1.6)	8.6 (0.1)	6.8 (0.1)
Ca	59.7 (0.7)	60.7 (6.1)	26.6 (0.2)	23.7 (0.5)
Fe	38.0 (0.3)	36.2 (0.6)	22.5 (0.7)	20.2 (0.4)
	-- mg kg^{-1} --			
Zn	72 (4)	70 (5)	76 (1)	64 (3)
Rb	70 (4)	70 (1)	26 (1)	29 (1)
Sr	314 (3)	308 (6)	275 (5)	257 (20)
Ba	607 (165)	615 (16)	<320 ---	268 (15)
Pb	15 (6)	14 (1)	24 (5)	21 (3)

Table 1-4. Mean mass balance calculations for samples analyzed by proton-induced x-ray emission (PIXE) and x-ray fluorescence (XRF).

Sample	Method	Oxide sum	LOI (900°C)	Total
		-------------------- g kg^{-1} --------------------		
NIST Flyash 1633a	PIXE	1044	33	1077
	XRF	1016	33	1049
IAEA Soil 5	PIXE	896	60	956
	XRF	951	60	1011
NRCC PACS-1	PIXE	895	153	1048
	XRF	891	153	1044
Ringold sediment	PIXE	838	126	964
	XRF	830	126	956
Pierce Muck	PIXE	588	469	1057
	XRF	547	469	1016

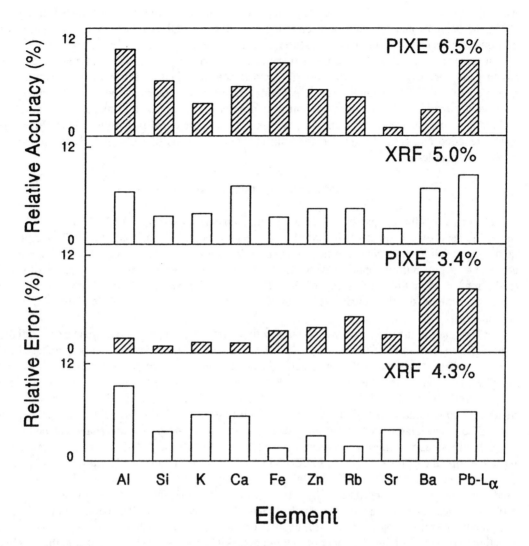

Fig. 1-25. Relative accuracies and relative errors obtained during analysis of three standard reference materials and two unknown sediments for major and minor elements by PIXE and XRF spectrometries. Mean values obtained by each technique for the elements listed are also shown to the right of each acronym.

handled by the same database). It is possible that the accuracy of the x-ray attenuation database is better than those of the proton-attenuation and ionization-cross-section databases. Another reason might be that the sample preparation technique, which involves dilution of the powdered sample in graphite powder, leads to greater difficulties in estimating the matrix composition and possibly to particle-size effects. The average relative precision of PIXE results, however, was very good (3.4%), particularly for the low-energy elements where the ionization cross sections for PIXE are most favorable. The relative precision of the results, therefore, suggests that the specimen preparation technique is reproducible and that the databases might be at fault. The worst precision was obtained for analysis of the Ba-Kα x-ray line at 32.07 keV, where the ionization cross section for protons is 100 000 times less than that for photons.

The relative precision for the XRF analyses was also good (4.3%), but clearly was better for the high-Z elements. This result can be explained by the much larger volume sampled by the higher-energy x-rays needed to excite the K-lines of these elements than with proton excitation and by the more favorable ionization cross sections for photon excitation. Both accuracy and precision were relatively poor (8.9 and 6.9% relative, respectively) for the analysis of Pb by PIXE or XRF using the Lα x-ray lines. This inaccuracy could be related to the relatively low concentrations of Pb in the samples and resultant poor counting statistics, as well as to the low ionization cross sections for L-shell x-ray production and the overlap of the Pb-Lα and As-Kα x-ray lines.

In general, the results of this limited experimental comparison of the PIXE and XRF techniques support the conclusions drawn from the theoretical considerations. However, even though the precision of the PIXE technique seems superior for the light elements, the poorer accuracy is somewhat surprising and suggests a potential problem with either the specimen preparation technique or the data reduction step. In any event, the accuracy and precision of both PIXE and XRF are good and can be relied on for bulk elemental analysis of soils, sediments, and other geological samples.

Economic and Logistical Comparisons

The costs and throughput associated with bulk elemental analysis by XRF, PIXE, PIGE, and SRIXE depend on a number of site-specific and macroeconomic factors and thus will vary from laboratory to laboratory. Nevertheless, some generalizations in the areas of capital costs, operating costs, and specimen throughput can be made about the different techniques and may prove of value to the reader. For this discussion, PIXE and PIGE will be considered together since they are typically applied simultaneously to the same specimen and require substantially the same equipment.

At first glance, the capital costs associated with the three techniques tend to scale exponentially in going from XRF to PIXE/PIGE to SRIXE. A first-rate XRF spectrometer can be purchased for roughly $150 000, whereas a Van de Graaff accelerator and associated beam line, sample analysis chamber, and electronics may run $1 000 000 to 1 300 000. An insertion device, beamline, and experimental station at a third-generation synchrotron facility may cost $3 000 000 to 6 000 000, in addition to the initial cost of the synchrotron facility itself ($\sim$ $500 000 000). The full capital costs of the accelerator and synchrotron facilities are rarely borne by the analytical end-user, however, inasmuch as these instruments are used for a variety of purposes. For example, the marginal capital cost for setting up a PIXE/PIGE analytical system may well be less than $100 000 for the sample chamber, detector, and electronics, if the accelerator is already in place and being used part-time for other purposes. Synchrotron beamlines may also be used for a number of different research and analytical activities, and the capital costs distributed among larger programmatic entities. Nevertheless, the capital costs are real, will fluctuate with interest rates and inflation, and tend to limit the number of individual facilities for each technique. Thus, the number of XRF facilities worldwide is probably on the order of 10^4-10^5, whereas PIXE/PIGE facilities and SRIXE facilities number about 10^2 and 10^1, respectively).

Estimates of operating costs for the various techniques depend on access costs and on sample throughput, which can be divided into components based on the labor required for specimen preparation, data collection, data reduction, and equipment maintenance and repair. Access costs will depend largely on the geographical location of the facility relative to the individual laboratory requiring the analysis. Thus, for XRF, these costs are generally nil, and for PIXE/PIGE they are relatively low, based on the number of these facilities available. Access to a synchrotron facility, on the other hand, may involve a substantial travel-time commitment on the part of the user as only a few of these facilities are in existence. Such access may also depend on participation in an organization formed to operate a particular beamline. Most synchrotron facilities, however, have set aside a substantial fraction of the total operating time to make it available for general users and that time is allocated in advance on the basis of proposals evaluated by a scientific peer-review process. Thus, the logistics of SRIXE spectrometry usually involve substantial advance planning to arrange for beamtime, transportation, and efficient specimen introduction to the analyzing beam during the hours allocated to a specific set of analyses.

Specimen preparation for any technique based on x-ray detection is similar and can be considered a constant across all the techniques. Data collection time, however, varies inversely with the ionization intensity of the technique and the detection limits desired. For a complete bulk elemental analysis at comparable detection limits, energy-dispersive XRF may require 2 h of irradiation per specimen, PIXE/PIGE on the order of 30 min, and white-radiation SRIXE only a few minutes. Because of the large databases and more complicated yield model, data reduction is more computationally intensive for PIXE/PIGE than for XRF or SRIXE. From a labor standpoint, however, the time involved in fitting standard spectra is probably comparable across the techniques and continuing improvements in the speed and capabilities of personal computers tend to reduce the computational costs to an insignificant level. The reliability and availability of the equipment associated with these techniques varies with the complexity of the equipment. The XRF spectrometer requires very little in the way of maintenance and is very reliable. The Van de Graaff accelerator is less reliable and may be down for maintenance and repairs several days to weeks of each year. The synchrotron facilities have regularly scheduled preventive maintenance that typically amount to 3-5 wk yr^{-1}, in addition to any time needed for facility modifications and unanticipated repairs.

The limiting factor in sample throughput for these elemental techniques, therefore, is the data-collection step. Under ideal conditions, and with comparable elemental sensitivities, energy-dispersive detection, and maximum utilization of the resource, about 12 to 18 samples a day can be analyzed by XRF, 50 to 100 by PIXE/PIGE, and perhaps as many as 300 by SRIXE. As many as 60 specimens per day can be analyzed by automated wavelength-dispersive XRF spectrometers. The actual costs to the individual requiring the data will vary widely according to the number of samples and geographic adjacency to the facility. In our own laboratories, however, typical fully burdened costs per specimen for analyses by energy-dispersive PIXE and XRF are comparable.

FUTURE DEVELOPMENTS

Although historically the need for quantitative mineralogical information about soils and sediments has arisen in the context of agriculture, mineral exploration, and large-scale construction projects (e.g., highways, dams, and buildings), an increasing share of this type of information is expected to derive from site-characterization activities for environmental remediation. Rapid, sensitive, accurate, and inexpensive bulk elemental analysis is a key component in the suite of mineralogical characterization activities and its use should grow hand in hand with the need for quantitative soil mineralogical information. Future developments in nondestructive bulk elemental analysis are expected to include the increased use of several new techniques, as well as the enhancement of existing techniques by improvements in the basic instrumentation used to collect spectral data and the software used to extract quantitative elemental data from the spectra. What follows is a summary of our speculations and those gleaned from the literature (Gilfrich, 1990; Vrebos & Kuiperes, 1991).

The need for spatially resolved elemental information will encourage further development of x-ray microprobes and their use in two- and three-dimensional mapping of elemental distribution in bulk soils and individual soil particles. For example, x-ray tomography of Fe-and Mn-nodules, weathered biotite and feldspar particles, and the regions surrounding plant roots in undisturbed soils and pores in metal-contaminated soils will become one of the standard tools used in characterization activities. Although currently linked primarily to synchrotron x-ray sources, tomography techniques might be extended to more conventional x-ray sources through the development of new focusing optics that rely on total internal reflectance of x-rays in tapered capillary tubes (Thiel et al., 1992).

The use of particle accelerators and synchrotron sources will increase, partly as a result of advantages in intensity, but also as a result of their versatility. The versatility, in particular will lead to the hybridization of several techniques and their simultaneous use on a single specimen. For example, the PIXE and PIGE techniques are already routinely applied simultaneously to the same specimen, and the addition of other accelerator-based techniques, such as Rutherford back-scattering, may become more common. The proton microprobe is already finding increased usage for sensitive (ppm level) multielemental analysis of individual mineral grains and inclusions.

Advances in instrumentation will include improved optical configurations and detector capabilities. X-ray focusing techniques have already been mentioned and are critical to the development of the microprobe. In addition, continued development of new crystals for wavelength dispersion (Gilfrich et al., 1982) produced by vacuum deposition of layers of atoms having very different electron densities (e.g., W and C) will lead to enhanced detection of low-energy x-rays such as C and O. The use of x-ray diffraction near 90° 2Θ to produce polarized radiation that has a much lower background than unpolarized x-rays will also improve detection limits (Hanson, 1986; Ryon & Zahrt, 1979). Improvements in x-ray detectors include (i) the development of new ultratransparent windows for detectors to extend the useful range in the soft x-ray region from Na to B (Bogert, 1988), (ii) continued development of position-sensitive detectors for rapid collection of wavelength-dispersed radiation, and (iii) superconducting tunnel-diode detectors with the eventual promise of 10- to 20-eV resolution rather than the present limit of about 180 eV with Si(Li) detectors (Twerenbold, 1986).

Last, the analytical software and databases used to estimate elemental concentrations from spectral intensities will continue to improve. One example of this trend is a new back-scatter/fundamental parameters code for XRF that combines single-peak intensity data from wavelength-dispersive systems with the continuous-spectrum type of data produced by energy-dispersive analysis (Arthur & Sanders, 1992). The resulting composite spectrum combines the high resolution and sensitivity offered by wavelength-dispersion for the light elements with the similar advantages of energy-dispersive detection for the heavier elements.

SUMMARY

Determination of the elemental composition of a soil sample is one of a suite of characterization activities that eventually lead to the identification and quantification of the mineral phases present. Nondestructive techniques for elemental analysis are generally used where rapid multielemental analysis of bulk samples is needed, and these techniques typically involve the measurement of characteristic x-rays or gamma-rays given off by specimens during exposure to a beam of high-energy x-rays or particles. These x-rays are produced by electronic transitions from high-energy atomic shells to fill vacancies created in shells of lower energy. The nondestructive x-ray based techniques reviewed in this chapter (XRF, PIXE, and SRIXE) can be distinguished largely by the source of the radiation or particles used to create inner-shell electron vacancies. The gamma-rays used for the PIGE technique, on the other hand, are produced by transitions from excited to ground states occurring in the nuclei of atoms as a result of collisions with high-energy particles.

Because the probabilities of analytical x-ray and gamma-ray production by photon and particle excitation and the extent to which these photons are attenuated by the specimen and detection systems are reasonably well known, it is possible, by using the appropriate yield equation, to estimate the concentrations of analytes present in the specimens based on the observed intensities of the analytical photons. This approach is an iterative process, however, because some of the fundamental parameters in the yield equation depend on the composition of the specimen. An estimate of the specimen composition is made, this composition is used to assign values to parameters in the yield equation, and the equation is used to calculate the concentrations of all elements present. The newly calculated specimen composition is then substituted for the initial estimate, and the process repeated until no difference between the initial and final compositions is found.

Theoretical comparisons among the x-ray-based techniques show the photon-excitation used in XRF and SRIXE to be more efficient than excitation by 2-MeV protons, especially for heavier elements. However, the relatively high intensity of a typical proton beam gives PIXE a significant x-ray yield advantage over conventional XRF for the light elements. The overwhelming incident x-ray intensities of synchrotron sources combine with the inherent efficiency of photon excitation to make SRIXE the most sensitive of the three techniques. The PIXE also differs from XRF and SRIXE in that the volume of the specimen that contributes analytical x-rays is limited by the penetrating ability of the protons and is essentially constant for x-rays of energy greater than

3 keV. In contrast, the sampling volume for XRF and SRIXE varies according to the energy of the incident x-ray.

A limited comparison of actual PIXE and XRF analyses of three geochemical reference materials and two unknown soils by our laboratories showed good agreement between the techniques and with published reference concentrations. Trends in the precision of the data showed PIXE to be the better technique for light elements and XRF for the heavier elements, in good agreement with the theoretical predictions. However, the accuracy of the PIXE data was slightly less than that of XRF, even for the light elements. This result was attributed to more uncertainty in the databases used to estimate PIXE yields than those used for XRF.

Although the capital costs associated with each of the techniques vary exponentially in going from XRF through PIXE/PIGE to SRIXE, the actual operating costs borne by the individual analyst will depend largely on geographical proximity to the facility and sample throughput. The number of facilities available varies inversely with the capital cost, whereas the sample throughput and logistical considerations scale directly with the capital cost. As a consequence, the selection of the appropriate technique will depend on the individual circumstances of the analyst. In our laboratories, where both PIXE and XRF are available, analytical costs per specimen are roughly comparable.

Future developments in nondestructive analytical techniques are expected to include an increased emphasis on spatially resolved data from intact specimens made possible by the availability of extremely bright x-rays from synchrotrons and advances in focusing optics for x-rays. The development of hybridized techniques (e.g., PIXE/PIGE) based at accelerators or synchrotrons will continue as new ways are found to extract information from the interactions of photons and particles with matter. And advances in x-ray optics and detectors will extend the range and sensitivity of the x-ray based techniques to lighter elements and lower detection limits.

Much remains to be done. As exemplified by the emergence of accelerator- and synchrotron-based nondestructive techniques over the last two decades, new ways of quantifying elemental concentrations in matter will continue to evolve and with them our ability to characterize, understand, and manipulate soil systems to our benefit.

ACKNOWLEDGMENTS

We are grateful to Don Baer, Mark Engelhard, Mark Middendorf, Alistair Leslie, Kirk Nielson, and Julie Olivier for their early work in developing the PIXE and XRF capabilities at the Pacific Northwest Laboratory (PNL), and to the Geosciences Department at PNL for partial support during the preparation of the manuscript. The manuscript benefited from careful reviews by Richard J. Arthur, James L. Campbell, Laurel K. Grove, A. D. (Tasos) Karathanasis, and one anonymous reviewer. Pacific Northwest Laboratory is operated for the U.S. Dep. of Energy by Battelle Memorial Institute under Contract DE-AC06-76RLO 1830.

REFERENCES

Ahlberg, M. S. 1977. Comparison of yield versus depth for particle-induced and photon-induced x-ray emission analysis. Nucl. Instrum. Meth. 146:465-467.

Alvarez, M. 1990. Glass disk fusion method for the x-ray fluorescence analysis of rocks and silicates. X-ray Spectrom. 19:203-206.

Anttila, A., R. Hanninen, and J. Raisanen. 1981. Proton-induced thick-target gamma-ray yields for the elemental analysis of the Z=3-9,11-21 elements. J. Radioanal. Chem. 62:293-306.

Arndt, U. W. 1986. X-ray position-sensitive detectors. J. Appl. Crystallogr. 19:145-63.

Arthur, R. J., and R. W. Sanders. 1992. Backscatter/fundamental-parameters analysis of unweighed samples using multi-target, multi-crystal regions of interest from WDXRF and EDXRF. Adv. X-ray Anal. 35:1101-1106.

Augustithis, S. S. 1988. Synchrotron radiation applications in mineralogy and petrology. Theophrastus Publi., Little Compton, RI.

Azaroff, L. 1974. X-ray spectroscopy. McGraw-Hill Publ. New York.

Bambynek, W. 1984. Private communication of material presented verbally at the international conference on x-ray and inner shell processes in atoms, molecules, and solids, Univ. of Leipzig. p. 11-12. *In* S. A. E. Johansson and J. L. Campbell (ed.) PIXE: A novel technique for elemental analysis. John Wiley & Sons, Chichester, England.

Berger, M. J., and J. H. Hubbell. 1987. XCOM. Photon cross sections on a personal computer. Report NBSIR 87-3597. U.S. Dep. of Commerce, Gaithersburg, MD.

Bertin, E. P. 1975. Principles and practice of x-ray spectrometric analysis. 2nd ed. Plenum Publ. Corp., New York.

Bertin, E. P. 1978. Introduction to x-ray spectrometric analysis. Plenum Publ. Corp., New York.

Bertin, E. P. 1981. X-ray spectrometric analysis: principles, instrumentation, practice, and applications. A detailed illustrated outline of a series of lectures presented during an integrated short-course on x-ray spectrometry at the Dep. of Physics, State Univ. of New York, Albany. June 1981. E.P. Bertin.

Bogert, J. R. 1988. Advances and enhancements in light element EDXRF. Adv. X-ray Anal. 31:449-454.

Boni, C., A. Caridi, E. Cereda, and G.M. Braga Marcazzan. 1990a. PIXE-PIGE analysis of thin fly-ash samples. Nucl. Instrum. Methods Phys. Res. B45:352-355.

Boni, C., A. Caridi, E. Cereda, and G.M. Braga Marcazzan. 1990b. A PIXE-PIGE setup for the analysis of thin samples. Nucl. Instrum. Methods Phys. Res. B47:133-142.

Briggs, D. 1983. Practical surface analysis by auger and x-ray photoelectron spectroscopy. John Wiley & Sons, New York.

Burkhalter, P. G., and W. J. Campbell. 1967. Comparison of detectors for isotopic x-ray analyzers. p. 393-423. *In* P. S. Baker and M. Gerrard (ed.) Proc. 2nd Symp. onLlow-Energy X- and Gamma Sources and Applications, Austin, TX. 27-29 March, ORNL-IIC-10. Vol. 1. Oak Ridge Natl. Lab., Oak Ridge, TN.

Bygrave, W. P. Treado, and J. Lambert. 1970. Accelerator nuclear physics: fundamental experiments with a Van de Graaff accelerator. High voltage Engineering Corp., Burlington, MA.

Cabri, L. J. 1988. Applications of proton and nuclear microprobes in ore deposit mineralogy and metallurgy. Nucl. Instrum. Methods Phys. Res. B30:459-465.

Cahill, T.A. 1980. Proton microprobes and particle-induced x-ray analytical systems. Annu. Rev. Nucl. Part. Sci. 30:211-252.

Campbell, J.L., J.A. Maxwell, W.J. Teesdale, and J.-X. Wang. 1990. Micro-PIXE as a complement to electron microprobe microanalysis in mineralogy. Nucl. Instrum. Methods Phys. Res. B44:347-356.

Campbell, J.L., J.-X. Wang, J.A. Maxwell, and W.J. Teesdale. 1989. An exact treatment of secondary and tertiary fluorescence enhancement in PIXE. Nucl. Instrum. Methods Phys. Res. B43:539-555.

Carlson, T. A. 1975. Photoelectron and auger spectroscopy. Plenum Press, New York.

Catlow, C. R. A., and G. N. Greaves. 1990. Applications of synchrotron radiation. Blackie, New York.

Chen, J.R., E.C.T. Chao, J.A. Minkin, J.M. Back, K.W. Jones, M.L. Rivers, and S.R. Sutton. 1990. The uses of synchrotron radiation sources for elemental and chemical microanalysis. Nucl. Instrum. Methods Phys. Res. B49:533-543.

Chen, M.H., and B. Crasemann. 1985. Relativistic cross sections for atomic K- and L-shell ionization by protons, calculated from a Dirac-Hartree-Slater model. At. Data Nucl. Data Tabl. 33:217-233.

Chen, M. H., B. Crasemann, and H. Mark. 1981. Widths and fluorescence yields of atomic L-shell vacancy states. Phys. Rev. A 24:177-182.

Clayton, E. 1986. PIXAN: The Lucas Heights PIXE analysis computer package. Aust. At. Energy Comm., Lucas Heights, Australia.

Cohen, D. D. 1987. Average L-shell fluorescence yields. Nucl. Instrum. Methods Res. Phys. B22:55-58.

Cohen, D. D., and M. Harrigan. 1985. K-shell and L-shell ionization cross sections for protons and helium ions calculated in the ECPSSR theory. At. Data Nucl. Data Tabl. 33:255-343.

Cohen, D.D., and M. Harrigan. 1986. Calculated L-shell x-ray-line intensities for proton and helium ion impact. At. Data Nucl. Data Tabl. 34:393-414.

Eastell, J., and J.P. Willis. 1990. A low dilution fusion technique for the analysis of geological samples. X-ray Spectrom. 19:3-14.

Fleming, S.J., and C.P. Swann. 1987. Color additives and trace elements in ancient glasses: specialized studies using PIXE spectrometry. Nucl. Instrum. Methods Phys. Res. B22:411-418.

Fraser, G.W. 1984. X- and γ-ray imaging using microchannel plates. Nucl. Instrum. Methods Phys. Res. A221:115-130.

Gihwala, D., and M. Peisach. 1992. Analysis using prompt gamma-ray emission. p. 307-348. In Z. B. Alfassi and M. Peisach (ed.) Elemental analysis by particle accelerators. CRC Press, Boca Raton, FL.

Gilfrich, J.V., D.J. Nagel, and T.W. Barbee, Jr. 1982. Layered synthetic microstructures as dispersing devices in x-ray spectrometers. Appl. Spectrosc. 36:58-61.

Gilfrich, J. V. 1990. New horizons in x-ray fluorescence analysis. X-ray Spectrom. 19:45-51.

Hanson, A. L. 1986. The polarization of x-rays scattered into 90°. Nucl. Instrum. Methods Phys. Res. A249:515-521.

Hubbell, J.H., W.J. Veigele, E.A. Briggs, R.T. Brown, D.T. Cromer, and R.J. Howerton. 1975. Atomic form factors, incoherent scattering functions, and photon scattering cross sections. J. Phys. Chem. Ref. Data 4:471-538.

Jaklevic, J.M., R.D. Giauque, and A.C. Thompson. 1990. Recent results using synchrotron radiation for energy-dispersive X-ray fluorescence analysis. X-ray Spectrom. 19:53-58.

Jenkins, R. 1974. An introduction to x-ray spectrometry. Heyden & Son, Ltd., London.

Johansson, S.A.E., and J.L. Campbell. 1988. PIXE: A novel technique for elemental analysis. John Wiley & Sons, Chichester, England.

Johansson, S.A.E., and T.B. Johansson. 1976. Analytical application of particle induced X-ray emission. Nucl. Instrum. Methods 137:473-516.

Johansson, T.B., K.R. Akselsson, and S.A.E. Johansson. 1970. X-ray analysis: Elemental trace analysis at the 10^{-12} g level. Nucl. Instrum. Methods Phys. Res. 84:141-143.

Jones, A.A. 1982. X-ray fluorescence spectrometry. p. 85-121. In A. L. Page et al. (ed.) Methods of soil analysis. Part 2. 2nd ed. Agron. Monogr. 9. ASA, Madison, WI.

Jones, A.A. 1991. X-ray fluorescence analysis. p. 287-324. In K. A. Smith (ed.) Soil analysis: Modern instrumental techniques. 2nd ed. Marcel Dekker, New York.

Jones, K.W., and B.M. Gordon. 1989. Trace element determinations with synchrotron-induced X-ray emission. Anal. Chem. 61:341A-356A.

Krause, M.O. 1979. Atomic radiative and radiationless yields for K and L shells. J. Phys. Chem. Ref. Data 8:307-327.

Leroux, J., and T.P. Thinh. 1977. Revised tables of x-ray mass attenuation coefficients. Corp. Sci. Claisse, Quebec.

Margaritondo, G. 1988. Introduction to synchrotron radiation. Oxford University Press, London.

Mitchell, I.V., and J.F. Ziegler. 1977. Ion induced x-rays. p. 311-484. In J. W. Mayer and E. Rimini (ed.) Ion beam handbook for material analysis. Acad. Press, New York.

Nielson, K.K., and R.W. Sanders. 1982. The SAP3 computer program for quantitative multielement analysis by energy dispersive x-ray fluorescence. PNL-4173. Pacific Northwest Lab., Richland, WA.

Norrish, K., and J.T. Hutton. 1969. An accurate x-ray spectrographic method for the analysis of a wide range of geological samples. Geochim. Cosmochim. Acta 33:431-453.

Osborn, J.C., and T.R. Welberry. 1990. A position-sensitive detector system for the measurement of diffuse x-ray scattering. J. Appl. Crystallogr. 23:476-484.

Paul, H., and J. Sacher. 1989. Fitted empirical reference cross sections for K-shell ionization by protons. At. Data Nucl. Data Tabl. 42:105-156.

Pineda, C.A. 1992. Thick target particle-induced x-ray emission. p. 279-305. In Z. B. Alfassi and M. Peisach (ed.) Elemental analysis by particle accelerators. CRC Press, Boca Raton, FL.

Potts, P.J., A.G. Tindle, and P.C. Webb. 1992. Geochemical reference material compositions. Rocks, minerals, sediments, soils, carbonates, refractories, and ores used in research and industry. CRC Press, Boca Raton, FL.

Raisanen, J. 1990. Experimental arrangements for the simultaneous use of PIXE and complementary accelerator based techniques. Nucl. Instrum. Methods Phys. Res. B49:39-45.

Reuter, W., A. Lurio, F. Cardone, and J. F. Ziegler. 1975. Quantitative analysis of complex targets by proton-induced X-rays. J. Appl. Phys. 46:3194-3202.

Rogers, P. S. Z., C. J. Duffy, and T. M. Benjamin. 1987. Accuracy of standardless nuclear microprobe trace element analysis. Nucl. Instr. Meth. Phys. Res. B22:133-137.

Russ, J.C. 1984. Fundamentals of energy dispersive x-ray analysis. Butterworths, London.

Ryan, C.G., D.R. Cousens, S.H. Sie, W.L. Griffin, G.F. Suter, and E. Clayton. 1990. Quantitative PIXE microanalysis of geological material using the CSIRO proton microprobe. Nucl. Instrum. Methods Phys. Res. B47:55-71.

Ryon, R.W., and J.D. Zahrt. 1979. Improved x-ray fluorescence capabilities by excitation with high intensity polarized x-rays. Adv. X-ray Anal. 22:453-460.

Salem, S.I., S.L. Panossian, and R.A. Krause. 1974. Experimental K and L relative x-ray emission rates. At. Data Nucl. Data Tabl. 14:91-109.

Schulze, D., and J.V. Smith (ed.). 1990. Synchrotron x-ray sources and new opportunities in the soil and environmental sciences: Workshop report. Proc. Workshop Argonne Natl. Lab., Argonne, IL. 8-10 January. ANL/APS-TM-7. Argonne Natl. Lab., Argonne, IL.

Scofield, J.H. 1973. Theoretical photoionization cross sections from 1 to 1500 keV. UCRL-51326. Univ. of California at Livermore, Lawrence Livermore Lab., Livermore, CA.

Scofield, J.H. 1974. Hartree-Fock values of L X-ray emission rates. Phys. Rev. A 10:1507-1510.

Smith, G.C. 1984. High accuracy gaseous x-ray detectors. Nucl. Instrum. Methods Phys. Res. A222:230-237.

Smith, J.V., and M. Manghnani (ed.). 1988. Synchrotron x-ray sources and new opportunities in the earth sciences: workshop report. Proc. Workshop Argonne Natl. Lab., Argonne, IL. 18-20 January. ANL/APS-TM-3. Argonne Natl. Lab., Argonne, IL.

Thiel, D.J., D.H. Bilderba, A. Lewis, E.A. Stern, and T. Rich. 1992. Guiding and concentrating hard x-rays by using a flexible hollow-core tapered glass fiber. Appl. Optics 31:987-992.

Thiesen R., and D. Vollath. 1967. Tables of x-ray mass attenuation coefficients. Verlag Stahleisen M. B. H., Dusseldorf, Germany.

Thinh, T.P., and J. Leroux. 1979. New basic empirical expression for computing tables of x-ray mass attenuation coefficients. X-ray Spectrom. 8:85-91 (errata published in 10:v).

Twerenbold, D. 1986. Giaver-type superconducting tunneling junctions as high-resolution x-ray detectors. Europhys. Lett. 1:209-214.

Veigele, W.J. 1973. Photon cross sections from 0.1 keV to 1 MeV for elements $Z=1$ to $Z=94$. At. Data Tabl. 5:51-111.

Vrebos, B.A.R., and G.T.J. Kuiperes. 1991. Trends in quantification in XRF. X-ray Spectrom. 20:5-7.

Wätjen, U. 1987. Currently used computer-programs for PIXE analysis. Nucl. Instrum. Methods Phys. Res. B22:29-33.

White-DePace, S., N.F. Gmur, R. Garrett, J. Jordan-Sweet, J. Phillips, J. Preses, and W. Thomlinson. 1988. National synchrotron light source experimenter's handbook. Brookhaven Natl. Lab., Upton, NY.

Winick, H., and S. Doniach (ed.). 1980. Synchrotron radiation research. Plenum Press, New York.

Zahorowski, W., J. Mitternacht, and G. Wiech. 1991. Application of a position-sensitive detector to soft x-ray emission spectroscopy. Meas. Sci. Technol. 2:602-609.

2   Dissolution and Elemental Analysis of Minerals, Soils and Environmental Samples

B. L. Sawhney and D. E. Stilwell
Connecticut Agricultural Experiment Station
New Haven, Connecticut

Quantitative analytical techniques for elemental analysis of minerals, soils, sediments, rocks, and environmental samples such as composts and incinerator ashes, etc., generally require that the samples be decomposed first into soluble forms and dissolved completely. Two common methods for the decomposition and solubilization of samples are fusion with an alkali flux and digestion in an acid or combination of acids. Classical techniques for elemental analysis mainly involved gravimetric determinations of elements where individual components were sequentially precipitated from solution and weighed. Principles of classical analytical methods have been discussed by Hildebrand (1919) and Washington (1930). In the 1950s, rapid schemes were developed where flame photometers, spectrophotometers and complexometric titrations were used. Details of these procedures are given by Jeffery (1970). Total elemental analysis is now carried out using modern instrumental techniques that permit rapid, multielement analysis of many major and minor elements. The most common instrumental methods for elemental analysis of solutions are Atomic Absorption Spectrometry (AAS) and Atomic Emission Spectrometry (AES). Procedures for sample preparation and analysis of rocks and minerals are described in detail by Johnson and Maxwell (1981). This chapter is concerned primarily with procedures for the decomposition and dissolution of samples, and the principles and applications of the two instrumental methods, AAS and AES.

DECOMPOSITION AND DISSOLUTION

Fusion with Alkali Salts

Common fluxes used for silica-rich materials such as minerals and soils include Na_2CO_3, NaOH, Na_2O_2 (or equivalent K salts), or $LiBO_2$. Fusion decomposes silicate constituents into compounds that are soluble in the acid used for dissolving the fused melt. For complete decomposition, it is essential that the sample be finely ground and mixed thoroughly with a large excess of the flux, followed by high-temperature (about 1000 °C) fusion. Although a high flux:sample ratio (generally 5:1) is used for silica-rich materials, an inductively coupled plasma (ICP) technique employing ratios as low as 1:1 has been reported (Burman, 1987). Fusion methods have certain advantages over the acid decomposition methods described below. They do not require hazardous acids, such as HF and $HClO_4$, nor any specialized Teflon apparatus. Also, Li and borate ions in solution suppress some matrix effects, especially when AAS is used for analysis (Potts, 1987). The principal disadvantage of fusion methods is the high salt content of the resultant solution, which can clog the burner aperture and cause erratic performance of the nebulizer. Moreover, a high salt content can produce scattered light effects resulting in high background signals.

A procedure commonly used for fusion of soils and silicate minerals (Jackson, 1958) described by Lim and Jackson (1982) is given below.

Pretreatment to Remove Organic Matter

Because soils almost always contain organic matter, it is suggested that the soil sample be ignited to remove organic matter before fusion. Place 1 g representative sample (sieved through a 250-mesh sieve) in a tared, ignited 20- to 30-mL Pt crucible. Place crucible (partly covered) on a Nichrome wire triangle or silica tube triangle and ignite with a low flame until organic matter is burned off, continue ignition to about 900 °C (cherry red crucible bottom). Cool, weigh, and repeat ignition until a constant weight is obtained.

Pretreatment of Soils High in Iron and Manganese Oxides

Because iron and manganese form alloys with Pt during fusion (Kanehiro & Sherman, 1965), soils containing more than 400 g Fe_2O_3 kg^{-1} or 10 g MnO_2 kg^{-1} should be pretreated with H_2O_2 and HCl to prevent damage to the Pt crucibles. Place 0.5 to 1 g of ignited soil from previous paragraph in 150-mL beaker, add 10 mL 2 M HCl and 30 mL H_2O_2, digest at low temperature on a hot plate and evaporate to dryness. Add 10 mL 15 M HNO_3 and 20 mL 12 M HCl, digest on a hot plate and evaporate to dryness. Add 20 mL 6 M HCl and dissolve soluble material using stirring rod. Filter through low-ash filter paper, wash beaker and paper several times with 2 M HCl and then with H_2O. Save filtrate for determination of Si and other elements. Place filter paper plus residue in Pt crucible and ignite as in previous section.

Fusion with Sodium Carbonate

To the ignited sample from the previous paragraphs, add 4 to 5 g anhydrous Na_2CO_3 and mix thoroughly using a spatula or glass rod. Brush adhering particles into the crucible. Add 1 g Na_2CO_3 on top of mixture, place covered crucible at a slight angle on a triangle and warm over a Meker burner by heating with a low flame for about 10 min. With the cover slightly ajar, gradually increase the heat to about 90% of full flame (approximately 900 °C; Pt turns cherry red) and maintain this heat for 15 to 20 min. Increase heat to full flame and maintain heat at 1000 °C (Pt turns bright cherry red) for 5 to 10 min. Heat any patches of fused material sticking to crucible sides so that the fused material forms a single mass. No effervescence should exist at this stage. Uncover the crucible and heat for a few minutes more. Remove crucible with a pair of Pt-tipped tongs and swirl to solidify melt along crucible sides. If any cracks or bubbles appear, remelt cake and again heat the crucible to 1000 °C for 15 min.

Cool the crucible to room temperature, then add water to the crucible, warm slightly to loosen the cake and transfer to a Pyrex or Teflon beaker. Wash the crucible and its cover and add washings to the beaker. Add 5 mL 6 M HCl to the crucible, heat with low flame or on a hot plate to disintegrate any remaining cake, transfer to the beaker, add a few more drops of HCl and transfer all material from the crucible to the beaker.

If some fused cake still sticks to the crucible, transfer the crucible to the beaker and cover with water. Heat beaker gently and disintegrate the cake with a flattened stirring rod. Remove crucible and wash any adhering particles into the beaker. All soil constituents are now present as chlorides, except Si which is present as soluble silicic acid or as insoluble particles. Dehydrate the Si to render it insoluble by evaporating the solution to dryness at a temperature below 130 °C. Resuspend the residue in water, filter, and wash filter paper several times. Oven dry the filter paper, ignite and weigh the SiO_2. Analyze filtrate for other elements by instrumental methods.

Fusion with Lithium Metaborate

For many years, lithium metaborate ($LiBO_2$) has been used as a flux for geological samples, especially for trace-element analysis. Varying flux:sample ratios (1:1 to 7:1) have been used (Burman, 1987). However, the use of high flux:sample ratios can lead to clogging of the nebulizer in the sample introduction train of the atomic spectrometer, causing readings to drift continuously.

Graphite crucibles have been used for fusion with $LiBO_2$, and are preferred by some because they are cheaper and easier to use than Pt when running large numbers of samples. When decomposing/dissolving geological samples using an automatic tunnel furnace for fusion, Govindaraju et al. (1976) have used sets of 200 graphite crucibles. Trace elements, which are often difficult to determine in geological samples, can be separated from the $LiBO_2$-fused matter by using preconcentration techniques such as ion exchange (Barnes & Genna, 1979).

Digestion with Acids

Acid digestion is probably the most widely used method for samples containing silicate, primarily because it avoids the introduction of extraneous salts. Acid digestion is also less time-consuming, and expensive Pt crucibles are not required. Hydrofluoric acid, in combination with $HClO_4$ and HNO_3, is the acid of choice for decomposing Si-rich materials because it is the only acid which readily dissolves silicates. Perchloric acid is added to the mixture to oxidize organic matter (OM). To prevent the explosive reaction of $HClO_4$ with OM, small amounts of HNO_3 are added first to oxidize the more reactive constituents of OM at low temperature. Because HF can dissolve Si from glassware even at low concentrations, digestion is carried out in Pt or plastic containers.

Open Digestion Systems

Decomposition of silicates with HF involves its reaction with Si to form gaseous SiF_4, which then escapes during digestion. Hence, the open-digestion method is not applicable for Si determinations. The procedure for digestion of soils described here is from Jackson (1974). Place 0.1 g finely divided soil sample (increase sample size to 0.5 g if chemical rather than instrumental analysis of some elements is intended) in a 30-mL Pt crucible. Wet the sample with a few drops of H_2SO_4, add 5 mL HF and 0.5 mL $HClO_4$ (for organic soils add 3 mL HNO_3 and 1 mL $HClO_4$) and heat on a hot plate until $HClO_4$ fumes evolve. Cool the crucible and add 5 mL HF. Digest the sample on a sand bath at 200 to 225 °C, with crucible nine-tenths covered, and evaporate the contents to dryness. If OM stains remain, heat with a Meker burner until OM is oxidized. Remove crucible, add 5 mL 6 M HCl and 5 mL H_2O and heat gently to a boil. If sample is not dissolved completely, evaporate solution to dryness and repeat digestion. After dissolution in HCl, dilute sample to 50 mL for elemental analysis. Because of volatilization losses, elements whose fluorides have boiling points lower than 200 to 225 °C (e.g., Si, B, As, Sb, and S) cannot be analyzed using the open system.

Closed Digestion Systems

Plastic containers of polypropylene or polytetrafluoroethylene (PTFE), which resist attack by HF, are used as closed systems to prevent volatilization losses. The following is a typical method for acid digestion of geological samples (Potts, 1987). To 0.2 g of finely divided rock, add 0.5 to 1 mL aqua regia as a wetting agent, then add 5 mL 40% HF to decompose the silicates and heat at 100 to 110 °C for 30 min. If the solution is clear, add 50 mL saturated boric acid solution. If the solution is not clear, digest for longer periods. An excess of boric acid is added to neutralize the remaining HF (by complexing it as HBF_4), suppress the hydrolytic decomposition of HBF_4, and dissolve any insoluble metal fluorides that may have formed. Neutralization of the HF also prevents contamination from glass containers and protects the glass nebulizer of the spectrometer from damage.

Introduction of acid digestion bombs (Bernas, 1968; Langmyhr & Paus, 1968) has enhanced the efficiency of the acid digestion methods. A PTFE beaker containing the sample and acids is capped with a PTFE lid, fitted tightly into an outer stainless steel pressure jacket and heated at 110 to 150 °C for only a few minutes. Bombs sealed and heated must never be filled to more than 10 to 20% of their free volume, nor should organic reagents and oxidizing agents be added as these conditions may cause the pressure bomb to explode.

The closed-digestion method described by Jackson (1974) for soils is as follows. Exchangeable cations are replaced with NH_4^+ by treating the soil with 1 M $(NH_4)_2CO_3$ and drying at 110 °C. Transfer 0.2 g of the NH_4^+-saturated sample to a 250-mL wide-mouth polypropylene bottle, and add 1 mL aqua regia to disperse the sample and decompose any carbonates. Add 10 mL HF (48%), cap bottle immediately, shake for 2 to 8 h depending on sample type. If sample is still colored after 2 h, indicating incomplete dissolution, heat the bottle at about 100 °C for 30 min. Cool, add 100 mL saturated boric acid solution and dilute to 200 mL for elemental analysis.

Bakhtar et al. (1989) have recently developed a two-step dissolution process in which soil OM is first oxidized using an H_2O_2 treatment, followed by an HCl treatment. In the second step, digestion of the residue with dilute HF is carried out at low temperature in closed fluorinated ethylene propylene (FEP) centrifuge tubes. Their procedure follows: Suspend 1.0 g finely ground (60-mesh) sample in 5 mL 30% H_2O_2 in an FEP centrifuge tube. When effervescence has ceased, gradually add an additional 10 mL H_2O_2, and shake slowly overnight (16-18 h). Place tube in a digestion block (90 °C) and evaporate to about 5 mL. If the sample is still dark, repeat H_2O_2 addition and heat again. Add 5 mL concentrated HCl and, leaving the tubes open, shake for 16 to 18 h. Cap loosely and digest at 90 °C for 4 h. Remove, cool, and centrifuge at 1100 rpm for 20 min. Decant supernatant into 100-mL calibrated plastic containers. Resuspend residue in 10 mL 6 M HCl, shake for 4 h, and digest overnight at 90 °C. Centrifuge and combine the supernatants in the plastic container. Repeat HCl treatment five times or until a colorless solution is obtained. This is the HCl fraction.

Suspend residue in about 5 mL distilled water, shake vigorously for 2 h, add 4 mL HF, shake for 1 h and digest overnight at 90 °C. Centrifuge, and decant supernatant into a new plastic container. Resuspend any remaining residue in 10 mL H_2O, shake for 1 h, centrifuge, decant supernatant into the plastic container and dilute to 100 mL. This is the HF fraction. If any organic residue is observed, add 1 mL HNO_3, cap the tube and digest for 2 h (90 °C). Remove caps and heat until reddish fumes disappear, add 2 mL H_2O and continue digestion until bubbling ceases. This solution is then transferred to the HCl fraction, diluted to 100 mL with 6 M HCl and saved for elemental analysis.

Analyses of 25 major and trace elements in U. S. Geological Survey standard rock samples obtained by the two-step method compared favorably with other methods. The advantages of this procedure are that a large number of samples (50-100) can be processed simultaneously and the loss of volatile constituents is minimized. On the negative side, the procedure requires numerous time-consuming steps and would be difficult to automate.

Microwave Dissolution

Sample dissolution by conventional procedures, such as fusion and acid digestion, is clearly time consuming and tedious. Recent developments in microwave technology, combining the rapid heating ability of microwaves with the use of sealed digestion vessels, offer an alternative that saves time and is less hazardous.

Abu-Samra et al. (1975) were the first to demonstrate the use of microwave energy as a heat source in the digestion of plant and animal tissue prior to metal analysis. Thereafter, most early investigations were concerned with specific applications, primarily involving biological materials, and used open or covered vessels in household microwave ovens. Nadkarni (1980) studied the dissolution of a variety of materials, including powdered coal, fly ash, oil shales, rock, sediment, biological samples and a number of standard reference materials (SRMs). Complete dissolution of some materials was observed in as little as 3 min. Determination of about 25 elements using ICP demonstrated that the microwave digestion method gave reproducible and accurate results for most materials. However, metals in nonporous organic solids such as heavy oils and cokes were not completely removed. Fischer (1986) analyzed a variety of geological materials using microwave-digestion and isotopic-dilution (ID) techniques. Control over pressure in the Teflon digestion containers was obtained by including two transfer ports in the container lids. After digestion, trace elements in the sample solutions were separated using chromatographic techniques and analyzed by thermal-ionization mass spectrometry. Because complete equilibrium between the sample and spike isotope is crucial to quantitative ID analysis, the satisfactory results obtained with the microwave-digestion technique attest to the complete dissolution of the various geological materials.

Hewitt and Reynolds (1990) have shown that the choice of microwave-digestion methodology depends not only on the type of material to be digested, but also on the analyte, as well as the analyte's origin (anthropogenic vs. natural deposits). They compared microwave digestion using HNO_3, or HNO_3-HF (8:2 v/v), to a certified hot-plate digestion. Using reference soils and sediments they determined that metals such as Cd, Cu, Pb, and Zn were efficiently digested by all methods. However, metals such as Ba, Cr, and Ni, contained in aluminosilicate lattices, could only be efficiently recovered from the HNO_3-HF digest. All the metals derived from anthropogenic sources were successfully extracted using HNO_3. Kemp and Brown (1990) and Liqiang and Wangxing (1989) have also developed HF-based microwave-digestion methods for silicate samples and carbonate rocks.

Kingston and Jessie (1986) described a closed-vessel microwave-digestion system whereby in situ measurement of the temperature and pressure in a closed PFA (perfluoroalkoxy ethylene) vessel is carried out during acid decomposition of organic samples. They discussed microwave power absorption by various acids and estimates of the time required during dissolution to reach target temperatures. Papp and Fischer (1987) compared microwave digestion with dry ashing and with wet digestion using HNO_3-$HClO_4$-HF for a number of peat samples. Microwave digestion took much less time and yielded good agreement with the other methods. Recently, the availability of microwave-transparent PFA containers for acid dissolution, coupled with advances in fiberoptic thermometry, have helped in the development of sophisticated microwave systems for the laboratory. Their use permits not only monitoring of acid-initiated sample decomposition but offers a means of studying rates of reactions in various systems. Kingston and Jessie (1988) provide a detailed discussion of microwave sample preparation for many types of materials, including dissolution methods for geological samples.

SEPARATION AND PRECONCENTRATION OF TRACE ELEMENTS

Separation and preconcentration of the analyte becomes necessary when it is present in the sample solution in quantities below the detection limit or when the matrix elements, usually present in high concentrations, interfere with its determination. The most commonly used methods for separation and concentration are solvent extraction, ion exchange, and chelation. In some instances, reverse-phase high-performance liquid chromatography (HPLC) has also been used. These procedures have been described by Minczewski et al. (1982).

Solvent extraction of metals generally requires the use of a buffer since the quantity of a metal chelate extracted is strongly dependent on the pH of the extraction solvent. Van Loon (1985) has discussed the use of different buffers including formate, acetate, borate, phosphate, and citrate for metal extraction from aqueous solutions. For samples containing silicates, Govindaraju and Mevelle (1983) used an acidic cation-exchange resin to separate a number of trace elements. Uchida et al. (1979) described a procedure for concentrating the Cu, Ni, Zn and Cd from silicates by using a chelating resin. Their method consisted of adding 2 mL of 2.5 M malonic acid to the HF digest of silicate minerals, adjusting the pH of the mixture to 6 with concentrated ammonia, centrifuging and passing the supernatant through the Chelax 100 resin. Metals retained by the resin were then eluted with 2 M HNO_3 and analyzed by AAS. Likewise, Miyazaki and Barnes (1981) used polydithiocarbamate chelating resin to separate Cr(VI) from Cr(III). A number of other chelating resins have been used for specific metals (Van Loon, 1985). More recently, Siriraks et al. (1990) combined chelation-ion chromatography as a method for trace-element analysis of complex environmental and biological samples.

In laboratories with high sample throughput, preconcentration is commonly automated using flow injection analysis (FIA) technology. In FIA, the sample is injected into a moving nonsegmented carrier stream. Microprocessor-controlled peristaltic pumps and valves are used to introduce the required reagents, buffers, eluents, etc. For preconcentration FIA, the sample and buffer are pumped through the preconcentration column. The analyte is then eluted, and detected by FAA or ICP-AES. Automated preconcentration FIA systems have been described by Knapp et al. (1987) and Bysouth et al. (1990). The FIA methods have been reviewed by Stewart (1983) and by Ruzicka (1983).

STANDARD SOLUTIONS AND REFERENCE MATERIALS

Standard solutions of an element are prepared using reagent-grade or high-purity chemicals. Generally, a stock solution containing 100 to 1000 mg L^{-1} is prepared from which calibration-standard solutions are then prepared as needed. Concentrated stock solutions are chosen because the amount of chemical required is large enough to permit accurate weighing and because concentrated solutions remain stable for longer periods of time. To prepare a metal standard, the high-purity metal or metal salt is dissolved in an acid, usually HCl or HNO_3, and diluted to an appropriate volume. The standard should contain at least 1% acid for stabilization of the metal solution and to minimize sorption losses on the container walls. Commercially prepared standard solutions are now available from a number of companies. Preparation of mixed-element standard solutions from single-element stock solutions to produce compatible mixtures for multielement analysis is discussed by McQuaker et al. (1979), Johnson and Maxwell (1981) and others. Precautions for quantitative analysis of trace elements, conditions for the storage of solutions, and treatments to avoid contamination and losses are given by Van Loon (1985). Typical levels of trace elements in soils are given by Bowen (1966) (Table 2-1) and should provide guidelines for preparing calibration standards for trace-element analysis of soil digests.

Table 2-1. Levels of selected trace elements in
        soils (Bowen, 1966).

Element	Mean	Range
	-------- $\mu g\ g^{-1}$ dry soil --------	
Ag	0.1	0.01-5
As	6	0.1-40
B	10	2-100
Be	6	0.1-40
Cd	0.06	0.01-0.7
Co	8	1-40
Cr	100	5-3 000
Cu	20	2-100
Fe	38 000	7 000-550 000
Hg	0.03	0.01-0.3
Li	30	7-200
Mn	850	100-4 000
Mo	2	0.2-5
Ni	40	10-1 000
Pb	10	2-200
Se	0.2	0.01-2
Sn	10	2-200
V	100	20-500

Use of internal standards and the spiking of sample solutions are techniques commonly employed for quality control in analytical chemistry. However, in multielement analysis of geological and environmental samples, use of standard reference materials (SRMs) is invaluable. Major sources of these reference materials in the USA are the National Institute of Standards and Technology, Gaithersburg, Maryland; the U.S. Geological Survey, Denver, Colorado; and the U.S. Environmental Protection Agency (USEPA), Cincinnati, Ohio.

ENVIRONMENTAL SAMPLES AND SOIL EXTRACTS

Environmental Samples

Environmental samples generally include industrial and municipal solid wastes requiring disposal and the soils and sediments that have been contaminated by these wastes. Among the many solid wastes are coal and municipal-incineration ashes, wastes from metal plating and electrical industries, and municipal sewage sludge. Public concern over the potential leaching of heavy metals from industrial and municipal wastes, and the resulting soil and groundwater contamination, has focused attention on the development of methods for analyzing these materials. This includes methods for total analysis (i.e., fusion and acid digestion) as well as those for determining the available or mobile fraction of an element.

Acid-digestion and fusion techniques have been used for many years for the elemental analysis of coal ash (Nadkarni, 1980), and, currently, are being used for the analysis of incineration ashes (Karstensen & Lund, 1989). Development and testing of various procedures to predict the leaching of metals from a variety of solid wastes have been reported by a number of investigators (Elseewi et al., 1982; Giordano et al., 1983; Frances & White, 1987; Boyle & Fuller, 1987; Loring & Rantala, 1988; Warren & Dudas, 1989; Lisk et al., 1989; Sawhney & Frink, 1991). Guidelines developed by the USEPA (1986) for evaluating solid wastes involve digestion with 1:1 HNO_3 (Method 3050), followed by oxidation using H_2O_2 and subsequent analysis for 23 elements using atomic spectrometric methods. Kimbrough and Wakakuwa (1989) have modified the procedure to include recovery of Sb and Ag from soil or sludge samples, followed by analysis of the digest by AAS or ICP. Although these guidelines are widely applied in the analysis of solid wastes, correlations of the amounts of elements extracted by this procedure with the total amounts present or their potential leachability have not been rigorously tested so far.

Soil Extracts

In many cases, the total concentration of an element does not provide a reasonable estimate of its plant availability or mobility in soils. Consequently, techniques for extracting and measuring the available/mobile fractions of an element are of considerable interest. For example, Lindsay and Norvell (1978) used DTPA to extract the available fraction of a number of heavy metals from soils. Similarly, Soltanpour and Workman (1979) used a mixture of 1 M NH_4HCO_3 and 0.005 M DTPA for the extraction of K, Zn, Fe, Cu, and Mn and analyzed the extract using ICP. Soltanpour et al. (1982) have discussed the use of various soil extractants, pretreatments of the soil extracts for analysis of certain elements (e.g., As, Se, and Hg) and their subsequent analysis by atomic spectroscopic techniques.

It is also worthwhile to note that, in addition to assessing the availability of an element, the extractant is often chosen to maximize the solution concentration of the analyte and bring it within the analytical limits of the detection technique.

ATOMIC SPECTROSCOPY

Atomic spectroscopy, the most common instrumental technique used for elemental analysis, includes absorption, emission, and fluorescence. In order to observe these atomic phenomena, the analyte must first be converted to a neutral atomic vapor or, in some cases, an ionic vapor. In analytical atomic spectroscopy this conversion is usually met by heating the sample to a high temperature (>2000 K). The electromagnetic radiation absorbed or emitted during orbital transitions by the electrons of the free analyte atoms or ions is the analytical signal of interest. A brief description of the theory, instrumentation and methods generally used in analytical atomic spectroscopy is given below. More detailed material including precision, sensitivity, and accuracy of the various techniques can be found in Billings and Angins (1973), Robinson (1981), Van Loon (1980, 1985), Winge et al. (1985), Welz (1986), Montaser and Golightly (1987), Boumans (1987), Ingle and Crouch (1988), Thompson and Walsh (1988), Moore (1989), Ottaway and Ure (1989), Burguera (1989), and Varma (1989).

Basic Principles

A simplified model for the absorption and emission of energy, based on the Bohr atom, is shown in Fig. 2-1 (top). When energy is absorbed by an electron orbiting the nucleus of the atom, it is raised from the lowest energy level, called the ground state, to some higher energy level, called the excited state. The source of absorbed energy can be electromagnetic (photons) or kinetic (collisional). When the energy absorbed during excitation comes from photon sources it is termed atomic absorption. Decay of the excited electron to a less energetic level can occur by emission of radiation of a characteristic wavelength. The emission of radiation is termed atomic emission when the excitation source is kinetic energy, and atomic fluorescence when the excitation source is photon absorption.

A simplified energy-level diagram depicting the excitation and emission events is shown in Fig. 2-1 (bottom). When the absorbed energy is sufficient to completely dissociate the electron from the atom, a positively charged ion can result; these energy levels for the ion are analogous to those

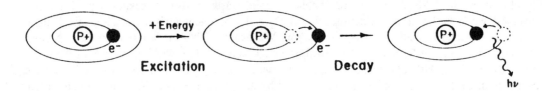

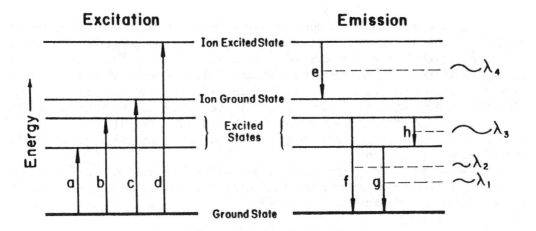

Fig. 2-1. Absorption and emission of energy resulting in excitation and decay of an orbital electron based on the Bohr atom (top), and energy-level diagram showing energy changes where a and b represent excitation; c, ionization; d, ionization/excitation; e, ion emission and f, g, and h, atom emission (Boss & Fredeen, 1989).

for the neutral atom. For a radiative transfer of energy, the difference in energy (ΔE) between the levels is given by

$$\Delta E = h\nu \text{ or } \Delta E = hc/\lambda, \qquad [1]$$

where h is Planck's constant, ν is the frequency, λ is the wavelength, and c is the speed of light. The energy transitions most useful for analytical work are found in the ultraviolet-visible portion of the electromagnetic spectrum (190-800 nm). Because the atoms are in the gaseous state, the emission spectrum for a given element consists of narrow lines, about 0.01 nm in width.

The principal components used in instruments for atomic spectroscopy are depicted in Fig.2-2. The flame, furnace, or plasma serves as the heat source (atomization device) which decomposes the sample into a gas containing free atoms. In atomic-absorption spectrometry (AAS), the lamp emits photons of a wavelength that is characteristic of the element of interest. The amount of light absorbed at this wavelength is proportional to the concentration of analyte atoms in the ground state, which, in turn, is proportional to the total concentration of analyte atoms sprayed into the atomization chamber. In inductively coupled plasma-atomic emission spectroscopy (ICP-AES), a high-resolution monochromator is used to distinguish the emission signal of a specific analyte. Quantitative evaluation is based on the principle that the radiative atomic emission of a specific transition is proportional to the analyte concentration. An important advantage of ICP-AES over AAS is that sequential, multielement analysis is possible by simply changing the wavelength sensed by the detector. Simultaneous analysis can be done either by using a detector array or by interfacing the ICP with a mass spectrometer. A major drawback of AES is spectral interference caused by the overlap of emission lines.

Atomic-fluorescence spectrometry (AFS) differs from AES in that the atom is excited by photons. Immediately afterwards, any incident radiation absorbed at wavelengths corresponding to the atomic-absorption lines is released as fluorescence of characteristic wavelength. The intensity of the fluorescence is proportional to the concentration of analyte. Because the excitation and emission wavelengths normally match, the detector is mounted perpendicularly to the excitation source (Fig. 2-2).

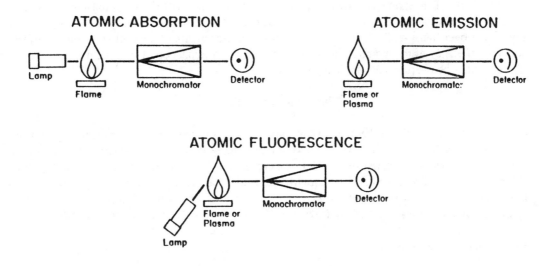

Fig. 2-2. Principal components of atomic absorption, emission and fluorescence instrumentation (Boss & Fredeen, 1989).

Temperature Effects

Consideration of the temperature stability of the excitation source is more important in AES than in AAS. Indeed, sensitivity to small changes in temperature has long been a serious drawback in flame AES. The importance of temperature stability for ICP-AES, and flame AES in particular, can be shown by application of the Boltzmann distribution to a collection of atoms at thermal equilibrium (Ingle & Crouch, 1988)

$$N*/N_0 = (g*/g)\exp(-\Delta E/kT)$$ [2]

where $N*/N_0$ is the fraction of atoms in the excited state ($N*$) compared to those in the ground state (N_0), ΔE is the energy difference between these states, k is the Boltzmann constant, T is the temperature in Kelvins, and $g*/g$ is the degeneracy factor (i.e., the number of states at each energy level). For example, using Eq. [2] for sodium atoms where $g*/g = 2$ and $\Delta E = 3.371 \times 10^{-19}$ J/atom, the fractions of atoms in the excited state in an acetylene-air flame at 2600 K, and in an ICP at 6500 K are

$$N*/N_0 = 1.68 \times 10^{-4} \qquad (2600\ K)$$
$$N*/N_0 = 4.68 \times 10^{-2} \qquad (6500\ K).$$

Thus, at these temperatures, less than 0.02 and 5% of the sodium atoms are excited in flame AES and ICP-AES, respectively. For a temperature increase of 10 K, the fractions of excited atoms are

$$N*/N_0 = 1.74 \times 10^{-4} \qquad (2610\ K)$$
$$N*/N_0 = 4.71 \times 10^{-2} \qquad (6510\ K).$$

Note that a 10 K temperature rise increased the excited state population in the flame by about 4% while the change was less than 1% in the plasma. Thus, two important advantages of the ICP source are greater sensitivity, due to the higher population of excited atoms, and greater stability of the signal with respect to temperature changes.

In AAS, modest temperature fluctuations have negligible effect on the absorption signal because: (i) the vast majority of the analyte atoms exist in the ground state favoring absorption, and (ii) the population of atoms in the ground state is not significantly affected by such fluctuations in temperature. As can be seen in the above example for Na atoms, at a flame temperature of 2600 K, 99.9833% of the atoms are in the ground state whereas at 2610 K, the percentage decreases to 99.9826. These attributes simplify AAS instrument design and provide AAS with significant advantages over flame AES. Nevertheless, because the temperature of the excitation source affects both the efficiency of the atomization process and degree of ionization of the analyte, temperature effects cannot be overlooked completely in AAS.

The Atomization Process

It has been stated above that the analyte must first be converted into atoms prior to detection. For flame AAS (FAAS) or ICP-AES, the sample is usually introduced continuously in the liquid form. The processes which transform solution samples into atomic vapors are outlined in Fig. 2-3. Other, more specialized methods include discrete liquid introduction, slurry, or gaseous sample introductions. The design, optimization and limitations of sample introduction in atomic spectroscopy have been reviewed by Browner (1983) and Browner and Boorn (1984a,b).

Liquid Sample Introduction

The first step in transporting the liquid sample to the flame or plasma is nebulization, i.e., conversion of the liquid into a fine mist or aerosol (Iveldi & Slavin, 1990). Connected to the

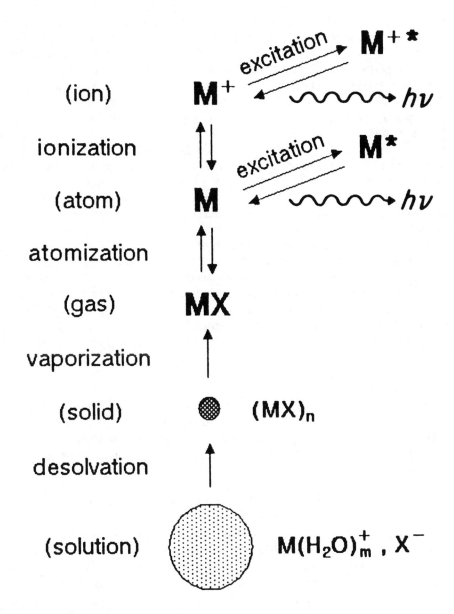

Fig. 2-3. Sequential processes in free-atom or ion formation for liquid sample introduction into a heat source, e.g., a flame, furnace, or plasma (Boss & Fredeen, 1989).

nebulizer is a spray chamber which removes large droplets, allowing only droplets of about 10 μm in diameter to pass into the flame or plasma. Depending on the design of the nebulizer and spray chamber, the aerosol transport efficiency, i.e., the percentage of the mass of nebulized solution reaching the flame or plasma, ranges from 1 to 90%. Experimental measurements (Smith & Browner, 1982) have shown, however, that 98 to 99.5% of the aerosol in a typical spray chamber is actually lost to waste. The two most common types of nebulizers are the cross-flow and V-groove types of pneumatic nebulizers. These nebulizers employ compressed gas to aspirate the sample (Meinhard, 1976). The inside diameter of the sample capillary in pneumatic cross-flow nebulizers is only about 0.1 mm and, therefore, is subject to clogging by samples high in particulate matter. Furthermore, the tip is prone to clogging by the evaporation of samples with high dissolved solids.

In the V-groove or Babington pneumatic nebulizers, sample flows down the groove until it is aspirated by the gas flowing through the small orifice (Wolcott & Sobel, 1982). The V-groove nebulizer is essentially free of clogging problems because the diameter of the solution delivery tube is about 1 mm. However, a peristaltic pump must be used to maintain a constant flow rate of solution. Other disadvantages include high sample-consumption rates and lower efficiencies. Walton and Goulter (1985) compared the performance of a pneumatic and a Babington nebulizer, and concluded that the solution detection limits were similar for the eleven elements studied.

Other types of nebulizers have also been developed. A high-pressure pneumatic nebulizer (Strasheim & Olson, 1983) was shown to exhibit good long-term stability as well as excellent resistance to salt build-up. A frit nebulizer (Layman & Lichte, 1982) has been designed to produce mean droplet diameters of the order of 1 μm, with transport efficiencies of 90%. A desirable feature of the frit nebulizer is its ability to function with sample flow rates as low as 0.01 to 0.03 mL min^{-1}. At these low flow rates, any organic solvent can be transported into the plasma and, for these reasons, the frit nebulizer is of interest for coupling liquid chromatography with ICP-AES.

Recently, ultrasonic nebulization has reemerged as a possible replacement for pneumatic nebulizers (Olson et al., 1977; Wiederin et al., 1990). The sample is fed continuously onto the ultrasonic transducer plate, resulting in high transport efficiencies and increased sample-residence times in the flame or plasma. Consequently, the detection limits of most elements are about one order of magnitude lower than with pneumatic nebulization. Disadvantages of ultrasonic nebulization include high salt build-up at the injector tube, memory effects, increased background and spectral interferences, the need for a specialized desolvation spray chamber, and high cost.

Hydride Sample Introduction

The detection limits of some environmentally important elements such as As, Sb, Se, Bi, Ge, Sn, Pb, and Te can be improved by about two orders of magnitude by first converting them into gaseous hydrides. Metal hydrides of these elements are produced in acidic solutions immediately upon addition of a reducing agent (NaBH$_4$). For example, the reaction of As(III) to produce arsine proceeds as follows:

$$BH_4^- + H^+ + H_3AsO_3 \rightarrow H_3BO_3 + H_2(g) + AsH_3(g) \qquad [3]$$

The detection limit is lowered because all of the analyte in the sample volume (10-50 mL) is introduced into the atomizer as the hydride in only a few seconds. Another advantage is that the analyte is removed from the sample matrix, thus eliminating the potential for matrix interferences. Manual and automated hydride generators are available. The principal disadvantage of this method is the increased analysis time and the complications of pH and oxidation state adjustments. Recently, Zhang et al. (1988) described a new generator for the determination of As, Sb, and Bi in geological samples and Novozamsky et al. (1988) have developed a procedure for generating methyl borate for the determination of B in soils. Hydride formation has been reviewed by Robbins and Caruso (1979) and by Nakahara (1983).

Free-Atom Formation

Once the sample enters the atomization device, four major processes (Fig. 2-3) occur: desolvation, volatilization, dissociation, and ionization. Desolvation results in the formation of a dry aerosol composed of molten or solid particles. The desolvation process varies according to droplet size, gas flow-rate, and temperature of the atomizer.

The next step, volatilization, converts the particles into gaseous constituents. Inefficient volatilization not only prevents optimum free-atom formation but also causes background problems due to light scattering. Because the metal oxides are less volatile than the metals or metal halides, a reducing (fuel-rich) environment is employed for oxide reduction as well as for minimizing oxide reformation in FAAS. With the ICP, however, the higher temperature is sufficient for oxide volatilization.

The volatilized materials then undergo dissociation into free atoms. The free atoms can recombine to form molecular species by reactions with solvent (O, OH) or with matrix dissociation products. Thus, molecular species such as ZnO, $CaOH$, $NaCl$, etc., can exist in equilibrium with the free atoms. Ionic species can also form by electron loss. Chemical interferences due to molecule formation are much more prevalent in flame systems, whereas ionic effects are more common in ICP systems. Specific factors and equations governing atom formation are discussed in detail by Ingle and Crouch (1988) and Boumans (1987).

Analytical Performance in Atomic Spectroscopy

Detection limits

Detection limit is generally considered as the concentration of an analyte that produces a net response [total signal minus background (BG) signal] equal to three times the standard deviation of the BG signal. Standard deviation is often estimated from 11 consecutive integrations of the signal obtained from the blank solution. In scanning instruments, the BG signal can be measured on either side of the analyte line and, in polychromators, it is measured using the blank. Although the detection limits for various instruments can be readily determined, these values are usually very low compared to those obtained from calibration curves and should be referred to as "instrumental detection limits" (IDLs). The numerical value of the detection limit depends not only on the value of the blank solution but also on the sensitivity, i.e., the slope of the calibration curve. The higher the sensitivity, the lower the detection limit and vice versa. Several instrumental factors, as well as sample injection rate and dissolution method, affect sensitivity. Detection limits for real samples are much higher than the IDLs because the samples often contain additional substances that give rise to chemical and spectral interferences (discussed later). It must be remembered also that practical detection limits vary with the time between calibration checks because of systematic drift. Various factors which affect detection limits are discussed in the books referenced earlier (Atomic Spectroscopy Section). Table 2-2 lists the detection limits (IDLs) of atomic spectroscopy techniques for common soil and mineral elements.

The detection limits listed in Table 2-2 taken from Perkin-Elmer (1988) were determined using elemental standards in dilute aqueous solutions. In addition to yielding the lowest detection limits of any of the AAS or AES techniques, graphite furnace atomic absorption (GFAA) spectrometry has the additional advantages of (i) requiring only a small amount of sample and (ii) possessing the capability to analyze solids (as slurries) directly. The advantages of ICP, on the other hand, include a large linear range for most elements and simultaneous, or near-simultaneous, multielement analysis of both major and minor elements. Moreover, using these atomic spectroscopy techniques, elements such as B and S can be analyzed quantitatively only by ICP. Despite the advantages of GFAA and ICP-AES, the most commonly used atomic spectroscopy method continues to be flame atomic absorption (FAA). The primary reason for this is much lower cost of the conventional FAA instruments.

Interferences

For quantitative determination of the concentration of an analyte, only the intensity of the signal from the analyte should be considered. To obtain a true measure of the intensity from the analyte alone, however, the total intensity measured by the spectrometer must be corrected for signals arising from interfering substances in the sample matrix.

In optical atomic spectroscopy there are two classes of interference, nonspectral and spectral. Nonspectral interferences affect the free-atom population in the atomizer and are caused by the chemical, ionization, and physical properties of the sample matrix. Spectral interferences are caused by broadband background absorption, or by overlapping elemental emission lines. In ICP-MS, interferences are caused by mass overlaps. The types of interference predominantly associated with a particular atomic spectroscopy technique, as well as typical methods of

Table 2-2. Detection limits of common elements in soils and
minerals (Perkin-Elmer, 1988).

Element	FAAS	GFAAS†	ICP-AES
	-------------------- $\mu g\ L^{-1}$ --------------------		
Al	30	0.04	4
As	100	0.2	20
B	700	20	2
Ba	8	0.1	0.1
Ca	1	0.05	0.08
Cd	0.5	0.003	1
Co	6	0.01	2
Cr	2	0.01	2
Cu	1	0.02	0.9
Fe	3	0.02	1
Hg	200	1	20
K	2	0.02	50
Mg	0.1	0.004	0.08
Mn	1	0.01	0.4
Mo	30	0.04	5
Na	0.2	0.05	4
Ni	4	0.1	4
P	50 000	30	30
Pb	10	0.05	20
Se	70	0.2	60
Si	60	0.4	3
Zn	0.8	0.01	1

†Graphite furnace atomic absorption spectrometry.

compensating for these interferences, are outlined in Table 2-3. Brief discussions of these interferences are presented in the following sections. Additional details can be found in the books referenced in the Atomic Spectroscopy Section.

Chemical Interferences. One type of chemical interference is caused by thermally stable molecular species, i.e., molecules containing the analyte atom, which fail to dissociate completely in the flame or furnace atomizer. This results in a lower concentration of free analyte atoms than would be detected in the absence of the chemical interferent. In FAA, chemical interferences can be reduced by using a higher temperature flame, or by adding a releasing agent. Chemical interferences are negligible in ICP-AES due to the high atomization temperature of the torch.

An example of chemical interference in FAA is the reaction of calcium with phosphates to form the thermally stable compound, calcium pyrophosphate ($Ca_2 P_2 O_7$). The formation of the pyrophosphate decreases the free Ca concentration compared to that obtained in solutions that are phosphate-free (i.e., calibration standards). Phosphate interference can be minimized by switching to the higher temperature nitrous oxide-acetylene flame which decomposes $Ca_2 P_2 O_7$. The high temperature flame, however, is more prone to ionization interferences (see below). An alternative approach is the addition of a releasing agent, such as lanthanum or strontium (about 1%), to

Table 2-3. Atomic spectroscopy interferences.

Technique	Type of interference	Method of compensation
FAA	Chemical/ionization	Releasing agent, nitrous oxide-acetylene flame, or ionization buffer
	Physical (matrix)	Dilution, matrix matching, or method of standard additions
GFAA	Physical (matrix)	See FAA, matrix modifiers
	Background absorption	Zeeman or continuum source background correction
ICP-AES	Physical (matrix)	See FAA
	Spectral	Spectral background correction or use alternative emission line
ICP-MS	Matrix	See FAA
	Mass overlap	Interelement correction, use alternate mass values, or increase mass resolution

preferentially bind the phosphate. Likewise, a chelating type of releasing agent, such as EDTA, can be added to complex the Ca, thereby releasing it from the phosphate.

Another type of chemical interference is that due to volatilization losses (particularly in the GFAA method). Highly volatile elements include Pb, Cd, Zn, As, and Se. Volatilization problems are particularly severe in matrices where metal halide formation is significant (Navarro et al., 1989). To prevent halide formation, and minimize volatilization losses in the GF atomizer, matrix modifiers are added to the sample. For example, in samples having a high NaCl content, NH_4NO_3 (Slavin & Manning, 1979, 1982) can be added to remove the NaCl by converting it to the volatile ammonium chloride and sodium nitrate, with the added benefit of producing less volatile metal nitrate. Other modifiers, such as phosphate (Bass & Holcombe, 1987), magnesium nitrate (Slavin et al., 1982), and Pd (Styris et al., 1991), also have been used to prevent volatilization losses.

Ionization Interferences.   Elements with low ionization potentials (e.g., Cs, Pb, K, and Na) can undergo significant ionization in the atomizer, thus decreasing the free-atom concentration. In the higher-temperature nitrous oxide-acetylene flame, determinations of calcium, the rare earths, and strontium can also suffer from severe ionization interferences.

The extent of ionization of the analyte depends on the concomitant elements and their concentrations. For example, the ionization of Ca is suppressed by K, a more easily ionizable element (EIE). Addition of an EIE floods the flame with electrons ($EIE \rightarrow EIE^+ + e^-$), thereby shifting the equilibrium to favor the Ca atom ($Ca^+ + e^- \rightarrow Ca$), in accordance with Le Chatelier's principle.

A diagnostic for ionization interferences in AAS is a positive deviation from linearity when absorbance versus concentration is plotted. To prevent inequality between samples and standards, a large excess of an EIE (approximately 0.1%) is added to all solutions to act as an ionization buffer.

Ionization effects can also occur in ICP-AES (Thompson & Ramsey, 1985; Thompson, 1987). In this case, however, analytes with large excitation potentials (e.g., Zn^+, 15.5 eV), rather than those with low excitation potentials (e.g., Li, 1.85 eV), are affected. This effect appears to be related to ionization of the matrix element in the plasma (Thompson, 1987). To prevent these interferences, the sample and standard matrices should be matched as closely as possible. If a constant matrix cannot be assumed from sample to sample, such as in soil analysis, then the method of standard additions (discussed below) is preferable.

Physical Interferences. Physical interferences are caused by a mismatch of the physical characteristics of the sample and standard. Differences in viscosity and surface tension affect the sample intake rate and nebulization efficiency, thus causing changes in the free-atom population in the atomizer. This can result in either positive or negative errors. The effects on the analytical signal are most noticeable in solutions containing differing concentrations of mineral acids, especially sulfuric and phosphoric acid with their large differences in viscosity and density (Fig. 2-4). To compensate for these effects, the major matrix components of the sample and the standards should be matched as closely as possible.

The method of standard additions (Van Loon, 1980; Bader, 1980) is useful in cases where the sample matrix is unknown, or difficult to match. Standard additions can also be used in methods development to determine which materials cause matrix interferences. In this method, aliquots of a standard are added to known (constant) amounts of the sample and diluted to a constant volume. Thus, any matrix interferent in the sample will also affect the standard. In the presence of interfering substances, the slope of the standard-additions plot will not be parallel to a standard plot derived from simple aqueous solutions. Although the method of standard additions is useful in many cases, it must be used with caution as it does not correct for any type of spectral interference.

Background-Absorption Interferences. Background interference is spectral interference that arises from incomplete atomization of the matrix materials in the flame or furnace atomizer. Undissociated molecular forms can exhibit broadband absorption spectra (Culver & Surles, 1975) and tiny solid particles in the optical path scatter the light. Both of these effects cause background absorption which must be measured and subtracted from the total signal. Background-correction techniques are discussed later in this chapter. Background interferences are minimal in ICP-AES due to the high temperature of the plasma.

Spectral Interferences. Spectral interferences are encountered mainly in ICP-AES, and present only minor difficulties in AAS. They are caused by stray light or overlapping spectral features.

Stray light is radiation from wavelengths outside the instrument bandpass that reaches the detector. It is caused by the intense emission of easily excited species (e.g., Ca or Na), and by imperfections in the instrument. Modern spectrometer designs minimize stray light. However, in trace analysis of analytes with emission or absorption lines near a strong concomitant-analyte emission line, some stray light is probable and matrix-matching procedures should be employed.

In ICP-AES, overlapping emission is due to the complex spectra arising from the high-temperature plasma. These overlaps are caused by spectral and molecular bands from the discharge atmosphere, the solvent, or other analytes in the sample. Spectral overlapping can be severe in samples containing large amounts of transition metals (particularly Fe, Cr, Ti and Mn), producing spectra consisting of hundreds of emission lines. Interferences of this type cause additive errors in the analysis, and are independent of the analyte concentration. As a result, spectral interferences are much more serious at low analyte levels. An example of a spectral

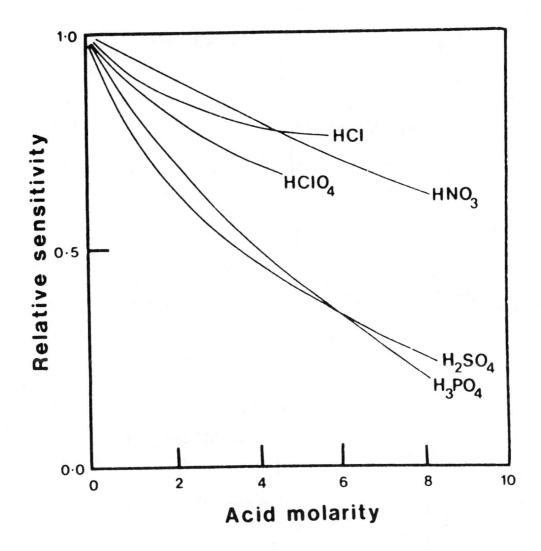

Fig. 2-4. Effect of different mineral acids and their concentration on the sensitivity of an analyte (Thompson, 1987).

interference is shown in Fig. 2-5 (top): the overlap of one wing of the nearby aluminum line causes a slope to occur in the background. A typical correction procedure would be to measure the emission signal on either side of the Cd 214.438-nm line, and then subtract the average from the line signal. In some cases, there is a direct spectral overlap of the emission lines as illustrated in Fig. 2-5 (bottom). In this situation, the use of an alternative emission line should be considered. If an alternative line is not available, the background can be corrected by using interelement correction techniques. In this method, the concentration of the interfering element is ascertained at another wavelength, and a predetermined correction factor is then applied. Spectral interferences in ICP-AES have been well documented (Boumans, 1987; Montaser & Golightly, 1987; Winge et al., 1985). The concentration equivalents of some of the more frequently encountered interferents in soil and environmental analysis are given in Table 2-4.

In atomic-absorption spectroscopy, spectral interferences are rare because of the wavelength-specific nature of the hollow-cathode lamp (HCL) (see discussion of light sources below). Nonetheless, some interferences do exist (Van Loon, 1980; Norris & West, 1974), particularly when multielement HCL lamps are used.

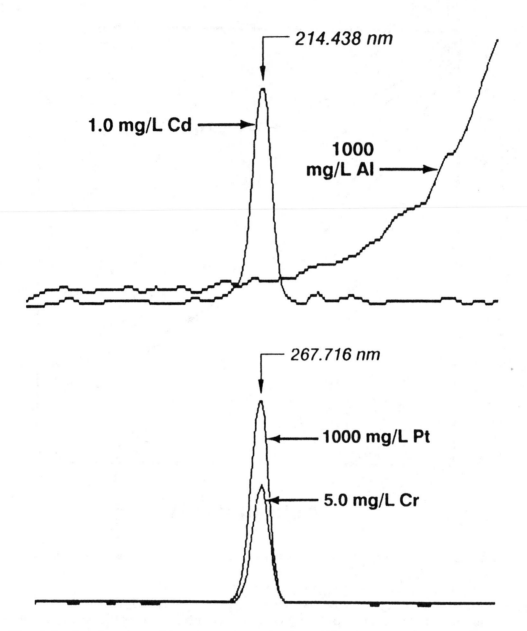

Fig. 2-5.   Interference at the 214.438-nm line of Cd by the spectrum of Al (top) and at the 267.716-nm line of Cr by direct spectral overlap from Pt (Boss & Fredeen, 1989).

ATOMIC-ABSORPTION SPECTROMETRY

Atomic-absorption spectrometers, due to their long history, relatively low cost, and versatility, are the most widely used instruments for elemental analysis in academic research and commercial laboratories. The principles of AAS are well established, being first set forth by Kirchhoff in 1860. However, the analytical usefulness of AAS for elemental analysis was not demonstrated until 1955. In that year, Walsh (1955) in Australia, and Alkemade and Milatz (1955) in Holland, independently published papers showing that radiation from a line-source emitter is absorbed by the atomic vapor, and that the absorption is proportional to analyte concentrations at the trace (mg L^{-1}) level. The first commercial instrument became available in 1959. In the 1960s, L'vov (1961), Massman (1968), and others (reviewed by Slavin & Manning, 1982), developed the electrothermal atomizer. The use of electrothermal atomizers in GFAA techniques lowered the

Table 2-4. Analyte concentration equivalents for selected elements at the 100 mg L^{-1} interferent level (U.S. Environmental Protection Agency, 1986).

Analyte	Wavelength	Interferent†							
		Al	Ca	Cr	Cu	Fe	Mg	Mn	Ni
	nm	---------------------------------- mg L^{-1} ----------------------------------							
As	193.696	1.3	--	0.44	--	--	--	--	--
Bo	249.773	0.04	--	--	--	0.32	--	--	--
Cd	226.502	--	--	--	--	0.03	--	--	0.02
Ca	317.933	--	--	0.08	--	0.01	0.01	0.04	--
Cr	267.716	--	--	--	--	0.003	--	0.04	--
Co	228.616	--	--	0.03	--	0.005	--	--	0.03
Cu	324.754	--	--	--	--	0.003	--	--	--
Fe	259.940	--	--	--	--	--	--	0.12	--
Pb	220.353	0.17	--	--	--	--	--	--	--
Mg	279.079	--	0.02	0.11	--	0.13	--	0.25	--
Mn	257.610	0.005	--	0.01	--	0.002	0.002	--	--
Mo	202.030	0.05	--	--	--	0.03	--	--	--
Ni	231.604	--	--	--	--	--	--	--	--
Se	196.026	0.23	--	--	--	0.09	--	--	--
Si	288.158	--	--	0.07	--	--	--	--	--
Zn	213.856	--	--	--	0.14	--	--	--	0.29

†Dashes indicate that no interference was observed even when interferent was introduced at the following levels: Al, Ca, Fe, and Mg at 1000 mg L^{-1} and Cr, Cu, and Mn at 200 mg L^{-1}.

detection limit for many elements by at least one order of magnitude, compared to flames. The graphite-furnace accessory became available commercially in the 1970s.

Instrumentation

The basic components of a typical commercial double-beam atomic-absorption spectrometer, outlined in Fig. 2-6, include: (i) the light source, (ii) the atomizer and sample-introduction unit (sample cell), and (iii) specific light detectors.

Light Sources

Because of the narrow absorption line-widths of free atoms (0.001-0.005 nm), all commercially available instruments employ line emission (typically <0.001 nm wide) for the light source. Under these conditions, Beer's law is valid, and the concentration of analyte atoms is directly proportional to the absorbance. In addition, the narrow line-source provides the maximum sensitivity with a minimum of spectral interference. Hollow-cathode lamps (HCLs) or electrodeless-discharge lamps (EDLs) are used as line sources in commercial instruments. Recently, however, Ng et al. (1990) have demonstrated that both continuous-wave and multiple-mode diode lasers can serve as line sources in AAS. These lasers are inexpensive, compact, tunable, and show promise for simultaneous background correction capabilities. However,

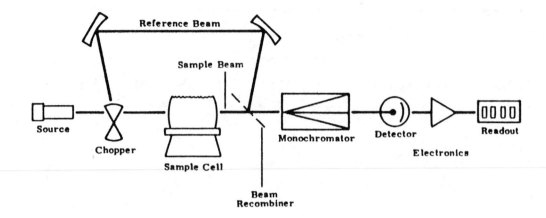

Fig. 2-6. Components of a double-beam atomic-absorption spectrometer, including beam chopper and reference-beam arrangement for background-interference correction (Beaty, 1988).

the wavelength range currently available (500-700 nm) limits this technique to elements with absorbance bands over this low energy-range (i.e., the alkali metals).

The HCL lamp is filled with an inert gas (Ne or Ar) under low pressure (0.1 -0.7 kPa). The cathode is constructed from the element of interest while the anode is made from a robust metal with low volatility, such as W. Upon application of a high voltage, the gas is ionized and the positively charged ions accelerate towards the cathode. Upon impact, the ion sputters (dislodges) some of the metal atoms from the surface. The sputtered atoms are then excited to emission states by collisional transfer of kinetic energy with the inert gas.

The HCL works well, except for some of the more volatile elements, such as As or Se, where low intensity and short life pose a problem. For these elements, EDLs exhibit superior performance. The EDL is essentially composed of the element of interest sealed under a low pressure (<1 kPa) of inert gas in a small quartz bulb, surrounded by an RF coil, and housed in a glass tube. The RF energy couples to the conducting metal in the bulb, vaporizing the metal and exciting the atoms to emit radiation. Since the metal is confined inside the bulb there are no volatile losses and the lifetime is much greater compared to the HCL. The EDLs, however, require a separate and rather expensive power supply.

The alternative to a line source is a continuum light source, such as a deuterium or tungsten lamp. Over the years, numerous papers have appeared on the use of continuum sources for AAS. The advantages of such sources include simple optical trains, single source requirements, simultaneous multielement detection, and simplified background correction. The disadvantages include poor source intensity and analyte sensitivity below 250 nm, and the need for a high-resolution monochromator and for wavelength modulation. These problems appear to prevent the commercial production of a continuum-source atomic-absorption instrument. Robinson (1981) illustrated the loss in sensitivity using continuum sources; for example, if the bandpass for a deuterium source is 0.1 nm, then the complete absorption of atoms with a line width of 0.005 nm will only result in a 5% decrease in the amount of radiation reaching the detector. Clearly, the analytical range is very poor, between 95 to 100% transmission. Despite these difficulties, research on continuum sources coupled with high-resolution optics remains active (Hsiech et al., 1990; O'Haver et al., 1988; Moulton et al., 1989; Jones et al., 1989).

Regardless of the source, the light must be modulated in order to obtain a stable output signal. In double-beam instruments, the mirror-coated chopper (Fig. 2-6) modulates the light-source output, serving to subtract out the source and background emissions from the atomizer. Fluctuations in source intensity are corrected by using a double-beam arrangement where the chopper modulates the reference beam 180° out of phase with respect to the sample beam. The double-beam arrangement results in a much more stable baseline than the single-beam design, and is standard in most commercial instruments. Much less common is electronic modulation of the light source, which is mechanically simple but shortens lamp life and reduces lamp stability.

Flame Atomizers

Flames are the most commonly utilized atomizers for routine use in AAS. The modern Lundegardh burner-nebulizer design for FAA is shown in Fig. 2-7. The elongated slotted burner head differs little from that first described by Walsh (1955). The major difference in modern design is the provision for premixing the combustion gases with the aerosol in the spray chamber. The most prevalent fuel-oxidizer combination is acetylene-air, producing a flame temperature of 2400 to 2700 K. For refractory elements such as B, Al, Si, V, Zn, and the rare earths, a hotter flame is needed and nitrous oxide is normally used to replace air as the oxidant. This not only produces a hotter flame but also maintains a more reducing environment, which minimizes oxide formation and solute vaporization interferences. However, the $N_2O-C_2H_2$ flame is prone to dangerous flashbacks, and should never be ignited directly.

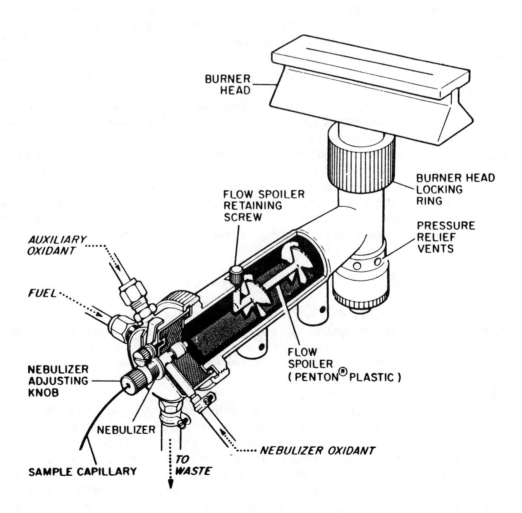

Fig. 2-7. FAA assembly for liquid sample introduction, including nebulizer, premix chamber and burner system (Perkin-Elmer Corp., 1988).

Graphite-Furnace Atomizers

Graphite-furnace atomic-absorption (GFAA) spectrometry is the ideal choice for ultratrace elemental analysis, and is particularly well suited when the sample size is small. For example, whereas the solution flow rate in conventional nebulizer systems ranges from 1 to 5 mL min^{-1}, with an analysis time of at least one minute, the total sample size introduced into the graphite furnace is only 5 to 50 μL. This small sample size is the primary reason that the absolute detection limit (ADL, or volume used in the determination times the solution detection limit), expressed in mass units, is generally 100 times less than the ADL for flame AA. Additional advantages include the capability of GFAA to analyze solid and slurry samples directly. One disadvantage is the slow sample throughput, typically one sample in three minutes. Background and chemical interferences are also common problems encountered with the GFAA. In addition, elements that readily form carbides (e.g., Hf, Zr, and Ta) cannot be analyzed by GFAA.

Basically, graphite-furnace atomizers are electrically heated tube-furnaces, as shown in Fig. 2-8 (top). The cylindrical tube is aligned in the optical path of the atomic-absorption spectrometer, and is held in place by contact cylinders which also provide electrical connections. To prevent overheating, the assembly is enclosed in a water-cooled housing. Inert gas flows around and through the tube to protect it from oxidation. The sample (5-50 μL) is introduced discretely through the hole in the furnace tube, and is subjected to at least three heating steps, as illustrated in Fig. 2-8 (bottom). In the first step, the drying or desolvation step, the furnace is heated to 100 to 150 °C and the solvent evaporated. The second step, the ashing or charring step, volatilizes inorganic and organic matrix components. The ashing temperature ranges from 250 to 1200 °C, depending on the analyte. Atomization occurs during the third step when the furnace temperature is increased to 2000 to 2700 °C. Usually, a final clean-up step is employed to remove any remaining residues by briefly increasing the temperature to about 2900 °C.

In early studies on GFAA, the samples were directly deposited onto the furnace wall. One of the problems with this approach is that it takes some time to achieve steady-state temperatures. Since the analyte vaporization temperature depends on the matrix constituents, slight variations in sample composition lead to poor precision in the final results. A vast improvement in the control of the analyte vaporization temperature occurs when the sample is deposited onto a L'vov platform (L'vov 1978). The L'vov platform is a thin graphite plate, usually with a groove to contain the liquid sample. The temperature of the platform lags behind the wall temperature, delaying sample vaporization until steady-state conditions are achieved. The use of the L'vov platform eliminates many of the matrix-dependent effects that occur as a result of wall atomization (Slavin & Manning, 1982).

Background Correction

For quantitative estimates of an analyte by atomic absorption, the background signal must be subtracted from the total absorption. Three approaches are used for background correction in commercial instruments: continuum-source, Zeeman-effect, and line-reversal. Background correctors compensate for spectral interferences caused by molecular broadband absorption and for light-scattering effects from particles. It should be emphasized that they do not correct for matrix, chemical, or spectral-overlap interferences.

In continuum-source background correction, the light from a deuterium lamp (for the UV range) or a tungsten lamp (for the visible range), is optically aligned along the path with the line source and detector, and the chopper (Fig. 2-6) is used to modulate the signal from the line (HCL) and continuum source. The background absorbs light over the entire spectral bandpass of the monochromator (0.1-0.5 nm) and, hence, affects the light emitted by the HCL and deuterium (or tungsten) lamp equally. Because atomic absorption lines are very narrow (0.001 nm), however, the absorbing line(s) specific to the analyte will have a negligible effect on the continuum radiation. Quantitative evaluations are performed by first measuring the total absorption of the HCL line,

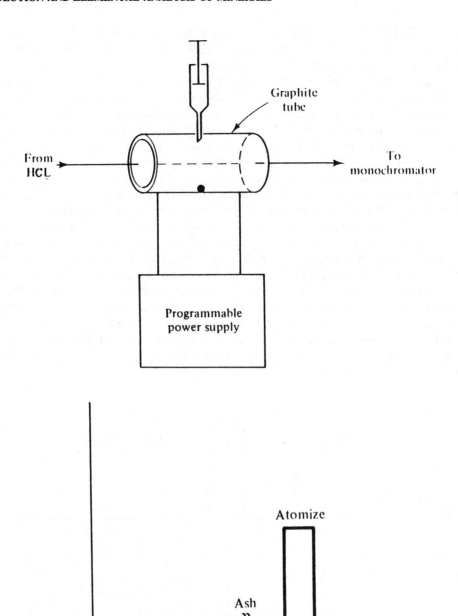

Fig. 2-8. Graphite furnace with programmable power supply (top) to permit discrete steps in sample atomization (bottom) (Ingle & Crouch, 1988).

which includes both the atomic (analyte-specific) and molecular (background) components, and then the broadband absorption due to the background. The difference between the two is the atomic absorption.

For the continuum-source background correction it is assumed that the background is constant (unstructured) over the spectral bandpass. However, in samples with complex matrices, the background is structured due to absorption by other elements, and by molecular (fine-structure) absorption arising from rotational and vibrational bands for small molecules. In instances where the background is structured, the continuum-source approach can overcorrect or undercorrect for background absorption. For this situation, which is normally much more prevalent in GFAA than FAA, the Zeeman-effect background corrector is superior.

The Zeeman (pronounced *ZAY-MON*) effect involves splitting the atomic spectral line into absorption lines for each electronic transition by application of a strong magnetic field (Brown, 1977; Stephens, 1980). In its simplest pattern, the line is split into a central pi (π) and two equally spaced sigma (σ) components (Fig. 2-9). The percentage component has the same wavelength as the original absorption line, and absorbs only radiation that is polarized parallel to the magnetic field. The σ components have wavelengths on either side of the percentage component and only absorb radiation that is polarized perpendicularly to the field. Consequently, light polarized perpendicularly to the magnetic field is used to remove the percentage component, thus, yielding a measure of the background absorption. Polarization of the source light is accomplished by placing a fixed polarizer between the light source and the atomizer. More-complex splitting patterns can occur (with Mn, As, and Cu for example) and these are called the anomalous Zeeman effect. The Zeeman-effect background corrector works by modulating a magnet which surrounds the atomizer (Liddell & Brodie, 1980). The total absorbance is read with the magnet off and the background with the magnet on, the difference being the analyte absorbance. In practice, the σ components may not be completely resolved from the percentage component, thus reducing the sensitivity of Zeeman-AAS. This loss of sensitivity ranges from about 2 to 50% (Van Loon, 1985; Slavin et al., 1983). On the other hand, Zeeman-AAS does correct for high levels of background absorption, and is demonstrably more accurate in the presence of structured background (Voellkopf & Grobenski, 1984; Letourneau & Joshi, 1987).

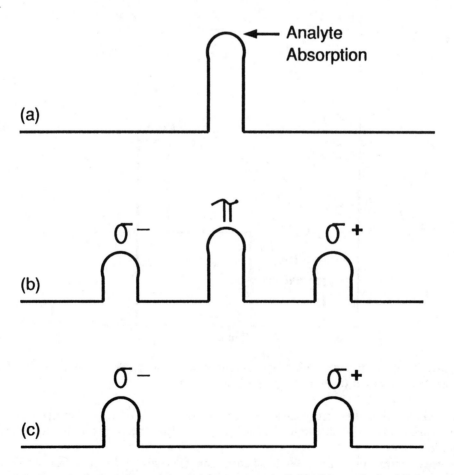

Fig. 2-9. Zeeman background correction: (*a*) no magnetic field; (*b*) unpolarized light, magnetic field on; and (*c*) polarized light, magnetic field on.

The line-reversal, or pulsed-HCL, background-correction technique is commonly called Smith-Hieftje background correction, after the originators of the method (Smith & Hieftje, 1983). For this correction, the HCL is first pulsed under normal current (5-20 mA) and then at high current (100-500 mA), each for about 0.5 ms. The lamp emission line is broadened at the high current and the excess of unexcited atoms produced at this current absorbs much of the emission in the center of the line. This phenomenon is called self-reversal. The background absorbs over the broad emission line during the high-current cycle and the total absorbance is measured during the low-current cycle. The Smith-Heiftje technique is most useful for volatile elements such as Pb and As. However, large losses in sensitivity (up to 90% for V) occur for less-volatile elements. Moreover, lamp lifetime is considerably reduced because of the high current. At present, there does not seem to be any compelling advantage to recommend this method over the others.

INDUCTIVELY COUPLED PLASMA-ATOMIC EMISSION SPECTROMETRY

The use of inductively coupled plasma-atomic emission spectrometry (ICP-AES) for elemental analysis is a relatively recent technique, commercial instruments becoming available only in the late 1970s. Babat is credited (Boumans, 1987) with generating the first sustained ICP, but the technique received little attention until after the landmark paper by Dickenson and Fassel (1969). In this paper, a toroidal ICP discharge arrangement was described. This arrangement improved the detection limits for many elements by two to three orders of magnitude compared to previous efforts which involved "teardrop" discharge. The ICP-AES designs developed in the 1970s were followed by the much more reliable commercial instruments of the 1980s. Today, modern ICP instruments can routinely determine 70 elements at concentrations below 1 mg L^{-1}.

Instrumentation

The major components of an ICP-AES instrument, as outlined in Fig. 2-10, are: (i) sample introduction system, (ii) ICP torch and RF generator, and (iii) spectrometer.

Sample Introduction

Methods for liquid sample introduction were discussed earlier. In ICP-AES, however, the analysis of samples dissolved in organic solvents requires special consideration. Deleterious effects caused by organic-solvent introduction include plasma instability, plasma cooling, spectral interferences by molecular-band emission, and rapid deterioration of the peristaltic pump tubing (Wiederon et al., 1990; Nygaard & Sotera, 1986). Ebdon et al. (1989) and Boumans and Lux-Steiner (1982) describe procedures for optimizing the operating conditions for sample analysis in a variety of organic solvents. These optimization methods include increasing the RF power and plasma-gas flow-rate substantially and decreasing the rate of sample intake.

Gaseous introduction of analytes via hydrides was also discussed earlier and other forms of sample introduction, such as direct-insertion or graphite-furnace, have been reviewed elsewhere (Routh & Tikkanen, 1987). Consequently, these topics will not be discussed here.

The Inductively Coupled Plasma Torch and Radio Frequency Generator

The most commonly employed ICP torches consist of three concentric quartz tubes, similar in design to those described by Fassel and associates in the early 1970s. Argon gas enters the torch through three inlets at different flow rates. Gas from the outer inlet, termed the coolant or plasma gas, flows tangentially up the outer tube at a rate of 7 to 15 L min^{-1} and serves the dual purpose of cooling the quartz sheath, to prevent melting, while maintaining the plasma. The gas flow through the middle concentric tube, normally 1 L min^{-1}, aids in aerosol introduction and keeps the plasma discharge away from the inlet tube, while gas flowing through the sample inlet carries the

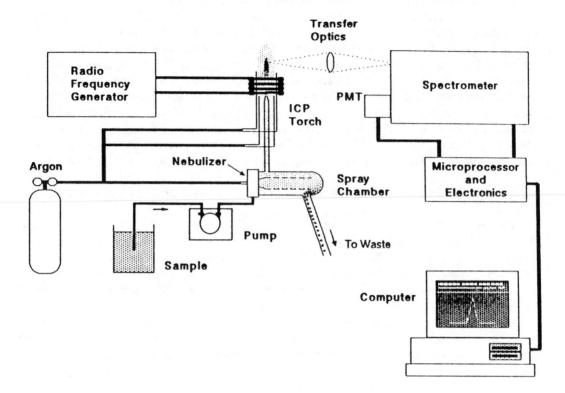

Fig. 2-10. Schematic diagram showing major components and layout of a typical ICP-AES instrument (Boss & Fredeen, 1989).

the sample aerosol. A small-diameter constriction near the top of the sample-inlet tube increases the gas velocity so that flow rates as low as 1 L min^{-1} are sufficient to pierce through the center of the plasma.

Energy to sustain the ICP discharge comes from radio frequency (RF) power (750-1500 W) produced by an RF generator (Greenfield, 1987) and applied to the load-coil surrounding the torch. The power is transferred to the conductive plasma by magnetic fields generated by the alternating current (27 MHz) passing through the coil. The Ar gas is initially rendered conductive by seeding with electrons supplied by a Tesla coil. The Ohmic resistance, caused by oscillation of the electrons in response to the high-frequency-induced alternating current, produces Joule heat. The high-energy electrons thus produced collide with Ar atoms in a process which sets up a chain reaction. The conductive plasma (consisting of Ar ions, atoms, and electrons) is sustained by the continual transfer of energy from the load-coil to the plasma by the inductively coupled process.

The ICP discharge appears as an intense white teardrop. The hottest portion of the plasma (10 000 K) is in the toroidal region. It takes about 2 ms for the analyte to travel from the bottom of the toroid to the normal observation height 15 to 20 mm above the coil (Fassel, 1979). These residence times and temperatures are about twice those produced by the best FAA conditions (i.e., the nitrous oxide-acetylene).

Spectrometers

The purpose of the spectrometer system is to separate and detect narrow (0.001-0.01 nm wide) analyte emission lines in the 170- to 900-nm range. To detect wavelengths below 200 nm, the spectrometer is housed under vacuum. Monographs (Strasheim, 1987; Olesik, 1987) are available which describe and compare a variety of the optical systems used in ICP-AES instruments. Only a brief comparative outline of simultaneous and sequential designs for multielement analysis will be discussed here.

A polychromic spectrometer is used for simultaneous multielement detection, whereas a monochromic spectrometer is used for the sequential analysis of multiple elements. In the dispersive Paschen-Runge type polychromator (above references), the entrance and exit slits, as well as the concave grating, are all in focus because they all are on the circumference of the Rowland circle. The Rowland-circle diameter (typically 1 m) is equal to the radius of curvature of the grating. In this arrangement, the number of elements that can be detected is only limited by the number of photomultiplier tube (PMT) phototransducers.

In spectrometers based on the sequential Paschen-Runge design, the PMT (or 2 PMTs) is (are) moved on a track mounted along the Rowland circle. The emission signal is detected, and the PMT is moved to the next position of interest, with a typical slew time of 1.5 seconds between elements.

The relative advantages and disadvantages of each design depend mainly on the intended application. With simultaneous systems, many elements can be detected in short times (1-15 s) and the precision and accuracy is generally better than that of the sequential analyzer. However, choice of elements is fixed, corrections for spectral interferences can be difficult, and the system is usually more expensive. On the other hand, sequential spectrometers provide spectral flexibility and ease in spectral-background interference analysis for complex samples. The latter is an important consideration when developing methods for environmental and geologic samples because of the high levels of Si, Al, Ca, and/or Fe, which exhibit complex ICP emission spectra.

Inductively Coupled Plasma-Mass Spectrometry

Since the initial report by Houk et al. (1980), inductively coupled plasma-mass spectrometry (ICP-MS) has developed into an important technique for elemental analysis. Numerous recent reviews (Horlick et al., 1987; Date & Gray, 1988; Dolan et al., 1990; Thompson & Walsh, 1988; Koppenaal, 1990) attest to the growing importance of this technique.

In ICP-MS, the ICP serves as the ion generator and the mass spectrometer (MS) serves as the ion measuring device. In this respect, ICP-MS is merely a new method of sample introduction for atomic mass spectrometry, a technique introduced in the early 1900s. Alternative ion-generation sources include spark, glow discharge, thermal (surface) ionization, laser microprobe, accelerator, and resonance ionization. Koppenaal (1990) has reviewed sample introduction methods for atomic mass spectrometry.

The unique capability of the ICP-MS sample introduction system rests with its capability to achieve detection limits of less than 0.1 ng L^{-1} in the direct analysis of solution samples. The detection limits are 100 to 1000 times better than those obtained by ICP-AES for over 60 elements. The large linear range associated with ICP-AES is also carried forward in ICP-MS. There are, however, some elements with poor detection limits (e.g., P, S, Si, K, C and the halogens). Nitrogen, oxygen, and the inert gases are not suited to analysis by ICP-MS due to poor ionization efficiencies (ranging from 0.04% for N to 10^{-9} % for He).

The ICP-MS instrument consists of an ICP torch, a quadrupole mass spectrometer and the interface between them. The ICP torch is positioned horizontally and is interfaced to the MS via a set of sampling cones (Fig. 2-11). The purpose of the cones is to reduce the pressure from ambient to about 10^{-4} Pa, as well as to focus the trajectory of the sample stream. A complete description of the instrument can be found in the articles listed above.

For detection by the MS the analyte must exist as an ion. Most elements present in the ICP are at least 90% ionized. Important exceptions include P, B, C and the halogens. The degree of ionization of an element is related to its ionization energy, as illustrated by the examples presented in Table 2-5. A complete tabulation of the degree of ionization for the elements can be found in Horlick et al. (1987).

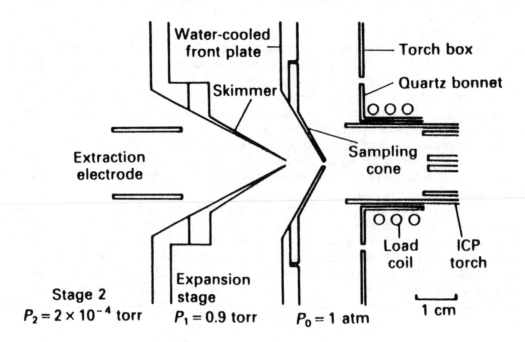

Fig. 2-11. Schematic diagram showing plasma sampling-interface with ICP-MS (Date & Gray, 1983).

Table 2-5. Degree of ionization in ICP for selected elements.

Element	Ionization potential	Degree of ionization
	eV	%
Cs	3.894	99.98
Na	5.139	99.91
Fe	7.870	96.77
B	8.298	62.03
Cd	8.993	85.43
Zn	9.394	74.50
P	10.486	28.79
Cl	12.967	0.46
Ar	15.759	0.01

Two types of mass-spectral interference are encountered in ICP-MS. One of these, isobaric spectral overlap, is the overlap among stable isotopes of elements such as ^{48}Ti:^{48}Ca, ^{58}Fe:^{58}Ni, and ^{64}Ni:^{64}Zn. In general, because the isotopic abundances for the elements are well established, these overlaps can be corrected readily. In the determination of titanium, for example, the overlap between ^{48}Ti and ^{48}Ca (0.19% relative abundance) can be corrected by measuring the ^{44}Ca (2.08% relative abundance) ion count and subtracting 9.13% of the ^{44}Ca counts from the ^{48}Ti signal. The other class of mass-spectral interference originates from background mass overlaps. These overlaps can cause serious errors and correction protocols are not always straightforward. Some of the more pronounced overlaps are ^{40}Ar with ^{40}Ca, ^{28}Si with $^{14}N^{14}N$, ^{32}S with $^{16}O^{16}O$, and ^{80}Se with $^{40}Ar^{40}Ar$. In most cases, adequate corrections for these mass overlaps can be made by subtracting the ion count for a blank from that of the sample. However, correcting for mass overlaps in complex matrices containing analytes at extremely low (sub $\mu g\ L^{-1}$) levels poses a more difficult problem, and one that is not as easily resolved.

The applications of ICP-MS for geological and environmental samples are similar to those for ICP-AES (Date & Gray, 1988; Dolan et al., 1990). Recently, Casetta et al. (1990) compared ICP-MS with several other techniques (including AAS) for the determination of rare earth and other trace elements in geological samples. They reported that the accuracy and precision of the ICP-MS technique were excellent and that the advantages of this technique included enhanced sensitivity and analytical speed. In addition, the capability for isotope measurement in ICP-MS can be used in isotope-dilution analysis and in stable-isotope tracer studies. Isotope-dilution analysis by ICP-MS is being used increasingly in the certification of reference materials (Koppenaal, 1990). Furthermore, ICP-MS stable-isotope tracer studies (Ting & Janghorbani, 1988; Ting et al., 1989) have the added advantage of eliminating the use of radioactive substances. The growth of ICP-MS as a generally available analytical tool is presently limited by its high cost.

Laser-Solid-Sampling Inductively Coupled Plasma-Mass Spectrometry

Although the use of a laser to vaporize solid samples can be coupled to any elemental analyzer (Denoyer et al., 1991), it is predominantly used in ICP-MS. Typically, a pulsed (10-20 Hz) Nd:YAG laser is focused (20-50 μm resolution) on an enclosed solid sample. The vaporized material is carried by argon gas into the plasma of the ICP-MS. A video camera and monitor are usually attached to the laser-solid sampler to provide for user control of lateral and depth profiling to determine the distribution of elements within the solid sample and on its surface.

One of the advantages of this method is that solids are directly analyzed, eliminating sample preparation procedures, which minimizes the opportunity for contamination. Moreover, the elimination of dilution steps and of a spray chamber result in lower detection limits compared to the conventional liquid-sample introduction methods. Laser-sampling ICP-MS has been used for the determination and distribution of elements in geological and soil samples (Broadhead et al., 1990). Details of the procedure and applications can be found in Denoyer et al. (1991).

SUMMARY AND CONCLUSION

Despite the availability of a variety of instrumental techniques for the total elemental analysis of soils, minerals, rocks and environmental samples (e.g., sludges, coal, and incineration ashes), atomic spectroscopic analyses following decomposition and dissolution of the sample remain the most common analytical procedures for quantitative analysis. Decomposition of samples can be attained either by fusion with alkali fluxes or digestion with HF in combination with other acids, e.g., $HClO_4$ and HNO_3. Use of pressurized plastic containers or bombs has substantially reduced the digestion time and the loss of certain volatile constituents that is encountered in open systems. Recent developments in the use of microwave energy as a heat source for the dissolution of solid samples, combining the rapid heating ability of microwaves and use of sealed vessels, has reduced the decomposition and digestion time from hours to minutes.

Although use of ICP-AES for multielement analysis is growing rapidly, FAA is still the most commonly used instrumental technique because of its simple operation and low cost. However, when many elements need to be determined in large numbers of samples, ICP-AES offers a distinct advantage, e.g., over 72 elements in the milligram per liter range can be determined rapidly by this technique. Because of its enhanced sensitivity, GFAA is the method of choice for trace-element analysis (especially for B, Cd, As, Pb, etc.). Moreover, it is particularly well adapted for the analysis of solid samples and slurries. Although still in its infancy, ICP-MS is recognized as an important analytical tool for the elemental analysis of soils, minerals, and environmental samples. However, despite its vast potential, general use of this technique is presently limited by its high cost. Likewise, the limited use of AFS stems not from some inherent weakness of the technique but rather from the fact that it is more costly and has no distinct advantage.

REFERENCES

Abu-Samra, A., J.S. Morris, and S.R. Koirtyohann. 1975. Wet ashing of some biological samples in microwave oven. Anal. Chem. 47:1475-1477.

Alkemade, C.T.J., and J.M.W. Milatz. 1955. Double-beam method of spectral selection with flames. J. Opt. Soc. Am. 45:583-584.

Bader, M. 1980. A systematic approach to standard addition methods in instrumental analysis. J. Chem. Ed. 57:703-706.

Bakhtar, D., G.R. Bradford, and L.J. Lund. 1989. Dissolution of soils and geological materials for simultaneous elemental analysis by inductively coupled plasma optical emission spectrometry and atomic absorption spectrometry. Analyst 114:901-909.

Barnes, R.M., and J.S. Genna. 1979. Concentration and spectrochemical determination of trace metals in urine with polydithiocarbamate resin and inductively coupled plasma-atomic emission spectrometry. Anal. Chem. 51:1065-1070.

Bass, D.A., and J.A. Holcombe. 1987. Mass spectral investigation of mechanisms of lead vaporization from a graphite surface used in electrothermal atomizer. Anal. Chem. 59:974-980.

Beaty, R.D. 1988. Concepts, instrumentation, and techniques in atomic absorption spectrophotometry. Perkin-Elmer Corp., Norwalk, CT.

Bernas, B. 1968. A new method for decomposition and comprehensive analysis of silicates by atomic absorption spectrometry. Anal. Chem. 40:1682-1686.

Billings, K., and E. Angins. 1973. Atomic absorption spectrometry in geology. Elsevier, New York.

Boss, C.B., and K.J. Fredeen. 1989. Concepts, instrumentation, and techniques in inductively coupled plasma atomic emission spectrometry. Perkin-Elmer Corp., Norwalk, CT.

Boumans, P.W.J.M. (ed.). 1987. Inductively coupled plasma emission spectroscopy. Part 1 andt 2. Wiley-Interscience, New York.

Boumans, P.W.J.M., and M.L. Lux-Steiner. 1982. Modification and optimization of a 50-MHz inductively coupled argon plasma with special reference to analysis using organic solvents. Spectrochim. Acta 37B:97-126.

Bowen, H.M. 1966. Trace elements in biochemistry. Acad. Press, London.

Boyle, M., and W.H. Fuller. 1987. Effect of municipal solid waste leachate composition on zinc migration through soils. J. Environ. Qual. 16:357-360.

Broadhead, M., R. Broadhead, and J. Hager. 1990. Laser sampling ICP-MS: Semi-quantitative determination of sixty-six elements in geological samples. At. Spectrosc. 11:205-209.

Brown, S. 1977. Zeeman effect-based background correction in atomic absorption spectrometry. Anal. Chem. 49:1269A-1281A.

Browner, R.F. 1983. Sample introduction for ICP's and flames. Trends Anal. Chem. 2:121-124.

Browner, R.F., and A.W. Boorn. 1984a. Sample introduction: The Achilles heel of atomic spectroscopy. Anal. Chem. 56:787A-798A.

Browner, R.F., and A.W. Boorn. 1984b. Sample introduction techniques for atomic spectroscopy. Anal. Chem. 56:875A-888A.

Burguera, J.L. 1989. Flow injection atomic spectroscopy. Marcel Dekker, New York.

Burman, Jan-Ola. 1987. Applications: Geological. p. 27-47. In P.W.J.M. Boumans (ed.) Inductively coupled plasma spectroscopy. Part 2. John Wiley & Sons, New York.

Bysouth, S.R., J.F. Tyson, and P.B. Stockwell. 1990. Use of masking agents in determination of lead in tap water by flame atomic absorption spectrometry with flow-injection pre-concentration. Analyst 115:571-573.

Casetta, B., A. Giaretta, and G. Mezzacasa. 1990. Determination of rare earth and other trace elements in rock samples by ICP-mass spectrometry: Comparison with other techniques. At. Spectrosc. 11:222-228.

Culver, B.R., and T. Surles. 1975. Interference of molecular spectra due to alkali halides in non-flame atomic absorption spectrometry. Anal. Chem. 47:920-921.

Date, A.R., and A.L. Gray. 1983. Development progress in plasma source mass spectrometry. Analyst 108:159-165.

Date, A.R., and A.L. Gray. 1988. Applications of inductively coupled plasma mass spectrometry. Blackie, Glasgow.

Denoyer, E., K. Fredeen, and J. Hager. 1991. Laser solid sampling for inductively coupled plasma spectrometry. Anal. Chem. 63:445A-447A.

Dickenson, G.W., and V.A. Fassel. 1969. Emission spectrometric detection of the elements at the nanogram per milliliter level using induction-coupled plasma excitation. Anal. Chem. 41:1021-1024.

Dolan, R., J. Van Loon, D. Templeton, and A. Paudyn. 1990. Assessment of ICP-MS for routine multielement analysis of soil samples in environmental trace element studies. Z. Anal. Chem. 336:99-116.

Ebdon, L., E.H. Evans, and N.W. Barnett. 1989. Simplex optimisation of experimental conditions in inductively coupled plasma atomic emission spectrometry with organic solvent introduction. J. Anal. At. Spectrom. 4:505-508.

Elseewi, A.A., A.L. Page, and C.P. Doyle. 1982. Environmental characterization of trace elements in fly ash. p. 39-47. In D.D. Hemphill (ed.) Trace substances in environmental health - XVI. Univ. Missouri, Columbia.

Fassel, V.A. 1979. Simultaneous or sequential determination of the elements at all concentration levels: The renaissance of an old approach. Anal. Chem. 51:1290A-1308A.

Fischer, L.B. 1986. Microwave digestion of geologic material: Application to isotopic dilution analysis. Anal. Chem. 58:261-263.

Frances, C.W., and G.H. White. 1987. Leaching of toxic metals from incinerator ashes. J. Water Pollut. Control Fed. 59:979-986.

Giordano, P.M., A.D. Behel, Jr., J.E. Lawrence, Jr., J.M. Solieau, and B.N. Bradford. 1983. Mobility in soil and plant availability of metals derived from incinerated municipal refuse. Environ. Sci. Technol. 17:193-198.

Govindaraju, K., and G. Mevelle. 1983. Geostandards and geochemical analysis. Spectrochim. Acta 38B:1447-1456.

Govindaraju, K., G. Mevelle, and C. Chouard. 1976. Automated optical emission spectrochemical bulk analysis of silicate rocks with microwave plasma excitation. Anal. Chem. 48:1325-1330.

Greenfield, S. 1987. Common radio frequency generators, torches, and sample introduction systems. p. 123-161. In A. Montaser and D.W. Golightly (ed.) Inductively coupled plasmas in analytical atomic spectrometry. VCH Publ., New York.

Hewitt, A.D., and C.M. Reynolds. 1990. Dissolution of metals from soils and sediments with a microwave-nitric acid digestion technique. At. Spectrosc. 11:187-192.

Hildebrand, W.F. 1919. The analysis of silicate and carbonate rocks. U.S. Geol. Surv. Bull. 700.

Horlick, G., S.H. Tan, M.A. Vaughan, and Y. Shao. 1987. Inductively coupled mass spectrometry. p. 361-398. In A. Montaser and D.W. Golightly (ed.) Inductively coupled plasmas in analytical atomic spectroscopy. VCH Publ., New York.

Houk, R.S., V.A. Fassel, G.D. Flesch, H.J. Svec, A.L. Gray, and C.F. Taylor. 1980. Inductively coupled argon plasma as an ion source for mass spectrometric determination of trace elements. Anal. Chem. 52:2283-2289.

Hsiech, C., S.C. Petrovic, and H.L. Pardue. 1990. Continuum-source atomic absorption spectroscopy with an Echelle spectrometer adapted to a charge injection device. Anal. Chem. 62:1983-1988.

Ingle, J.D., Jr., and S.R. Crouch. 1988. Spectrochemical analysis. Prentice-Hall, Englewood Cliffs, NJ.

Ivaldi, J.C., and W. Slavin. 1990. Cross-flow nebulisers and testing procedures for ICP nebulisers. J. Anal. At. Spectrom. 5:359.

Jackson, M.L. 1958. Soil chemical analysis. Prentice-Hall, Englewood Cliffs, NJ.

Jackson, M.L. 1974. Soil chemical analysis-Advanced course. 2nd ed. M.L. Jackson, Madison, WI.

Jeffery, P.G. 1970. Chemical methods of rock analysis. Pergamon Press, New York.

Johnson, W.M., and J.A. Maxwell. 1981. Rock and mineral analysis. 2nd ed. John Wiley & Sons, New York.

Jones, B.T., M. Mignardi, B.W. Smith, and J.D. Winefordner. 1989. High resolution continuum source atomic absorption spectrometry in an air-acetylene flame with photodiode array detection. J. Anal. At. Spectrom. 4:647-651.

Kanehiro, Y., and G.D. Sherman. 1965. Fusion with sodium carbonate for total elemental analysis. p. 952-958. *In* C.A. Black et al. (ed.) Methods of soil analysis. Part 2. Agron. Monogr. 9. ASA and SSSA, Madison, WI.

Karstensen, K.H., and W. Lund. 1989. Multi-element analysis of a city waste incineration ash reference sample by inductively coupled plasma atomic emission spectrometry. J. Anal. At. Spectrom. 4:357-359.

Kemp, A.J., and C.J. Brown. 1990. Microwave digestion of carbonate rock samples for chemical analysis. Analyst 115:1197-1199.

Kimbrough, D.E., and J.R. Wakakuwa. 1989. Acid digestion for sediments, soils and solid wastes. A proposed alternative to EPA SW 846 method 3050. Environ. Sci. Technol. 23:898-900.

Kingston, H.M., and L.B. Jessie. 1986. Microwave energy for acid decomposition at elevated temperatures and pressures using biological and botanical samples. Anal. Chem. 58:2534-2541.

Kingston, H.M., and L.B. Jessie (ed.). 1988. Introduction to microwave sample preparation: Theory and practice. ACS, Washington, DC.

Knapp, G., K. Muller, M. Strunz, and W. Wegscheider. 1987. Automation in element pre-concentration with chelating ion exchangers. J. Anal. At. Spectrom. 2:611-614.

Koppenaal, D.W. 1990. Atomic mass spectrometry. Anal. Chem. 62:303R-324R.

Langmyhr, F.J., and P.E. Paus. 1968. The analysis of inorganic silicious materials by atomic absorption spectrometry and the hydrofluoric acid decomposition technique. Part I. Anal. Chim. Acta 43:397-408.

Layman, L.R., and F.E. Lichte. 1982. Glass frit nebulizer for atomic spectroscopy. Anal. Chem. 54:638-642.

Letourneau, V.A., and B.M. Joshi. 1987. Comparison between Zeeman and continuum background correction for GFAA on environmental samples. At. Spectrosc. 8:145-149.

Lim, C.H., and M.L. Jackson. 1982. Dissolution for total elemental analysis. p.1-12. *In* A.L. Page et al. (ed.) Methods of soil analysis. Part 2. 2nd. ed. Agron. Monogr. 9. ASA and SSSA, Madison, WI.

Liddell, P.R., and K.G. Brodie. 1980. Application of a modulated magnetic field to a graphite furnace in Zeeman effect AAS. Anal. Chem. 52:1256-1260.

Lindsay, W.L., and W.A. Norvell. 1978. Development of a DTPA soil test for zinc, iron, manganese and copper. Soil Sci. Soc. Am. J. 42:421-428.

Liqiang, X., and S. Wangxing. 1989. Rapid decomposition method for silicate samples. Fresenius Z. Anal. Chem. 333:108-110.

Lisk, D.J., C.L. Secor, M. Rutzke, and T.H. Kuntz. 1989. Element composition of municipal refuse ashes and their aqueous extracts from 18 incinerators. Bull. Environ. Contam. Toxicol. 42:534-539.

Loring, D.H., and R.T.T. Rantala. 1988. An intercalibration exercise for trace metals in marine sediments. Mar. Chem. 24:13-18.

L'vov, B.V. 1961. The analytical use of atomic absorption spectra. Spectrochim. Acta 17:761-770.

L'vov, B.V. 1978. Electrothermal atomization. Spectrochim. Acta 33B:153-193.

Massman, H. 1968. Atomic absorption analysis with a heated graphite cell. Spectrochim. Acta 23B:215-226.

McQuaker, N.R., P.D. Kluckner, and G.N. Chang. 1979. Calibration of an inductively coupled plasma-atomic emission spectrometer for the analysis of environmental materials. Anal. Chem. 51:888-895.

Meinhard, J.E. 1976. Pneumatic nebulizers, present and future. ICP Inf. Newslett. 2:163-168.

Minczewski, J., J. Chwastowska, and R. Dybczynski. 1982. Separation and preconcentration methods in inorganic trace elements. John Wiley & Sons, New York.

Miyazaki, A., and R. Barnes. 1981. Complexation of some transition metals, rare earth elements, and thorium with a polydithiocarbamate chelating resin. Anal. Chem. 53:299-304.

Montaser, A., and G.W. Golightly (eds.) 1987. Inductively coupled plasmas in analytical atomic spectrometry. VCH Publ., New York.

Moore, G.L. 1989. Introduction to inductively coupled plasma atomic emission spectroscopy. Elsevier, New York.

Moulton, G.P., T.C. O'Haver, and J.M. Harnly. 1989. Continuum source atomic absorption spectrometry with a pulsed source and a photo-diode array detector. J. Anal. At. Spectrom. 4:673-674.

Nadkarni, R.A. 1980. Multitechnique multielement analysis of coal and fly ash. Anal. Chem. 52:929-935.

Nakahara, T. 1983. Applications of hydride generation techniques. Prog. Anal. At. Spectrosc. 6:163-223.

Navarro, J.A., V.A. Granadillo, O.E. Parra, and R.A. Romero. 1989. Determination of lead in whole blood by GFAA spectrometry with matrix modification. J. Anal. At. Spectrom. 4:401-406.

Ng, K.C., A.H. Ali, T.E. Barber, and J.D. Winefordner. 1990. Multiple mode semiconductor diode laser as a spectral line source for graphite furnace atomic absorption spectroscopy. Anal. Chem. 62:1893-1895.

Norris, J.D., and T.S. West. 1974. Some applications of spectral overlap in AA spectrometry. Anal. Chem. 45:1423-1425.

Novozamsky, I., R. van Eck, J.J. van der Lee, V. Houba, and G. Ayaga. 1988. Continuous-flow technique for generation and separation of methyl borate from iron-containing matrices with subsequent determination of boron by ICP-AES. At. Spectrosc. 9:97-99.

Nygaard, D.D., and J.J. Sotera. 1986. Analysis of electronic-grade organic solvents by ICP emission spectrometry. Spectroscopy 1:8-12.

O'Haver, T.C., J. Carrol, R. Nichol, and D. Littlejohn. 1988. Extended range background correction in continuum source atomic absorption spectrometry. J. Anal. At. Spectrom. 3:155-157.

Olesik, J.W. 1987. Spectrometers. p. 466-535. *In* P.W.J.N. Boumans (ed.) Inductively coupled plasma emission spectroscopy. Part 1. Wiley-Interscience, New York.

Olson, K.W., W.J. Haas, Jr., and V.A. Fassel. 1977. Multielement detection limits and sample nebulization efficiencies of an improved ultrasonic nebulizer. Anal. Chem. 49:632-637.

Ottaway, J.M., and A.M. Ure. 1989. Practical atomic absorption spectrometry. Pergamon Press, New York.

Papp, C.S.E., and L.B. Fischer. 1987. Application of microwave digestion to the analysis of peat. Analyst 112:337-338.

Perkin-Elmer Corp. 1988. The guide to techniques and applications of atomic spectroscopy. Perkin-Elmer Corp., Norwalk, CT.

Potts, P.J. 1987. A handbook of silicate rock analysis. Blackie, London.

Robbins, W.B., and J.A. Caruso. 1979. Development of hydride generation methods in atomic spectroscopic analysis. Anal. Chem. 51:889A-899A.

Robinson, J.W. 1981. Atomic absorption spectroscopy. p. 730-800. *In* P. Elving et al. (ed.) Treatise on analytical chemistry. Vol. 7. Wiley-Interscience, New York.

Routh, M.W., and M.W. Tikkanen. 1987. Introduction of solids into plasmas. p. 431-486. *In* A. Montaser and D.W. Golightly (ed.) Inductively coupled plasmas in analytical atomic spectrometry. VCH Publ., New York.

Ruzicka, J. 1983. Flow injection analysis from test tube to integrated microconduits. Anal. Chem. 55:1041A-1053A.

Sawhney, B.L., and C.R. Frink. 1991. Potential leachability of heavy metals from incinerator ash. Water Air Soil Pollut. 57-58:289-296.

Siriraks, A., H.M. Kingston, and J.M. Riviello. 1990. Chelation ion chromatography as a method for trace elemental analysis in complex environmental and biological samples. Anal. Chem. 62:1185-1193.

Slavin, W., G.R. Carnrick, and D.C. Manning. 1982. Magnesium nitrate as a matrix modifier in the stabilized temperature platform furnace. Anal. Chem. 54:621-624.

Slavin, W., G.R. Carnrick, D.C. Manning, and E. Pruszkowska. 1983. Recent experiences with the stabilized temperature platform furnace and Zeeman background correction. At. Spectrosc. 4:69-86.

Slavin, W., and D.C. Manning. 1979. Reduction of matrix interferences with lead determination with the L'vov platform and the graphite furnace. Anal. Chem. 51:261-265.

Slavin, W., and D.C. Manning. 1982. Graphite furnace interferences: A guide to the literature. Prog. Analyt. At. Spectrosc. 5:243-340.

Smith, D.D., and R.F. Browner. 1982. Measurement of aerosol transport efficiency in atomic spectroscopy. Anal. Chem. 54:533-537.

Smith, S.B., Jr., and G.M. Hieftje. 1983. A new background correction method for atomic absorption spectrometry. Appl. Spectrosc. 37:419-424.

Soltanpour, P.N., J.B. Jones, and S.M. Workman. 1982. Optical emission spectrometry. p. 29-66. *In* A.L. Page et al. (ed.) Methods of soil analysis. Part 2. 2nd ed. Agron. Monogr. 9. ASA and SSSA, Madison, WI.

Soltanpour, P.N., and S.M. Workman. 1979. Modification of the $NH_4 HCO_3$-DTPA soil test to omit carbon black. Commun. Soil Sci. Plant Anal. 10:1411-1420.

Stephens, R. 1980. Zeeman modulated atomic absorption spectroscopy. CRC Crit. Rev. Anal. Chem. 9:167-221.

Stewart, K.K. 1983. Flow injection analysis. Anal. Chem. 55:931A-940A.

Strasheim, A. 1987. Instrumentation for optical emission spectrometry. p. 69-121. *In* A. Montaser and D.W. Golightly (ed.) Inductively coupled plasmas in analytical atomic spectrometry. VCH Publ., New York.

Strasheim, A., and S.D. Olsen. 1983. Correlation of the analytical signal to the characterized nebulizer spray. Spectrochim. Acta 38B:973-975.

Styris, D., L. Prell, and D. Redfield. 1991. Mechanisms of palladium-induced stabilization of arsenic in EAAS. Anal. Chem. 63:503-507.

Thompson, M. 1987. Analytical performance of ICP-AES. p. 163-199. *In* A. Montaser and D.W. Golightly (ed.) Inductively couple plasmas in analytical atomic spectroscopy. VCH Publ., New York.

Thompson, M., and J.N. Walsh. 1988. Handbook of inductively coupled plasma spectroscopy. 2nd ed. Blackie, London.

Thompson, M., and M.H. Ramsey. 1985. Matrix effects due to calcium in ICP-AES. Analyst 110:1413-22.

Ting, B.T.G., and M. Janghorbani. 1988. Optimisation of instrumental parameters for the precise measurement of isotope ratios with ICP-MS. J. Anal. At. Spectrom. 3:325-336.

Ting, B.T.G., C.S. Mooers, and M. Janghorbani. 1989. Isotopic determination of selenium in biological materials with ICP-MS. Analyst 114:667-674.

Uchida, H., T. Uchida, and C. Iida. 1979. Determination of major and minor elements in silicates by inductively coupled plasma emission spectrometry. Anal. Chim. Acta 108:87-92.

U.S. Environmental Protection Agency. 1986. Test methods for evaluating solid wastes. USEPA 846, U.S. Gov. Print. Office, Washington, DC.

Van Loon, J.C. 1980. Analytical atomic absorption spectroscopy. Acad. Press, New York.

Van Loon, J.C. 1985. Selected methods of trace metal analysis: Biological and environmental samples. John Wiley & Sons, New York.

Varma, A. 1989. CRC handbook of furnace atomic absorption spectroscopy. CRC Press, Boca Raton, FL.

Voellkopf, U., and Z. Grobenski. 1984. Interference in the analysis of biological samples using the stabilized temperature platform furnace and Zeeman background correction. At. Spectrosc. 5:115-122.

Walsh, A. 1955. Application of atomic absorption spectra to chemical analysis. Spectrochim. Acta 7:108-117.

Walton, S.J., and J.E. Goulter. 1985. Performance of a commercial maximum dissolved solids nebulizer for ICP spectrometry. Analyst 110:531-534.

Warren, C.J., and M.J. Dudas. 1989. Leachability and partitioning of elements in ferromagnetic fly ash particles. Sci. Total Environ. 83:99-111.

Washington, H.S. 1930. The chemical analysis of rocks. 4th ed. John Wiley & Sons, New York.

Welz, B. 1986. Atomic absorption spectrometry. VCH Publ., New York.

Wiederin, D.R., R.S. Houk, R.K. Winge, and A.P. D'Silva. 1990. Introduction of organic solvents into ICP's by ultrasonic nebulization with cryogenic desolvation. Anal. Chem. 62:1155-1160.

Winge, R.K., V.A. Fassel, V.J. Peterson, and M.A. Floyd. 1985. Inductively coupled plasma emission spectroscopy: An atlas of spectral information. Elsevier, New York.

Wolcott, J.F., and C.B. Sobel. 1982. Fabrication of a Babington type nebulizer for ICP sources. Appl. Spectrosc. 36:685-686.

Zhang, L., S. Xiao-quan, and N. Zhe-ming. 1988. An oblique section hydride generator for simultaneous determination of arsenic, antimony and bismuth in geological samples by inductively coupled plasma-atomic emission spectrometry. Fresenius Z. Anal. Chem. 332:764-768.

3   Quantitative Oxidation-State Analysis of Soils

J. E. Amonette
Pacific Northwest Laboratory
Richland, Washington

F. A. Khan and A.D. Scott
Iowa State University
Ames, Iowa

H. Gan and J. W. Stucki
University of Illinois
Urbana, Illinois

The oxidation states of Fe, Mn, Cr, and other transition metals in soil minerals play an important role in many soil processes. For example, the weathering of minerals often includes the oxidation of Fe(II) in micas, producing expandable clay minerals and releasing K^+ and other ions to solution. Cyclic natural processes, such as wetting and drying, result in seasonal variability of the oxidation state of the soil, setting up conditions for in situ oxidation or reduction of constituent minerals. Redox (i.e., oxidation and reduction) processes are also linked to the activity of microorganisms that may colonize the surfaces of soil minerals, the rhizosphere, and other soil microenvironments, and thus may affect the fertility and the pathogenic properties of soils. The mobility of metals (e.g., Cr, Mn, U) and organic compounds in soils and sediments depends strongly on their speciation, and, hence, on their oxidation state, and may be influenced by interactions with the redox-sensitive species in soil minerals and organic matter. The ability to quantify the amounts and the reactivity of redox-sensitive species in soil solids is crucial to understanding the processes involved in plant nutrition, pedogenesis, and contaminant migration in soils.

The oxidation state of structural cations also influences the physical-chemical properties of the host crystalline mineral. Striking examples are found in the effect of Fe oxidation state on the swelling pressure, surface area, layer charge, cation fixation, and other properties of smectites. Manipulation and measurement of the oxidation state of structural cations thus provides another way to probe the fundamental forces governing the surface properties of clay minerals.

Because of these important effects on mineral properties and the need to identify the charges on structural cations for structural formula determinations, considerable emphasis has been placed on developing methods for the quantitative determination of elemental oxidation states in minerals. Collection and interpretation of element-specific oxidation-state data is more difficult in soils, however, because they are complex mixtures of minerals and organic compounds. In some instances, element-specific redox data are less relevant than is a measure of the "total reductive capacity" or "total oxidative capacity" of a soil. As will be seen, some methods that are nominally element-specific actually provide this broader type of information when applied to soils or other complex systems that contain a variety of redox couples.

Nevertheless, there is also a need for element-specific oxidation-state analysis of soils as well as of minerals. Although numerous wet-chemical methods for measuring the total amounts of various transition elements in soils have been published (Page et al., 1982), the specific and quantitative differentiation of their oxidation states has been largely ignored. The reasons for this neglect rest largely in the complexity of the substrate. A method that works for one soil may not be appropriate for analysis of another because the set of possible interferences is different. Recently developed direct spectroscopic methods such as Mössbauer, x-ray photoelectron (XPS), and x-ray absorption near-edge structure (XANES) avoid many such matrix-dependent problems and are capable of resolving elemental oxidation states, but they are semiquantitative at best. Thus, quantitative analysis of soil oxidation states continues to rely on a suite of wet-chemical techniques. The legitimacy of the data obtained, in turn, depends on the analyst's ability to select the appropriate sampling and analytical techniques after considering the potential interferences presented by a particular soil matrix and the uses for which the data are intended.

This chapter provides an overview of the various techniques commonly used to quantify the oxidation states of specific metals in soil solids. The focus is on methods for quantifying Fe(II) and Fe(III) because Fe is the dominant redox-sensitive metal in most soils and sediments. Some attention, however, is also given to analysis for Mn and Cr oxidation states and for overall soil redox properties (as distinct from oxidation states of specific metals), and it is hoped that the reader can extend the principles that are presented to other situations when necessary. The chapter begins with a general discussion of the principles that underly the various methods for measuring oxidation states in soil solids, followed by more detailed descriptions of techniques originally developed for analysis of minerals. The chapter concludes with a comparison of the results obtained when three representative methods for the determination of Fe oxidation states in minerals are applied to the analysis of bulk soils.

GENERAL PRINCIPLES

Sampling

The procedures followed during collection and storage of soil samples before redox analysis can have a large bearing on the quality and relevance of the results obtained. Many soil systems are poised at redox levels that are more reducing than systems in equilibrium with atmospheric O_2. Exposure of samples from these systems to air (e.g., by the common practice of sieving and air drying) will result in their oxidation and a loss of information that may be relevant to the problem being studied. Before field sampling occurs, therefore, the analyst must decide whether the air-oxidizable fraction of the sample is important to the problem. If measurement of the redox conditions present at the time of sample collection is deemed important, then steps may be taken to preserve the sample until laboratory analysis is possible. These steps ideally include storage of the sample under an inert atmosphere such as N_2 or Ar and chilling or freezing the sample immediately after it is collected. In many instances, however, especially where the sample will be analyzed in the laboratory within a few hours of collection, preservation in a plastic bag from which the air has been squeezed is adequate. As an added precaution, a field analysis of the redox state may be conducted for later comparison with the laboratory results.

Other factors to consider in sampling include sample size, numbers of samples, bulk vs. selected mottles, concretions, etc. Obviously the samples collected should represent the problem that is being addressed and a large enough number should be collected to provide statistical certainty to the results obtained. Loveland (1988) presents an excellent discussion of the factors to consider when sampling soils for Fe analysis.

Standards

Calibration of the analytical results to well-characterized standards is particularly important for redox-state analyses because of the analytical difficulties that are typically encountered. Two types of standards may be employed. The first type is a readily soluble standard consisting of a single compound containing the element of interest in the appropriate oxidation state. The advantage of this type of standard is that it is easy to work with and offers essentially no complications as a result of its matrix. Examples of primary standards include ferrous ethylenediammonium sulfate for Fe(II) and potassium dichromate for Cr(VI). These standards are of high purity, have high equivalent weights to minimize errors in weighing, react readily, and do not readily absorb moisture or change oxidation state when stored in the solid form.

The second type of standard is a standard reference material, typically a finely ground soil, rock, or mineral powder that contains the analyte in a matrix that is close to that of the samples being analyzed. Because it involves homogenizing a heterogeneous solid, this type of standard is more difficult to prepare reliably and is best obtained from organizations such as the National Institute of Standards and Technology (NIST) with expertise in preparing and characterizing standard powders. The advantage of using standard reference materials is that matrix effects are also included in the analysis thus affording a more realistic test of the applicability of the analytical method.

In recent years, the development of a suite of geochemical reference standards covering a variety of mineral matrices has been coordinated by an international working group under the auspices of the Centre de Recherches Petrographiques et Geochimiques in Nancy, France. This group publishes *Geostandards Newsletter* that summarizes the results of analyses on their standard reference materials by laboratories worldwide (contact Dr. K. Govindaraju, Geostandards, CRPG, B. P. 20, 54501 Vandoeuvre Cedex, France, for further information). Other sources of standard reference materials include NIST (U.S. Department of Commerce, Gaithersburg, MD, 20899), the U.S. Geological Survey (contact Dr. Stephen Wilson, USGS Reference Materials Program Coordinator, USGS, Box 25046, MS 973, Denver Federal Center, Denver, CO, 80225), and numerous other organizations worldwide. Potts et al. (1992) provide the addresses of more than 30 organizations that issue geochemical reference materials along with the analytical values recommended by these organizations. More than 700 reference materials are listed, including silicate rocks, minerals, sediments, soils, carbonates and ores. These materials typically have been analyzed for Fe oxidation states; a few of the ores listed have also been analyzed for Mn oxidation states. Of the soil reference materials listed, however, only two (the GXR series distributed by the USGS on behalf of the Association of Exploration Geologists) report values for Fe(II) and Fe(III) separately from total Fe, and none list values for Mn or Cr oxidation states. In practice, therefore, rock, mineral, or ore standards have been used for the calibration of redox-state analyses of soils and other geological materials.

Analytical Techniques

Elemental oxidation states can be determined either by direct spectroscopic measurement of essentially unaltered specimens or by indirect chemical methods that involve decomposition of the specimen and subsequent measurement of the oxidation state by colorimetry or titrimetry. The direct spectroscopic techniques include XPS, XANES, and Mössbauer spectroscopy. The first two spectroscopic techniques involve the irradiation of a specimen with x-rays and the detection of either (i) the energies and intensities of the photoelectrons emitted (XPS) or (ii) the relationship between x-ray absorption intensity and incident x-ray energy in traversing the x-ray absorption edge of the element of interest (XANES). Generally speaking, these two techniques probe the energies of transitions between the inner-shell electrons and the valence-shell orbitals whose occupancy determines the oxidation state of the atom of interest. An introduction to the physics of x-ray absorption and production is provided in Chapter 1 (Amonette & Sanders, 1994); detailed discussions of the XPS and XANES techniques are given by Hochella (1988) and Brown et al. (1988), and in Chapter 7 (Cocke et al., 1994).

Mössbauer Spectroscopy

Mössbauer spectroscopy is a well-known, nondestructive technique for identifying Fe(II) and Fe(III) in mineral samples. The technique relies on the Mössbauer effect (Mössbauer, 1958a,b), which involves the resonant absorption and recoil-free emission of γ-rays by specific nuclei. Thus, in contrast to XPS and XANES, Mössbauer spectroscopy directly probes nuclear energy levels rather than electronic energy levels. The nuclear energy levels, however, are sensitive to the electronic energy levels and to the magnetic and electrical fields present in the solid. As a result they contain information about the oxidation state of the atom whose nucleus is sampled. Although approximately half of the elements in the Periodic Table exhibit the Mössbauer effect, the method is not practical in most cases because of: (i) the very short half-lives of the relevant parent nuclei that serve as γ-ray sources for these elements, (ii) insufficient recoil-free fractions in the absorber nuclei, or (iii) the short lifetime of the excited state of the decay product. Of all the elements, Fe is the best suited for study by this method. Specimens can be analyzed as liquids or solids, at ambient pressure or in vacuo, and at temperatures ranging from hundreds of degrees Celsius to near absolute zero. The two common oxidation states of Fe have distinctly different fingerprints in the Mössbauer spectrum but, because their absorption envelopes overlap, spectrum decomposition is necessary to extract relative peak areas for the two states.

The uncertainty associated with peak decomposition decreases the quantitative resolution of the Mössbauer method. Another complication is that the area ratio may not be an accurate measure of relative abundance. This is because the observed absorbance depends on the recoil-free fractions of each state. Recoil-free fractions are temperature-sensitive down to 120 K, are somewhat difficult to determine precisely (Lear, 1987; Rancourt, 1989, 1991a,b; de Grave & van Alboom, 1991), and may vary for different minerals in the same specimen (Whipple, 1974a). Furthermore, Mössbauer spectroscopy is not suited for analysis of large numbers of specimens. Two spectra at most can be collected simultaneously with a single spectrometer, and several hours to days of data collection may be necessary, depending on the activity of the ^{57}Co source, to collect spectra of sufficient quality for quantitative analysis.

The great advantages of Mössbauer spectroscopy, however, are that it: (i) is specific to Fe and thus is free from the interferences (e.g., sulfides, and organic matter) that chemical methods cannot avoid, (ii) can usually be performed nondestructively, (iii) gives a semiquantitative result rather easily, and (iv) reveals other information about the environment of the Fe in the sample. Excellent introductions to the technique are given by Hawthorne (1988) and, specifically for applications to Fe oxides, by Murad (1988). More detailed treatments are given in several texts (Bancroft, 1973; Gibb, 1977; Cranshaw et al., 1986).

Chemical Methods

As noted by Fudali et al. (1987), ". . . no other major element determinations are more demanding in terms of human skill and care, than Fe^{2+} and Fe^{3+} wet chemical determinations." Nevertheless, chemical methods are by far the most common methods for oxidation state determination. Because the specimen preparation requirements are similar for all the chemical methods, the methods can be distinguished generally on the basis of how the sample is decomposed, and how the oxidation state(s) are measured.

Specimen Preparation. Preparation of the specimens usually involves decreasing the particle size to silt-sized or smaller by grinding. In this way, the total surface area of the specimen is increased significantly, and both the total time required to decompose the specimen and the risk of altering the redox state of the specimen during decomposition are decreased. Nevertheless, the benefits gained from grinding must be weighed against the potential for altering the specimen's redox state by the grinding process itself; for this reason, grinding should be kept to a minimum. Whipple et al. (1984) suggest grinding to pass a 200-mesh sieve (75 μm) when poorly soluble minerals are present (e.g., ilmenite and magnetite), and note that grinding to only pass a 70-mesh sieve (212 μm) results in a specimen that is possibly too coarse. To preserve the relative contents of oxidized and reduced species in the specimen, grinding is best done when the specimen is wetted with acetone (Fitton & Gill, 1970; French & Adams, 1972). Acetone is preferred to alcohols because of its higher volatility and lower propensity to remain in the specimen and act as a reductant when the specimen is decomposed. Because grinding implements made from reductants like carbides, borides, nitrides, or metallic Fe may contaminate the specimen and react with oxidized species [e.g., Fe(III)] in the specimen during decomposition (Ritchie, 1968; Whipple et al., 1984), they should be avoided in favor of those made from alumina, mullite, porcelain, silica, or agate, which do not contain redox-sensitive elements.

Aside from the potential errors induced by grinding (or not grinding), the major difficulty associated with oxidation-state analysis by chemical methods is in preserving the ratio of oxidized to reduced species during decomposition. In specimens that already contain compounds that are not at redox equilibrium [e.g., sulfides, organic matter, Mn oxides, or Ti(III) when analyzing for

Fe(II) and Fe(III)], these effects are difficult to eliminate, if they can be eliminated at all. In some instances, the results of separate analyses can be used to correct for these errors. In others, the analyst can hope to minimize changes in redox state arising from sources, like atmospheric oxygen or light, that are external to the specimen and thus amenable to control.

Sample Decomposition. Samples are decomposed either by alkali-salt fusion or by acid digestion (usually HF-H_2SO_4). The former has the advantage of effecting complete dissolution on a wide range of poorly soluble minerals, but it requires special set-up and apparatus (such as Pt-Au crucibles). It also may pose a greater risk for altering the redox status of metal ions during digestion because of the relatively high temperatures, lengthy handling times, and the dependence of metal oxidation states on O_2 fugacity. Based on a study of these factors, a review of the high-temperature phase-equilibria work of Muan (1958, 1963) at low O_2 fugacities, and data demonstrating the tendency for Fe to alloy with the Pt in Pt crucibles at high temperatures, Donaldson (1969) concluded that Fe(II) determinations involving fusion decomposition were not reliable. Ayranci (1992), however, seems to have eliminated much of this unreliability by decomposition in a Pt-Au alloy crucible, using a specially designed Ar-purged induction oven that allows rapid heating and cooling of the specimen and minimizes the formation of an Fe-Pt alloy during the procedure.

For nonrefractory samples (i.e., samples that are easily brought into solution), acid digestion by dilute solutions of HF-H_2SO_4 is simpler than fusion, has a higher sample throughput, and requires only apparatus normally available in a wet-chemistry laboratory. In addition, many refractory (i.e., difficultly soluble) samples can be dissolved in concentrated H_3PO_4 solutions at temperatures of 300 to 325 °C or in anhydrous mixtures of HF and H_2SO_4 (Kiss, 1974, 1987). Acid-digestion techniques are susceptible to error from oxidation of the metals involved, but this error can be largely avoided by adding stabilizing agents [e.g., 1,10-phenanthroline or vanadium(V)], using short decomposition times, purging with fumes released by concentrated H_2SO_4 and HF mixtures or with N_2 or CO_2 to displace atmospheric oxygen above the sample, and maintaining a low (<2) pH. French and Adams (1972) presented data suggesting that purging with acid vapors arising from hot, premixed HF and H_2SO_4 was more effective than purging with bottled inert gas. Whipple (1974b) notes that bottled N_2 or CO_2 may contain trace amounts of reductants, which can be removed by slow bubbling through a solution of $K_2Cr_2O_7$ in concentrated H_2SO_4. Similarly, Kiss (1967) removes trace amounts of oxidants from these gases by passing the gases through an acrylic cylinder containing a 15% solution of pyrogallol in 50% KOH. Another source of interference commonly encountered in digestion methods that use HF is the strong complexation of Mg, Ca, Al, Fe(II), and Fe(III) by free F^-. The general effect of this interference is to lower the results obtained for Fe(II) and Fe(III) either by sequestration or precipitation of the Fe; the effect is minimized by the removal of free F^- from the digested solution after decomposition.

Oxidation-State Measurement. Most methods for differentiating metal oxidation states in mineral digestates fall within two general categories: (i) titrimetric, in which one species [e.g., Fe(II)] is determined by reaction with a known, titrated amount of oxidant or reductant; and (ii) colorimetric, which involves the chelation of the cationic species [usually Fe(II)] by a chromogenic compound (e.g., 1,10-phenanthroline) followed by spectrophotometric determination of the concentration. Selection of the appropriate method will depend on the types of interferences expected from a particular sample and on the type of information needed.

Titrimetric methods rely on an oxidant sufficiently strong to oxidize the reduced form of the analyte [e.g., Cr(VI) for Fe(II)] or on a reductant sufficiently strong to reduce the oxidized form [e.g., ferrocene for Fe(III)]. The strengths of the oxidants and reductants relative to the analyte couple can be assessed, to a first approximation, by their standard reduction potentials (Table 3-1). These standard potentials, however, are measured for a specific set of conditions (1 molal aqueous concentrations, 1 atmosphere gas pressures) and may not represent the actual potentials encountered

Table 3-1. Standard reduction potentials (E^o) and formal reduction potentials ($E^{o\prime}$) in various acid media for selected aqueous half-cell reactions.

Reaction	Medium[†]	E^o	$E^{o\prime}$	Source[‡]
		---- V ----		
$MnO_4^- + 8 H^+ + 5 e^- = Mn^{2+} + 4 H_2O$	----	1.51	----	Va84
$Mn^{3+} + e^- = Mn^{2+}$	----	1.54	----	Va84
	$(4\ F\ H_3PO_4)$	----	1.31	Se84
$I^+Cl^- + e^- = \tfrac{1}{2} I_2 + Cl^-$	----	1.51	----	Ba74
$H_2Ce(SO_4)_3 + e^- = Ce^{3+} + 2 H^+ + 3 SO_4^{2-}$	$(4\ F\ H_2SO_4)$	----	1.43	Sc69
$IO_3^- + 6 H^+ + 5 e^- = \tfrac{1}{2} I_2 + 3 H_2O$	----	1.20	----	Va84
$HCrO_4^- + 7 H^+ + 3 e^- = Cr^{3+} + 4 H_2O$	$(8\ F\ H_2SO_4)$	----	1.35	Sc69
	$(6\ F\ H_2SO_4)$	----	1.30	Sc69
	$(4\ F\ H_2SO_4)$	----	1.15	Sc69
	$(2\ F\ H_2SO_4)$	----	1.11	Sc69
	$(1\ F\ H_2SO_4)$	----	1.03	Di74
$VO_2^+ + 2 H^+ + e^- = VO^{2+} + H_2O$	$(8\ F\ H_2SO_4)$	----	1.30	Gr63
	$(4\ F\ H_2SO_4)$	----	1.14	Sc69
	$(2\ F\ H_2SO_4)$	----	1.07	Sc69
	$(1\ F\ H_2SO_4)$	----	1.02	Sc69
$[Fe(III)\text{-}phen]^{3+} + e^- = [Fe(II)\text{-}phen]^{2+}$	----	1.15	----	Va84
	$(4\ F\ H_2SO_4)$	----	1.03	Sc69
$[Fe(III)\text{-}bipy]^{3+} + e^- = [Fe(II)\text{-}bipy]^{2+}$	----	1.10	----	Sc69
	$(4\ F\ H_2SO_4)$	----	1.00	Sc69
Diphenylaminesulfonate(ox) + H^+ + e^- = Diphenylaminesulfonate(red)	----	0.84	----	Di74
$Ag^+ + e^- = Ag(m)$	----	0.80	----	Va84
$Fe^{3+} + e^- = Fe^{2+}$	----	0.77	----	Va84
	$(4\ F\ H_2SO_4)$	----	0.69	Di74
	$(4\ F\ HCl)$	----	0.66	Sc69
	$(4\ F\ H_3PO_4)$	----	0.44	Fu63
$[Fe(III)\text{-}ferricene]^+ + e^- = Fe(II)\text{-}ferrocene$	----	0.40	----	Va84
$Ti(OH)^{3+} + H^+ + e^- = Ti^{3+} + H_2O$	$(4\ F\ H_2SO_4)$	----	0.20	Se84

[†] Media are not specified for standard potentials; F = formal concentration (mol L^{-1}).

[‡] Compilations from which data were obtained: Va84 = Vanysek (1984); Se84 = Serjeant (1984, p. 543-548); Ba74 = Banerjee (1974); Sc69 = Schilt (1969, p. 107-109, 118); Di74 = Diehl (1974, p. 213, 241); Gr63 = Grady (1963, p. 191); Fu63 = Furman (1963, p. 2294).

under the conditions of analysis. Potentials measured under "nonstandard" conditions are termed "formal" reduction potentials. The formal reduction potential of the Fe(III)/Fe(II) couple in aqueous solution varies significantly from its standard potential (0.77 V) when complexation of Fe(III) by phosphate ions or of Fe(II) by 1,10-phenanthroline (phen) molecules occurs (Fig. 3-1, top). The competition between Fe(II) and protons for the phen ligand, and the sulfate and phen ligands for Fe(II) results in a decrease in the formal potential of the Fe-phen couple with increasing concentration of H_2SO_4. The opposite trend in formal potential with acid concentration can be seen for the oxidants vanadate (Fig. 3-1, bottom) and bichromate (Table 3-1). Much of the increase in the formal potential for these oxidants can be attributed to the higher concentrations of protons pushing the half-cell reactions to the right (i.e., le Chatelier's principle).

Titrimetric methods are subject to several possible sources of error. The primary source of error arises from the presence of ions in the specimen (other than the analyte) having formal reduction potentials below those of the oxidant or above those of the reductant. These ions may react just like the analyte and cause errors that are proportional to their concentration. Another problem arises from indistinct titration end-points. The reliability of the titrated volume depends

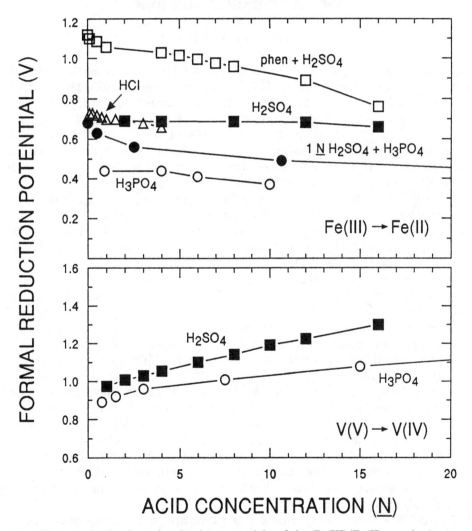

Fig. 3-1. Changes in the formal reduction potentials of the Fe(III)/Fe(II) couple (top) and the V(V)/V(IV) couple (bottom) with acid concentration and ligand type [data compiled from Furman (1963, p. 2294), Grady (1963, p. 191), Schilt (1969, p. 107, 109), Diehl (1974, p. 210, 241), Goldman (1975, p. 7197), and Serjeant (1984, p. 543-548)].

strongly on having a sharp end-point and a skilled operator. Much of the error (and the tedium) associated with titration end-points can be overcome by the use of an automated titration system and potentiometric determination of the end-point.

Colorimetric methods rely on use of a chelating agent that is sufficiently specific for the analyte to avoid interferences from other common ions present in the sample, and sufficiently chromophoric to yield a strong, easily measured absorbance peak. The strength of the analyte:chelate complex and the intensity of its absorption of light at a specific wavelength are given by the formation constant (K_f) and the molar absorptivity (ϵ) of the complex, respectively. Errors in colorimetric methods may arise from (i) other transition-metal ions that form colored complexes with absorbance peaks overlapping that of the chromophore, (ii) strongly acidic solutions in which protons compete with the metal cations for the chromophore ligand, (iii) complexing anions that compete with the chromophore ligand for the metal cation, thus preventing quantitative complexation of the metal species by the chromophore, and (iv) photochemical reduction of oxidized-metal/chromophore complexes. Elimination of photochemical reduction may require the use of a darkened room during portions of the analysis or the use of opaque sample containers.

SPECIFIC CHEMICAL METHODS

Iron(II) and Iron(III)

Numerous quantitative methods have been developed for determining Fe(II) and Fe(III) both in nonrefractory silicate minerals (Pratt, 1894; Sarver, 1927; Wilson, 1955, 1960; Clemency & Hagner, 1961; Reichen & Fahey, 1962; Meyrowitz, 1963; Ungethüm, 1965; Van Loon, 1965; Schafer, 1966b; Peters, 1968; French & Adams, 1972; Murphy et al., 1974; Banerjee, 1974; Whipple, 1974b; Kiss, 1977; Begheijn, 1979; Stucki, 1981; Loveland, 1988; Komadel & Stucki, 1988; Johnson & Maxwell, 1989; Amonette & Scott, 1991) and in such refractory minerals as magnetite, chromite, garnet, and tourmaline (Rowledge, 1934; Hey, 1941; Seil, 1943; Groves, 1951; Ito, 1962; Kleinert & Funke, 1962; Cheng, 1964; Novikova, 1966; Donaldson, 1969; Meyrowitz, 1970; Kiss, 1987; Ayranci, 1992). Schafer (1966a) provides an excellent review of the historical development of techniques for analysis of Fe(II) in minerals.

The successful adaptation of any of these methods to analysis of Fe oxidation state in a soil obviously requires consideration of the effects of soil-specific properties on the accuracy and precision of the method. For example, soils generally differ from minerals in that they contain organic matter, which is a moderately reducing substance in its own right. Methods that rely on the oxidimetric titration of Fe(II) to Fe(III) will probably also oxidize organic matter that is present and, consequently, will obtain high results. Methods that rely on complexation of the Fe(II) by a chromogen may suffer spectral interferences from complexation of other metal species by functional groups in the organic matter, competition from these functional groups for complexation of the Fe, and reduction of structural Fe(III) by the organic matter on decomposition of the specimen (Pruden & Bloomfield, 1969; Begheijn, 1979). Soils may contain a mixture of nonrefractory and refractory minerals and, thus, the harsher decomposition techniques used for refractory minerals may be needed if a complete analysis is desired. Furthermore, some of the Fe in soil may be present in mixed-valence Fe oxides (e.g., green rust), as structural Fe in clay-sized layer silicates, or bound to exchange sites and, in any of these instances, be very sensitive to oxidation by air once the soil is sampled. Because this fraction of the soil-Fe reservoir is the most reactive or "labile" fraction (and possibly the most important), samples should be either stored immediately under an inert atmosphere or frozen to minimize changes in the relative amounts of Fe(II) and Fe(III) before analysis.

Pruden and Bloomfield (1969) presented a particularly discouraging assessment of the possibility of accurately determining Fe oxidation states in soils. Their assessment was based on results showing substantially higher Fe(II) and lower total-Fe values than expected in a "humose meadow soil" that had been amended with known amounts of freshly precipitated Fe(OH)$_3$ before HF digestion and colorimetric analysis. Using the same Fe method with lateritic soils from

Hawaii, however, Walker and Sherman (1961) showed relatively little effect of organic matter on Fe(II) values. Presumably, the different results of Walker and Sherman (1961) and Pruden and Bloomfield (1969) stemmed from differences in the form and availability of Fe(III) and the type and amount of organic matter present in the soils analyzed. Mitsuchi and Oyama (1963) reported eliminating the reducing effect of organic matter by digesting the specimens at temperatures below 80 °C. Brinkman (1977) and Begheign (1979) minimized the effects of organic matter by drastically decreasing the time required for sample decomposition and limiting the decomposition temperature to about 65 °C. And Ungethüm (1965) described a titrimetric method using Ag(I) as the oxidant that showed no interference from additions of a standard humic acid, sucrose, tartaric acid, or hydrazine sulfate. Thus it seems that there is a diversity of opinion regarding the extent to which the difficulties involved in analysis of soils can be overcome.

Titrimetric Methods

The titrimetric methods have relied on standardized solutions of oxidants such as permanganate, dichromate, and vanadate to oxidize the Fe(II) present. Early methods used permanganate or dichromate to titrate the HF-H_2SO_4 digests after decomposition was complete. As it became clear that much of the error in these determinations arose from oxidation of the Fe(II) by atmospheric O_2 that occurred after decomposition and before titration, investigators added oxidants to the specimen before digestion and allowed them to react immediately as the Fe(II) was released by the HF. With this approach, however, a new source of error arose in that permanganate and dichromate solutions are not particularly stable in hot HF-H_2SO_4 acid solutions. Furthermore, permanganate is subject to interferences from Mn(II) and, possibly, Cr(III). Of these three oxidants, then, the V(V)/V(IV) couple has formal redox potentials closest to those of the Fe couple (Table 3-1), the vanadate titer is particularly stable in strong acid solutions at moderate temperatures (Wilson, 1955; Peters, 1968; Whipple, 1974b; Amonette & Scott, 1991), and its reduced form, vanadyl, is more resistant to oxidation by oxygen than Fe(II). These properties, along with a lack of interference from Mn(II) and Cr(III) and easy adaptation to batch analyses with large numbers of samples (Amonette & Scott, 1991) have combined to make vanadate the oxidant of choice.

Three other oxidants, iodine monochloride, iodate, and silver ion, have been proposed for Fe(II) analyses (Van Loon, 1965; Banerjee, 1974; Ungethüm, 1965), based on their stability in strong acids. Very precise results were reported for the iodate method of Van Loon (1965) and the iodine-monochloride method of Banerjee (1974). However, these two methods do not seem well suited for rapid batch analyses of samples. Because the oxidant in the silver ion method (Ungethüm, 1965) has the lowest formal redox potential (+0.78 V in fluoride solution), it seems least subject to interferences from other ions and organic matter and the method may prove to be a valuable technique for analysis of soils.

The titrimetric methods for Fe(II) are not specific to Fe(II), but rather include in their measurement all substances that can be oxidized by the particular reagent used. For titration with V(V) or Ag(I), these substances include such ions as Ti(III) and V(III), as well as many sulfides and organic matter. With a stronger oxidant, such as permanganate, Mn(II) and Cr(III) must be added to this list. Oxidizing species, such as Mn(III) and Mn(IV), may also be present in the specimens and these will tend to lower the results for Fe(II) obtained with titrimetric measurements by oxidizing some of the Fe(II) as it is released during specimen decomposition.

Although the high fluoride content of the digests helps to sharpen the end-point when diphenylamine or its sulfonate are used as indicators, it tends to favor the oxidation of Fe because the Fe(III)-fluoride complexes are more stable than those with Fe(II). This effect is countered by inclusion of a stable oxidant [e.g., V(V) in 4.5 M H_2SO_4] in the digest to immediately react with the Fe(II) on release from the solid and preserve the titer of Fe(II) until the final titration.

Other consequences of high fluoride levels in these digests can affect the titrimetric results. Amonette (1988) noted a very strong effect of gibbsite [Al(OH)$_3$] in depressing Fe(II) levels measured by V(V) titration and, after testing several hypotheses, attributed the effect to the formation of Al-Fe(II)-Fe(III)-Mg-F-OH complexes that prevented oxidation by V(V). Addition of equivalent amounts of Al as $AlCl_3$ or as the aluminosilicate mineral muscovite, however, had no effect on the Fe(II) results (Amonette & Scott, 1991). Murphy et al. (1974) also noted a decrease in Fe(II) and Fe(III) values in analyses of Mg-rich rocks, if the Mg-Fe(II)-Fe(III)-F precipitates that formed were not dissolved completely (i.e., the digest remained cloudy). Langmyhr and Kringstad (1966) and Croudace (1980) observed several complex aluminofluorate solids in HF digests of aluminosilicate minerals, some of which corresponded to the mineral ralstonite (Pabst, 1939; Pauly, 1965; Croudace, 1980). Most of these effects of fluoride can be explained by incomplete dissolution of the complex fluoride precipitates that form during specimen decomposition and can be eliminated by addition of sufficient H_3BO_3 to react with the fluoride and dissolve the precipitates.

The most precise titrimetric methods rely on sharp end-points. Factors that yield a sharp end-point include (i) a stable, reversible, soluble indicator with a formal reduction potential midway between that of the oxidizing couple [e.g., V(V)/V(IV)] and the Fe(II)/Fe(III) couple; (ii) a high contrast between the colors of the oxidized and reduced forms of the indicator; and (iii) moderately low activation energies for the reduction or oxidation of the indicator and the Fe(II)/Fe(III) couple. Vogel et al. (1989, p. 360-368) discuss many of the properties of oxidation-reduction indicators as well as the importance of formal reduction potentials relative to standard reduction potentials in redox titrimetry. Diphenylaminesulfonic acid (Sarver & Kolthoff, 1931a) meets these requirements and is particularly well suited for oxidimetric titrations of the Fe(II)/Fe(III) couple. However, the end-point with this indicator may shift slightly, depending on the acid matrix and the amount of Fe(II) present (Sarver & Kolthoff, 1931b; Toni, 1962; Schafer, 1966b; Amonette & Scott, 1991). For best results, therefore, the effects of the matrix on the indicator end-point must be assessed and all standardizations performed using the same matrix as in the actual analysis. The end-points obtained with diphenylaminesulfonic acid are improved somewhat by the addition of reagents to lower the formal reduction potential of the Fe(II)/Fe(III) couple. Phosphoric acid forms a moderately strong complex with Fe(III) and yields a formal reduction potential near 0.6 V. Fluoride complexes with Fe(III) also have similar effects (Schafer, 1966b), but too much fluoride can yield an indistinct end-point (Amonette & Scott, 1991).

Several titrimetric methods in which the end-points are determined by potentiometry have been described (Schafer, 1966b; Kiss, 1967, 1977; 1987; Beyer et al., 1975). The advantages of this kind of approach are that (i) the reaction equivalence-point and the titration end-point generally coincide, (ii) the end-point is sharp, and (iii) the use of an automatic titrator is facilitated, thus eliminating some of the tedium and operator bias associated with indicator solutions and manual titrimetry. In addition, with the technique of Beyer et al. (1975), simultaneous determination of Fe(II) and Fe(III) by direct-current polarography followed by a determination of total Fe using alternating-current polarography is possible on a single specimen.

All the titrimetric methods discussed so far have involved the oxidation of Fe(II) to Fe(III) by a standardized oxidant. The reduction of Fe(III) by a standardized reductant has also been used to determine Fe oxidation states in minerals (Clemency & Hagner, 1961; Murphy et al., 1974). The method of Clemency and Hagner (1961) involves the coulometric generation of Ti(III) to reduce Fe(III). The titration is automated, with the end-point being determined by loss of the blue color associated with the methylene blue indicator. Ferrocene [dicyclopentadienyliron(II)] was used by Murphy et al. (1974). The method relies on the loss of the red color associated with the Fe(III)-SCN complex to determine the end-point and the addition of a nonpolar surfactant to disperse the ferrocene in a reactive form during the titration.

Colorimetric Methods

Although numerous chelating agents for Fe(II) have been suggested, the active portion of the structure of the most effective ligands consists of an -N-C-C-N- group and forms a five-membered ring with Fe atoms. Chromophoric properties are derived from the delocalized electrons present when aromatic rings (e.g., phenyl groups, pyridines, diazines, and triazines) form part of the ligand structures. Thus, although ethylenediaminetetraacetic acid (EDTA) is a very effective chelate for Fe(II) and Fe(III), it has very weak chromophoric properties because of the lack of aromatic rings in its structure. A comparison of the structures and molar absorptivities (ϵ) for several commonly used Fe(II) chromogens (Fig. 3-2, Table 3-2) shows the rough correlation between aromaticity and chromogenicity. The increase in chromogenic character, however, usually adds more steps to the synthesis of these ligands and increases their costs accordingly.

The formation constants for these ligands with Fe(II) also depend on their structures (Fig. 3-2, Table 3-2). The ligands based on the bipyridine structure (bipy and ferrozine) do not bind Fe(II) nearly as strongly as those based on the phenanthroline structure (phen and bathophen), presumably because the phenanthroline structure rigidly fixes the five-membered ring in a single plane and at the optimum size for Fe(II), whereas the bypyridine structure is more flexible. Comparing pK_a values (Table 3-2), the bipyridine-based ligands are more acidic than the phenanthrolines and thus will form complexes at a lower pH. On the other hand, the forward rate constant (k_f) values suggest that phenanthrolines form their complexes much faster (assuming that bathophen behaves as phen). The reverse rate constant for these ligands is essentially the same (log k_r = 4 as calculated from β_3 and k_f values).

In selecting a chromogen for analysis of Fe(II) in samples decomposed by an HF-H$_2$SO$_4$ mixture, key factors to consider include the formation constant, molar absorptivity, pK_a, and cost. The most important of these properties is the formation constant. The chromogen ligand must be able to compete effectively with fluoride ions for complexation of the Fe(II) ion. On this basis alone, the phenanthroline-based reagents are clearly better (log β_3 in Table 3-2) The molar

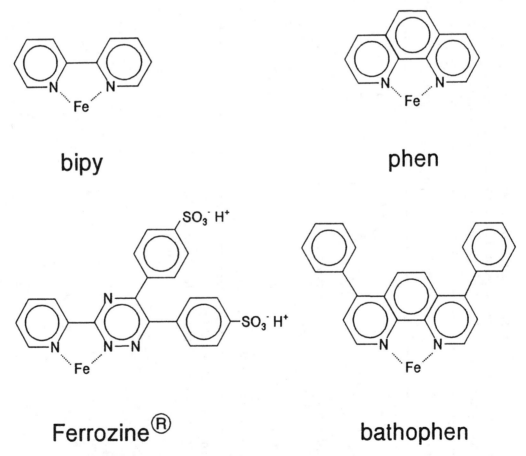

Fig. 3-2. Chemical structures of several chromophoric ligands for Fe(II) (see Table 3-2 for IUPAC names).

Table 3-2. Logarithms of the Fe(II)-trischelate formation (β_3) and forward rate (k_f) constants at 25 °C and 0.1 M ionic strength, the negative logarithm of the acid dissociation constant (K_a), and the molar absorptivity (ϵ_{max}), wavelength of maximum absorbance (λ_{max}), and cost of several Fe(II) chromogens [chemical property data as compiled by Schilt (1969, p. 28-36, 43, 55) and Thompsen and Mottola (1984)].

IUPAC Name	Abbreviation	log β_3	log k_f	pK_a	ϵ_{max}	λ_{max}	Cost†
			L mol⁻¹ s⁻¹		L mol⁻¹ cm⁻¹	nm	$ mmol⁻¹
2,2'-bipyridine	bipy	17.2	13.1	4.33	8 650	522	0.12
3-(2-pyridyl)-5,6-bis(4-phenylsulfonic acid)-1,2,4-triazine	Ferrozine‡	15.7	11.5	3.2	27 900	562	4.46
1,10-phenanthroline	phen	21.4	17.4	4.96	11 100	510	0.50
4,7-diphenyl1-1,10-phenanthroline	bathophen	21.8	--	4.84	22 400	533	14.53

† 1993 Catalog, Aldrich Chemical Co.
‡ Ferrozine is a registered trademark of Hach Chemical Co.

absorptivity of the chromogen largely determines the precision and detection limits that are possible. In this regard, the more highly substituted ligands (ferrozine and bathophen) seem to have an advantage. Only bathophen, however, combines a high molar absorptivity with a high formation constant and this advantage is offset by its considerably greater cost relative to phen. Thus, bathophen is best reserved for measurements that require high precision or very low detection limits.

The pK_a of the chromogens is roughly equivalent to the pH above which the ligand is available to complex Fe(II). The bipyridine-based reagents have lower pK_a s than the phenanthrolines and thus can form complexes in solutions that are one to two pH units more acidic. In Fe(II) analysis of natural waters, this property may have some advantage, but in acid digests where the pH is between 0 and -1, substantial neutralization is necessary before any of the chromogens will complex Fe(II) stoichiometrically. Based on this discussion, the authors recommend phen and bathophen as the best chromogens for use in Fe(II) analyses of HF-H_2SO_4 digests of soils or minerals.

In contrast to the titrimetric methods, whose specificity is based on reduction potential, the specificity of phen for Fe(II) is based on ion size, charge, and electronic structure. It comes as little surprise then that phen forms strong complexes with other metal ions having properties similar to those of Fe(II). Only the complexes of Ru(II), Rh(I), Cu(I), Cr(III), Pd(II), and Fe(III), however, have sufficient chromophoric character to interfere directly in the Fe(II)-phen spectrum (see Schilt, 1969, p. 36 and p. 58-59) and of these, only Fe(III) is of any concern in routine soil analysis (Table 3-3).

Two separate complexes of Fe(III) with phen are theoretically possible. The blue [Fe(III)-$phen_3$]$^{3+}$ complex, which is analogous in structure to the strongly absorbing [Fe(II)-$phen_3$]$^{2+}$ complex, is formed only by oxidizing its reduced analog using a strong oxidant (Baxendale & Bridge, 1955). With time, however, the blue complex converts to a yellow, dinuclear, oxo-bridged complex, [$phen_2$ LFe(III)-O-Fe(III)L$phen_2$]$^{n+}$, where L is a monodentate ligand and $n = 4$ for

Table 3-3. Wavelength of maximum absorbance (λ_{max}) and the molar absorptivity at this wavelength (ϵ_{max}) for visible-light absorption bands of selected metal-phen complexes (Schilt, 1969, p. 36; David et al., 1972).

Complex	λ_{max}	ϵ_{max}
	nm	L mol^{-1} cm^{-1}
Fe(II)-$phen_3$	510	11 100
Fe(III)-$phen_3$	590	600
Fe(III)$_2$ -$phen_4$ OL$_2$ †	360	9 200
Ru(II)-$phen_3$	448	18 500
Co(II)-$phen_3$	425	100
Ni(II)-$phen_3$	519	12
Cu(II)-$phen_2$	700	60
Cu(I)-$phen_2$	435	7 000

† $phen_2$ LFe(III)-O-Fe(III)L$phen_2$.

uncharged L (Harvey and Manning, 1952; Harvey et al., 1955; David et al., 1972). The yellow dinuclear complex also forms directly when Fe(III) and phen are in solution, and has a molar absorptivity at 510 nm that is about 2% of that of the $[Fe(II)\text{-phen}_3]^{2+}$ complex (Harvey et al., 1955). Thus, the yellow complex is the one most likely to interfere in the Fe(II)-phen absorbance reading.

The extent of this interference for analysis of HF-H_2SO_4 digests mixed with H_3BO_3 and citrate is unclear. Harvey et al. (1955) reported a tenfold decrease in absorbance by the yellow complex when 500 μg mL^{-1} of F$^-$ was present in solution, and Stucki and Anderson (1981) saw no evidence for the complex in Fe(III) standard solutions prepared in an HF-H_2SO_4-H_3BO_3-citrate matrix. Novak and Arend (1964), on the other hand, were trying to measure trace levels of Fe(II) in hematites and noted that "slight colour from the iron(III) increases the blank value." These authors recommended chloroform extraction of the Fe(II)-phen complex before determination of the absorbance, but it is not clear from their description whether interference by the yellow Fe(III)-phen complex was avoided by this procedure or, for that matter, what acid was used to dissolve the hematite. It seems prudent, therefore, to prepare and analyze a separate set of Fe(II) standards that contain appropriate levels of Fe(III) to ensure accurate analysis of specimens that contain high amounts of Fe(III) relative to Fe(II).

In addition to the potential for direct spectral interference by Fe(III)-phen complexes, the presence of other ions that prevent stoichiometric complexation of Fe(II) by phen [i.e., the formation of the trischelate, $Fe(phen)_3^{2+}$] may result in low values for Fe(II). As noted earlier, high concentrations of fluoride ions can depress the Fe values by forming strong, relatively inert complexes with Fe(II) and Fe(III) ions. The presence of large concentrations of such cations as Co(II), Ni(II), Zn(II), Hg(II), or Tl(III) together with limited quantities of phen may also result in lower Fe(II) values by limiting the amount of phen available for complexation of the Fe(II).

Because of the large difference (approximately 10^7) in their stabilities, the Fe(III)-phen complexes are metastable with respect to Fe(II)-phen. Fortunately, the rate at which reduction occurs is insignificant as long as the samples are not treated with a reductant or irradiated with ultraviolet (UV) light. Even the low intensities of UV light emitted by indoor light fixtures have been shown to produce a substantial photoreductive effect (Stucki & Anderson, 1981). Stucki (1981) avoided this problem by performing the analyses under red light in a dark room with the help of an automatic dilutor. In a further adaptation of the technique, Komadel and Stucki (1988) took advantage of the phenomenon, using intense UV light from a mercury vapor lamp to rapidly reduce all Fe(III)-phen to Fe(II)-phen for determinations of total Fe in a separate aliquot of the sample digest.

As with the titrimetric methods, most of the colorimetric techniques involve direct determination of Fe(II). If determination of Fe(III) is also desired, it is usually taken as the difference between the Fe(II) value and a value for total Fe obtained either by another technique (always for titrimetric techniques) or by reducing the Fe(III) in a separate aliquot of the sample digest and repeating the Fe(II) colorimetric analysis. A method that involves direct colorimetric determination of Fe(III) in aqueous solution, however, has been described by Dinsel and Sweet (1963). In this technique, anthranilic acid is used to precipitate the Fe(III) salt, which is then extracted into 1-pentanol and the absorbance measured at 475 nm. The Fe(II) remaining in the aqueous sample is oxidized to Fe(III) with H_2O_2 and the precipitation-extraction-measurement procedure repeated. With this technique, then, Fe(III) and Fe(II) are separated and each is determined directly. The molar absorptivity of the Fe(III)-anthranilate complex in 1-pentanol, however, is 1300 L mol^{-1} cm^{-1} [about an order of magnitude lower than that for Fe(II)-phen], and the competition of fluoride for Fe(III) is expected to pose a significant interference based on the results of Harvey et al. (1955) with Fe(III)-phen complexes.

Manganese(II), Manganese(III), and Manganese(IV)

After Fe, Mn is the next most abundant element in soil minerals to exhibit multiple oxidation states. Even though Mn exists both in oxide phases and as an octahedrally coordinated cation in silicate minerals, methods for the determination of Mn oxidation states have focused on the Mn in oxides. The reasons for this stem from (i) the relatively easy and selective dissolution of Mn oxides by reducing agents in mild acid solutions, and (ii) the reduction of Mn(III) and Mn(IV) by large amounts of Fe(II) released when silicates are dissolved by strong $HF\text{-}H_2SO_4$ acid solutions. Implicit in the use of reductive dissolution treatments, therefore, is the assumption that most, if not all, the Fe released by the treatment was originally present as Fe(III) coprecipitated with Mn oxides.

Recent theoretical calculations (Sherman, 1990) suggest that the octahedrally coordinated Mn atoms in silicates cannot be trivalent if located adjacent to an octahedral site containing Fe(II). Trivalent Mn is possible when other cations or trivalent Fe atoms are in the adjacent sites but, for the most part, Mn(II) can be considered the dominant oxidation state of Mn in silicates. Complete verification of this assumption, however, must come from direct spectroscopic measurements by XPS or XANES. The XANES work of Mande and Deshpande (1990) and Manceau et al. (1992) and the x-ray emission work of Urch and Wood (1978) with Mn oxides provide examples of this approach.

Numerous spectrophotometric methods for the determination of total Mn in aqueous solutions have been developed (Chiswell & Rauchle, 1990; Kessick et al., 1972), yet the adaptation of these for the analysis of geochemical specimens (typically, Mn nodules and sediments from ocean floors) has not proven particularly successful (Murray et al., 1984). The most robust methods, therefore, have relied on titrimetric techniques, in which the oxides are dissolved in the presence of a known excess of a weak reducing agent (iodide, oxalate, or ferrous iron), the amounts of reductant consumed and of total Mn and Fe released are determined, and from these data an estimate of the average oxidation state of the Mn oxides is calculated. Thus, it is not possible by titrimetric techniques to distinguish between a sample containing 2 mmol of Mn(III) and another sample containing 1 mmol each of Mn(II) and Mn(IV). In these titrimetric methods (Freeman & Chapman, 1971; Grill, 1978; Murray et al., 1984), all the Fe released by dissolution is assumed to be Fe(III)--the presence of significant amounts of Fe(II) relative to Mn(III) or Mn(IV) in the fraction dissolved can lead to low or negative results for the oxidized fraction of Mn. Murray et al. (1984) compared the three titrimetric methods using standard Mn oxide samples and achieved identical results with the oxalate and iodimetric methods. However, results obtained with the method of Grill (1978) involving Fe(II) as the reductant were unreasonably low, possibly as a result of using colorimetric analysis to measure total Fe and the Fe(II) remaining after the reduction of Mn(III) and Mn(IV). In short, the routine wet-chemical analysis of Mn oxidation states in soils faces many of the problems inherent to Fe oxidation state analysis, is potentially subject to large interferences from Fe(II), and, consequently, is even more difficult to perform accurately than Fe oxidation-state analysis.

Chromium(III) and Chromium(VI)

Chromium in soil minerals is usually Cr(III) and behaves in an analogous fashion to Fe(III) because of the similar ionic radii of the two cations (Rai et al., 1986, 1988; Amonette & Rai, 1990). Hexavalent chromium may also occur in the solid phase under certain circumstances (Hauff et al., 1983; Duesler & Foord, 1986; Amonette & Rai, 1987), but typically Cr(VI) occurs as an anion that is weakly adsorbed to oxide surfaces and the edges of phyllosilicate minerals (Zachara et al., 1988). Determination of the oxidation state of Cr in soils has therefore generally involved an extraction procedure in which Cr(VI) is complexed by s-diphenyl carbazide and the absorbance measured at 540 nm. Bartlett and Kimble (1976a,b) and Bartlett and James (1979) used this technique to study the behavior of Cr(III) and Cr(VI) in soils containing organic matter [which reduces Cr(VI)] and Mn oxides [which oxidize Cr(III)]. The s-diphenyl carbazide reagent has also been used for the determination of Cr oxidation states in the solid phase (Pilkington & Smith, 1967). More recently, x-ray absorption spectroscopy has been used to study the molecular mechanisms of Cr(III) oxidation to Cr(VI) by manganese oxides (Manceau & Charlet, 1992).

Nonspecific Measurements

In some instances, it may be desirable to measure the reductive or oxidative *capacity* of a soil or soil fraction in addition to, or instead of, the oxidation state of a particular element. With titrimetric methods, this capacity is indeed what is generally measured, albeit within certain prescribed limits. Measurements of overall reductive capacities before and after certain extractions or treatments to a soil can help identify the availability of reductants to participate in soil reactions, and, in combination with element-specific measurements, to isolate reaction mechanisms.

For example, the coarse-grained sediments at the Hanford Site in Washington State contain relatively large quantities of Fe(II), much of which is present in hornblende-type minerals and not readily accessible to the environment surrounding the particles. At the same time, the clay fraction also contains large amounts of Fe, which may be in either oxidation state and which is relatively sensitive to environmental conditions. As part of a screening process to identify zones in these sediments where facultatively anaerobic microorganisms might be found, a test for the "labile" reductive capacity of a sediment was devised using the method of Amonette and Scott (1991). In addition to a set of samples analyzed by the usual method, a second set of samples was treated once with H_2O_2 at 65°C in a pH 5 acetate buffer to oxidize any "H_2O_2-labile" reductants present in the clay fraction before proceeding with the analysis. Because even trace amounts of unreacted H_2O_2 would interfere in the analysis by forming a blood-red, relatively inert complex with V(V), the samples were incubated for a week until all the H_2O_2 had decomposed. Subtraction of the results obtained with the second set of samples from those obtained with the untreated samples yielded values for the "H_2O_2-labile V(V)-reducing" capacity of the sediments. The zones where this capacity was greatest corresponded to other measurements that indicated reducing conditions and aided in the selection of samples for microbial analysis.

EXPERIMENTAL COMPARISON

To put the relative strengths and weaknesses of titrimetric, colorimetric, and spectroscopic techniques for Fe oxidation-state analysis in perspective, we used three representative methods to analyze samples of the A, B, and C horizons of four Iowa soils. We selected the vanadate method of Amonette and Scott (1991), the phen method of Komadel and Stucki (1988), and Mössbauer spectroscopy for this comparison, which complemented a similar study comparing the three methods when applied to nonrefractory mineral samples (Amonette et al., manuscript in preparation). The results from the mineral analyses showed general agreement among the three methods, although the vanadate method was considerably more precise than the others.

Materials and Methods

Samples were collected from three horizons of four Iowa soils (Table 3-4) as part of a study evaluating the moisture regimes of soils having the same parent material, but different landscape positions (Khan, 1991). The Clarion and Canisteo soils shared a common parent material (late Wisconsinan glacial till), as did the Sharpsburg and Nira soils (loess). After collection, the samples were air dried, dry ground to pass a 100-mesh sieve, and then separate splits were taken for analysis by the various techniques. Five replicate analyses for Fe(II) content were performed using the titrimetric method of Amonette and Scott (1991), and five replicates for Fe(II) and total Fe were performed using the colorimetric method of Komadel and Stucki (1988). Separate HF-H_2SO_4 digests of the soils were also prepared in triplicate by the method of Bernas (1968) and the total Fe present measured by atomic absorption spectrometry. The organic C contents of the samples were determined by the method of Mebius (1960) as modified by Nelson and Sommers (1982). Inorganic C contents were determined by the method of Boellstorff (1978). All results were reported on an air-dry soil weight basis.

Table 3-4. Description and classification of soils used in this study.

Soil	Depth in profile	Horizon	Classification	Textural class
	-- cm --			
Clarion	0-30	Ap	Typic Hapludoll	Loam
	54-83	Bw		Clay loam
	135-150	C		Loam
Canisteo	0-22	Ap	Typic Endoaquoll	Silty clay loam
	83-103	Bg		Loam
	135-155	Cg		Loam
Sharpsburg	0-18	Ap	Typic Argiudoll	Silty clay loam
	60-86	Bt		Silty clay loam
	172-193	C		Silty clay loam
Nira	0-15	Ap	Typic Hapludoll	Silty clay loam
	43-58	Bw		Silty clay loam
	102-137	C		Silty clay loam

A Mössbauer spectrum was obtained for each soil sample at 80 K in a bath-type cryostat (sample cooled by He exchange gas surrounded by a liquid-N_2 reservoir) using a Ranger Scientific MS-900 spectrometer in triangular wave-form mode (which gives mirror-image spectra) over 1024 channels. The γ-ray source was ^{57}Co in a 10% Rh matrix (DuPont-New England Nuclear). Each sample was packed in powder form to a uniform absorber thickness (5 mg Fe/cm^2 in the plane normal to the beam) between two Mylar windows. Spectra were folded into 512 channels to combine both halves of the mirror-image pattern generated by the triangular wave form, which resulted in a flat background. The curve-fitting algorithm assumed Lorentzian line shapes. One Fe(II) and two Fe(III) doublets were required to achieve a satisfactory fit. The areas and widths of both peaks in each doublet were constrained to be equal, then the curve deconvolution was optimized through numerical iteration, until the statistical χ^2 was at a minimum. Velocities were measured relative to Fe foil observed under similar conditions. Isomer shift (IS) and quadrupole splitting (QS) parameters were calculated from the following relations (Lear et al., 1988):

$$IS = 0.5(V_1 + V_2 - 2V_0) \tag{1}$$

and

$$QS = |V_2 - V_1| \tag{2}$$

where V_i is the velocity position of the respective peaks, in order of ascending velocity; and V_0 is the velocity of the center position of the Fe foil spectrum.

As a first approximation, the Fe(II):Fe(III) ratio in a Mössbauer absorber is proportional to the relative areas of deconvoluted peaks assigned to either Fe(II) or Fe(III), viz.,

$$F2/F3 \approx k\,(A2/A3) \tag{3}$$

where $F2/F3$ is the Fe(II):Fe(III) ratio; $A2/A3$ is the ratio of the areas of all deconvoluted peaks assigned to either Fe(II) or Fe(III), respectively; and k, is the proportionality constant. The value of k is affected by a number of factors including: (i) effective thickness of the sample; (ii) texture (orientation of the Fe electric field gradient relative to the γ-ray); (iii) temperature; (iv) choice of theoretical line-shape for curve fitting; and (v) possible variations in the recoil-free fractions for Fe(II) and Fe(III), which depend on bond strengths, and the effect of the matrix on these fractions (Rancourt, 1989, 1991a,b; de Grave & van Alboom, 1991). A minimum level of Fe in each oxidation state (perhaps 5% of total Fe) must also exist in the sample. Lear (1987, p. 83) found for ferruginous smectite that k was constant below 120 K. If all soils are similar in their mineralogy and Fe site occupancy, and if sample packing is reasonably comparable among all samples, the most significant factor influencing k will be the recoil-free fractions, f_2 and f_3, of Fe(II) and Fe(III), respectively. Under this condition, Eq. [3] can be rewritten as,

$$F2/F3 \approx (A2/A3)(f_3/f_2) \qquad [4]$$

RESULTS AND DISCUSSION

Values of Fe(II) and total Fe determined by the various wet-chemical methods, and the organic and inorganic C contents of the soil samples are reported in Table 3-5. In general, the soils contained low levels of Fe(II) relative to total Fe, at least in part because they were air dried prior to analysis rather than being preserved in a field-moist condition. The total Fe contents of the Clarion and Canisteo soils were comparable (170-280 mg kg^{-1}) and generally less than those found in the Sharpsburg and Nira soils (280-370 mg kg^{-1}).

The Fe(II) values obtained by the colorimetric method were precise and were consistent across horizons for a given soil (Table 3-5). The titrimetric values for Fe(II), however, did not agree well with the colorimetric values and were less precise, particularly in samples from the A horizon where organic C content was highest. As the organic C content decreased, the precision of the titrimetric results improved and the discrepancy with the colorimetric results decreased. The relationship between organic C content and the apparent Fe(II) content as measured by the titrimetric and colorimetric methods is shown in Fig. 3-3. Nearly all the variability in the titrimetric data ($r^2 = 0.97$) could be explained by a linear dependence on the organic C content, whereas the colorimetric data were essentially independent ($r^2 = 0.15$) of organic C.

It seems clear, then, that the organic matter in these soils was able to reduce V(V) [thus causing falsely high values for Fe(II)] but had essentially no effect on the Fe(II) values obtained when phen was used as a chromogen. Calculations made under the assumption that the average oxidation state of the C was initially 0 showed that only about 3% of the total amount of organic C present in each soil was oxidized by the V(V) treatment. Thus, the conditions encountered in the titrimetric method with V(V) are not nearly as harsh as those used in the Mebius (1960) and related techniques (Nelson & Sommers, 1982), where Cr(VI) in fuming H_2SO_4 solutions at elevated temperatures completely oxidizes any C that is present. Nevertheless, the interference from organic C is quite pronounced in the titrimetric method [10 g organic C kg^{-1} $\approx$ 5.5 g Fe(II) kg^{-1}] and its use for soils is not recommended when Fe-specific data are required. If a measure of the "labile" reductive capacity of the soil is required, however, the titrimetric method is more appropriate than an Fe-specific method.

A comparison of the total-Fe data obtained by the colorimetric and atomic absorption methods (Table 3-5, Fig. 3-4) showed excellent agreement for the Sharpsburg and Nira soils, but a consistent difference for the Clarion catena soils (i.e., Clarion and Canisteo). These results could be due to differences in moist weight of the soils when analyzed, although this seems unlikely. Alternative explanations involve either the presence of an interfering element in the Clarion-catena parent material (Cr, Rh, or Pd to give high phen absorbances; Ni to give low Fe atomic absorption results) or incomplete dissolution of these soils (blockage of light by colloidal material to give higher phen absorbances; incomplete release of Fe to give lower atomic absorption values). Because of the more predictable relationship between the Mössbauer Fe(II) values obtained using atomic absorption for total Fe (see below) and the colorimetric Fe(II) values, we speculate that the colorimetric results for total Fe are too high.

Table 3-5. Means and standard deviations for the Fe(II) and total Fe contents of twelve soil samples as measured by titrimetric, colorimetric, and atomic absorption methods, and the organic and inorganic C contents.

Soil	Fe(II)		Total Fe		C	
	Titrimetric	Colorimetric	Colorimetric	Atomic absorption	Organic (as C)	Inorganic (as CO_3)
			---------- g kg^{-1} ----------			
Clarion Ap	10.7 ± 0.9	1.5 ± 0.1	24.1 ± 0.3	21.3 ± 0.7	20	--†
Clarion Bw	5.6 ± 0.7	1.6 ± 0.1	28.1 ± 0.4	25.1 ± 0.6	7	--
Clarion C	3.2 ± 0.1	1.8 ± 0.1	23.1 ± 0.2	20.7 ± 0.3	2	89
Canisteo Ap	29.9 ± 1.6	2.9 ± 0.2	26.1 ± 0.5	21.3 ± 0.9	50	18
Canisteo Bg	4.1 ± 0.6	2.7 ± 0.1	19.9 ± 0.2	17.3 ± 0.9	3	46
Canisteo Cg	3.6 ± 0.1	2.1 ± 0.1	20.3 ± 0.5	16.9 ± 0.4	2	130
Sharpsburg Ap	11.8 ± 0.4	2.0 ± 0.1	27.9 ± 0.2	27.9 ± 0.6	23	--
Sharpsburg Bt	3.8 ± 0.5	1.6 ± 0.1	35.7 ± 0.5	35.1 ± 0.8	6	--
Sharpsburg C	2.4 ± 0.1	2.1 ± 0.1	29.4 ± 0.2	33.5 ± 0.8	1	--
Nira Ap	16.8 ± 1.5	1.3 ± 0.1	31.4 ± 0.2	30.9 ± 0.7	23	--
Nira Bw	4.5 ± 0.4	1.0 ± 0.1	37.1 ± 0.6	36.5 ± 0.8	4	--
Nira C	2.0 ± 0.1	1.1 ± 0.1	30.1 ± 1.0	30.1 ± 0.3	1	--

† -- = below detection.

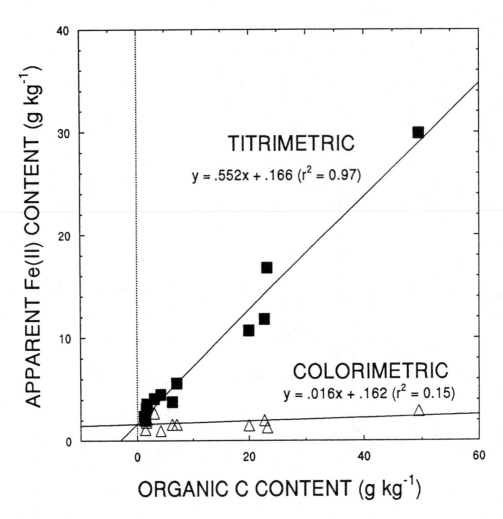

Fig. 3-3. Relation between the apparent Fe(II) content as measured by titrimetric and colorimetric methods and the organic C content of 12 Iowa soil samples.

A typical Mössbauer spectrum of the soils analyzed in this study is shown in Fig. 3-5, and the corresponding parameters for all of the soils are given in Table 3-6. These parameters were calculated from the curve-fitting results of the experimental patterns. The principal strong absorbance at about 0.3 mm s^{-1} derives primarily from Fe(III), but the small peak at about 2.5 mm s^{-1} indicates the presence of Fe(II) as a minor component. Peak pairs 1 and 5, and 2 and 4, made up the two doublets assigned to Fe(III). The Fe(II) doublet consisted of Peaks 3 and 6, with an area that ranged from about 4% of the total peak area in the Nira B-horizon sample to about 16% in the Canisteo C-horizon sample.

As explained above, the determination of the Fe(II):Fe(III) ratio from the Mössbauer spectrum requires an estimate of the proportionality constant, k, in addition to a measurement of $A2/A3$, the ratio of the Fe(II) and Fe(III) peak areas (see Eq. [3]). Assuming that $k \approx f_3/f_2$, the ratio of the recoil-free fractions of Fe(III) and Fe(II), and that f_3/f_2 is about one for the conditions under which the spectra were collected (de Grave & van Alboom, 1991), then the $F2/F3$ ratio for the Clarion A horizon estimated by the Mössbauer technique would be about 0.12. This value is considerably higher than 0.07 to 0.08, which is the value obtained when the colorimetric data in Table 5 are used to calculate an $F2/F3$ ratio for the Clarion A horizon. Either a considerable amount of oxidation of Fe(II) occurred during the colorimetric procedure or the value for k is not appropriate for this sample and the Mössbauer analytical conditions. It is also possible that Eq. [4] is invalid, even as a first approximation, for these soils and that additional factors besides the recoil-free

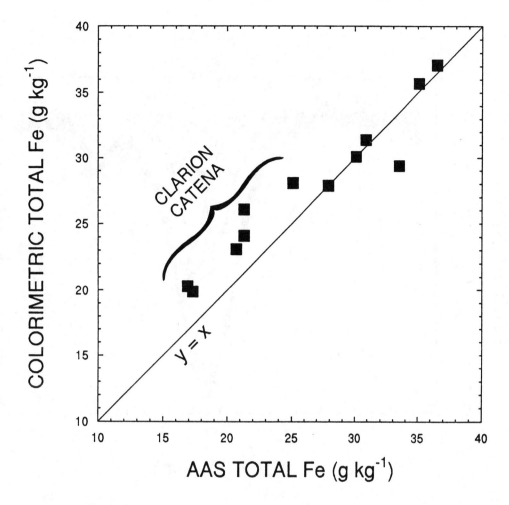

Fig. 3-4. Relation between total Fe values as determined by a colorimetric method and by atomic absorption spectroscopy for 12 Iowa soils.

fraction ratio (such as sample thickness, texture, and curve-fitting algorithm) are important components of k.

A plot of the Mössbauer $A 2/A 3$ ratio data as a function of the colorimetric $F 2/F 3$ ratio data should yield a line through the origin whose slope is equal to $1/k$ if the two methods measure the same relative amounts of Fe(II) and Fe(III). Such a plot offers a way to independently determine k and, in this particular situation, two such plots can be made, depending on which value of total Fe is used to calculate the Fe(III) content for use in the ratio. When the ratio data calculated using the atomic absorption (AAS) values for total Fe are plotted (Fig. 3-6), a line with a slope of about 1.43 and a correlation coefficient (r^2) of 0.92 is obtained (ignoring, for the time being, the outlying Canisteo A and B horizon data). The k value for the soils in this plot, therefore, would be 0.70 and is reasonably constant. A similar plot (not shown) using the colorimetric values for total Fe in the ratio calculations yielded a k value of 0.57 and a somewhat poorer correlation coefficient ($r^2 = 0.84$). These results suggest that the total-Fe data obtained by AAS yield Fe(III) values that are more highly correlated with the Mössbauer data than do the colorimetric total-Fe data.

As a further check on this hypothesis, the k values obtained from the ratio plots (e.g., Fig. 3-6) were substituted along with the appropriate total Fe values into the equation

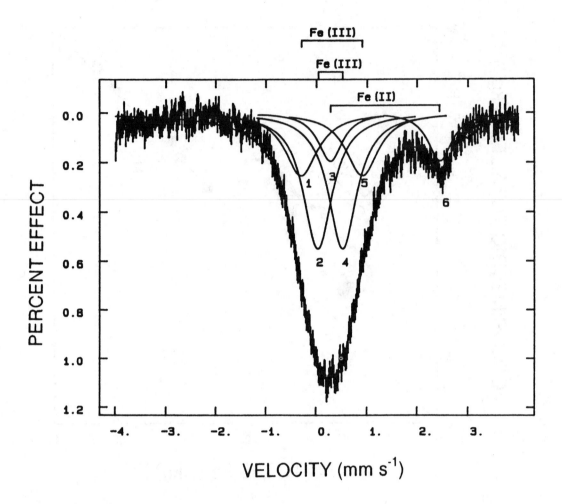

Fig. 3-5. A typical Mössbauer spectrum for the soils analyzed.

$$Fe(II) = Fe_{TOT} [1/(1 + (A3/(kA2)))] \qquad [5]$$

where Fe(II) is the Mössbauer-estimated Fe(II) content of the sample. These Mössbauer values for Fe(II) were then plotted versus the direct colorimetric Fe(II) values (Fig. 3-7 and 3-8). As can be seen (again ignoring the outlying Canisteo A and B data), a reasonable 1:1 correlation (slope = 0.79, r^2 = 0.72) between the colorimetric and Mössbauer values was obtained using the AAS total-Fe data (Fig. 3-7), whereas, when the colorimetric total-Fe data were used (Fig. 3-8), considerably more scatter in the data was observed and an r^2 of only 0.54 was calculated. Based on these results, therefore, it seems likely that the AAS total-Fe data for the Clarion-catena soils give a truer estimate of the Fe content than the colorimetric total-Fe data.

Throughout this data analysis, the data points corresponding to the Canisteo A and B horizons have been clearly different from those of the other 10 samples. One possible explanation of these results is the presence, particularly in the Canisteo A horizon, of siderite or ankerite [i.e., carbonate phases containing Fe(II)]. These phases tend to form in reducing, carbonatic systems having pH values between 8 and 10 (see, for example, Stumm & Morgan, 1981, p. 448). The Canisteo soil is the only one of the soils analyzed to have significant carbonate contents in the A and B horizons and a pH near 8. Because it has an aquic moisture regime, it would be expected to have reducing conditions and high levels of CO_2 during wet periods that could result in the

Table 3-6. Mean Mössbauer parameters obtained for analysis of Fe in 12 soil samples (IS = isomer shift, QS = quadrupole splitting, HWHM = half-width at half-maximum for each deconvoluted peak in the doublet, $A\,2/A\,3$ = ratio of sum of Fe(II) peak areas to sum of Fe(III) peak areas).

Soil	Peaks	Oxidation state	IS	QS	HWHM	Relative area	$A2/A3$	χ^2
			--------- mm s^{-1} ---------					
Clarion Ap	3,6	II	1.38	2.15	0.37	0.10	0.1109	0.851
	1,5	III	0.30	1.16	0.33	0.17		
	2,4	III	0.29	0.50	0.38	0.74		
Clarion Bw	3,6	II	1.36	2.20	0.27	0.07	0.0799	0.724
	1,5	III	0.30	1.16	0.28	0.14		
	2,4	III	0.29	0.48	0.42	0.79		
Clarion C	3,6	II	1.34	2.18	0.28	0.14	0.1655	1.007
	1,5	III	0.30	1.16	0.32	0.25		
	2,4	III	0.26	0.49	0.36	0.61		
Canisteo Ap	3,6	II	1.30	2.45	0.27	0.05	0.0547	0.648
	1,5	III	0.32	1.28	0.25	0.08		
	2,4	III	0.31	0.47	0.43	0.87		
Canisteo Bg	3,6	II	1.33	2.23	0.34	0.15	0.1788	0.645
	1,5	III	0.24	1.18	0.23	0.11		
	2,4	III	0.28	0.45	0.40	0.74		
Canisteo Cg	3,6	II	1.34	2.17	0.29	0.16	0.1962	0.732
	1,5	III	0.28	1.20	0.39	0.29		
	2,4	III	0.25	0.49	0.34	0.55		
Sharpsburg Ap	3,6	II	1.36	2.24	0.55	0.10	0.1060	0.926
	1,5	III	0.33	1.25	0.32	0.12		
	2,4	III	0.53	0.86	0.46	0.78		
Sharpsburg Bt	3,6	II	1.39	2.24	0.28	0.05	0.0571	0.914
	1,5	III	0.33	1.23	0.29	0.16		
	2,4	III	0.33	0.46	0.41	0.79		
Sharpsburg C	3,6	II	1.38	2.16	0.39	0.07	0.0799	0.874
	1,5	III	0.32	1.22	0.24	0.10		
	2,4	III	0.32	0.48	0.41	0.83		
Nira Ap	3,6	II	1.37	2.14	0.25	0.07	0.0705	0.770
	1,5	III	0.31	1.10	0.26	0.15		
	2,4	III	0.31	0.50	0.35	0.78		
Nira Bw	3,6	II	1.39	2.21	0.20	0.04	0.0438	1.136
	1,5	III	0.32	1.23	0.25	0.11		
	2,4	III	0.31	0.48	0.39	0.85		
Nira C	3,6	II	1.40	2.22	0.17	0.06	0.0593	0.671
	1,5	III	0.33	1.25	0.28	0.18		
	2,4	III	0.30	0.48	0.38	0.77		

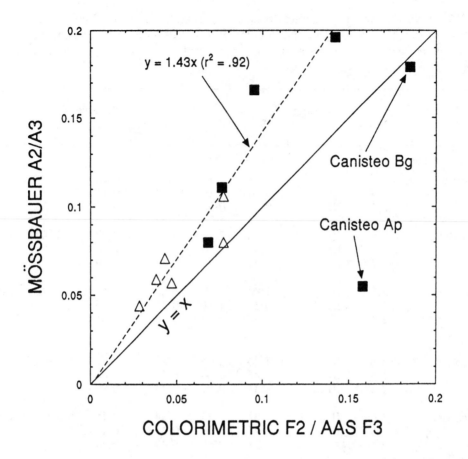

Fig. 3-6. Relation between the Mössbauer Fe(II)/Fe(III) peak area ratio and the Fe(II)/Fe(III) concentration ratio as determined by colorimetry and atomic absorption (for total Fe) for 12 Iowa soils. (Clarion-catena soils are represented by filled squares, Sharpsburg and Nira soils by triangles; Canisteo Ap and Bg outliers are not included in regression analysis).

precipitation of the ferrous carbonate phases. Furthermore, this soil is well known for its ability to cause Fe chlorosis in soybean during these wet periods.

For the carbonate hypothesis to be valid, the recoil-free fraction of Fe(II) in the carbonate would have to be significantly smaller than that in a silicate mineral. The slightly larger quadrupole splittings seen for the Canisteo A and B horizons suggest that the Fe(II) is in a more distorted octahedral environment than in the silicates. Whether this supports the carbonate hypothesis remains to be seen. As yet, however, no positive identification of a ferrous carbonate-containing phase in this soil has been made.

We conclude by noting that more work on the technique is required before the Fe oxidation states can be quantitatively differentiated in soils using Mössbauer spectroscopy. The technique is still useful for qualitative estimates when the sample must be recovered after analysis, but, as we have demonstrated, interpretation of the results from a potentially mixed-phase system is not always straightforward.

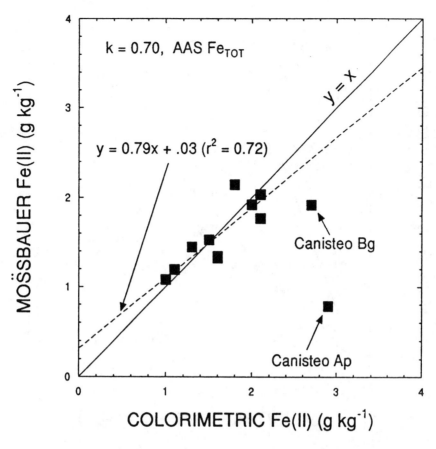

Fig. 3-7. Relation between Fe(II) contents of 12 Iowa soil samples as determined by Mössbauer spectroscopy (using total Fe values from atomic absorption spectroscopy and a proportionality constant value of 0.70) and by a colorimetric method (Canisteo Ap and Bg outliers are not included in regression analysis).

FUTURE DEVELOPMENTS

Because the redox status of soil systems is one of the key factors controlling the mobility, availability, and toxicity of various chemical species in soils, the need to identify and quantify specific oxidized or reduced species will continue to be strong. In combination with soil organic matter, soil minerals serve to buffer the electron activity and to record the redox chemistry [e.g., by the formation of Fe(III) oxides or Fe(II) carbonates] in a particular system. In some instances, measurements of a specific oxidant or reductant will be needed (e.g., for determination of reaction mechanisms), but the trend is more towards an operational definition of oxidizing or reducing capacity with respect to a specific species. Various wet-chemical techniques for determining the redox status of Fe species are fairly well developed but methods for determination of the redox status of other important elements, such as Mn, Cr, and Ti, still leave much to be desired. The increasing power of direct spectroscopic techniques (e.g., XPS, XANES, Auger) offers some hope for avoiding the possible errors associated with sample decomposition, but often the results are semiquantitative and the expense greater. For the time being, then, wet-chemical techniques will continue to be the dominant means of obtaining redox information. Improvements in these techniques will be largely the product of i) the increased use of automated dilutions, titrations, and

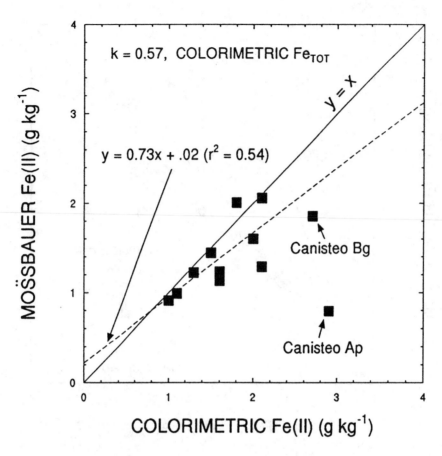

Fig. 3-8. Relation between Fe(II) contents of 12 Iowa soil samples as determined by Mössbauer
spectroscopy (using total Fe values from a colorimetric method and a proportionality constant
value of 0.57) and by a colorimetric method (Canisteo Ap and Bg outliers are not included in
regression analysis).

extractions; (ii) the incorporation of gravimetric in lieu of volumetric measurements; (iii) the
development of new chromogens and of new combinations of chromogens with oxidants or
reductants for increased specificity; and (iv) an increased reliance on electronic data collection and
handling.

SUMMARY

This chapter surveys the various spectroscopic and wet-chemical methods that have been
developed to determine redox status of soils and soil minerals. The dominant redox-sensitive
element in most soils is Fe and it may be analyzed by x-ray based spectroscopic techniques (e.g.,
XAS, XANES, Auger), by Mössbauer spectroscopy, and by wet-chemical techniques involving
decomposition of the specimen and either titration of the Fe(II) species with a suitable oxidant or
complexation of Fe(II) by a chromogenic chelate and analysis by colorimetry. The spectroscopic
techniques, while offering nondestructive analysis of specimens, are generally less accurate,
precise, or accessible for routine characterization than the wet-chemical methods. The wet-
chemical methods require careful attention to specimen preparation, decomposition, and oxidation-
state quantification, and they are subject to numerous types of interferences. The chief
interferences in titrimetric methods are other oxidized or reducing substances (e.g., Mn oxides,
organic matter) that can react with Fe or with the titrant to give false readings. Interfering species
in the colorimetric techniques include other complexing agents (e.g., F⁻), and other metal-

chromogen complexes or organic species that may absorb light at the same wavelengths as the analytic complex. Inasmuch as soils inevitably contain at least some organic matter, titrimetric methods generally yield "total reductive capacity" measurements that include some of the reductive ability of the organic matter in addition to that for the mineral Fe, whereas colorimetric methods are much more specific to Fe. In the opinion and experience of the authors, phen-based colorimetric reagents are the best chromogens to use when Fe-specific data for soils are needed. For overall assessments of reductive capacity, on the other hand, titrimetric techniques using V(V) are preferred.

An experimental comparison of a colorimetric (phen), a titrimetric [V(V)], and a spectroscopic (Mössbauer) method for quantitative determination of the Fe oxidation state in soils showed a strong interference by organic matter in the titrimetric method, resulting in poor precision and erroneously high values for Fe(II). The results of the Mössbauer analyses showed that other factors in addition to the recoil-free fractions of Fe(II) and Fe(III) were important in determining the appropriate proportionality factor for conversion of Fe(II) and Fe(III) peak areas into relative concentrations. These results suggested that Mössbauer analysis of mixtures of Fe-containing phases is difficult because additional peak deconvolutions are required and different recoil-free fractions must be accounted for in the spectrum interpretation. Thus, the most reliable technique for use in soils was the colorimetric method using phen, which showed little influence from organic matter and excellent precision for Fe(II). Slightly more variability was encountered when total Fe was measured by this method, and a direct measurement by atomic absorption spectrometry may be advisable in some instances. Of the three methods investigated, the phen method is relatively accurate and is the only one that gives both Fe(II) and total Fe content simultaneously.

ACKNOWLEDGMENTS

The manuscript benefited from careful reviews by Laurel K. Grove and three anonymous technical reviewers. Pacific Northwest Laboratory is operated for the U.S. Department of Energy by Battelle Memorial Institute under Contract DE-AC06-76RLO 1830.

REFERENCES

Amonette, J.E. 1988. The role of structural iron oxidation in the weathering of trioctahedral micas by aqueous solutions. Ph.D. diss. Iowa State Univ., Ames (Diss. Abstr. 88-25372).

Amonette, J., and D. Rai. 1987. Ba(S,Cr)O$_4$ solid solution as a possible phase controlling Cr(VI) levels in soils. p. 165. *In* Agronomy abstracts. ASA, Madison, WI.

Amonette, J.E., and D. Rai. 1990. The identification of noncrystalline (Fe,Cr)(OH)$_3$ by infrared spectroscopy. Clays Clay Miner. 38:129-136.

Amonette, J.E., and R.W. Sanders. 1994. Nondestructive techniques for bulk elemental analysis. p. 1-48. *In* J. Amonette and L. W. Zelazny (ed.) Quantitative methods in soil mineralogy. SSSA Misc. Publ. SSSA, Madison, WI.

Amonette, J.E., and A.D. Scott. 1991. Determination of ferrous iron in non-refractory silicate minerals--1. An improved semi-micro oxidimetric method. Chem. Geol. 92:329-338.

Ayranci, B. 1992. Analysis of the oxidation states of iron in silicate rocks and refractory minerals by fusion disintegration. Kontakte (Darmstadt) 1992:16-20.

Bancroft, G.M. 1973. Mössbauer spectroscopy. An introduction for inorganic chemists and geochemists. McGraw Hill, New York.

Banerjee, S. 1974. Direct determination of ferrous iron in silicate rocks and minerals by iodine monochloride. Anal. Chem. 46:782-786.

Bartlett, R.J., and J.M. Kimble. 1976a. Behavior of chromium in soils: I. Trivalent forms. J. Environ. Qual. 5:379-383.

Bartlett, R.J., and J.M. Kimble. 1976b. Behavior of chromium in soils: II. Hexavalent forms. J. Environ. Qual. 5:383-386.

Bartlett, R., and B. James. 1979. Behavior of chromium in soils: III. Oxidation. J. Environ. Qual. 8:31-35.

Baxendale, J.H., and N. K. Bridge. 1955. The photoreduction of some ferric compounds in aqueous solution. J. Phys. Chem. 59:783-788.

Begheijn, L.Th. 1979. Determination of iron(II) in rock, soil, and clay. Analyst 104:1055-1061.

Bernas, B. 1968. A new method for decomposition and comprehensive analysis of silicates by atomic absorption spectrometry. Anal. Chem. 40:1682-1686.

Beyer, M.E., A.M. Bond, and R.J.W. McLaughlin. 1975. Simultaneous polarographic determination of ferrous, ferric, and total iron in standard rocks. Anal. Chem. 47:479-482.

Boellstorff, J.D. 1978. Procedures for the analysis of pebble lithology, heavy minerals, light minerals and matrix calcite-dolomite of tills. In G. R. Hallberg (ed.) Standard procedures for evaluation of Quaternary materials in Iowa. Tech. Inf. Ser. no. 8. Iowa Geol. Surv., Iowa City, IA.

Brinkman, R. 1977. Surface-water gley soils in Bangladesh: Genesis. Geoderma 17:111-144.

Brown, G.E, Jr., G. Calas, G.A. Waychunas, and J. Petiau. 1988. X-ray absorption spectroscopy: Applications in mineralogy and geochemistry. Rev. Miner. 18:431-512.

Cheng, K.L. 1964. The determination of ferrous iron in ferrites. Anal. Chem. 36:1666-1667.

Chiswell, B., and G. Rauchle. 1990. Spectrophotometric methods for the determination of manganese. Talanta 37:237-259.

Clemency, C.V., and A.F. Hagner. 1961. Titrimetric determination of ferrous and ferric iron in silicate rocks and minerals. Anal. Chem 33:888-892.

Cocke, D.L., R.K. Vempati, and R.H. Loeppert. 1994. Analysis of soil surfaces by x-ray photoelectron spectroscopy. p. 205-235. In J. Amonette and L.W. Zelazny (ed.) Quantitative methods in soil mineralogy. SSSA Misc. Publ. SSSA, Madison, WI.

Cranshaw, T.E., B.W. Dale, G.O. Longworth, and C.E. Johnson. 1986. Mössbauer spectroscopy and its applications. Cambridge Univ. Press, Cambridge, England.

Croudace, I. W. 1980. A possible error source in silicate wet-chemistry caused by insoluble fluorides. Chem. Geol. 31:153-155.

David, P.G., J.G. Richardson, and E.L. Wehry. 1972. Photoreduction of tetrakis(1,10-phenanthroline)-μ-oxodiiron(III) complexes in aqueous and acetonitrile solution. J. Inorg. Nucl. Chem. 34:1333-1346.

de Grave, E., and A. van Alboom. 1991. Evaluation of ferrous and ferric Mössbauer fractions. Phys. Chem. Miner. 18:337-342.

Diehl, H. 1974. Quantitative analysis--elementary principles and practice. 2nd ed. Oakland St. Sci. Press, Ames, IA.

Dinsel, D.L., and T.R. Sweet. 1963. Separation and determination of iron(II) and iron(III) with anthranilic acid using solvent extraction and spectrophotometry. Anal. Chem. 35: 2077-2081.

Donaldson, E.M. 1969. Study of Groves' method for determination of ferrous oxide in refractory silicates. Anal. Chem. 41:501-505.

Duesler, E.N., and E.E. Foord. 1986. Crystal structure of hashemite, $BaCrO_4$, a barite structure type. Am. Miner. 71:1217-1220.

Fitton, J.C., and R.C.O. Gill. 1970. The oxidation of ferrous iron in rocks during mechanical grinding. Geochim. Cosmochim. Acta 34:518-524.

Freeman, D.S., and W.G. Chapman. 1971. An improved oxalate method for the determination of active oxygen in manganese dioxide. Analyst 96:865-869.

French, W.J., and S.J. Adams. 1972. A rapid method for the extraction and determination of iron(II) in silicate rocks and minerals. Analyst 97:828-831.

Fudali, R.F., M.D. Dyar, D.L. Griscom, and H.D. Schreiber. 1987. The oxidation state of iron in tektite glass. Geochim. Cosmochim. Acta 51:2749-2756.

Furman, N.H. 1963. Potentiometry. p. 2269-2302. In I. M. Kolthoff and P. J. Elving (ed.) Treatise on analytical chemistry. Part I. Vol. 4. Sect. D-2. John Wiley & Sons, New York.

Gibb, T.C. 1977. Principles of Mössbauer spectroscopy. Chapman and Hall, London.

Goldman, J.A. 1975. Titrimetry: Oxidation-reduction titration. p. 7191-7225. In I. M. Kolthoff et al. (ed.) Treatise on analytical chemistry. Part 1. Vol. 11. Sect. I-2. John Wiley & Sons, New York.

Grady, H.R. 1963. Vanadium. p. 177-272. In I. M. Kolthoff et al. (ed.) Treatise on analytical chemistry. Part II. Analytical chemistry of the elements. Vol. 8. Sect. A. John Wiley & Sons, New York.

Grill, E.V. 1978. The effect of sediment-water exchange on manganese deposition and nodule growth in Jervis Inlet, British Columbia. Geochim. Cosmochim. Acta 42:485-494.

Groves, A.W. 1951. Silicate analysis. 2nd ed. George Allen and Unwin, Ltd., London.

Harvey, A.E., Jr., and D.L. Manning. 1952. Spectrophotometric studies of empirical formulas of complex ions. J. Am. Chem. Soc. 74:4744-4746.

Harvey, A.E. Jr., J.A. Smart, and E.S. Amis. 1955. Simultaneous spectrophotometric determination of iron(II) and total iron with 1,10-phenanthroline. Anal. Chem. 27:26-29.

Hauff, P.L., E.E. Foord, and S. Rosenbloom. 1983. Hashemite, $Ba(S,Cr)O_4$, a new mineral from Jordan. Am. Miner. 68:1223-1225.

Hawthorne, F.C. 1988. Mössbauer spectroscopy. Rev. Miner. 18: 255-340.

Hey, M.H. 1941. The determination of ferrous iron in resistant silicates. Miner. Mag. 26:116-118.

Hochella, M.F., Jr. 1988. Auger electron and x-ray photoelectron spectroscopies. Rev. Miner. 18:573-637.

Ito, J. 1962. A new method of decomposition for refractory minerals and its application for the determination of ferrous iron and alkalies. Sci. Pap. Coll. Gen. Educ., Univ. Tokyo 11:47-68.

Johnson, W.M., and J.A. Maxwell. 1989. Rock and mineral analysis, 2nd ed. Robert Krieger, Malabar.

Kessick, M.A., J. Vuceta, and J.J. Morgan. 1972. Spectrophotometric determination of oxidized manganese with leuco crystal violet. Environ. Sci. Technol. 6:642-644.

Khan, F.A. 1991. Relationships of saturated zones and distribution of selected chemical and mineralogical properties in the Clarion catena. Ph.D. diss. Iowa State Univ. Ames (Diss. Abstr. 92:12152).

Kiss, E. 1967. Chemical determination of some major constituents in rocks and minerals. Anal. Chim. Acta 39:223-234.

Kiss, E. 1974. Synthesis of a sulphonated ferroin reagent for chelating iron(II) in strong acid. Spectrophotometric determination of the oxidation state of iron in silicates. Anal. Chim. Acta 72:127-144.

Kiss, E. 1977. Rapid potentiometric determination of the iron oxidation state in silicates. Anal. Chim. Acta 89:303-314.

Kiss, E. 1987. Integrated scheme for micro-determination of iron oxidation states in silicates and refractory minerals. Anal. Chim. Acta 193:51-60.

Kleinert, P., and A. Funke. 1962. The determination of iron(II) and total iron in nickeliferous ferrites. Z. Chem. 2:155-157.

Komadel, P., and J.W. Stucki. 1988. The quantitative assay of minerals for Fe^{2+} and Fe^{3+} using 1,10-phenanthroline. III. A rapid photochemical method. Clays Clay Miner. 36:379-381.

Langmyhr, F.J., and K. Kringstad. 1966. An investigation of the composition of the precipitates formed by the decomposition of silicate rocks in 38-40% hydrofluoric acid. Anal. Chim. Acta 35:131-135.

Lear, P.R. 1987. The role of iron in nontronite and ferrihydrite. Ph.D. diss. Univ. of Illinois, Urbana.

Lear, P.R., P. Komadel, and J.W. Stucki. 1988. Mössbauer spectroscopic identification of iron oxides in nontronite from Hohen Hagen, Federal Republic of Germany. Clays Clay Miner. 36:376-378.

Loveland, P.J. 1988. The assay for iron in soils and clay minerals. p. 99-140. In J. W. Stucki et al. (ed.) Iron in soils and clay minerals. D. Reidel, Dordrecht, Holland.

Manceau, A., and L. Charlet. 1992. X-ray absorption spectroscopic study of the sorption of Cr(III) at the oxide-water interface--I. Molecular mechanism of Cr(III) oxidation on Mn oxides. J. Coll. Inter. Sci. 148:425-442.

Manceau, A., A.I. Gorshkov, and V.A. Drits. 1992. Structural chemistry of Mn, Fe, Co, and Ni in manganese hydrous oxides: Part I. Information from XANES spectroscopy. Am. Miner. 77:1133-1143.

Mande, C., and A.P. Deshpande. 1990. A study of manganese oxides by EXAFS spectroscopy. Phys. Stat. Sol.(B) 158:737-742.

Mebius, L.J. 1960. A rapid method for determination of organic carbon in soils. Anal. Chim. Acta 22:120-124.

Meyrowitz, R. 1963. A semimicroprocedure for the determination of ferrous iron in nonrefractory silicate minerals. Am. Miner. 48:340-347.

Meyrowitz, R. 1970. New semimicroprocedure for determination of ferrous iron in refractory silicate minerals using a sodium metafluoborate decomposition. Anal. Chem. 42:1110-1113.

Mitsuchi, M., and M. Oyama. 1963. Total ferrous and ferric iron in soils. J. Sci. Soil Manure Jpn. 34:23-27.

Mössbauer, R. 1958a. Kernresonanzabsorption von gammastrahlung in Ir[191]. Naturwissenschaften 45:538-539.

Mössbauer, R. 1958b. Kernresonanzfluoresent von gammastrahlung in Ir[191]. Z. Phys. 151:124-143.

Muan, A. 1958. Phase equilibria at high temperatures in oxide systems involving changes in oxidation states. Am. J. Sci. 256:171-207.

Muan, A. 1963. Silver-palladium alloys as crucible material in studies of low-melting iron silicates. Bull. Am. Cer. Soc. 42:344-347.

Murad, E. 1988. Properties and behavior of iron oxides as determined by Mössbauer spectroscopy. p. 309-350. In J. W. Stucki et al. (ed.) Iron in soils and clay minerals. D. Reidel, Dordrecht, Holland.

Murphy, J.M., J.I. Read, and G.A. Sergeant. 1974. A method for the direct titrimetric determination of iron(III) in silicate rocks. Analyst 99:273-276.

Murray, J.W., L.S. Balistrieri, and B. Paul. 1984. The oxidation state of manganese in marine sediments and ferromanganese nodules. Geochim. Cosmochim. Acta 48:1237-1247.

Nelson, D.W., and L.E. Sommers. 1982. Total carbon, organic carbon, and organic matter. p. 539-579. In A. L. Page et al. (ed.) Methods of soil analysis. Part 2. 2nd ed. Agron. Monogr. 9. ASA and SSSA, Madison, WI.

Novak, J., and H. Arend. 1964. The photosensitivity of the complex of iron(III) with 1,10-phenanthroline. Talanta 11:898-899.

Novikova, Y.N. 1966. The possibility of determination of iron(II) in rocks and minerals by fusion with sodium-metafluoborate decomposition. Zh. Anal. Khim. 23:1057-1059.

Pabst, A. 1939. Formula and structure of ralstonite. Am. Miner. 24:566-576.

Page, A.L., R.H. Miller, and D.R. Keeney. 1982. Methods of soil analysis. Part 2. 2nd ed. Agron. Monogr. 9, ASA and SSSA, Madison, WI.

Pauly, H. 1965. Ralstonite from Ivigtut, South Greenland. Am. Miner. 50:1851-1864.

Peters, von A. 1968. Ein neues verfahren zur bestimmung von eisen(II)oxid in mineralen und gesteinen. Neues Jahrb. Miner. Monatsh. 1968(3/4):119-125.

Pilkington, E.S., and P.R. Smith. 1967. Spectrophotometric determination of chromium in ilmenite. Anal. Chim. Acta 39:321-328.

Pratt, J.H. 1894. On the determination of ferrous iron in silicates. Am. J. Sci. 48:149-151.

Pruden, G., and C. Bloomfield. 1966. The effect of organic matter on the determination of iron(II) in soils and rocks. Analyst 94:688-689.

Rai, D., J.M. Zachara, L.E. Eary, D.C. Girvin, D.A. Moore, C.T. Resch, B.M. Sass, and R.L. Schmidt. 1986. Geochemical behavior of chromium species. Rep. EA-4544. Electric Power Res. Inst., Palo Alto, CA.

Rai, D., J.M. Zachara, L.E. Eary, C.C. Ainsworth, J.E. Amonette, C.E. Cowan, R.W. Szelmeczka, C.T. Resch, R.L. Schmidt, and D.C. Girvin. 1988. Rep. EA-5741, Electric Power Res. Inst., Palo Alto, CA.

Rancourt, D.G. 1989. Accurate site populations from Mössbauer spectroscopy. Nucl. Instrum. Methods B58:199-210.

Rancourt, D.G. 1991a. Spectral analysis: Getting the most true information from Mössbauer spectra. Workshop on Mössbauer spectroscopy lecture notes, Houston, TX. 5 October. Clay Miner. Soc., Boulder, CO.

Rancourt, D.G. 1991b. Mössbauer spectroscopy of phyllosilicates. Workshop on Mössbauer spectroscopy lecture notes, Houston, TX. 5 October. Clay Miner. Soc. Boulder, CO.

Reichen, L.E., and J.J. Fahey. 1962. An improved method for the determination of FeO in rocks and minerals including garnet. Geol. Surv. Bull. 1144-B:B1-B5.

Ritchie, J.A. 1968. Effect of metallic iron from grinding on ferrous iron determinations. Geochim. Cosmochim. Acta 32:1363-1366.

Rowledge, H.P. 1934. New method for the determination of ferrous iron in refractory silicates. J. R. Soc. W. Aust. 20:165-199.

Sarver, L.A. 1927. The determination of ferrous iron in silicates. J. Am. Chem. Soc. 49:1472-1477.

Sarver, L.A., and I.M. Kolthoff. 1931a. Diphenylamine sulfonic acid as a new oxidation-reduction indicator. J. Am. Chem. Soc. 53:2902-2905.

Sarver, L.A., and I.M. Kolthoff. 1931b. Indicator corrections for diphenylamine, diphenylbenzidine and diphenylamine sulfonic acid. J. Am. Chem. Soc. 53:2906-2909.

Schafer, H.N.S. 1966a. The determination of iron(II) oxide in silicate and refractory materials--I. A review. Analyst 91:755-762.

Schafer, H.N.S. 1966b. The determination of iron(II) oxide in silicate and refractory materials--II. A semi-micro titrimetric method for determining iron(II) oxide in silicate materials. Analyst 91:763-790.

Schilt, A.A. 1969. Analytical applications of 1,10-phenanthroline and related compounds. Pergamon Press, Oxford, England.

Serjeant, E.P. 1984. Potentiometry and potentiometric titrations. Chemical analysis. Vol. 69. John Wiley & Sons, New York.

Seil, G.E. 1943. The determination of ferrous iron in difficultly soluble materials. Ind. Eng. Chem. Anal. 15:189-192.

Sherman, D.M. 1990. Molecular orbital (SCF-Xα-SW) theory of Fe^{2+}-Mn^{3+}, Fe^{3+}-Mn^{2+}, and Fe^{3+}-Mn^{3+} charge transfer and magnetic exchange in oxides and silicates. Am. Miner. 75:256-261.

Stucki, J.W. 1981. The quantitative assay of minerals for Fe^{2+} and Fe^{3+} using 1,10-phenanthroline. II. A photochemical method. Soil Sci. Soc. Am. J. 45:638-641.

Stucki, J.W., and W.L. Anderson. 1981. The quantitative assay of minerals for Fe^{2+} and Fe^{3+} using 1,10-phenanthroline: I. Sources of variability. Soil Sci. Soc. Am. J. 45:633-637.

Stumm, W., and J.J. Morgan. 1981. Aquatic chemistry. An introduction emphasizing chemical equilibria in natural waters. John Wiley & Sons, New York.

Thompsen, J.C., and H.A. Mottola. 1984. Kinetics of the complexation of iron(II) with ferrozine. Anal. Chem. 56:755-757.

Toni, J.E.A. 1962. Titration of uranium with potassium dichromate--determination of disproportionality effects. Anal. Chem. 34:99-102.

Ungethüm, H. 1965. Eine neue methode zur bestimmung von eisen(II) in gesteinen und mineralen, insbesondere auch in bitumenhaltigen proben. Z. Angew. Geol. 11:500-505.

Urch, D.S., and P.R. Wood. 1978. The determination of the valency of manganese in minerals by x-ray fluorescence spectroscopy. X-ray Spectosc. 7:9-11.

Van Loon, J.C. 1965. Titrimetric determination of the iron(II) oxide content of silicates using potassium iodate. Talanta 12:599-603.

Vanysek, P. 1984. Electrochemical series. p. D156-D163. In R. C. Weast et al. (ed.) CRC handbook of chemistry and physics. CRC Press, Boca Raton, FL.

Vogel, A.I., G.H. Jeffery, J. Bassett, J. Mendham, and R.C. Denney. 1989. Vogel's textbook of quantitative chemical analysis. 5th ed. Longman Sci. and Tech. Harlow, England.

Walker, J.L., and G.D. Sherman. 1961. Determination of total ferrous iron in soils. Soil Sci. 93:325-328.

Whipple, E.R. 1974a. Quantitative Mössbauer spectra and chemistry of iron. Ph.D. diss. Massachussetts Inst. Technol., Cambridge.

Whipple, E.R. 1974b. A study of Wilson's determination of ferrous iron in silicates. Chem. Geol. 14:223-238.

Whipple, E.R., J.A. Speer, and C.W. Russell. 1984. Errors in FeO determinations caused by tungsten carbide grinding apparatus. Am. Miner. 69:987-988.

Wilson, A.D. 1955. A new method for the determination of ferrous iron in rocks and minerals. Bull. Geol. Surv. Eng.. 1955(9):56-58.

Wilson, A.D. 1960. The micro-determination of ferrous iron in silicate minerals by a volumetric and a colorimetric method. Analyst 85:823-827.

Zachara, J.M., C.E. Cowan, R.L. Schmidt, and C.C. Ainsworth. 1988. Chromate adsorption by kaolinite. Clays Clay Miner. 36:317-326.

4   Structural Formulae of Layer Silicates

B. Číčel and P. Komadel
Slovak Academy of Sciences
Bratislava, Slovakia

Layer silicates are natural inorganic compounds of variable chemical composition. Their structure is built from two types of building units, tetrahedra $[TO_4]^{n-}$ and octahedra $[MO_6]^{m-}$, connected together to form two-dimensional sheets. The layer structure is comprised of tetrahedral and octahedral sheets joined face-to-face in ratios of 1:1 (e.g., kaolinite), 2:1 (e.g., mica), or 2:1:1 (chlorite). The central atoms in the tetrahedra are usually Si, Al, or Fe(III). The choice of central atom in the octahedral sheet is much more variable, e.g., Al, Fe(III), Fe(II), Mg, Cr(III), Ni(II), Cu(II), etc. Different combinations of central atoms in the octahedral and tetrahedral sheets may result in a net negative charge on the layer structure which is balanced by interlayer cations.

In order to classify and compare minerals, a systematic set of chemical compositions for each group of minerals has been developed. One way this is accomplished is by determining the type of site (i.e., tetrahedral, octahedral, interlayer) each central atom occupies, from which a structural formula for each mineral may be deduced.

Structural formulae provide a statistical picture of the occurrence of individual central atoms (Si, Al, Mg, etc.) in the given type of position (tetrahedra, octahedra, interlayer). The occupancy factors, known from structural analysis, can be calculated easily from the structural formula. For example, the structural formula of montmorillonite from Kriva Polanka (Yugoslavia) is

$$[Si_{7.70} Al_{0.30}][Al_{3.09} Fe_{0.25} Mg_{0.68}][Ca_{0.47}]O_{20}(OH)_4 , \qquad [1]$$

and the occupancy factors for tetrahedra and octahedra are $(Si_{0.96} Al_{0.04})$ and $(Al_{0.77} Fe_{0.06} Mg_{0.17})$, respectively.

The basic requirement for determining structural formulae is a knowledge of the structure of the mineral. Real ideas about structures of layer silicates appeared in the works of Pauling (1930a,b) (muscovite, kaolinite, and chlorite), Jackson and West (1930, 1933) (muscovite), Gruner (1932a,b, 1934) (kaolinite, dickite, and vermiculite), Hofmann et al. (1933) (montmorillonite), and Kazancev (1934) (vermiculite). Several refinements of these structures were published later, and today are the bases for current confidence in these structures. Formula units for several types of layer silicates are given in Table 4-1.

Two different methods commonly are used to locate the positions of the central atoms in the structure: (i) calculation of structural formulae from bulk chemical analyses, and (ii) detection of the distribution of central atoms within the mineral structure by chemical methods such as acid dissolution and by spectral methods such as infrared (IR), magic-angle spinning nuclear magnetic resonance (MAS-NMR), and Mössbauer spectroscopies.

Table 4-1. Formula units for layer silicate minerals.

Layer	Arrangement	Formula unit
1:1	Dioctahedral	$T_4 M_4 O_{10} (OH)_4$
	Trioctahedral	$T_4 M_6 O_{10} (OH)_8$
2:1	Dioctahedral	$T_8 M_4 I_x O_{20} (OH)_4$
	Trioctahedral	$T_8 M_6 I_x O_{20} (OH)_4$
2:1:1	di - di	$T_8 M_4 O_{20} (OH)_4 M_4 (OH)_{12}$
	di - tri	$T_8 M_4 O_{20} (OH)_4 M_6 (OH)_{12}$
	tri- di	$T_8 M_6 O_{20} (OH)_4 M_4 (OH)_{12}$
	tri- tri	$T_8 M_6 O_{20} (OH)_4 M_6 (OH)_{12}$

STRUCTURAL FORMULAE FROM BULK CHEMICAL ANALYSES

The reliability of structural formula calculations depends on several basic assumptions and rules.

1. The sample used for chemical analysis must be monomineralic. The presence of other minerals will always lead to inaccurate results, which obviously would significantly affect the calculated distribution of atoms in the structure (Kelley, 1945; Číčel, 1981).

2. The chemical analysis must be correct and exact. The influence of inaccurate chemical analysis on calculated distributions of central atoms has been investigated from the first days of their use (Kelley, 1945) even until today (Giaramita & Day, 1990).

3. The main central atoms in the tetrahedral sites are Si and Al. Silicon exists only in tetrahedra, and Al fills the rest of the tetrahedral sites (Marshall, 1935; Ross & Hendricks, 1945; Stevens, 1946; Kelley, 1955; Hofmann et al., 1956; Bates, 1959; Nyrkov, 1961). If the measured quantity of Si and Al is insufficient to fill all the tetrahedra, the remaining sites are filled by Fe.

4. The other cations are assigned either to the octahedral sheet (1:1 silicates) or to the octahedral sheet and the interlayer space (2:1 and 2:1:1 silicates).

5. At least one type of site, i.e., tetrahedral or octahedral, in the structure must be fully occupied.

The calculation of a structural formula is a *calculation*, not a *determination,* of the probable distribution of central atoms within the structure. The calculation algorithm is based on the above-stated assumptions 1 through 5, and is valid only to the extent that these assumptions are fulfilled.

Structural formulae may be calculated from bulk chemical analysis in either of two ways. The first depends on the assumption that the O network is perfect, and that the central atoms compensate for the network's negative charge -- 18 O^{2-} in 1:1 silicates, 24 O^{2-} in 2:1 silicates, or 36 O^{2-} in 2:1:1 silicates. Among those who used this method were Marshall (1935), Ross and Hendricks (1945), Stevens (1946), McConnell (1951), Kelley (1955), Nyrkov (1961), Köster (1977), and others. The second method assumes that the central atoms occupy all of the tetrahedral and octahedral positions in the unit cell. This was used by Hofmann et al. (1956), Borneman-Starynkevic (1964), Grim and Guven (1978), and others. These methods are referred to hereafter as "Method I" and "Method II," respectively.

These methods differ in the way of filling the central atom positions in the tetrahedra and octahedra. They may also differ in the way that H_2O or structural OH^- is distributed, based on dehydroxylation (McConnell, 1951). Examples for both these methods of assigning central atoms

For silicates, which always contain Si^{4+} and some other formally mono-, bi-, or trivalent cations, Eq. [11] applies, viz.,

$$\left| \sum_{i=1}^{n} \frac{\partial F}{\partial A_i} \, dA_i \right| < \sum_{i=1}^{n} \left| \frac{\partial F}{\partial A_i} \, dA_i \right| . \qquad [11]$$

For this reason the sensitivity of F to the admixture of other layer silicates, and also feldspars, zeolites, or volcanic glass, is always lower than to the admixture of free oxides such as quartz, cristobalite, and goethite. The sensitivity to free oxides in the sample increases with increasing V_i and with decreasing E_i.

Method II

The other structural-formula calculation method that uses bulk-chemical data is based on the assumption that all central atoms are bound in a fixed number of positions derived from the type of structure (e.g., 12 for dioctahedral micas, 18 for di-tri- and tri-dioctahedral chlorites, etc.). From this assumption, the condition $dF/dA_i = 0$ applies, regardless of the presence or absence of admixtures. Only an unacceptable distribution of central atoms in the structure would indicate the incorrectness of the calculation. The procedure is as follows:

Step 1. Divide the mass concentration (A_i) of each oxide by its equivalent mass (E_i = molecular mass divided by the number of metal atoms in the oxide formula).

Step 2. Sum the mass equivalents to equal C.

Step 3. Calculate the conversion factor, M, from $M = K/C$, where K is the number of central atom positions to be included in the calculation (e.g., 8 or 10 for 1:1 minerals; 12 or 14 for 2:1 minerals; etc., the higher number of each pair corresponding to trioctahedral minerals).

Step 4. The coefficients, f_i, in the structural formula are then obtained by multiplying the mass equivalents (A_i/E_i) by the conversion factor M.

Step 5. Assume that the number of tetrahedral central atoms in the unit cell is 4 for 1:1, and 8 for 2:1 and 2:1:1 layer silicates.

A comparison of the two methods made by Grim and Guven (1978) is given in Table 4-2. The first structural formula (Sample 7) presented in Table 4-2 gives identical values for both methods of calculation. In both cases $F_{(o)}$ = 4.00. When calculated using Method I, the second structural formula (Sample 12) indicates that too many central atoms were assigned to the octahedral sheet, whereas the formula calculated using Method II seems to be formally acceptable. The structural formula for Sample 151 calculated using Method I shows the presence of at least 5% of free silica, while using Method II leads again to adequate f_i values. The mean of 152 analyses presented by Grim and Guven (1978) indicates that most of the samples contained free Al and/or Fe oxides. The mean value of $F_{(o)}$ = 4.14 calculated by Method I can be achieved only if Fe oxides comprise at least 6% of the sample mass. The comparable structural formula calculated by Method II indicates nothing about the presence of admixed minerals.

Effect of Free Oxides

The oxides SiO_2, Al_2O_3, Fe_2O_3, MgO, and CaO not bound in smectite are frequently present in the fine fractions separated from bentonites, in spite of no identification by XRD. Using calculation Method I, the addition of 10 g "free" silica per kilogram increases f_{Si} in smectites by 0.04, while $F_{(o)}$ decreases by 0.01 or 0.02. A value of f_{Si} greater than 8.00 unambiguously indicates the presence of free SiO_2 in the system. To reach $f_{Si} > 8.00$, more than 200 g free SiO_2 per kilogram is needed for beidellitic samples, and more than 20 g kg^{-1} is needed for montmorillonites. For $f_{Si} < 8$, the only indication of free silica is the decrease of $F_{(o)}$. Once $F_{(o)}$ is less than

Table 4-2. Structural formulae for smectites calculated using (i) the method of fixed anionic charges, and (ii) the method of fixed number of central atom positions (Grim & Guven, 1978).

Sample†	Method	Tetrahedral ions		Octahedral ions				
		Si	Al	Al	Fe^{3+}	Fe^{2+}	Mg	$F_{(o)}$
7	1	7.98	0.02	2.92	0.16	--	0.92	4.00
	2	7.98	0.02	2.92	0.16	--	0.92	(4.00)
12	1	6.66	1.34	3.56	0.46	--	0.50	4.52
	2	6.38	1.62	3.08	0.44	--	0.48	(4.00)
151	1	7.84	0.16	2.88	0.66	0.06	0.26	3.86
	2	7.94	0.06	3.00	0.66	0.06	0.26	(4.00)
Mean	1	7.68	0.32	3.06	0.34	0.02	0.72	4.14
	2	7.60	0.40	2.94	0.34	0.02	0.72	(4.00)

†7, Chambers, Arizona (Robertson et al., 1968); 12, Nashville, Arkansas (Ross & Hendricks, 1945); 151, Humber River, New South Wales, Australia (Grim & Kulbicki, 1961); Mean = mean of 152 analyses.

3.96 for a montmorillonite, a test for the presence of free SiO_2 (i.e., amorphous SiO_2, cristobalite, quartz) in the sample is recommended.

The influence of increasing amounts of free silica on f_{Si}, $f_{Al(o)}$, is given in Fig. 4-1. Because Al may occur in both tetrahedral and octahedral sites, the effect of free silica on $f_{Al(o)}$ is complex. With increasing SiO_2 content, part of Al is formally "transferred" from the tetrahedra into the octahedra. This takes place until $f_{Si} = 8.00$.

The changes in calculated coefficients caused by the presence of free alumina and/or ferric oxides were discussed by Kelley (1945) and Číčel (1981). Changes of f_{Si} caused by the presence of 10 g free alumina per kilogram in smectites, for example, are -0.07 to -0.09 depending on the Al/Fe ratio in the octahedral sheet. The same quantity of free Fe_2O_3 causes f_{Si} to change between -0.045 and -0.057 in smectites. The sum of coefficients of atoms in octahedra ($F_{(o)}$) increases by 0.025 to 0.030 for free Al_2O_3, and by 0.015 to 0.020 for free Fe_2O_3, at levels of 10 g kg^{-1}.

Effect of Layer Silicates

The presence of kaolinite has little effect on the calculated tetrahedral substitution of smectites. Changes in f_{Si} caused by 10 g kaolinite per kilogram vary between 0.008 and 0.016. Even smaller changes are calculated if illite is the admixed component. Changes in f_{Si} due to 10 g illite per kilogram usually range between 0.000 to 0.005. The sum $F_{(o)}$ increases by about 0.002 for each 10 g kaolinite per kilogram increment, and is constant if illite is added.

The effects of 10 g free Al_2O_3 per kilogram on f_{Si} in the structural formulae of three smectites, and the amounts of Fe_2O_3, kaolinite, and illite required to produce the same effect on f_{Si} as 10 g Al_2O_3 per kilogram, are given in Table 4-3.

Evaluation of Bulk Chemical Methods

Mathematical analyses of the methods for calculating structural formulae, as well as the results of recalculation of synthetic mixtures of natural samples and admixtures, indicate that if assumptions i) and ii) above are valid, the results obtained by Methods I and II are identical (Grim

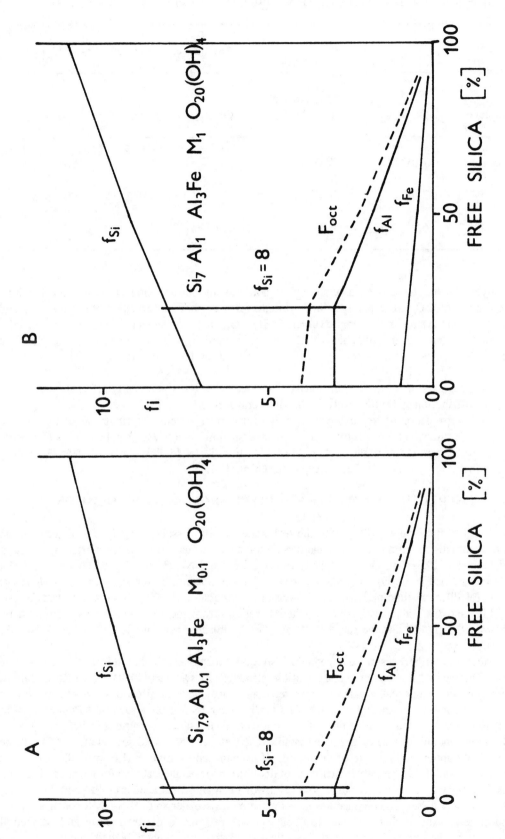

Fig. 4-1. Changes in f_{Si}, $f_{Al(o)}$, f_{Fe}, and $F_{(o)}$ as a function of free silica present in the sample.

Table 4-3. Changes in f_{Si} caused by the presence of 10 g free Al_2O_3 per kilogram sample, and equivalent amounts of other compounds causing the same effect on f_{Si} (Číčel, 1981).

| Sample | f_{Si} | | Amounts equivalent to 10 g Al_2O_3 kg^{-1} | | |
	Original	Original with 10 g Al_2O_3 kg^{-1}	Fe_2O_3	Kaolinite	Illite
			---------------------- g kg^{-1} -------------------		
Kuzmice	7.91	7.81	20	70	100
Branany	7.07	6.98	18	100	200
Garfield	6.91	6.83	17	120	300

& Guven, 1978). Once these assumptions do not hold, the results obtained by Method I, with a fixed number of O atoms per structural formula, provide more information about the type and amount of admixtures in the sample (Číčel, 1981). One must remember that the calculation of a structural formula from bulk-chemical data is only another way of representing the chemical composition of the sample in terms of a probable structure. To summarize:

1. Quantities of up to 10% of other layer silicates do not significantly affect the calculated distribution of central atoms in the structural formula.
2. Determination of free oxides is essential for correct structural formula calculation.
3. The sum of coefficients of atoms in octahedral coordination, $F_{(o)}$, can be used as a measure of the correctness of the structural formula. For dioctahedral 2:1 layer silicates, $F_{(o)}$ values between 3.96 and 4.05 are acceptable (Číčel, 1981).

STRUCTURAL FORMULAE DERIVED FROM ACID DISSOLUTION

Brindley and Youell (1951) used the technique of acid dissolution (ADT) to determine the central atom distribution in chlorite, and dissolution curves of smectites were reported by Karsulin and Stubican (1954) and Osthaus (1954, 1956). Číčel and Novák (1977) and Novák and Číčel (1978) described an improved method of ADT. They extracted 500-mg samples with $6M$ HCl at 96 °C for different time periods. After solid-phase separation the liquids were analyzed for Al, Fe, Mg, and Ca. The leaching of octahedral cations from the solid phase was monitored by following the gradual disappearance of the $Si_{(t)}$-O-$Al_{(o)}$ infrared band at 520 cm^{-1} (Číčel & Novák, 1977).

Číčel et al. (1990) described a method for structural formulae calculation from acid dissolution data. The method is based on the assumption introduced by Osthaus (1954), i.e., that the rate of dissolution of exchangeable cations is greater than that of octahedral cations, which, in turn, is greater than that of tetrahedral cations. This method gave reasonable structural formulae even in instances where calculation from bulk-chemical analysis failed because of the presence of substantial amounts of free oxides in the sample (Číčel et al., 1992, Madejová et al., 1992). There is no doubt about the high rate of dissolution of exchangeable cations. However, the validity of the assumption of substantially higher rates of dissolution for octahedral than for tetrahedral cations was investigated only very recently. The results of Luca and MacLachlan (1992) and Tkáč et al. (1993) on acid dissolution of smectites investigated by Mössbauer and MAS-NMR spectroscopies, respectively, showed that the acid treatment appears to remove the tetrahedral and octahedral cations from the smectite structure at about the same rate. The change of the slope of the dissolution curves after substantial portions of the cations were dissolved (Osthaus, 1954) is caused by the low smectite content remaining in the reaction mixture, and not by a change of sites from which the cations are being dissolved.

Portions of cations indicated as (s), ($o+t$), and (n) can be calculated from the dissolution curves (Fig. 4-2). Natural samples frequently contain readily soluble forms of Al and Fe oxides, and exchangeable Mg, Ca, K, and Na. Those are dissolved or exchanged for protons during the first few minutes of the dissolution process. This appears as a single step on the dissolution curve (Fig. 4-2). The readily soluble portion of the atom is given by the intercept on the 1-α axis at t = 0. This portion contains the exchangeable and all other readily soluble forms of the given atom (e.g., Fe bound in hydrated oxides such as goethite), but it could also include a portion of the octahedral cations (i.e., those which are readily soluble due to their position on the edges of the clay particles).

The readily soluble fraction of aluminum in most smectites is low, at most 5% of the total Al present in the sample (Čičel et al., 1992). Because of problems in quantifying Al from ^{27}Al MAS-NMR spectra of Fe-bearing materials (Morris et al., 1990), there is little, if any possibility of directly determining whether this readily soluble Al is bound in the smectite structure or as a free oxide.

Readily soluble Mg is also frequently found in homoionic forms (usually Ca or Na) of smectites that contain no exchangeable cations. Chen et al. (1990) reported the solubilization of octahedral Mg from Na-montmorillonite suspensions containing NaCl. Therefore, the readily soluble Mg in homoionic smectites could be bound in the smectite structure and/or in extraneous, readily soluble phases. Identification of the extraneous phases is difficult because they occur in such low concentrations (usually no more that 15% of the total Mg in the sample).

The situation for readily soluble Fe is similar to that for Mg, i.e., it can occur in the octahedral sites, and/or in extraneous, readily soluble phases. In contrast to Al and Mg, however, low-temperature Mössbauer spectroscopy is sensitive enough to provide information on Fe compounds present in the sample. Therefore the combined results of ADT and Mössbauer spectroscopy give reasonable compositions for readily soluble Fe phases. Komadel et al. (1993) showed a good correlation between the amounts of iron identified as goethite by Mössbauer spectroscopy and the amounts of readily soluble Fe extracted from the fine fractions of some Czech smectites.

Sometimes "insoluble" minerals ($t_{1/2}$ > 100 h) are present in the sample as well, for which the curve of 1-α vs. t stabilizes at a positive value. Absence of the most sensitive infrared band for octahedral cations, $Si_{(t)}$-O-$Al_{(o)}$ at about 520 cm^{-1}, is used to identify the dissolution time, (t_n), after which all smectite has dissolved and only "insoluble" minerals remain. The portions of cations bound in these minerals, (n), are calculated from the chemical analysis of the liquid phase at t_n. The most common "insoluble" phases that occur with smectites are kaolinite, quartz, anatase, and volcanic glass.

The $o+t$ fractions of atoms are obtained by subtracting $s+n$ from one. Magnesium is assumed to have no tetrahedrally coordinated atoms. Presence of tetrahedral Fe, and the Fe t:o ratio, can be estimated from Mössbauer spectra. In the absence of Mössbauer data, the common assumption for montmorillonites is that Fe occurs only in the octahedral sites. From knowledge of the amounts of octahedral Mg and Fe, as well as tetrahedral Fe, the distribution of Al between octahedral and tetrahedral sites can be calculated by computer optimization of $F_{(o)}$ = 4.00. The structural-formula coefficient for silicon is then derived from the condition F_t = 8.00. A sample calculation of the structural formula for SWy-1 montmorillonite from acid-dissolution data is given in the Appendix.

Advantages and Limitations of the Acid-Dissolution Technique

The ADT provides the quantitative distribution of atoms bound in various operationally defined fractions of the sample and, in particular, it enables an estimation of the fractions of atoms bound in smectite phases. However, more accurate results are obtained when ADT is used in combination with spectroscopic techniques.

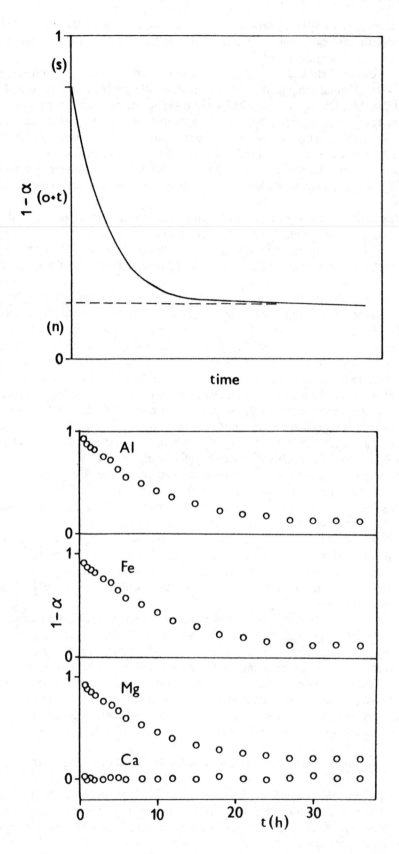

Fig. 4-2. Acid dissolution curves for structural cations in layer silicates: (top) typical dissolution
curve of an octahedral cation, and its fractions (s), (o+t), and (n); (bottom) dissolution
curves of SWy-1 montmorillonite in 6M HCl at 95 °C.

The ADT is unsuitable for mixtures of layer silicates and mixed-layer minerals. The actual procedure prevents its application to nonswelling dioctahedral layer silicates such as kaolinite, mica, and some chlorites, because of the long dissolution time ($t_{1/2} > 20$ h) required for these minerals. Vermiculites, illites, swelling chlorites, trioctahedral micas and chlorites have not been tested. The method is time consuming. The amount of sample needed is 5 to 10 g, plus the amount for chemical analysis.

Present methods do not allow distinctions to be made between octahedrally bound, but readily soluble, portions of Al and Mg in smectites and these elements in other readily soluble phases. The ADT is also unable to directly distinguish between octahedral and tetrahedral aluminum, but the problem is partly solved in the algorithm for calculating the structural formula by assuming that the sum of negative charges in octahedra and tetrahedra equals the charge of exchangeable cations, and the sum of occupied tetrahedral and octahedral positions is 8.00 and 4.00, respectively.

Despite these limitations, the ADT is the only method that provides an estimate of the site distribution of several atoms bound in the structure of smectites. Its application to other minerals remains open.

INFRARED IDENTIFICATION OF OCTAHEDRAL CATIONS

Slonimskaya et al. (1986) were the first to apply IR spectroscopy to structural formula analysis, using celadonites and glauconites. Madejová (1990) and Madejová et al. (1992) recently extended the technique to smectites. The method is based on the principle that the vibrational stretching frequency (ν) of the structural OH groups in dioctahedral 2:1 layer silicates varies with the two octahedral cations to which the OH is coordinated. These OH-stretching frequencies occur in the region 3400 to 3700 cm^{-1}. The integrated intensity of each band is given by the number of proper absorption centers, and by the absorption coefficients. Then the distribution of central atoms in octahedral sites can be calculated from the experimental values of integrated intensities if the assignment of the individual absorption bands is known. These assignments are summarized in

Table 4-4. OH-stretching frequencies and coordinating pairs of central atoms in octahedra for dioctahedral micas (Slonimskaya et al., 1986) and smectites (Madejová, 1990).

Micas		Smectites	
ν	Atom pair	ν	Atom pair
cm^{-1}		cm^{-1}	
		3681	AlMg†
3640	AlAl	3636	AlAl
3620	AlAl	3619	AlAl
3605	MgAl	3602	MgAl
3580	Fe^{2+}Al	3582	Fe^{3+}Al
3560	MgFe^{3+}	3563	MgFe^{3+}
		3555	Fe^{3+}Fe^{3+}
3545	Fe^{3+}Fe^{3+}		
3534	Fe^{3+}Fe^{3+}	3535	Fe^{3+}Fe^{3+}
3528	Fe^{2+}Fe^{3+}	3525	Fe^{3+}Fe^{3+}
3505	MgMg		
3495	Fe^{2+}Fe^{2+}		

†Assigned by Farmer (1974).

Table 4-4 for dioctahedral micas (Slonimskaya et al., 1986) and smectites (Madejová, 1990). Samples were prepared using the alkali-halide pressed-disk technique. Pressed disks were heated at 170 to 200 °C to remove most of the adsorbed water. The IR spectra of leucophyllite and montmorillonite and their deconvolution into individual absorption bands are given in Fig. 4-3.

Quantitative evaluation after Slonimskaya et al. (1986) is as follows. Because each OH group coordinates to two central atoms, 100% of the integral intensity represents 200% of the concentration of octahedral central atoms. The concentrations are expressed in the form of structural formula coefficients, so

$$f_{Al} + f_{Fe(III)} + f_{Fe(II)} + f_{Mg} = 2 \qquad [12]$$

for $O_{10}(OH)_2$. Then the amount, f_i, of each central atom in octahedral sites is deduced from that atom's contribution, W_{ij}, to the integral intensity of the bands associated with structural OH in the coordination sphere of that atom. For example,

$$f_{Al} = 2W_{Al-Al} + W_{Al-Fe(III)} + W_{Al-Fe(II)} + W_{Al-Mg}. \qquad [13]$$

Slonimskaya et al. (1986) compared the structural-formula coefficients obtained from IR spectra and those from chemical analysis, and found agreement to within 0.05 atomic units per half-cell, i.e., within 2.5%. Madejová (1990) studied the variability in the IR determinations of f_i for Al and Fe(III) in a montmorillonite from Jelšový Potok (CSFR) and obtained standard deviations between 0.04 and 0.08. These values were between 1 and 2% of $F_{(o)}$, but for individual atoms (i.e., f_i) the relative error was on the order of 10% and the 95% confidence interval was ± 0.08 atomic units.

A comparison of central-atom distributions in the octahedral sheets of some 2:1 layer silicates, as calculated from bulk chemical analyses, ADT, and IR spectra (Table 4-5), reveals a general qualitative agreement between the IR method and chemical methods. Correction of structural formulae by adjusting the results of the chemical method to be in better agreement with results from the IR method, however, is not generally recommended (Slonimskaya et al., 1986).

Advantages and Limitations of the Infrared Method

The distribution of central atoms in octahedra calculated from an IR spectrum is given exclusively by the short-range ordering in the structure of the layer silicate. This is the dominant advantage of the method. The composition of the octahedral sheet can be obtained directly from the spectrum and only a small amount of sample (less than 10 mg) is needed.

This method has yet to be applied to minerals other than 2:1 layer silicates. Most clay minerals absorb in the range 3700 to 3500 cm^{-1}. Therefore, this method cannot be applied to a mixture of clay minerals to distinguish between them. Admixtures of other minerals common in fine fractions of bentonites, such as quartz, cristobalite, anatase, goethite, carbonates, and volcanic glass, do not interfere in this method.

DETERMINATION OF TETRAHEDRAL ALUMINUM BY MAGIC-ANGLE SPINNING NUCLEAR MAGNETIC RESONANCE SPECTROSCOPY

The spectral parameters of nuclear magnetic resonance (NMR) are sensitive to the local environment of the coordination polyhedra of the studied nuclei. The most important among these parameters, the chemical shift, depends on the coordination number of the central atom (e.g., AlO_n and/or SiO_n), and on the kind and number of atoms joined directly with it. These general features of NMR, combined with the sophisticated magic-angle spinning (MAS) technique, experimental procedures, and mathematical treatment, permit the determination of tetrahedral to octahedral Al ratios $[Al_{(t)}/(Al_{(t)} + Al_{(o)})]$, and, consequently, $f_{Al(t)}$ from ^{27}Al MAS-NMR spectra (Woessner, 1989). Excellent agreement between the predicted and calculated tetrahedral Al contents was obtained by Nadeau et al. (1985) for Unterruptsroth beidellite using the $Al_{(t)}/Si$ ratio calculated from the formula

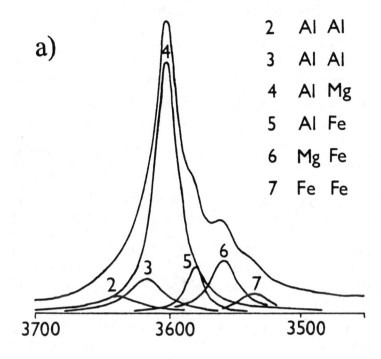

a)

2	Al	Al
3	Al	Al
4	Al	Mg
5	Al	Fe
6	Mg	Fe
7	Fe	Fe

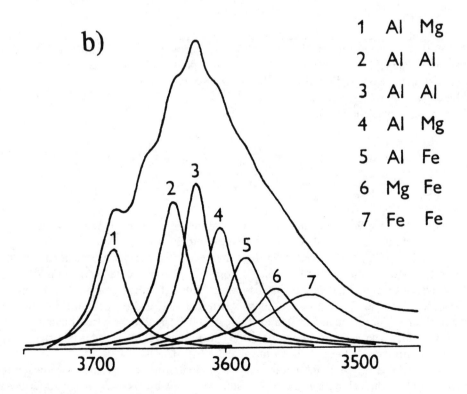

b)

1	Al	Mg
2	Al	Al
3	Al	Al
4	Al	Mg
5	Al	Fe
6	Mg	Fe
7	Fe	Fe

Fig. 4-3. Deconvolution of IR spectra of *a*) leucophyllite (Slonimskaya et al., 1986) and *b*) montmorillonite (Madejova et al., 1992).

Table 4-5. Comparison of structural formulae calculated from bulk chemical analyses (bulk), acid dissolution data (acid), and infrared spectroscopic data (IR).

		Tetrahedral ions		Octahedral ions				
Sample†	Method	Si	Al	Al	Fe³⁺	Fe²⁺	Mg	$F_{(o)}$
1	Bulk	7.88	0.12	0.10	2.30	0.72	0.82	3.94
	IR			0.14	2.28	0.72	0.86	
2	Bulk	7.88	0.12	2.20	0.34	0.14	1.28	3.96
	IR			2.14	0.38	0.10	1.38	
3	Bulk	7.28	0.72	0.16	2.16	0.24	1.32	3.88
	IR			0.50	1.92	0.26	1.32	
4	Bulk	7.70	0.30	2.99	0.38	--	0.63	4.00
	Acid	7.79	0.21	3.14	0.34	--	0.52	
	IR‡			3.06	0.42	--	0.52	
5	Bulk	7.95	0.05	3.07	0.40	--	0.49	3.96
	Acid	7.68	0.32	3.06	0.45	--	0.49	
	IR¶			3.08	0.45	--	0.47	
6	Bulk	7.28	0.72	1.53	1.92	--	0.59	4.04
	Acid	7.24	0.76	2.03	1.49	--	0.48	
	IR¶			2.04	1.45	--	0.51	

†1, Celadonite 69G (Slonimskaya et al., 1986); 2, leucophyllite 31 (Slonimskaya et al., 1986); 3, glauconite 655 (Slonimskaya et al., 1986); 4, montmorillonite, Jelšový Potok, CSFR; 5, montmorillonite SWy-1, Wyoming; 6, beidellite, Rokle, CSFR.
‡Madejová (1990).
¶Madejová, unpublished data.

$$\frac{2Si^2 + (Si^1 - 4Si^2)/3}{Si^0 + Si^1 + Si^2} \qquad [14]$$

where Si^0, Si^1, and Si^2 represent the intensities of the NMR peaks of Si atoms with, respectively, zero, one, and two tetrahedral Al sites as nearest neighbors.

Woessner (1989) was correct when he commented that the disagreement in several samples (Table 4-6) between the formulae calculated from bulk-chemical data and from NMR data is most likely due to unrepresentative structural formulae from chemical analyses. As noted by Číčel (1981), samples with $F_{(o)} < 4.00$ probably contain free SiO_2 and/or some minerals rich in Si. The data for Umiat, Alaska, montmorillonite (Table 4-6) indicate the presence of about 30 g free silica per kilogram. The corresponding "decrease" in tetrahedral Al is approximately -0.10. The calculated difference between bulk-chemical and NMR data is -0.19.

The calculated $F_{(o)} > 4.00$ for Almeria, Spain, montmorillonite (Table 4-6 indicates free Al_2O_3 and/or Fe_2O_3. The value for $F_{(o)}$ of 4.21 could be caused by about 70 g free Al_2O_3 per kilogram

Table 4-6. Comparison of tetrahedral Al contents ($f_{Al(t)}$) of several montmorillonites obtained from chemical analyses (CA) and from ^{27}Al MAS-NMR data at 11.74 T (Woessner, 1989).

Sample	$f_{Al(t)}$			
	CA	NMR	Δ†	$F_{(o)}$
Tatatilla	0.20	0.23	- 0.03	4.01
Crook County	0.12	0.19	- 0.07	4.00
Arizona	0.21	0.00	+ 0.21	4.00
Clay Spur	0.32	0.32	0.00	4.00
Belle Fourche	0.20	0.22	- 0.02	4.04
Polkville	0.00	0.00	0.00	4.00
Umiat	0.16	0.35	- 0.19	3.96
Otay	0.05	0.00	+ 0.05	4.00
Almeria	0.66	0.14	+ 0.52	4.21

†Δ= CA - NMR

of about 100 g free Fe_2O_3 per kilogram. The expected "calculated" increase of $f_{Al(t)}$ is approximately 0.45. The measured difference is 0.52. For Apache County, Arizona, montmorillonite, $F_{(o)}$ equals 4.00 and $f_{Al(t)}$ equals 0.21 by chemical analysis, whereas $f_{Al(t)}$ by NMR is 0.00. This can be explained by the combination of effects of free oxides present in the sample. The presence of 10 g free SiO_2 per kilogram increases f_{Si} by 0.04, while the same concentration of free Al_2O_3 decreases f_{Si} by -0.07 to -0.09. In other words, the effect of free Al_2O_3 is opposite and about twice as strong as the effect of the same amount of free SiO_2. Therefore, simultaneous presence of "x" g of free Al_2O_3 and "2x" g of free SiO_2 results in the same f_{Si} and $f_{Al(t)}$ as when these phases are absent.

No data on the reproducibility of $f_{Al(t)}$ from NMR spectra were given by either Goodman and Stucki (1984), Nadeau et al. (1985), or Woessner (1989). But an average rms (root mean square) error of 7% was reported by Kinsey et al. (1985).

Increasing Fe content leads to increasing NMR signal loss. Morris et al. (1990) reported recently that up to 35% of the Al present in SWy-1 montmorillonite does not give rise to an NMR signal. The presence of a ^{27}Al NMR signal in an Fe-containing clay depends upon the distance between Fe and Al. In essence, all Al atoms within 0.6 nm of Fe atoms will give signals so broad that they will be not observed. Accordingly, in an Fe-containing clay some Al atoms are not seen. The presence of ^{27}Al NMR signals from montmorillonites containing high quantities of Fe thus appears to be a consequence of separate domains of Fe-substituted and Al-substituted octahedral sites (Morris et al., 1990).

Advantages and Limitations of Magic-Angle Spinning
Nuclear Magnetic Resonance Spectroscopy

The principal advantage to using MAS-NMR for structural formula analysis is that the quantity of tetrahedral Al can be obtained directly and, if Si and Al are the only tetrahedral central atoms, the composition of the tetrahedral sheet can be calculated. Among the other advantages, a relatively small amount of sample is needed (<1 cm^3) and the method is nondestructive.

A major disadvantage is the specialized and expensive equipment required, which includes the need for high magnetic fields and high-speed sample spinning. The relative contributions to the tetrahedral-Al signal by the various 2:1 layer silicates present in a sample mixture are not discernible. In Fe-containing clays some Al atoms are not seen, and increasing Fe content leads to increasing signal loss.

DISTRIBUTION OF IRON FROM IRON-57 MÖSSBAUER SPECTROSCOPY

Iron-57 Mössbauer spectroscopy is frequently used for structural studies of clay minerals (Amonette et al., 1994; see Chapter 3). It allows distinction among Fe atoms in different oxidation states, coordination geometries, and site symmetries. All this information is valuable for structural-formula calculation.

Some of the Fe present in nominally monomineralic layer-silicate specimens has been found to actually occur in oxide or hydroxide minerals present in the specimens rather than in the layer-silicate structure (see, for example, Goodman & Nadeau, 1988; Goodman et al., 1988; Lear et al., 1988; Murad, 1988; Vandenberghe et al., 1990). Murad (1987) found between 2.9 and 13% of the Fe in three nontronite samples to be in the form of goethite. For example, the Garfield, Washington, nontronite studied by Murad contained 4.6% of the total Fe as goethite, and 5.8% of the layer-silicate Fe in tetrahedral coordination. Taking these data into consideration, he modified the structural formula given by Stucki et al. (1984) as

$$[Si_{7.12} Al_{0.88}][Al_{0.19} Fe_{3.72} Mg_{0.11}]Na_{0.93} O_{20} (OH)_4 \qquad [15]$$

to be

$$[Si_{7.20} Al_{0.59} Fe_{0.21}][Al_{0.49} Fe_{3.39} Mg_{0.11}]Na_{0.94} O_{20} (OH)_4 . \qquad [16]$$

An acceptable correlation has been found between the readily acid-soluble fraction of Fe, Fe(s), and the Fe bound in goethite, Fe(g). These values were, respectively, 65 and 67% for a Strimice (CSFR) beidellite, and 67 and 70% for a Cerný Vrch (CSFR) montmorillonite (Komadel et al., 1993).

Mössbauer spectra of 2:1 phyllosilicates often give evidence of tetrahedral Fe, but reported amounts sometimes vary widely for the same mineral. Cardile (1989) reviewed the data for smectites, and found that the amount of tetrahedral Fe reported was affected by the method of clay preparation, the sedimentation method, and the interlayer cation. If Mössbauer spectroscopy is used to estimate Fe_t content, some standard procedures must be adopted to allow comparisons to be carried out (Cardile, 1989; Rancourt et al., 1992).

Some studies reported good agreement between Mössbauer spectra and chemical analysis for Fe(II)/total Fe ratios. About 85% Fe(II) by Mössbauer, and about 92% by chemical analysis were found in reduced NG-1 nontronite, Germany (Komadel et al., 1990). The respective values for reduced SWa-1 nontronite, Washington, were 88 and 95%, and for reduced Garfield nontronite, Washington, 82 and 90%. The consistently lower values obtained from Mössbauer spectra could be caused by partial reoxidation of the samples when they were transferred from the freeze-dryer to sample holder, and/or by non-equal recoil-free fractions of Fe(II) and Fe(III).

Iron(II) in smectites is also readily soluble in HCl. Studies in our laboratory with the Hroznetín smectite (CSFR) have yielded Fe(II):Fe(total) ratios of 0.45, 0.39-0.41, and 0.40 for analyses by acid dissolution (i.e., $Fe_{(s)}$), by wet chemistry, and by Mössbauer spectrometry.

Advantages and Limitations of Mössbauer Spectroscopy

The main advantage of Mössbauer spectroscopy for structural formula determinations is the information about Fe distribution in the sample. Sample preparation is simple, and small amounts of sample (usually less than 100 mg) are needed. The technique is very sensitive. The identification of magnetically ordered Fe oxides is possible even at concentrations below 10 g kg^{-1}.

Among its limitations are the lack of quantification of tetrahedral Fe, as described above. For exact calculations the recoil-free fractions of different Fe sites are needed. And, the accumulation of spectra is slow (several hours or days).

FUTURE DEVELOPMENTS

Because of large uncertainties in the distribution of atoms within layer silicates, and among layer silicates and the free oxide phases that may be associated with them, the introduction of new methods of calculation is unlikely to improve the quality of the structural formulae that are estimated. This improvement is more likely to come from the incorporation of phase- and site-specific information about the distribution of atoms in the specimen. To obtain this supplemental information, the specimen may need to be analyzed by various spectroscopic (e.g., FTIR, MAS-NMR, Mössbauer, and x-ray absorption), microscopic (e.g., HRTEM, atomic force), and selective dissolution techniques in addition to the standard bulk elemental techniques. We expect the major development in structural formula estimation, therefore, to be an increase in the use of phase- and site-specific data, obtained from a variety of specialized analyses, to constrain and improve the formulae calculated using bulk elemental data.

SUMMARY

The most frequently used structural formulae are calculated from the chemical analyses of the samples studied. Formulas produced in this manner are the results of a *calculation,* not a *determination*, of the probable distribution of central atoms within the structure. The accuracy of the results depends on the fulfilment of several assumptions, and is negatively affected by the presence of other minerals in the sample.

No single method is available for the *determination* of the distribution of all, or even most, of the central atoms in the structure of clay minerals. However, a combination of acid-dissolution, infrared, MAS-NMR, and Mössbauer techniques provides information on even small amounts of admixtures in the sample, and on the distribution of most atoms present. Structural formulae, calculated using data from these techniques, are closer to the real composition of the mineral, than the structural formulae calculated from bulk-chemical data.

REFERENCES

Amonette, J.E., F.A. Khan, H. Gan, J.W. Stucki, and A.D. Scott. 1994. Quantitative oxidation-state analysis of soils. p. 83-113. In J.E. Amonette and L.W. Zelazny (ed.) Quantitative methods in soil mineralogy. SSSA Misc. Publ. SSSA, Madison, WI.

Bates, T.F. 1959. Morphology and crystal chemistry of 1:1 layer silicates. Am. Mineral. 44:78-114.

Berry, F.J., M.H.B. Hayes, and S.L. Jones. 1986. Investigations of intercalation in inorganic solids with layered structures: Iron-57 Mössbauer spectroscopy studies of size fractionated and iron-exchanged montmorillonite clays. Inorg. Chim. Acta. 122:19-24.

Borneman-Starynkevic, I.D. 1964. How to calculate formulas of minerals. (In Russian.) Nauka, Moscow, USSR.

Brindley, G.W., and R.F. Youell. 1951. A chemical determination of the tetrahedral and octahedral aluminum in a silicate. Acta Crystallogr. 4:495-497.

Cardile, C.M. 1989. Tetrahedral iron in smectite: a critical comment. Clays Clay Miner. 37:185-188.

Chen, J.S., J.H. Cushman, and P.F. Low. 1990. Rheological behavior of Na-montmorillonite suspensions at low electrolyte concentration. Clays Clay Miner. 38:57-62.

Číčel, B. 1981. The influence of mineral impurities on the calculated statistical distribution of atoms in unit cell formulae of smectites. p.35-40. In J. Konta (ed.) Proc. 8th Conf. Clay Mineral. and Petrol., Teplice. 1979. Charles Univ., Prague, Czechoslovakia.

Číčel, B., P. Komadel, E. Bednáriková, and J. Madejová. 1992. Mineralogical composition and distribution of Si, Al, Fe, Mg and Ca in the fine fractions of some Czech and Slovak bentonites. Geol. Carpathica, Ser. Clays 43:3-7.

Číčel, B., P. Komadel, and J. Hronský. 1990. Dissolution of the fine fraction of Jelšový Potok bentonite in hydrochloric and sulphuric acids. Ceramics-Silikaty 34:41-48.

Číčel, B., I. Novák, and I. Horváth. 1981. Mineralogy and crystal chemistry of clays. (In Slovak.) Veda, Bratislava, Czechoslovakia.

Číčel, B., I. Novák, and F. Pivovarnček. 1965. Dissolution of montmorillonite in HCl and its application. (In Slovak.) Silikaty 9:130-139.

Číčel, B., and I. Novák. 1977. Dissolution of smectites in hydrochloric acid--I. Half-time of dissolution as a measure of reaction rate. p.163-171. In J.Konta (ed.) Proc. 7th Conf. Clay Mineral. and Petrol,, Karlovy Vary. 1976. Charles Univ., Prague, Czechoslovakia.

Farmer, V.C. 1974. The layer silicates. p. 331-363. In V.C. Farmer (ed.) The infrared spectra of minerals. Miner. Soc., London.

Giaramita, M.J., and H.W. Day. 1990. Error propagation in calculations of structural formulas. Am. Mineral. 75:170-182.

Goodman, B.A., and P.H. Nadeau. 1988. Identification of oxide impurity phases and distribution of structural iron in some diagenetic illitic clays as determined by Mössbauer spectroscopy. Clay Miner. 23:301-308.

Goodman, B.A., P.H Nadeau, and J. Chadwick. 1988. Evidence for the multiphase nature of bentonites from Mössbauer and EPR spectroscopy. Clay Miner. 23:147-159.

Goodman, B.A., and J.W. Stucki. 1984. The use of nuclear magnetic resonance (NMR) for the determination of tetrahedral aluminum in montmorillonite. Clay Miner. 19:663-667.

Gregor, M., and B. Číčel. 1969. Bentonite and its use. (In Slovak.) Publ. House of the Slovak Acad. Sci., Bratislava, Czechoslovakia.

Grim, R.E., and N. Guven. 1978. Bentonites. Elsevier, Amsterdam, the Netherlands.

Grim, R.E., and G. Kulbicki. 1961. Montmorillonite: high temperature reactions and classification. Am. Mineral. 46:1329-1369.

Gruner, J.W. 1932a. Structure of kaolinite. Z. Krystallogr. 83:75-78.

Gruner, J.W. 1932b. Structure of dickite. Z. Krystallogr. 83:394-404.

Gruner, J.W. 1934. Structures of vermiculites and their collapse by dehydration. Am. Mineral. 19:557-575.

Hofmann, U., K. Endell, and D.Wilm. 1933. Structure und Quellung von Montmorillonit. Z. Krystallogr. 86:340-348.

Hofmann, U., A. Weiss, G. Koch, A. Mehler, and A. Scholz. 1956. Intracrystalline swelling, cation exchange and anion exchange of minerals of the montmorillonite group and of kaolinite. Clays Clay Miner. 4:273-287.

Jackson, W.W., and J. West. 1930. The crystal structure of muscovite $KAl_2(AlSi_3)O_{10}(OH)_2$. Z. Krystallogr. 76:211-227.

Jackson, W.W., and J. West. 1933. The crystal structure of muscovite $KAl_2(AlSi_3)O_{10}(OH)_2$. Z. Krystallogr. 85:160-164.

Karsulin, M., and V.I. Stubican. 1954. Uber die Structur und die Egenschaften syntetischer Montmorillonite. Monatsh. Chem. 85:343-358.

Kazancev, V.P. 1934. Structure and properties of vermiculite. (In Russian.) Zap. Vseros. Min. Obsc. Ser.2. 63:464-480.

Kelley, W.P. 1945. Calculating formulas for the fine grained minerals on the basis of chemical analysis. Am. Mineral. 30:1-26.

Kelley, W.P. 1955. Interpretation of chemical analysis of clays. Clays Clay Miner. 1:92-94.

Kinsey, R.A., R.J. Kirkpatrick, J. Hower, K.A. Smith, and E. Oldfield. 1985. High resolution aluminium-27 and silicon-29 nuclear magnetic resonance spectroscopic study of layer silicates, including clay minerals. Am. Mineral. 70:537-548.

Komadel, P., P.R. Lear, and J.W. Stucki. 1990. Reduction and reoxidation of nontronite: Extent of reduction and reaction rates. Clays Clay Miner. 38:203-208.

Komadel, P., J.W. Stucki, and B. Číčel. 1993. Readily HCl-soluble iron in the fine fractions of some Czech bentonites. Geol. Carpathica, Ser. Clays 44:11-16.

Konta, J. 1957. Clay minerals of Czechoslovakia. (In Czech.) Nakl. Cs. Akad. Ved, Prague, Czechoslovakia.

Köster, H.M. 1977. Strukturformeln und Kationenladungen von 2:1 Schichtsilikaten. Clay Miner. 12:45-54.

Lear P.R., P. Komadel, and J.W. Stucki. 1988. Mössbauer spectroscopic identification of iron oxides in nontronite from Hohen Hagen, Federal Republic of Germany. Clays Clay Miner. 36:376-378.

Luca V., and D.J. MacLachlan. 1992. Site occupancy in nontronite by acid dissolution and Mössbauer spectroscopy. Clays Clay Miner. 40:1-7.

Madejová J. 1990. IR spectroscopic study of the structure of dioctahedral smectites. Ph.D. diss. Inst. Inorg. Chem., Slovak Acad. of Sci., Bratislava, Czechoslovakia.

Madejová J., P. Komadel, and B. Číčel. 1992. Infrared spectra of some Czech and Slovak smectites and their correlation with structural formulas. Geol. Carpathica, Ser. Clays 43:9-12.

Marshall, C.E. 1935. Layer lattice and base-exchange clays. Z. Krystallogr. 91:433-449.

McConnell, D. 1951. The crystal chemistry of the montmorillonite II. Clay Miner. Bull. 1:179-188.

Morris, H.D., S. Bank, and P.D. Ellis. 1990. ^{27}Al NMR spectroscopy of iron-bearing montmorillonite clays. J. Phys. Chem. 94:3121-3129.

Murad, E. 1987. Mössbauer spectra of nontronites: structural implications and characterization of associated iron oxides. Z. Pflanzenernaehr. Bodenkd. 150:279-285.

Murad, E. 1988. Properties and behavior of iron oxides as determined by Mössbauer spectroscopy. p. 309-350. In J.W. Stucki et al. (ed.) Iron in soils and clay minerals. D. Reidel, Dordrecht, the Netherlands.

Nadeau, P.H., V.C. Farmer, W.J. McHardy, and D.C. Bain. 1985. Compositional variations of the Unterrupsroth beidellite. Am. Mineral. 70:1004-1010.

Novák, I., and B. Číčel. 1978. Dissolution of smectites in hydrochloric acid--II. Dissolution rate as a function of crystallochemical composition. Clays Clay Miner. 26:341-344.

Nyrkov, A.A. 1961. Methods of structural formulae calculation of hydromicas. (In Russian.) Mineral. Sbor. Lvov. Geol. Obsc. 15:386-395.

Osthaus, B.B. 1954. Chemical determination of tetrahedral ions in nontronite and montmorillonite. Clays Clay Miner. 2:404-417.

Osthaus, B.B. 1956. Kinetic studies on montmorillonites and nontronite by acid dissolution technique. Clays Clay Miner. 4:301-321.

Pauling, L. 1930a. Structure of micas and related minerals. Proc. Natl. Acad. Sci. USA 16:119-123.

Pauling, L. 1930b. Structure of the chlorites. Proc. Natl. Acad. Sci. USA 16:578-582.

Rancourt, D.G., M.Z. Dang, and A.E. Lalonde. 1992. Mössbauer spectroscopy of tetrahedral Fe^{3+} in trioctahedral micas. Am. Mineral. 77:34-43.

Robertson, H.E., A.H. Weir, and R.D. Wood. 1968. Morphology of particles in size fractionated Na-montmorillonites. Clays Clay Miner. 16: 239-247.

Ross, C.S., and S.B. Hendricks. 1945. Minerals of the montmorillonite group. U.S. Geol. Surv. Pap. 205(B). U.S. Govt. Print. Office, Washington, DC.

Slonimskaya, M.V., G. Besson, L.G. Daynyak, C. Tchoubar, and V.A. Drits. 1986. Interpretation of the IR spectra of celadonites and glauconites in the region of OH-stretching frequencies. Clay Miner. 21:377-388.

Stevens, R.E. 1946. A system for calculating analyses of micas and related minerals to end members. U.S. Geol. Surv. Bull. 950:101-119.

Stucki, J.W., D.C. Golden, and C.B. Roth. 1984. Effects of reduction and reoxidation of structural iron on the surface charge and dissolution of dioctahedral smectites. Clays Clay Miner. 32:350-356.

Tkáč, I., P. Komadel, and D. Müller. 1993. Acid treated montmorillonites - a ^{29}Si and ^{27}Al MAS NMR study. Clay Miner. (In press.)

Vandenberghe R.E., E. De Grave, C. Landuydt, and L.H. Bowen. 1990. Some aspects concerning the characterization of iron oxides and hydroxides in soils and clays. Hyp. Int. 53:175-196.

Woessner, D.E. 1989. Characterization of clay minerals by ^{27}Al nuclear magnetic resonance spectroscopy. Am. Mineral. 74:203-215.

APPENDIX

Definitions of Symbols and Values

A_i concentration of ith oxide in the system

B $$\sum_{i=1}^{n} \frac{A_i V_i}{E_i}(o) + \sum_{i=1}^{n} \frac{A_i V_i}{E_i}(ex) - \sum_{i=1}^{n} \frac{A_i V_i}{E_i}(t)$$

C $$\sum_{i=1}^{n} \frac{A_i}{E_i}(o) + \sum_{i=1}^{n} \frac{A_i}{E_i}(t)$$

E_i formula mass of the oxide of ith element divided by number of metal atoms in the formula

f_i coefficient of ith element in structural formula

F sum of f_i

$F_{(o)}$ sum of f_i in octahedral sheet in structural formula

$F_{(t)}$ sum of f_i in tetrahedral sheet in structural formula

Fe(g)   iron bound in goethite

K number of central atoms per structural formula

M conversion factor, $M = O/N$, $M = K/C$, or $M = O(o)/B$

N $$\sum_{i=1}^{n} \frac{A_i V_i}{E_i}$$

O number of negative charges per structural formula

$O(o)$ number of negative charges per octahedral sheet in structural formula

V_i formal valency of ith element

α mass fraction of oxide (central atom) dissolved

t time

t_N time of dissolution after which is all smectite dissolved

$t_{1/2}$ half-time of dissolution (t for which $\alpha = 0.5$)

(s)   part of sample readily soluble in acid

(o)   part of sample bound in octahedra

(t)   part of sample bound in tetrahedra

(ex)   exchangeable cations

(n)   insoluble part of sample ($t_{1/2} > 100$ h)

W_{ij} integral intensities of IR absorption bands

Calculations of Structural Formulae of Layer Silicates

Assuming Fixed Number of Oxygen Atoms

Horní Bříza (CSFR) Kaolinite

Chemical Analysis (Konta, 1957):

Oxide	A_i	A_i/E_i	$(A_i/E_i)(V_i)$	f_i†
	g kg^{-1}			
SiO_2	447.6	7.452	29.810	3.93
Al_2O_3	392.5	7.701	23.102	0.07
				3.99
MgO	3.6	0.089	0.179	0.05

$$N = 53.091 \quad F = 8.04$$

$O = 28$

† $f_i = (A_i/E_i)M$ where $M = O/N = 0.52740$

Structural Formula: $[Si_{3.93} Al_{0.07}][Al_{3.99} Mg_{0.05}]O_{10}(OH)_8$

$F_{(o)} = 4.04$, $F_{(t)} = 4.00$

Jelšový Potok (CSFR) Montmorillonite

Chemical Analysis:

Oxide	A_i	A_i/E_i	$(A_i/E_i)(V_i)$	f_i†
	g kg^{-1}			
SiO_2	591.3	9.841	39.365	7.70
Al_2O_3	214.1	4.200	12.599	0.30
				2.99
Fe_2O_3	38.5	0.482	1.446	0.38
MgO	32.5	0.806	1.613	0.63
CaO	31.1	0.555	1.109	0.43
K_2O	3.2	0.068	0.068	0.05

$$N = 56.200 \quad F = 12.48$$

$O = 44$

† $f_i = (A_i/E_i)M$ where $M = O/N = 0.78292$

Structural formula: $[Si_{7.70} Al_{0.30}][Al_{2.99} Fe_{0.38} Mg_{0.63}][Ca_{0.43} K_{0.05}]O_{20}(OH)_4$

$F_{(o)} = 4.00$, $F_{(t)} = 8.00$

Assuming Fixed Number of Central Atom Positions

Horní Bříza (CSFR) Kaolinite

Chemical Analysis (Konta, 1957):

Oxide	A_i	A_i/E_i	f_i†
	g kg⁻¹		
SiO_2	447.6	7.452	3.91
Al_2O_3	392.5	7.701	0.09
			3.95
MgO	3.6	0.089	0.05
	$C =$	15.242	

$K = 8$

† $f_i = (A_i/E_i)M$ where $M = K/C = 0.52487$

Structural Formula: $[Si_{3.91} Al_{0.09}][Al_{3.95} Mg_{0.05}]O_{10}(OH)_8$

Jelšový Potok (CSFR) Montmorillonite

Chemical Analysis:

Oxide	A_i	A_i/E_i	f_i†
	g kg⁻¹		
SiO_2	591.3	9.841	7.70
Al_2O_3	214.1	4.200	0.30
			2.99
Fe_2O_3	38.5	0.482	0.38
MgO	32.5	0.806	0.63
CaO	31.1	----(0.555)	0.43
K_2O	3.2	----(0.068)	0.05
	$C =$	15.329	

$K = 12$

† $f_i = (A_i/E_i)M$ where $M = K/C = 0.78283$

Structural Formula: $[Si_{7.70} Al_{0.30}][Al_{2.99} Fe_{0.38} Mg_{0.63}][Ca_{0.43} K_{0.05}]O_{20}(OH)_4$

From Acid Dissolution Data

The structural formula for Ca-saturated SWy-1 montmorillonite (Wyoming, USA) is calculated from acid dissolution data (Fig. 4-2) assuming equal dissolution rates for octahedral and tetrahedral cations. The readily soluble portions, $Al_s = 0.05$, $Fe_s = 0.06$, and $Mg_s = 0.05$, are considered to be bound in the octahedra, probably on the edges of the particles. No Fe oxides were identified in SWy-1 by Mössbauer spectroscopy (Berry et al., 1986). The insoluble portions, $Al_n = 0.13$, $Fe_n = 0.11$, and $Mg_n = 0.21$, were determined from the analysis of the liquid phase after 36 h dissolution. $Al_{o+t} = 1 - 0.13 = 0.87$, $Fe_t = 0$, $Fe_o = 1 - 0.11 = 0.89$, $Mg_o = 1 - 0.21 = 0.79$. $Ca_{ex} = 0.99$, $Ca_n = 0.01$.

Component	A_i	A_i/E_i	$A_i V_i/E_i$	f_i†
	g kg^{-1}			
SiO_2	670.5			7.68
Al_2O_3 (t)	18.4	0.361		0.32
Al_2O_3 (o)	175.3	3.439	10.317	3.06
Fe_2O_3 (o)	40.2	0.503	1.509	0.45
MgO(o)	22.0	0.546	1.092	0.49
CaO(ex)	25.4	0.453	0.906	0.40

$$\text{Sum} = 13.824$$
$$- \quad 0.361$$
$$B = 13.463$$

$O(o) = 12$

†$f_i = (A_i/E_i)(M)$ where $M = O(o)/B = 0.89133$

Structural Formula: $[Si_{7.68} Al_{0.32}][Al_{3.06} Fe_{0.45} Mg_{0.49}]Ca_{0.40} O_{20} (OH)_4$

Silica bound in smectite:

$A_{Si} = (f_{Si})(E_{Si})/(M) = (7.68)(60.01)/(0.89133) = 517.1$ g kg^{-1}

Silica not bound in smectite: $670.5 - 517.1 = 153.4$ g kg^{-1}

Quartz was identified by XRD and MAS-NMR spectroscopy in this sample.

5   Light Microscopic Techniques in Quantitative Soil Mineralogy

L. R. Drees
Texas A&M University
College Station, Texas

M. D. Ransom
Kansas State University
Manhattan, Kansas

Optical methods are among the oldest of techniques to analyze and characterize the objects and phenomena around us. The light microscope is simply an extension of the human eye. It allows for greater resolution of smaller sized particles. Russ (1990, p. 14) has stated that "the light microscope has been perhaps the most important scientific instrument of all time in both biological and materials research, again emphasizing the reliance that we place upon images to learn about the things that interest us." We would only add soils and minerals to the above list of research topics. With suitable accessories, the light microscope permits the measurement of unique optical properties of minerals that reflect their chemical composition and crystal structure. It also permits observation of secondary features, such as grain size, shape, spatial relationships to other minerals or pedological features, and alterations due to weathering. Optical microscopy is one of the few techniques that can examine each mineral separately rather than as a homogeneous composite. One of its benefits is that in thin sections, the constituents are assumed to be in their natural setting, undisturbed by either sampling or preparation procedures. Optical techniques are also applicable to nonmineral constituents in soils (e.g., organic matter), noncrystalline or poorly crystalline components (e.g., some of the Fe oxides and amorphous silica), and voids. Because the light microscope has unique functions, its use is appropriate for solving pedological problems where its power can be fully utilized.

In today's environment, pedologists need to be more quantitative concerning mineralogical composition, pore size and distribution, and the spatial distribution of soil constituents and pedological features. Quantitative statements are needed (i) for appropriate taxonomic classifications of soils, (ii) to document the effects of tillage and land management, and (iii) to help understand fluid-flow dynamics for hazardous wastes which could endanger water quality. Although descriptive micromorphology is important, quantitative statements are more convincing and acceptable than their qualitative analogs for solving many soil-related problems.

Quantitative mineralogy has been given several names. Modal analysis is probably the more common term used and applies to mineral composition (Chayes, 1956) as well as to pedological features (Anderson & Binnie, 1961; Milfred et al., 1967). Modal analysis is a term used in the geological sciences, from which pedologists have obtained most of their microscopic techniques and terminology. The term "quantitative microscopy" is applied to many disciplines and is also concerned with size, shape, and spatial distribution of constituents. Image analysis is the measurement of geometric features that are exposed on two-dimensional images. This is a broad term that refers to the extraction of visual information from spatial data, and ranges from the analysis of satellite images to high-resolution TEM micrographs. Currently the term is most often applied to computerized analysis of video images. However, any technique that measures geometric features (e.g., minerals or pedological features) of two-dimensional images (e.g., grain mounts, polished blocks, or thin sections) qualifies as image analysis. In this chapter, image analysis will be discussed in the context of computer-assisted image analysis. A more detailed discussion of microscopic image analysis is given by Russ (1990).

Although there are a number of light modes and specialized equipment and accessories for optical microscopy, the emphasis herein will be on transmitted light in the petrographic microscope with crossed and plane-polarized light. These light modes are most often used for quantitative analysis and computer-assisted image analysis.

HISTORICAL DEVELOPMENTS

Analysis and Preparation

Published reports of mineral analysis by light microscopy date back to the late eighteenth century. The early years of the nineteenth century saw the first published work by P.L. Cordier (see Crook, 1907). Polarized light microscopy had its true beginning with the development of the Nicol prism in 1828. This was about the same time that thin sections of minerals, bone, teeth, and fossil plants were first being prepared. In the following years, a few individuals examined thin sections with the polarizing microscope, but it was by no means a common tool. In 1849, Henry C. Sorby was the first to successfully prepare thin sections of sandstones, limestones, and other geologic materials and is considered the father of microscopic petrography. His contribution was not in preparing thin sections but in employing these thin sections to arrive at geological conclusions. Sorby was the first to use thin sections to study the structure of rocks and the structural relationships of their constituents (Higham, 1963). The microscope was little used in the geologic or earth sciences at that time, and optical petrography made little progress despite the work of Sorby and others. By 1867, David Forbes had published several papers using thin sections, and is reported to have made a collection of over 2000 sections (Holmes, 1930). Loewinson-Lessing (1954) marks 1870 as the beginning of the microscopic era of petrology with the publication of F. Zerkel's work on basalts. By the early 1870s, several books on the optical properties of rock-forming minerals had been published. Optical properties were defined and quantified for most of the major minerals. From this period on, optical microscopy of minerals and the use of thin sections have been major tools for attacking problems in geology and soil science.

Among the earlier publications on optical properties of soil-forming minerals was that of McCaughey and Fry (1913). Although this publication emphasized detrital grain mounts, the full range of optical properties of soil minerals was discussed. A later publication (Fry, 1933) expands further on microscopic methods of analysis for soil laboratories. Publication of *Micropedology* by Kubiena (1938) brought the microscope into the forefront of pedological research and established micromorphology as a science.

Because soil peds are generally friable, they are not easily cut for thin sections. Ross (1924, 1926) introduced the use of synthetic resins (Bakelite, kollolith) for impregnation of friable materials to retain their natural fabric. These resins were used by Redlich (1940) and Swanson and Peterson (1942) to prepare thin sections of soil samples for the quantitative evaluation of soil structure and porosity. With the exception of modern polyester resins (Ashley, 1973) and epoxy resins (Innes & Pluth, 1970), sample-impregnation techniques have changed little.

Measurement Techniques

DeHoff and Rhines (1968) indicate that ". . . the principal distinction between a science and an art lies in the insistence of the science upon quantitative relationships . . ." This distinction represents a transition from descriptive to quantitative analysis. In microscopy, quantitative analysis is a means of drawing inferences about three-dimensional objects by analyzing their two-dimensional analogues. Several measurement techniques have been, and continue to be used to place optical microscopy on a more quantitative base.

Areal Method

The earliest method of quantifying volume percentages of minerals was that of Delesse in 1847 (Holmes, 1930; Jones, 1977). He studied polished blocks, traced the outline of minerals onto paper, then transferred the pattern to tinfoil, and cut around the boundaries. The tinfoil pieces were sorted into mineralogical classes and weighed to yield relative percentages (Fig. 5-1). The volume percentage of a mineral was determined by comparing the weight of each cut pattern representing a mineral species (W_A) to the total weight of cut pieces (W_T)

$$\text{Percentage mineral} = (W_A/W_T) \times 100 \qquad [1]$$

This, and other methods equated measured weights or areas to volume equivalents. To measure the volume of clay skins, Buol and Hole (1961) projected the microscope image onto a ground glass, traced the image on translucent paper, and then cut and weighed each piece. They measured clay skin volume as a function of horizonation and genesis of translocated clay. Johannsen (1919) and Fry (1933) used a camera lucida attached to a microscope to project the image. The periphery of each mineral was traced with a planimeter to yield relative proportions. Despite the awkwardness of area measurements, recent developments in computer image analysis utilize this principle.

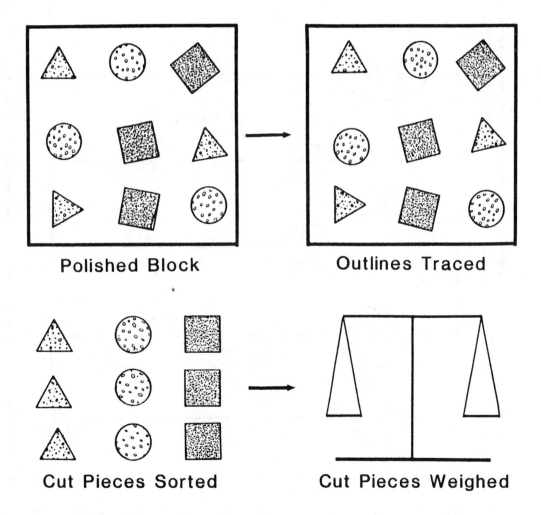

Polished Block   **Outlines Traced**

Cut Pieces Sorted   **Cut Pieces Weighed**

Fig. 5-1. Illustration of Delesse's method of areal analysis.

Linear Method

In 1898, A. Rosiwal developed a much simpler method of measuring the volume percentage of minerals (Holmes, 1930). This technique measures lengths of line segments rather than areas (Fig. 5-2), and samples a portion of the mineral area rather than the whole area, as in the Delesse method. It measures the frequency with which a mineral is encountered during the transect. The proportion of the length of the line crossing a mineral species (L_A) is compared to the total length of the traverse (L_T). The percentage of a mineral in the specimen is given by

$$\text{Percentage mineral} = (L_A / L_T) \times 100 \qquad [2]$$

The theoretical basis and statistical accuracy for this method are discussed by Lincoln and Rietz (1913).

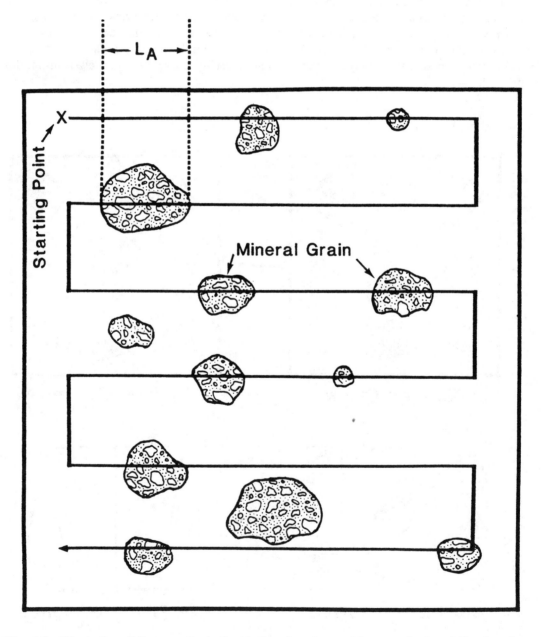

Fig. 5-2. Illustration of linear method of analysis. Area is equal to sum of mineral line lengths (L_A) divided by total line length.

The linear technique gained acceptance with the development of a recording stage micrometer for the microscope (Shand, 1916). The importance of the stage micrometer was in providing a convenient and practical means to make quantitative measurements. This early micrometer could only measure one constituent, requiring repeated traverses for each mineral species to be quantified. Later developments (Wentworth, 1923; Hunt, 1924) made this a standard technique for quantifying mineral grains in thin sections by allowing up to six constituents to be measured at a time. The linear method was recommended for soil laboratories (Fry, 1933) and its accuracy estimated to be about 1 to 2% (Thompson, 1930). Larsen and Miller (1935) recommended that the total length of the traverse be 100 times the mean grain size, with 15 traverses about 1 mm apart. Using this technique, Swanson and Peterson (1940) quantified differences in porosity between two contrasting soils. To quantify porosity and pore dimensions, Redlich (1940) found it necessary to survey a total length of 200 mm with 1 mm between traverses. Chayes (1956) designates the time period between Delesse and the Shand micrometer as the golden age of descriptive petrology.

Point-Count Method

Point counting, as proposed by Glagolev (1934), modified the linear method by taking equally spaced points along the transect (Fig. 5-3). Point counting is a stepwise scanning procedure that permits the sampling points to be spread uniformly and systematically over the entire area without accumulating more data points than necessary to attain the level of precision desired. The statistical validity and precision of point counting has been well established (Chayes & Fairbairn, 1951; Chayes, 1956). A major weakness of the areal and linear methods was the lack of an easily understood analytical demonstration of their validity and precision in thin-section analysis. The point-counting technique has gained wide favor not only because of its ease, but because statistical confidence limits and expected errors can be easily calculated. This method also gained popularity with the development of microscope stage attachments for the rapid counting of points and the independent movement of the sample in equal increments (Hurlbut, 1939; Chayes, 1949). One of the latest aids in point counting is the use of voice recognition by a computer (Dunn et al., 1985). The computer can recognize more than 100 voice commands, thus freeing the operator from manual manipulation of counters. The computer also activates movement of the microscope stage and stores the point data for later plotting or processing.

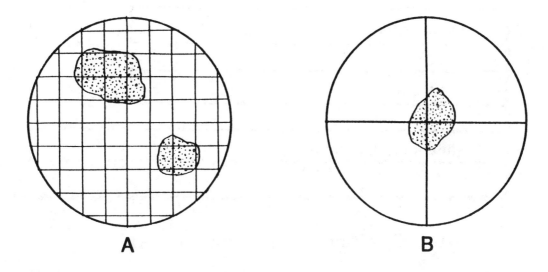

A   **B**

Fig. 5-3. Illustration of point-count method. (*A*) eyepiece grid reticule; (*B*) minerals at cross-hairs of microscope.

Although point counting is more suitable to statistical analysis, Hilliard (1968) derived theoretical coefficients of variation for the several methods mentioned (areal, linear, and point counting). Figure 5-4 illustrates that the systematic point-count method is markedly superior to other techniques in estimating volume percentages. The calculated values assumed a random distribution of equally sized spheres. In a more practical test, Weibel (1979) prepared five model sections with a component having an areal fraction of 20%. Seven individuals then made estimates of the areal fraction using several methods (Table 5-1). The variation for all methods is less than that illustrated in Fig. 5-4, but shows the accuracy of the systematic point-counting method. In addition, point counting took the shortest time, about one-quarter to one-sixth that of areal measurements (Table 5-1).

Table 5-1. Coefficient of variation and average time required for measuring the area of a component (area = 20%) by different methods (data from Weibel, 1979).

Method	Coefficient of variation	Average time
	%	s
Areal analysis		
planimeter	3.8	323
circle fitting	3.3	148
cut out and weigh	2.6	461
Linear analysis	7.5	105
Point count		
random	6.2	79
systematic	4.7	73

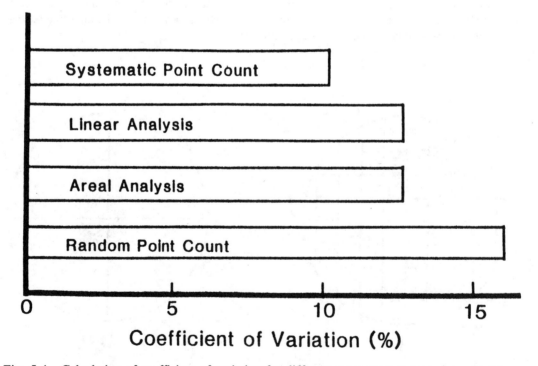

Fig. 5-4. Calculation of coefficient of variation for different measurement procedures based on 100 observations of randomly dispersed, equidimensional spheres (modified from Hilliard, 1968).

OPTICAL METHODS OF QUANTIFICATION

Quantification Considerations

Sampling

In analyzing a soil sample for any property, the aim is to obtain data from the sample in order to make quantitative statements about a soil horizon or unique soil zone. As stated by Snedecor and Cochran (1967), "it is the sample that we observe but it is the population which we seek to know." Optical examination, by its very nature, constitutes a small sample of a larger population. A typical thin section may be <0.05 cm^3 whereas the impregnated block from which it was obtained may be 500 cm^3 and may only represent a small portion of the vertical dimensions of the horizon. The magnitude of difference between the volume of the impregnated block and the thin section may be on the order of 10 000 or greater. Moreover, the pedon from which the block was sampled also has a rather limited horizontal extent on the landscape, typically ranging from 1 to 10 m^2. Soil microscopists should keep in mind that thin sections and grain mounts are essentially two-dimensional objects from which to extrapolate to three-dimensional bodies. Thus, a critical problem in measurement reliability is caused by improper sampling. The field sample and subsequent laboratory subsamples must be representative and unbiased. An unbiased sample is one taken and prepared without regard to assumptions relating to the results. For a thin section to produce reliable measurements, it must represent the population as a whole. A thin section cut parallel or normal to bedding features or translocated features in the horizon may yield different volume estimates of constituents depending on the orientation of the cut section. If soil horizons exhibit horizontal bedding or vertical features, such as illuviated clay or slickensides, the thin section should be cut perpendicularly to these features (Fig. 5-5). Optical microscopic techniques are well established, and quantification has a strong statistical basis. Although these techniques are capable of producing valid results, they cannot compensate for inadequate sampling, poor sample preparation, or improper experimental design.

In order for mineralogical analysis by microscopic techniques to be meaningful, the following constraints must be considered:

1. The mineral must be identifiable.
2. Minerals must be uniformly distributed and independent from one another.
3. The area selected for measurement must be representative of the specimen as a whole.
4. Minerals must occur in random orientation.

Size Limitations

The light microscope, although having many advantages in mineral analysis, has some limitations. In order to quantify a mineral species in either a thin section or in grain mounts, the mineral must be identifiable so that it can be assigned to the appropriate mineralogical class. The wavelength of light defines the physical limits of resolution of an optical microscope. Particles smaller than about 2 μm can be easily observed but not easily identified. In addition, two particles separated by distances of less than 1 or 2 μm often appear as a single entity. Thus, quantitative optical analysis must be limited to minerals or pedological features generally >10 μm in size that can be easily identified. The inability to identify small-sized minerals, especially clays, is one of the principal deficiencies of optical microscopy. Bullock and Thomasson (1979) put the lower limit for pore measurements at 30 μm with the use of image analyses. Other techniques discussed in this book are more suited for the analysis of individual particles <5 μm in size (Gilkes, 1994; see Chapter 6). At the other extreme, it becomes impractical to attempt characterization or mineral quantification of soils that contain nodules, lithics (rock fragments), or gravels that are larger than 1 cm in diameter. When constituents become so large that they occupy most of the microscope slide, multiple samples, rather than multiple sections from the same sample, may have to be examined to obtain a representative volume.

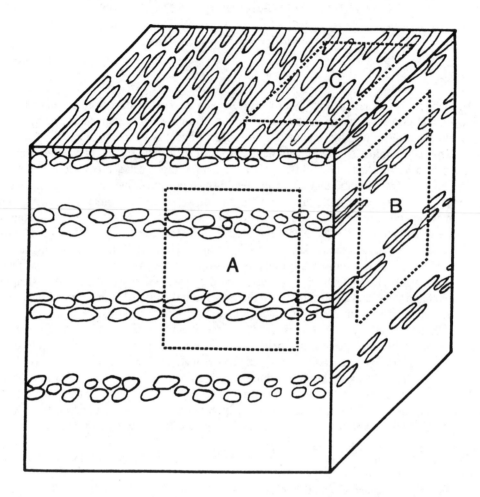

Fig. 5-5. Example of soil block with strongly lineated and bedded structure. Thin sections cut either along sides *A*, *B*, or *C* yield different areal proportions of constituents (modified from Hutchison, 1974).

Operational Limitations

Although size is one factor that hinders mineral identification, other factors also come into play that likewise hinder identification. These additional factors are due largely to mineral orientation, opacity, and weathering. Accurate mineral identification in grain mounts or thin sections requires familiarity with the sample. Familiarity includes sight recognition of common minerals, knowledge of likely mineral constituents, and the ability to distinguish optical properties between mineral species. A trained microscopist can usually identify most common and accessory minerals. If a true unknown is encountered, especially in trace amounts, optical identification may be difficult, if not impossible. In thin sections, definitive optical properties such as refractive index for principle directions of light vibration in the crystal or interference figures may be difficult to determine. In grain mounts, crystal habit may preferentially orient minerals such that the optic axis is parallel to the stage, making interference figures difficult to determine.

Opacity due to grain coatings or mineralogy often precludes the determination of optical properties, especially in grain mounts. In many cases, reflected light microscopy is an important aid. Grain coatings are usually not a problem in thin sections unless the mineral grain is smaller than the thickness of the section. Microcrystalline grains also pose greater difficulties in grain mounts than in thin sections for positive mineral identification.

Mineral weathering can result in alteration products which can confound optical analysis and the determination of optical properties. For example, biotite (Harris et al., 1985) and feldspar grains commonly exhibit weathering, especially along cleavage planes. Pyroxenes, amphiboles (Velbel, 1989) and olivines are also commonly weathered and exhibit alteration products. There can be a range of mineral alterations within a given sample, from unaltered to pseudomorphed grains. Even grains with the optical appearance of biotite may be multiphased, consisting of combinations of kaolinite, vermiculite, goethite, etc. Along this weathering continuum there are no definitive break points to stop naming one phase and begin naming another. Weathering and alterations may cause mineral pseudomorphs, where one mineral may take on the form and habit of another, such as calcite after gypsum (Drees & Wilding, 1987). In such instances, being able to determine optical properties is more important than form recognition.

These limitations are not included to discourage one from using optical microscopy for mineral quantification, but to make the reader aware of some of the realities of mineral identification. It is only when we understand the practical limitations of a given technique that we can have more confidence in our ability to make quantitative statements.

Grain Mounts

The quantitative determination of loose mineral grains by microscopy was exemplified by Fleet (1926), who counted all the grains in a heavy mineral suite instead of a representative sample. He felt that ". . . certain difficulties arise if mental estimates of relative proportions are made unsupported by actual counting of grains." A benefit of using loose grains is that they can be fractionated by size, density, or magnetic properties. Thus, particular isolates can be selected that may yield important information otherwise unavailable if the bulk sample were examined. For example, heavy minerals, which often constitute a very small proportion of a soil, can be concentrated and examined optically to determine mineralogical distributions (Cady et al., 1986). An added advantage is that grain mounts can be used to identify rock fragments, quartz, polycrystalline quartz, and chert which may not be distinguishable by other techniques.

Although grain mounts are easier to prepare than thin sections, results are usually in number percentages, which cannot be easily converted to area, volume, or weight percentages unless all grains are of the same size and do not have preferred orientations. For example, the only way to convert number percentages to area or volume percentages is to measure the area of each grain, which is a difficult and time-consuming task with the light microscope. With the introduction of computer-assisted image analysis, however, area determination is a simple and rapid procedure. The use of computer-assisted image analysis will be discussed later.

Sample Preparation

In order for grain mounts to be suitable for counting, individual minerals should be of a common size and distributed uniformly on the microscope slide such that they are neither touching nor superimposed on each other. Mounting media may be an immersion oil or permanent cement. Mounting media or cements should have a refractive index of about 1.54, but other immersion oils may be used if refractive index is useful for mineral identification. For heavy minerals, an immersion oil with a refractive index of about 1.7 would be more appropriate. Transmitted cross-polarized light is best used for most common light minerals such as quartz, feldspars, carbonates, etc. Plane-polarized light aids in identifying grains that are pleochroic. Because many heavy minerals are opaque, reflected light is often better suited for their identification. Reflected light is also suitable for other grains that are opaque due to thickness or coatings. Most texts on light microscopy list the optical properties used for mineral identification.

Grain mounts have some advantages and limitations in comparison with thin sections. Many grains that are platy or elongated will usually be in an orientation that may facilitate identification. For example, micas usually lie with their *c*-axis perpendicular to the plane of the slide. This orientation will exhibit some unique optical characteristics of the mineral. On the other hand, identification usually becomes more difficult as the grains become very large or very small. Galehouse (1971) recommends that grains be <0.5 mm in diameter, especially for heavy minerals. In most instances, particles in the medium-sand to coarse-silt range (0.5-0.05 mm) are most suitable for identification and counting. Cady (1950) mounted grains of 0.1 to 0.05 mm and made over 1000 grain counts per slide. The aspect of statistical validity is covered under the discussion of thin-section analysis.

Many mineral grains look similar under the microscope and time-consuming procedures must be used to determine the optical properties of each mineral for proper identification. A particular problem is distinguishing untwinned plagioclase from quartz. Staining procedures (see discussion under Thin Sections) are often used in grain mounts and in thin sections. Loose grains, however, cannot be easily stained. To stain grain mounts, the grains need to be attached firmly to a glass slide but have sufficient free surface exposed to take the stain. The mounting cement must not react with the etching solution or the chemical stains. A thin layer of epoxy with grains sprinkled over the surface works well (Langford, 1962). The epoxy must be able to hold the grains firmly, but not engulf the grains due to surface tension forces. Gross and Moran (1970) suggest using a thin layer of black roofing tar to hold the grains in place. The minerals must be examined in reflected light, but the dark background offers greater contrast with both stained and unstained minerals. These procedures have proved adequate for feldspar identification for family placement in Soil Taxonomy (Soil Survey Staff, 1975). Additional details in the use, preparation, and interpretations of grain mounts are presented by Cady et al. (1986).

Grain Shape Limitations

In measuring individual grains for quantitative mineralogy, we often assume that volume percentages are equal to areal proportions. This assumption holds true if all grains are equidimensional or do not exhibit any preferred orientation. In loose grain mounts, platy grains, such as mica, tend to settle with their flat face parallel to the glass slide, thus exposing the maximum cross-sectional area to the viewer (Harris & Zelazny, 1985). Because platy grains may have the same projected areas as cylindrical or cubic grains, the actual volumes of platy minerals will vary depending on the diameter-to-thickness ratio, resulting in an overestimation of their volume proportion (Rebertus & Buol, 1989). The diameter-to-thickness ratio of mica varies with particle size, being about 15 for very fine sand (0.1-0.05 mm) and up to 75 for very coarse (1-2 mm) sand (Harris et al., 1984). A simplistic model with tetragonal crystals illustrates the differences in volume and area proportions with different width to height ratios for different percentage composition (Fig. 5-6). Similar relationships have been discussed by Harris et al. (1984), Harris & Zelazny (1985), and Rebertus and Buol (1989). An alternate procedure, which eliminates much of the bias due to preferred orientation, is to thoroughly mix the grains with epoxy in a small sample cup and cut a thin section when hardened (Hagni, 1966). This is also helpful for identifying larger grains whose optical properties may be obscured by thickness or opacity. The optical properties of each mineral are more readily discerned, and staining can be used for mineral identification.

Mineral Quantification

There are several types of grain-mount counting techniques that should be considered, for they do not all yield the same results. The counting is rather easy, but interpretations may be difficult. The Fleet and ribbon methods involve counting the individual grains and assigning each to a specific mineralogy class. In the ribbon method, only mineral grains within a certain band width are counted (Fig. 5-7A), as opposed to all the observed grains in the Fleet method. In practice, the microscope slide is usually moved under the objective, and all grains within the band are

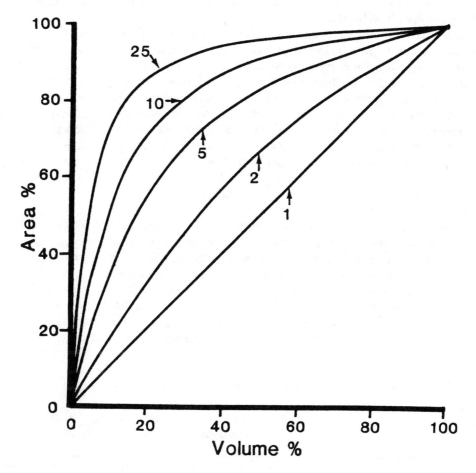

Fig. 5-6. Relationship between area and volume with tetragonal grains having width to thickness ratios of 1, 2, 5, 10 and 25.

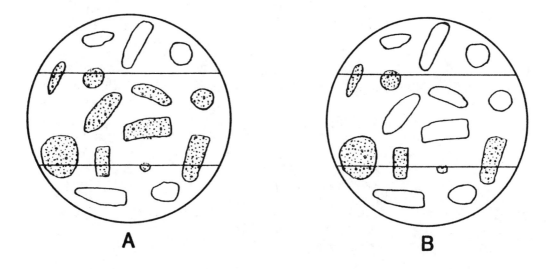

Fig. 5-7. Methods of grain counts. (*A*) Ribbon method - all grains between lines are counted; (*B*) Line method--only minerals touching line are counted (modified from Galehouse, 1971).

counted. If the band width is larger than the largest grain, the method is considered to yield number percentage (Galehouse, 1969, 1971).

The line method is probably used more often because it is somewhat easier to keep track of the minerals counted. In this method all mineral grains that intersect a line or the microscope crosshairs during a linear traverse of the slide are counted (Fig. 5-7B). This results in a number frequency, which is different than the number percentage above because the sample is biased toward larger grains. The line procedure underestimates the area occupied by large grains and overestimates the area of smaller grains (Galehouse, 1971; Rebertus & Buol, 1989). Regardless of the method used, grain-mount analysis should be treated with caution. The various counting methods for thin sections and grain mounts are summarized in Table 5-2. A more comprehensive discussion of number frequency vs. number percentage is presented by Galehouse (1969).

Thin Sections

Thin-section analysis of soils is the only technique that permits the soil scientist to study, on a microscopic scale, an undisturbed portion of the living soil in its natural environment. The spatial distribution of minerals, pores, and pedological features can also be viewed and measured. Prior to the beginning of any type of quantitative analysis, it is advisable to examine the entire section at low magnification. Such examinations may reveal artifacts, areas of grain plucking (void artifact), or possible unusual features such as the filling of krotovina channels. It is helpful to the microscopist if a profile description is available to ensure that the section observed is representative of the soil horizon. Preliminary examination may reduce errors and increase the reliability of the results.

Table 5-2. Counting methods for mineralogical analysis (modified from Galehouse, 1971).

Method	What measured	What determined	What implied	Can be converted to
	-----------------------Thin section or polished block-----------------------			
Areal	Area	Area %	Volume %	Weight % or chemical analysis
Linear	Length of lines	Area %	Volume %	Weight % or chemical analysis
Point	Grid of points	Area %	Volume %	Weight % or chemical analysis
	------------------------------Grain mounts------------------------------			
Fleet	All individuals	Number %	--	--
Ribbon	All individuals within band	Number %	--	--
Line	Individuals along line	Number frequency	--	--

Sample Preparation

Due to the friable nature of most soil materials, resin impregnation is a necessity. The common impregnation agents are polyester resin (Ashley, 1973) and epoxy (Innis & Pluth, 1970). The mechanics of impregnation and sample preparation will not be detailed herein as they are adequately described by Murphy (1986) and Cady et al. (1986). One prime requisite is that the section be prepared to a uniform thickness, usually 25 to 30 μm. Uniform thickness is critical as the interference color of minerals in cross-polarized light is a function of thickness, as well as of orientation and birefringence. Variations in thickness within a section or between sections could cause difficulties in mineral identification, especially with automatic image analysis. Artifacts such as grinding powder, lint, air bubbles, or grain plucking due to poor impregnation also complicate mineral identification. Although the human eye-brain combination can quickly compensate for variations in thickness, illumination, or artifacts, computer-assisted image analysis is not quite as forgiving and may lead to erroneous results.

Differential Mineral Staining

Although each mineral has unique optical properties, it is not convenient and fast to measure the optical properties at each point along a transect. A trained soil microscopist can readily identify most common minerals at first glance. There are instances, however, of minerals that are often difficult to distinguish from one another: calcite and dolomite, untwinned plagioclase feldspars and quartz, and K-feldspars and some plagioclase feldspars. Chemical staining techniques are especially useful in differentiating various minerals that appear similar. For staining to be effective, the grain must have an exposed surface. Grains that are embedded within the thin section cannot take a stain. Thus, incident or reflected light may prove more efficient than transmitted light. Likewise, a polished block is easier to prepare than a thin section and may be more efficient in identifying certain constituents.

Most feldspar staining begins by etching with vapors of HF, which is hazardous and should be handled with caution. Morris (1985) suggests the use of a glass-etching cream (HF in a paste matrix) which is used by artists and is available in many art supply and hobby stores. The senior author has used this product and found it to be acceptable and without the hazards associated with HF vapors. Excellent reviews on staining techniques are found in Friedman (1971) and Hutchison (1974). Feldspar-staining techniques are further detailed by Norman (1974) and Houghton (1980), while Friedman (1959) and Poole and Thomas (1975) discuss staining of sulfate minerals. Carbonate-staining methods are given by Friedman (1959), Lindholm and Finkelman (1972), and Warne (1962). The procedures are fairly well documented and are not repeated here. As an example, x-ray diffraction analysis of the sand and silt fractions indicated both calcite and dolomite present in a soil horizon. Initial optical examination of thin section of the horizon revealed only a single carbonate phase which was interpreted as dolomite (refer to Color Plate 5-8A). When the section was treated with Alizarin red-S to stain calcite, it was observed that calcite occurred as a syntaxial overgrowth on dolomite in optical continuity with it (refer to Color Plate 5-8B). Staining is equally effective in differentiating feldspar species, especially when the only purpose is to distinguish quartz from feldspar for family placement in Soil Taxonomy (Soil Survey Staff, 1975). Staining techniques are also helpful for identifying fine-grained material whose optical properties cannot be easily determined.

Holmes Effect

The discussion to this point has considered thin sections as two-dimensional representations of the three-dimensional horizon. For many mineral grains and pedological features, this may be a true statement. As minerals or features become smaller than the section thickness, however, pores or mineral grains may not be recognized or resolved because they may be underlain, overlain, or partly occluded by other constituents (Fig. 5-9). The distortion caused by the finite thickness of the thin section results in differences between the projected area and the true area proportion and is termed the "Holmes effect" (Chayes, 1956). The overall effect results in an overestimation of

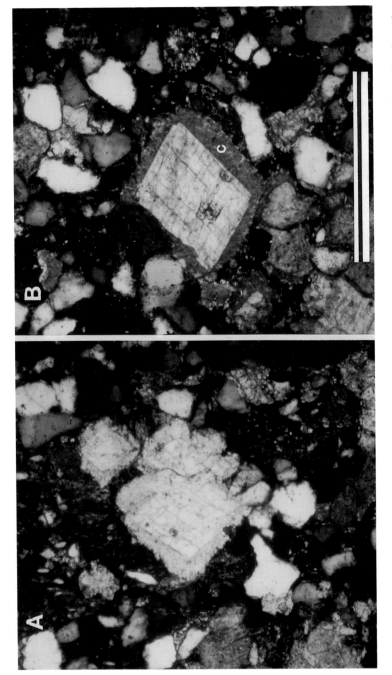

Color Plate 5-8. Syntaxial calcite overgrowth on dolomite. (*A*) Cannot distinguish calcite from dolomite in section; (*B*) stained portion shows distribution of calcite (*C*) around dolomite (*D*). Crossed polarized light; bar length = 0.5 mm.

fine particles and an underestimation of pores (Halley, 1978; Murphy & Kemp, 1984), with the magnitude of the error increasing as particle size decreases. Halley (1978) also indicates that the error increases with tighter packing, decreased sorting, and increased shape irregularity. The problem of overestimation of grains and underestimation of pores arises because any portion within the 30-μm slice occupied by a grain is counted as a grain, regardless of the pore space above and/or below it (Fig. 5-9). For a pore to be recognized and counted, light must pass unimpeded through the entire section. The edge effect, or amount of actual pore space not observed, is a function of grain size. As grains become larger, there is less potential for void spaces to be obscured (Fig. 5-10). Correction factors for the Holmes effect for different sizes of opaque spheres in a transparent matrix measured with transmitted light are given by Chayes (1956). Moreover, the ratio of grain size to thin-section thickness governs the size of the bias. As the thickness of the thin section approaches zero, the bias is eliminated, both for pores and mineral grains.

If these effects pose a problem in quantification of pores or mineral grains, the section can be examined in reflected light. In reflected light, the thin section or polished block is truly two dimensional, having zero thickness. The use of fluorescent dyes (Ruzyla & Jezek, 1987; Puentes & Wilding, 1990) in reflected-light fluorescence microscopy is well suited to show the actual pore space and pore distribution for counting and quantification. Likewise, chemical staining of soil minerals aids in their identification in the reflected-light mode. In reflected light, the image observed is the true areal portion of the mineral, pore, or pedological feature and is proportional to volume percentages.

Point-Count Statistics

As discussed earlier, the most widely used technique of mineral quantification in grain mounts or thin sections is point counting. An excellent discussion of this technique is given by Galehouse (1971). One of the primary benefits of point and other counting techniques, aside from their ease, is that number frequency is more suitable for statistical analysis than either area or volume percentages. Statistical calculations can yield the probable error at a given confidence level and the number of counts needed to achieve a certain probable error. Statistical analyses by point counting techniques are given by Van der Plas and Tobi (1965), Galehouse (1971), Brewer (1976, see Chapter 3), and Jones (1987, see Chapter 6).

If a binomial distribution is assumed, then the standard deviation (α) is

$$\alpha = \sqrt{P(100 - P)/N}, \qquad [3]$$

where P is the frequency percentage of the selected mineral and N is the total number of points or grains counted. The probable error about the mean is calculated from

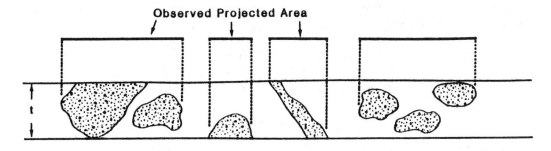

Fig. 5-9. Examples of bias due to the "Holmes effect" when particles become smaller than the thickness of the thin section (t). The observed projected area is greater than the actual area.

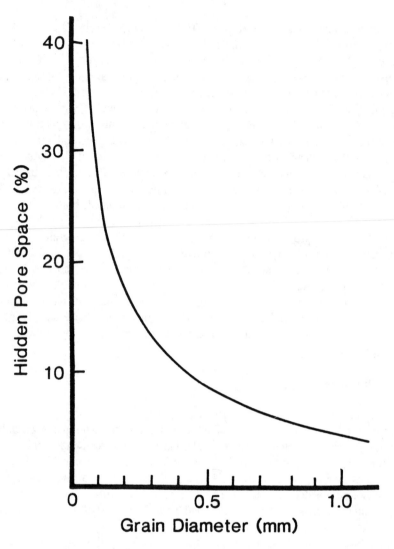

Fig. 5-10. Effect of grain size on proportion of pore space not observed in thin section. Curve based on cubic packing of spheres (modified from Halley, 1978).

$$e = t\sqrt{P(100 - P)/N},$$ [4]

where e is the absolute error in percentage, and t is the t distribution (1.96 for 95% confidence level, 0.6745 for 50% confidence level). For example, if we estimate 20% of a mineral in 200 counts, then the probable error is

$$e = t\sqrt{(20 \times 80)/200} = t \times 2.83.$$

At the 95% confidence level, the error is

$$e_{95} = 1.96 \times 2.83 = 5.5\%$$

and at the 50% level, the error is

$$e_{50} = 0.6745 \times 2.83 = 1.9\%.$$

Thus, there is a 50% chance that the true value lies between 18.1 and 21.9% and a 95% chance that the true value is between 14.5 and 25.5%. A graphical solution of the above equation is given by Van de Plas and Tobi (1965). The probable errors about the mean decrease as $1/\sqrt{N}$. Thus, to decrease the error by a factor of two requires four times the number of counts. The number of counts can be derived by rearranging Eq. [4] to give

$$N = t^2 P (100 - P)/e^2 \qquad [5]$$

A more useful calculation is to estimate the relative probable error (E) on P, the percentage of the selected constituent, where $E = e/P$. Thus, Eq. [5] becomes

$$N = t^2 P (100-P)/PE^2 . \qquad [6]$$

In this equation, the total number of counts (N) is inversely proportional to the percentage of the mineral being measured; as P becomes small, N becomes large. This presents a more realistic value for the accuracy of point or grain counting and illustrates the potential errors involved. Table 5-3 lists relationships between confidence interval, point counts, probable error, and mineral percentages based on Eq. [6]. A variation of this equation was used by Brewer (1976, Table 3.1) to arrive at similar numbers.

Table 5-3. Relationship between number of grains, confidence interval (CI), probable percentage error and estimated percentage of mineral in sample.

Percentage mineral in sample	5% Probable error		10% Probable error		20% Probable error	
	50% CI	95% CI	50% CI	95% CI	50% CI	95% CI
	----------------------------------Number of grains----------------------------------					
1	18 016	152 127	4 504	38 032	1 126	9 508
2	8 917	75 295	2 229	18 824	557	4 706
5	3 458	29 196	864	7 299	216	1 825
10	1 638	13 829	409	3 457	102	864
20	728	6 146	182	1 536	45	384
30	424	3 584	106	896	27	224
40	272	2 305	68	576	17	144
50	181	1 536	46	384	11	96
60	121	1 024	30	256	--	64
70	78	658	20	165	--	41
80	45	384	11	96	--	24
90	20	171	5	43	--	11

An alternate means of evaluating point-count statistics is by comparing coefficients of variation (CV) of different mineral percentages based on a range of counts. Coefficients of variations are a convenient means of comparing variability because values are independent of the mean. Coefficients of variation were calculated from Eq. [3] and are presented in Fig. 5-11. Based on these calculations, several hundred counts are sufficient to estimate the mean for major constituents

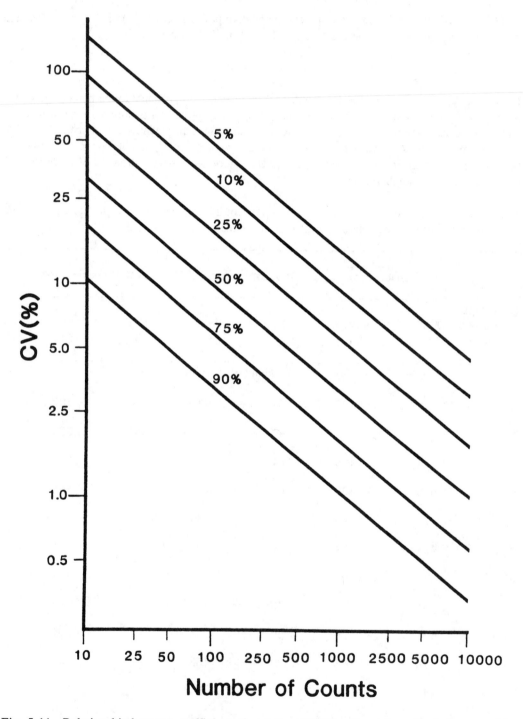

Fig. 5-11. Relationship between coefficient of variation (CV%) and number of points counted for constituents comprising various proportions of the sample.

(>50%) with a CV of <10%. On the other hand, almost 2000 counts are required to estimate minor constituents (5%) with the same degree of variance about the mean (Fig. 5-11). An algorithm based on random numbers indicates that CV values for random point counts are about twice those given in Fig. 5-11. This is in agreement with other data showing that random point-counting is more variable than systematic point-counting (Table 5-1, Fig. 5-4).

Table 5-3 and Fig. 5-11 indicate that the microscopist cannot make reliable mineral estimates when a mineral comprises a small fraction of the sample. The analytical error increases as the number of grains decreases. Quantitative analysis does not become impossible because of this decrease, it simply becomes increasingly unprofitable (Chayes, 1956). One advantage of optical microscopy is the ability to observe minerals in concentrations below the detection limits of other analytical instruments. We cannot be as precise in estimating minerals in low concentrations as for those that are major constituents. We may have to be satisfied with a lower confidence interval or greater probable error. A balance is needed between accuracy and the investment of time to make the measurements. The difference between an accuracy of 5 to 10% or 1% may correspond to an analysis taking 30 min or 30 h. At low mineral concentrations, sample variability may be a more significant source of error than counting statistics. Brewer (1976) indicates that ". . . low percentages are less useful for interpretative work unless the mineral species is a specific 'marker' mineral, in which case simply presence or absence is usually diagnostic."

Statistical errors are important considerations in mineral quantification. As shown in Table 5-3, statistical errors can be reduced, but they cannot be eliminated. Nonstatistical errors are mostly because of mineral misidentification but contribute to the degree of reliability we can put in the data. The tabulated statistics in Table 5-1 assume that the microscopist does not make mistakes. Mistakes will be made, but these errors are usually much less than the aforementioned statistical errors. Errors of misidentification can be largely eliminated by adequate training and practice.

Quantification Examples

In a comparison of quantitative mineralogy by x-ray diffraction and optical microscopy, Davis and Walawender (1982) used 500 points per section. They initially made 200 counts, calculated the weight percentage of each major mineral, and made additional 100 point counts until replicate values were within 5% of the original value. They found excellent agreement between the two methods for five igneous rocks. Friedman (1960) also found excellent agreement between wet chemical analyses and chemical analysis derived from thin sections. Friedman (1960) concludes that petrologists should make more chemical analyses with the petrographic microscope.

Quantitative microscopic techniques can also provide a means of characterizing soil structure and fabric (Anderson & Binnie, 1961). The determination of pedological features is fraught with difficulties because of feature heterogeneity and feature recognition. Point counts of 1000 (Eswaran, 1968), 1600 (Miedema & Slager, 1972), and 6000 (Murphy, 1983) have been used to provide a quantitative measure of pedological features in a soil horizon. One of the pedological features for which quantitative information is desirable is illuvial clay. Concentrations of illuvial clay above 7%, however, are usually rare (Miedema & Slager, 1972). Because of the low proportion, accurate quantification becomes difficult; Miedema and Slager (1972) suggest that more than 1% of illuvial clay should be present for quantitative microscopic determination. By counting 6000 points over a 25- by 25-mm section, Murphy (1983) was able to quantify the pore space but had reservations about the reliability of quantifying illuvial clay. These reservations were due to difficulties in illuvial clay recognition, orientation of illuvial clay, differentiation from other features and soil heterogeneity.

The difficulty of quantifying illuvial clay has been aptly discussed by McKeague et al. (1978, 1980). These studies focused on illuvial clay recognition as observed in thin sections, and the concepts and definition of illuvial clay among soil microscopists. In one study (McKeague et al., 1980), 10 individuals made at least 5000 point counts per section over an area of 25 by 25 mm. The coefficient of variation for the 10 individuals ranged from 39 to 64% for the six sections studied. The highest CV's (>60%) were for the lowest (1.1%) and the highest (8.5%) mean percentage of illuvial clay. The authors question the validity of making estimates of illuvial clay unless guidelines are developed for the recognition of illuvial clay in thin sections. It should again be emphasized that these errors and reservations by various authors arise because of feature recognition and do not fault the underlying principle of point counting.

An alternate procedure is to prepare multiple sections for quantitative analysis. In a measure of pedological features in argillic horizons (glaebules, pores, papules, argillans, skeletal grains), Milfred et al. (1967) concluded that an estimate of individual constituents with a standard error of 10% would require 21 cores, two thin sections per core, and 1000 points per section. This study illustrates the heterogeneity within soil horizons, and the difficulty in accurately identifying pedological features. Redlich (1940) measured soil structure and porosity and found that 24 sections from a humus-treated soil, but only 15 sections from an untreated soil were required to reduce the mean error to about 10%. For macroporosity determinations by image analysis, Puentes et al. (1992) found that variability decreased more rapidly by analyzing more samples rather than increasing the sections per sample or frames per section. These studies indicate that variability is greater between soil samples than between sections obtained from the same sample, and that error is reduced by analyzing multiple cores rather than multiple thin sections from a single core (Swanson & Peterson, 1940; Redlich, 1940; Milfred et al., 1967; Murphy & Banfield, 1978). Grossman (1964) suggested a technique of comparing composite thin sections for pedological analyses as a means of obtaining a large number of random units with a few sections. Such techniques may yield more statistically accurate results, but spatial relationships may be lost. Although the technique has some apparent advantages, it has not been widely used.

Quantification Recommendations

It is not possible to give exact specifications for mineral quantification by the point-count procedure although point counting is a proven technique with well-understood statistics. Precise data from point counts require an appropriate number of observations across the specimen at equally spaced intervals. The network of points may be a grid reticule inserted in the microscope eyepiece (Fig. 5-3A), a grid superimposed on a micrograph, or more often, minerals intersecting the crosshairs in the microscope as the slide is moved in equally spaced intervals (Fig. 5-3B). The increment between points should approximate the mean grain diameter, although Van der Plas and Tobi (1965) suggest that the point increment be larger than the largest grain that is included in the analyses. Although this may be a good rule of thumb, minerals and pedological features in soils seldom occur in a uniform size distribution, and situations occur where this rule of thumb is not advisable. For example, in a fine-grained sample, incrementing at small intervals may result in a large number of counts, but such a small area is covered that many features may be missed. Likewise, a large increment may cover the entire slide but provide insufficient points to give a true picture of the mineralogical composition. Points should be spaced sufficiently far apart that the constituents counted are independent of one another. Point-counting stages can usually be adjusted for interval spacings of 0.1 to 1.0 mm. If it is not practical to increment at the average grain diameter, measurement intervals can be calculated to cover the entire slide in 500 to 1000 points. For a 25- by 25-mm slide, an increment of 0.5 mm in a rectangular grid arrangement will yield 2500 points, more than sufficient for most detailed analyses. An alternate procedure would be to make transects in one direction at 0.5-mm intervals, but make the transect lines 1.0 mm apart. A balance is needed between number of points obtained, statistical accuracy desired, and sampling a representative portion of the slide. Most references suggest 300 to 600 points as desirable for maximum efficiency and minimal investment of time. For detailed analyses, about 1000 to 1500 points are suggested.

Quantitative results are accurate if the section is truly representative of the horizon or portion of the soil from which it was extracted. For many soils, minerals in the sand and silt fraction are usually randomly distributed, and a single thin section may be adequate for quantification. Recommendations become more difficult for quantifying pedological features. These features, such as illuvial argillans, stress cutans, infillings of root or animal channels, slickensides, ped surfaces, or effects due to tillage and cultivation, are not uniformly distributed, at least at the scale of the microscope slide (commonly < 1000 mm^2). Because some of these features occur at intervals greater than the dimension of the microscope slide, they may be missed entirely or observed in concentrations greater than their true distribution. In addition, many of these features may have preferential orientations in the soil profile. Movement and translocation of constituents usually occur vertically, but horizontal bedding may also be present, and their expression on the slide will be dependent on the orientation of the cut section in reference to the feature (Fig. 5-5). Preliminary guidelines suggest that vertical and horizontal sections be prepared, 500 counts be made on each section, and the results compared to evaluate differences caused by orientation. If large differences are noted, or if the horizon is known to be heterogeneous, additional sections or cores may be required for accurate quantification. Whether additional sections are prepared depends on whether the additional data are worth the investment of time and effort. This may seem like a tedious and arduous task, but optical microscopy is the only technique available to identify and quantify many pedological features.

COMPUTER-ASSISTED IMAGE ANALYSIS

Introduction

The petrographic microscope has been a standard laboratory instrument for nearly a century. The coupling of a video camera to the optical microscope in the early 1950s (Inoue, 1986) opened the door for video microscopy. Linking of the personal computer (PC) with image acquisition and processing boards and enhanced video technology in the 1980s has ushered in the era of computerized image analysis (Russ & Russ, 1984) and has improved the performance capabilities of the light microscope.

It should be emphasized that no amount of data processing and manipulation can extract information that was not present in the original image. Thus, the most critical component of an image-analysis system is a good microscope with clean lenses and good optics producing a clear optical image. Ideal optics and the most modern computer system cannot overcome inadequate sample preparation. Sample preparation is more critical in computer-assisted image analysis because the system cannot compensate as easily as the human eye and mind for variations in mineral orientation, sample thickness, illumination, artifacts, or portions of the specimen not injected by a stain or dye.

It is the ability to store the video image as digital data and then process the data mathematically, rather than optically, that gives a new dimension to the term "image analysis." With digital "image processing (enhancement)," the signal-to-noise ratio can be altered, contrast improved, edge-features and highlight-details enhanced, and resolution improved. Enhancement procedures do not yield descriptive information about the sample but manipulate the image such that it is more useful to the observer. Since the digital image is stored as a mathematical abstraction, it can be processed and manipulated in ways that cannot be duplicated by a physical device such as a lens, prism or filter. The manipulations only affect the gray level values of individual pixels, not their spatial distribution. A digital "image analyzer," however, is designed to extract statistical data concerning morphometric parameters, such as the number of particles, area, shape, perimeter and mathematical relationships between image components. The distinction between image processing and image analysis is important because commercial image analyzers are usually designed to emphasize one procedure over the other, although they may incorporate features of each.

The substitution of electronic image analysis and image processing for the human eye and mind permits unbiased analysis, independent of expected results or classification. Although image analyzers do not possess the same discriminating powers or pattern recognition capabilities as the human eye and brain, in many applications their speed and accuracy are advantageous. The ideal image analyzer is described as ". . . able to produce area analyses, line analyses, point analyses and particle counts, as desired. It would also be able to produce accurate information about each of any number of selected minerals within a specimen and it would be capable of doing this almost instantaneously" (Jones, 1987). He lists some additional attributes of the ideal analyzer, but concludes that there is no such system, and there likely will not be one.

Pattern Recognition

Computer software can be designed to recognize the mathematical expression of the form or pattern of images. This is accomplished by complex algorithms which analyze the size, perimeter, shape factors, spatial arrangement or orientation of structural elements composing the image. This capability has been used in satellite imagery for the identification of landscape characteristics and vegetation features, and in industry for the recognition of deformed parts during manufacture. In the quantification of minerals, pattern recognition is of limited value because soil minerals seldom have unique shapes in grain mounts or thin section. Physical and chemical weathering usually obliterates the original morphology of most minerals. In addition, the two-dimensional image observed in thin section may not be representative of the mineral morphology in three dimensions because a section cut through a three-dimensional object will always have one less dimension than the original object. For example, a cubic mineral will exhibit a number of one or two dimensional shapes depending upon the orientation of the object when it is cut (Fig. 5-12). The problem is

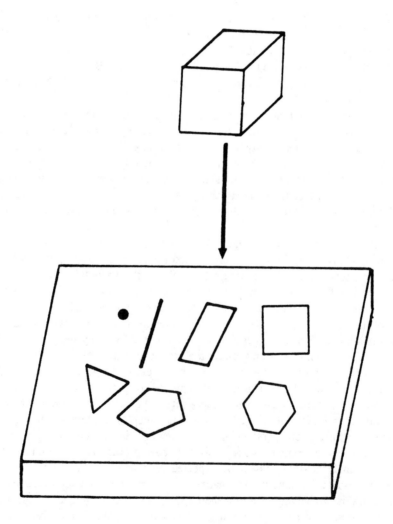

Fig. 5-12. Example illustrating the various one- and two-dimensional shapes obtained from random slices through a cube of uniform size (modified from Jones, 1987).

compounded if another mineral having a tetrahedral shape is added to the cubic mineral. Each mineral section will produce some two-dimensional shapes that are common. This commonality makes it nearly impossible to assign a two-dimensional shape to a three-dimensional object. Due to the loss of one dimension during preparation of the specimen, information critical for shape or pattern recognition is often lost.

Pattern recognition can be used successfully, however, in selected cases, such as the discrimination of porosity. In this use of pattern recognition, the pores are made visible with a dye to facilitate identification. By using a dye, the image becomes simpler: pores and nonpores. Computer software programming is then used to put limits on size, area, or shape factors of pores for characterization purposes. The application of pattern recognition for soil porosity and variable selection for characterization of pores is given by Ringrose-Voase and Bullock (1984).

Types of Image Analyzers

Two basic computerized image-analysis systems are in use, both using a video camera, an image board, and a computer. They are the semiautomatic or interactive analyzer and the automatic system. The main difference between the two is that the microscopist must make the decisions of identification and measurement in the semiautomatic system whereas the computer does much of the decision making in the fully automatic system. The pedologist must decide upon the most suitable image-analysis system based upon research needs, the measuring task at hand, and cost factors. Criteria for selecting an image-analyzing system, ranging from manual to fully automatic, are given in Table 5-4. Walter and Berns (1986) list about 25 commercially available image-processing systems and their specifications. This information may be dated because of rapid advances in image technology, but still provides a good base for comparison of features.

Semi-Automatic (Interactive) Image Analysis

In this system, the image is seen in the eyepiece of the microscope and is also observed on a graphics monitor. The image may be derived from thin sections, polished blocks, fluorescent images, or photographs. The minerals or features of interest are outlined or discriminated by the microscopist using a light pen or digitizing tablet and cursor. The computer, automatically measures various parameters of the feature that have been selected by the operator. Some of the parameters that may be selected in this mode or the fully automatic mode are given in Table 5-5 and are also discussed by Murphy et al. (1977a), Bullock and Murphy (1980), Delgado and Dorronsoro (1983), and Buckingham et al. (1988). Most analysis systems also output histograms; population statistics such as number frequency of objects, mean size, and means and standard deviations of all measured parameters. In this system we have come full circle, back to the method of areal analysis proposed by Delesse about 150 yr earlier. But the computer has made area measurements feasible, rapid, and accurate.

Although this analysis is much faster than manual methods, it still requires all the skill and experience of the microscopist to make decisions concerning the sample. The microscopist has the major decision role in assigning minerals or pedologic features to classes based on their optical properties. In the semiautomatic system, minerals that differ slightly in optical characteristics can be distinguished by the microscopist based on such characteristics as interference color, refractive index, twinning, pleochroism, etc. The operator can also compensate for variations in thickness and mineral orientation that give rise to differences in optical characteristics. The operator can also ignore artifacts such as air bubbles, grinding compound, strained embedding media, scratches, lint, or other debris on the slide. Because the operator is intimately involved in mineral recognition, identification is enhanced in this technique. Although slower and more tedious than automatic systems, Schafer and Teyssen (1987) felt that the high reliability of the results was worth the extra workload. On the other hand, any bias by the operator will be retained in this method.

Table 5-4. Criteria for selecting an image-analyzing system (modified from Jones, 1987).

Section criterion	Image-analyzing system		
	Automatic	Semiautomatic	Manual
Signals used to discriminate minerals	Optical or electron optical	Optical	Optical
Typical measuring device	TV system	Planimeter, digitizing tablet	Microscope
Specimen preparation requirements	Very demanding specimen must be polished and flat	Less demanding than for automatic measuring systems	Less demanding than for automatic or semiautomatic systems
Quality of mineral discrimination	Fair	Excellent	Excellent
Speed at which information is collected	Very fast; about 1 s per raster	Slow; about 1 s per observation	Slower than semiautomatic systems
Kind of information obtained	Zero-, one- and two-dimensional	Zero-, one- and two-dimensional	Zero-dimensional
Nature of material that should be measured	Simple materials, where optical discrimination is easy	Complex materials, where mineral discrimination is very difficult	Single, very difficult
Types of specimen that should be analyzed	Large numbers of routine, simple specimens	Small numbers of difficult specimens	Individual complex materials (for research purposes)
Analyzing costs per specimen	Small	Larger	Very Large

Automatic Image Analysis

Fully automatic image-analysis systems, such as the Quantimet (Cambridge Instruments, Deerfield, IL) (Bradbury, 1977), are based on television scanning technology. They are capable of doing many of the analyses listed above, but one component can be distinguished or discriminated from another based in the strength of the optical signal from the sample. Most typical systems process images with a resolution of 512 x 512 pixels (picture elements, the smallest part of the digital image). Using solid-state cameras, some high-resolution systems use 1024 x

Table 5-5. Variables easily measured and calculated by image-analysis systems (modified from Buckingham et al. 1988).

Variable	Description
Area	The size in area units of individual objects
Area fraction	The percentage of the field occupied by individual objects or a class of objects
Perimeter	The length of the boundary around the object
Form factor	A ratio of area to perimeter [$4\pi Area/(perimeter)^2$]
Equivalent spherical diameter	The diameter of a sphere whose bisected cross-sectional area is equal to the area of the feature
Feret lengths	The projected lengths of an object onto the X and Y coordinates
Orientation	The angular orientation in degrees of an object's long axis
Aspect ratio	An object's long-axis length divided by its perpendicular short-axis length

x 1024, or even 4096 x 4096, pixels. To distinguish an image, each pixel must have a range of intensities from black to white (gray levels). Commonly 256 gray levels per pixel are adequate, but some systems may use more or less depending on the requirements of the analysis to be performed. This is a wider brightness range than the human eye (20-30 levels) or photographic film (Russ, 1990). The ability to differentiate gray levels and to determine the number of pixels having a certain gray level is one of the most powerful tools in digital image analysis, especially in enhancing resolution of low-contrast images. These systems are usually equipped with a cursor or light pen to manually alter the image, erase artifacts, group constituents together, or select features of interest for measurement.

Images may be captured in black and white or color, although color images require considerably more computer storage space. The benefit of color is that many minerals and pedological features can be differentiated on the basis of color even though they may have matching gray tones.

Problems in Mineral Identification

The automatic system has many advantages, especially in speed. On the other hand, no optical procedure based solely on gray-tone differentiation can fully determine compositional differences, especially in a sample as complex as most soil systems. In crossed polarized light, crystal structure, mineral orientation, and sample thickness determine the gray level of all anisotropic minerals. Most common minerals (e.g, quartz) will exhibit gray tones ranging from black to white depending upon orientation (refer to Color Plate 5-13). Thus, cross-polarized light is usually inadequate to differentiate minerals in the automatic mode. Quartz and gypsum often give similar gray-level images, as do quartz and untwinned feldspars. Likewise, in plane polarized light, differentiating minerals based on gray tones is often difficult because many minerals have no color and appear clear. Some of the difficulties of analyzing something as complex as soils have been discussed by Amstutz and Giger (1971). Despite these difficulties, favorable comparisons have been noted between manual and automatic image analysis (Jones, 1977). In plane-polarized light, opaque or colored minerals such as biotite can be easily differentiated from the remainder of the mineral matrix (refer to Color Plate 5-14). Likewise, in a sample from an Oxisol, the Fe-rich matrix can be easily differentiated from the quartz skeletal grains and pores (refer to Color Plate 5-15).

To illustrate the potential problems involved in computer image analysis, consider a granite composed primarily of quartz and K-feldspar. The grain is shown in crossed (refer to Color Plate 5-16A) and plane (refer to Color Plate 5-16B) polarized light. In crossed polarized light, both the quartz (q) and feldspar (f) grains exhibit a range of gray tones (refer to Color Plate 5-16A) that make identification difficult. In plane polarized light (refer to Color Plate 5-16B), all portions of the grain have the same gray level, which does not facilitate mineral differences. Color Plate 5-16C is a corresponding x-ray microradiograph of the same grain, which shows the differences in composition based on x-ray absorbance, not optical properties. In Color Plate 5-16C, f is brighter than q because of greater x-ray absorbance. Image analysis of the microradiograph gives the correct proportions of quartz (53%) and feldspar (47%).

Although difficulties abound, Allard and Sotin (1988) present a technique for determining modal composition of rocks based on image analysis. However, their material had few mineral components and compositional phases could be differentiated by gray-level tones. They indicated that where two minerals have the same brightness range, staining techniques or thicker sections should be used. The selection of a different light source may also enhance mineral contrast. Amor and Block (1967) describe a technique for depositing a black layer over certain transparent particles (silica, alumina, and glass) for image analysis. The increased contrast facilitates identification of grains that are clear or transparent and not easily differentiated in light microscopy. Chemical stains and dyes are also effective means to enhance the contrast of minerals that exhibit similar optical characteristics.

In instruments using color-imaging procedures, any feature seen as a different color can be isolated and measured. Even so, groupings of components may include biotite and ferriargillans, quartz and gypsum, and all opaques. It is in color-image analysis that staining procedures can play an important role. The image analyzer can differentiate the red stain of calcite from unstained dolomite (refer to Color Plate 5-8B) and stained feldspar from quartz. Even the most sophisticated automatic image analyzer must at some time require the intervention of the microscopist to decide which mineral or property should be evaluated. Only the trained pedologist can truly assess the value of the results.

Automatic systems require careful sample preparation. Although they have the potential to be more precise, automatic systems are more sensitive to artifacts introduced during preparation or differences in color or image intensity caused by variable sample thickness. Algorithms are available, however, to partly compensate for gray-level differences and systematic variances in sample thickness or illumination, but not random variances in thickness from poor impregnation, grain plucking, or poor sample preparation.

Applications of Image Analysis

Although computer-assisted image analysis may be considered as an extension of the optical microscope in making qualitative and quantitative statements about a sample, it has been most extensively utilized in specific areas that optimize its enhanced capabilities. Image analysis has not been extensively used for mineral identification and quantification due to some of the difficulties outlined above, but it has been used in the analysis of soil porosity and structure. It is nearly impossible to use image analysis to identify and quantify soil minerals unless the size, shape, orientation, and distribution of soil pores is also considered.

Soil Porosity

Although pores are not minerals they are considered as important compositional phases of soils (Swanson & Peterson, 1942, Anderson & Binnie, 1961) and rocks (Ehrlich et al., 1984). Pores are among the easier compositional phases to differentiate, largely because of staining techniques which distinguish them from non-pore components. In plane polarized light, and especially in

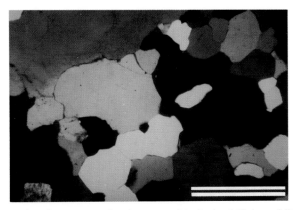

Color Plate 5-13. Example of quartz grains in crossed polarized light exhibiting a range of gray tones, bar length = 1 mm.

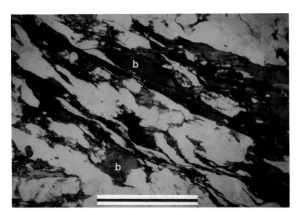

Color Plate 5-14. Brown biotite (b) from a biotite schist as seen in plane polarized light. Light colored minerals are quartz and feldspar, bar length = 1 mm.

Color Plate 5-15. Oxisol horizon exhibiting strong red pigmentation and clear quartz grains and pores in plane polarized light, bar length = 1 mm.

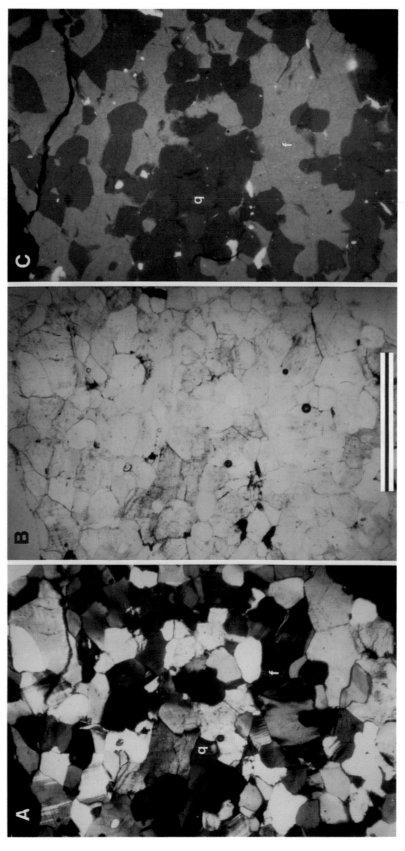

Color Plate 5-16.   Granitic grain composed of quartz (*q*) and feld spar (*f*).   (*A*) Cross-polarized light; (*B*) plane polarized light; (*C*) x-ray microradiograph.   Bar length = 1 mm.   [*A* and *C* modified from Drees & Wilding (1983) with permission of Elsevier Science Publishers].

reflected-fluorescence illumination, stained pores stand out against the background (Fig. 5-17). A number of studies have used staining and fluorescent dyes to analyze porosity in reservoir rocks (Ehrlich et al., 1984; Ruzyla, 1986) and soils (Murphy et al., 1977a,b; Ringrose-Voase & Bullock, 1984; Puentes & Wilding, 1990). Uvitex OB fluorescent dye (CIBA-GEIGY, New York) has been successfully used with polyester resins. Yanguas and Dravis (1985) and Ruzyla and Jezek (1987) list a number of dye formulas for use with epoxy resins. Although the procedure is not strictly optical, Bisdom et al. (1983) used backscattered scanning-electron images at low magnification (30 to 60 x) to differentiate pores and mineral grains.

Image analysis not only permits the quantification of porosity, but can quantify pore size, shape, orientation, and distribution, factors that affect infiltration, root penetration and air transfer. The first use of computer-image analysis of thin sections for the characterization of soil pore patterns and shape was by Jongerius (Jongerius et al., 1972; Jongerius, 1974). Image analysis permitted the separation and classification of pores according to shape. Using methylene blue as a dye, Bouma et al. (1977) were able to partition pores and vughs into three shape groups and three area classes for characterization and quantification. Murphy et al. (1977a) and Bullock and Murphy (1980) used Uvitex OB fluorescent dye to enhance pore recognition to discriminate pores from other pedological features (see Fig. 5-17). The combination of the fluorescent dye and high-contrast negatives facilitated pore recognition by image analysis. The authors made the distinction between two types of measurements, "basic" measurements (e.g., area, size, perimeter, counts, and Feret diameters) and "derived" measurements (e.g., shape, orientation, spherosity, form or irregularity, and spatial distribution). In a companion paper (Murphy et al., 1977b), the authors

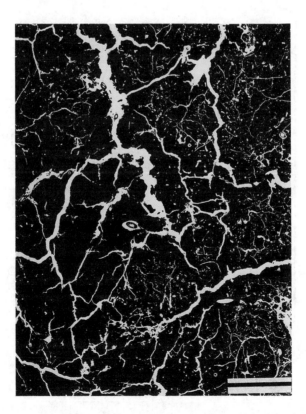

Fig. 5-17. Example of fluorescent Uvitex OB dye on a polished block to define pore pattern. Bar length = 20 mm. (Puentes & Wilding, 1990).

illustrate the range of usefulness of the image-analysis technique. The authors also indicated that image analysis permitted the rapid measurement of size, shape, and orientation of pores. Image analysis was more accurate and at least 20 times faster than point counting. Ringrose-Voase and Bullock (1984) have developed a computer algorithm for pattern recognition in order to describe pore patterns quantitatively. Depending on shape parameters, they were able to classify voids into channels, fissures and packing voids/vughs. In a later paper, Ringrose-Voase (1987) examined two models for pore recognition and quantification, but in the broader context of relating pore patterns to soil structure.

Quantitative pore distributions may be reported as cumulative curves (Fig. 5-18) or as histograms (Fig. 5-19). In each of these investigations, porosity was compared among similar soils under different management systems, and showed significant differences in pore-size distribution and in mean pore-size (Fig. 5-18 and 5-19). Murphy et al. (1977b) showed quantitative differences in pore size in two examples, one due to wetness and the other due to compaction. Such differences illustrate the utility of quantitative microscopy in characterizing practical soil problems, such as the impact of soil management systems on porosity, fluid flow, drainage, air transfer and rooting potential.

If thin sections are taken so as to maintain orientation, vertical and horizontal Feret diameters are a good measure of pore orientation. To show differences in pore orientation between treatments, star plots using polar coordinates can be drawn (Fig. 5-20). This example shows that soil samples from the cultivated field exhibited elongated pores with a strong horizontal orientation, whereas those from the pasture had randomly distributed equant pores. The use of star plots is discussed in theory (Murphy et al., 1977a) and in practice (Murphy et al., 1977b; Puentes & Wilding, 1990). Murphy et al. (1977a) caution that orientation measurements are sensitive to slight

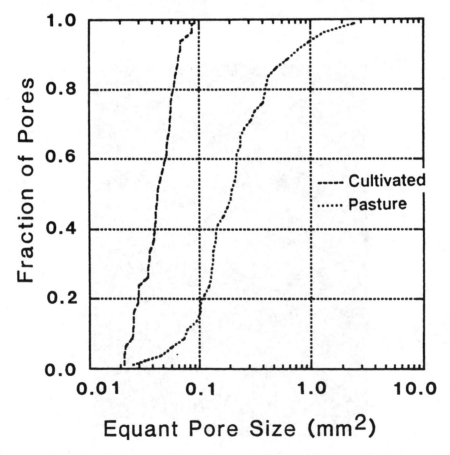

Fig. 5-18. Cumulative pore-size distribution for equant pores at continuous-cropping and pasture sites in Texas (modified from Puentes & Wilding, 1990).

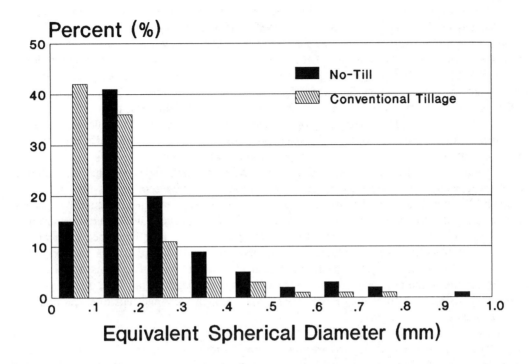

Fig. 5-19. Histograms showing pore-size distribution of nonplanar pores from no-till and conventional tillage management in Kentucky (Drees, unpublished data).

Percentage of macroporosity as a function of angular orientation

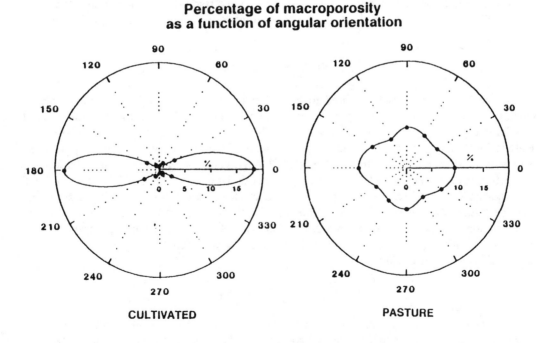

Fig. 5-20. Star plots showing differences in angular orientation of pores in continuous cropping (A) and pastures (B) for surface horizon. The distance from the center to the radius circle indicates total length in millimeter/100 mm^2 of vertical section [reprinted from Puentes and Wilding (1990) with permission of Int. Congr. Soil Sci.].

changes in orientation. Also, pores often exhibit a bimodal distribution pattern, with larger pores exhibiting orientation while smaller pores are more uniform in shape. In such cases, pores should be separated into size classes prior to analysis.

Pore-size distributions derived from thin sections by image analysis compares favorably with other techniques (Lawrence, 1977; Bui et al., 1989). However, Bullock and Thomasson (1979) found that image analysis measures the maximum pore diameter, whereas techniques based on water retention measure the diameter of the smallest point of each pore. Part of this problem is due to difficulties in measuring small pores (<30 μm) in thin section. Porosity, as defined by image analysis, is more closely linked to effective porosity than to total porosity. Ehrlich et al. (1984) prefers the term "Total Optical Porosity" to distinguish it from porosity measured by other techniques. Even so, porosity defined by optical means is dependent upon the magnification used. The term "Total Optical Porosity" may have little meaning unless the magnification used to measure pores is standardized, or at least reported along with the porosity values.

The ability to store digital data and mathematically manipulate the data has also allowed scientists to analyze porosity patterns that would have been tedious or impossible by manual methods. Macdonald et al. (1986a,b) were able to quantitatively reconstruct the three-dimensional microscopic pore structure in the Berea sandstone. The pore network was determined from 78 serial sections by successively grinding off 10-μm increments of a 1- by 1-mm block and analyzing photomicrographs of the polished block. The authors digitized the image using a graphics tablet, although a fully automatic system could have been used. This type of analysis offers a more accurate picture of the pore network than a two-dimensional image. In a single thin or polished section, the visible pore space only represents sections through pores and fails to reveal their true three dimensional distribution.

In a similar study, Scott et al. (1988a,b) analyzed the pore-structure pattern in a cracking clay soil. Their blocks were larger, measuring 72- by 48-mm in size, and impregnated with a fluorescent dye. The block was ground down in 50-μm increments and each surface photographed to reveal the changing crack pattern. This process produced a three-dimensional pattern based on a sequence of two-dimensional slices through the block. Scott et al. (1988a) concluded that only 10 to 20 sections at 50-μm intervals were necessary to define the type of crack pattern. The interval spacing between successive slices was larger in this study than in the one by MacDonald et al. (1986 a,b) because the cracks were of a larger dimension than the smaller pores in the Berea sandstone.

FUTURE TRENDS

Pedologists and mineralogists need to be visionary in their approach to problem solving; looking to the future rather than relying on past accomplishments. Predictions about future trends in microscopy, however, are difficult at best. Some aspects of light microscopy and other analytical techniques will change. Maximum utilization of the changes and new developments can only be made if we are also willing to change our thoughts and concepts about traditional ways of conducting research and the role and function of traditional research tools. This does not imply that we abandon our present microscopes or analytical tools, but that we are willing to accept the fact that future advances will allow us to look at soil samples differently, in ways that we do not now envision, and extract information heretofore unobtainable. Some of the techniques discussed exist now but have not been utilized to their full extent, while others are experimental and their potential has yet to be explored.

Fluorescence Microscopy

Fluorescence microscopy utilizes the property of minerals or pedological features to emit light when excited by an outside energy source. These energy sources may be heat (thermoluminescence), electrons (catholuminescence), x-rays (x-ray luminescence) or visible or ultraviolet light (UV fluorescence). Thin sections or polished blocks are usually used in fluorescence microscopy. Cathodoluminescence has been mostly applied to the study of compositional variations and diagenesis of carbonates (Solomon & Walker, 1985; Machel, 1985; Marshall, 1988). Fluorescence microscopy using UV light has also been used to study carbonates (Dravis & Yurewicz, 1985). Although these studies have been nonquantitative visual examinations, computer-assisted image analysis may have a potential in quantifying carbonate distributions as well as other minerals and pedological features.

Computer-Assisted Image Analysis

The computer processing and analysis of microscopic images has only been available for a relatively short period of time. Despite this short-term availability, the proliferation of computers, software, and the continued reduction of costs has made computer-assisted image analysis a common research tool in many laboratories. The entire field of computers and image processing is rapidly changing. The development of computer hardware, applications software, and processing algorithms is continuing at an accelerated pace. The description of the accelerated pace of development, processing speed, and capabilities of image analysis is similar to a statement made by Dott (1983) concerning geologic time: ". . . that which seems impossible by human standards becomes possible, and the improbable becomes inevitable." The mineralogical complexity of many soils, and the inability of currently available computer-assisted image analysis to successfully discriminate among minerals with similar optical properties has limited the use of fully automatic image analysis in quantitative mineralogy. As development costs decrease and computer speed and storage capacity increase, image analysis will be an asset to the soil mineralogist. Future techniques such as superimposing color images in crossed, plane and circularly polarized light modes may facilitate mineral or feature discrimination. We have just begun to explore the capabilities of image analysis by computers in soil micromorphology and mineral quantification. It should be remembered, however, that a computer, no matter how fast or great the storage capacity, cannot replace a trained microscopist. An image-analysis system will never be better than the operator who must insure that the decisions are correct and valid. Image analysis, like the optical microscope, is only a tool to be used in solving problems in an expeditious manner.

Linking computer-assisted image analysis and mathematical models (discussed below) is the field of mathematical morphology, which attempts to study the form and structure of image components (Serra, 1982). This is more encompassing than form recognition, as mentioned earlier, and allows the separation of features based on gray tones and their size, shape, and spatial arrangement. In mineral analysis and quantification, this approach permits the segmentation of objects based on morphological properties. Although these properties are not optical, and are fraught with many of the same problems as form recognition, mathematical morphology has the potential to greatly improve the performance of image analysis.

Mathematical Concepts

Mathematical models are becoming more common because of the increasing power of personal computers and the ability to capture images as numerical abstractions that can be processed and analyzed in a number of ways. Stereology attempts to develop structural relationships between objects. Stereology has been defined by Weibel (1979) as "a body of mathematical methods relating three-dimensional parameters defining the structure to two-dimensional measurements obtained on sections of the structure." Because stereology is mathematically based, techniques can be used to study three-dimensional structures and the geometric aspects of these structures, regardless of their size or composition. These techniques have been used successfully in the material sciences (DeHoff & Rhines, 1968; Underwood, 1970) and biology (Weibel, 1979) and have been introduced into the geologic (Amstutz & Giger, 1971; Jones, 1987) and soils literature (Ringrose-Voase & Nortcliffs, 1987; McBratney & Moran, 1990).

Another mathematical approach applicable to computer processing is Markov models. Markov models do not treat measured entities as random events, but examine the frequency or probability of transition from one state or one feature to another. For example, the probability of transition from one mineral to another, from matrix to pore, or matrix to argillan to pore can be estimated and modeled. A Markovian model has an element of randomness, but also an element of predictability in time or space (Lin & Harbaugh, 1984). The distribution and shape of an individual mineral or pedological feature influences the probability of a transition to another mineral or feature. Markov models may be used in one dimension (time or linear distance), two dimensions (thin-section slide or map), or three dimensions (solid objects). Markov models have been used in the analysis of soil structure (Hewitt & Dexter, 1981), but have not been used to any great extent in soil mineral investigations. Stereology, Markov models, mathematical morphology and other mathematical concepts will play increasing important roles in the quantification of soil constituents as the concepts become better known and computer software becomes available to utilize these concepts and interface with computer analysis.

New Microscopic Techniques

The resolution of the optical microscope (about 300 nm) is limited by the wavelength of visible light and the ability of lenses to diffract light. Other instruments such as the scanning electron (SEM) and transmission electron (TEM) microscopes have greatly extended the level of resolution by using electrons and magnetic lenses to replace visible light and glass lenses. These are by no means the only forms of microscopy. Newer instruments take different approaches in magnifying an image and extracting unique kinds of information from a sample. Confocal microscopy, or confocal laser-scanning microscopy is one of the emerging microscopic techniques and is quickly finding its way into biological and materials research (Keeler, 1991). McCrone (1991) says that "confocal microscopy is one of the most important developments in light microscopy since its invention about 400 years ago." Suppliers of confocal microscopes indicate that the resolution is on the order of 0.1 μm, considerably better than light microscopy, and well within the range to study individual clay minerals. The ability to study individual clay particles has been one of the disadvantages of the light microscope. Presently this technique has been utilized for examining the surface topography of opaque materials and the surface and subsurface topography of translucent samples. The sample is examined in reflected light and the scanning nature of the light beam greatly reduces the depth of field problems inherent in normal light microscopy. Images generated in confocal microscopy are ideally suited to computer enhancement and analysis. Confocal microscopy is also well suited to the science of stereology. In suitable situations, computer software can arrange successive confocal images into three-dimensional reconstructions, which could facilitate volume measurements of individual objects. The selection of the appropriate laser frequency may also facilitate fluorescence microscopy.

A new family of scanning-probe microscopes is vastly increasing the resolution of surface detail, often at the atomic level. Scanning-tunneling microscopes (STM) measure the current flow (tunneling current) between the sample and a probe tip positioned a few tenths of a nm above the sample (Binnig & Rohrer, 1985). The probe tip is scanned in an X-Y raster mode and measures variations in tunneling current in the Z direction. Three-dimensional changes in tunneling current may be due to topographic relief or charge-density variations yielding both geometric and electronic information about the sample surface (Kuk & Silverman, 1989). Hansma et al. (1988) point out that due to the sensitivity of the tunneling current between the probe tip & sample, contours better than 1/100 of an atomic diameter can be revealed. Lateral resolution, however, may be considerably less.

The STM can only analyze conducting and semiconducting specimens and, thus, is of limited usefulness for soil materials. The atomic-force microscope (AFM), on the other hand, is well suited for nonconductive specimens typical of soils. The AFM operates in a similar scanning mode to the STM, but records repulsive forces between the atomic cloud of surface atoms and the electron cloud of the probe tip. Other scanning-probe microscopes include the magnetic-force microscope (MFM), which senses magnetic characteristics of the sample, and the laser-force microscope (LFM) (Wickramasinghe, 1989). Each member of this family of scanning-probe microscopes interacts with the sample surface differently to extract different types of information, often with a resolution of a few nm, the unit-cell dimensions of clay minerals. They also can be operated at ambient conditions, eliminating the high vacuums required by SEM and TEM. The first application of this technology to soils was by Hartman et al. (1990), in which the AFM was used to examine the basal surfaces of montmorillonite and illite. The resulting image comprised 90 unit cells in area and showed hexagonal rings of O atoms as spots.

The scanning-probe technique has also been applied to transmission and reflected light microscopy. In the scanning near-field optical microscope (SNOM or NSOM) an extremely small aperture (about 10 nm in diameter) illuminated by an argon laser source (λ = about 500 nm) is scanned across the sample (Pohl et al, 1988). Sensitive photomultiplier tubes record the transmitted or reflected light. Resolution, which is determined by the aperture diameter, is about 10 to 20 nm laterally, but as small as 1 nm vertically, a considerable improvement over conventional light microscopes.

Scanning-probe microscopes do not all rely on electron tunneling, photons can also be used. Incident light at a critical angle to the interface between materials of different refractive indices will cause total internal reflection. With total internal reflection an evanescent field is generated parallel to the sample surface, but decays exponentially normal to the total reflection surface (Reddick et al, 1989; Guerra, 1990). In the photon scanning-tunneling microscope (PSTM or PTM), spatial variations in the sample surface are measured by recording the photon intensity in the evanescent field by an optical fiber probe tip. Vertical resolution is reported to be <1 nm with lateral resolution of about 150 nm.

Scanning-probe microscopes require a personal computer for probe positioning, rastering, and processing and storage of digital data. The digital image can be processed for real-time video display or analyzed by a variety of image-analysis software programs. Although the scanning-probe technology is only several years old, technological and computer control advancements have brought the price down to where these microscopes are competitive with a good research petrographic microscope. Because these microscopes are just being introduced to the scientific community on a commercial scale, their potential is still largely unknown. The problem with these microscopes is not in the technology, but in the interpretation of the images. They should be able to play important roles in problems involving surface-chemical reactions and clay-mineral interactions. Improved resolution of mineral surfaces should be able to substantiate or refute many of our ideas and hypotheses about the atomic structure of clay minerals and their chemical reactions (e.g., interactions involving organics and pesticides). If individual clay minerals can be identified, clay mineral quantification may be more exacting.

CONCLUSIONS

Many aspects of soil science are becoming more quantitative. One only has to read *Soil Taxonomy* (Soil Survey Staff, 1975) to appreciate the transition from descriptive morphology to quantitative analyses. Minerals and pedologic features in soils are among the components that need to be quantified, and many techniques are currently available which attempt to reach that goal. Quantification is not an end unto itself. The purpose is to solve practical problems and gain insight into soil dynamics and soil genesis.

Optical techniques are not a panacea, however. Like all quantification techniques, there are technical and practical problems which reduce reliability and increase error. The magnitude of these nonstatistical errors, however, cannot be easily calculated. They have been discussed herein, not so much to show the potential problems with optical measurements, but to make the microscopist aware of the difficulties and limitations so that appropriate action can be taken to minimize the error involved and maximize the reliability of measured data.

Optical techniques are among the oldest for soil mineral quantification. Through the years, advances have been made which make quantification both easier and more reliable. Optical analysis of thin sections allows the microscopist to view mineral grains and pedological features in situ, undisturbed by preparatory or analytical procedures. The point-counting technique is favored because of its ease and statistical validity. Even though new techniques and instruments continue to be added to our arsenal of investigative tools, the petrographic microscope will not be superseded by other forms of microscopy, but will continue to be at the forefront of mineralogical research and analysis.

ACKNOWLEDGMENTS

Contribution no. TA 30063 from the Texas Agricultural Experimental Station and Contribution no. 91-180-B from the Kansas Agricultural Experimental Station.

REFERENCES

Allard, B., and C. Sotin. 1988. Determination of mineral phase percentages in granular rocks by image analysis on a microcomputer. Comp. Geosci. 14:261-269.

Amonette, J.E., F.A. Khan, H. Gan, J.W. Stucki, and A.D. Scott. 1994. Quantitative oxidation-state analysis of soils. p. 83-113. *In* J Amonette and L. W. Zelazny (ed.) Quantitative methods in soil mineralogy. SSSA Misc. Publ. SSSA, Madison, WI.

Amor, A.F., and M. Block. 1967. A technique for staining curtain types of transparent particles, to facilitate size analysis by television scanning microscopy. J. R. Micros. Soc. 88:601-605.

Amstutz, G.C., and H. Giger 1971. Stereological methods applied to mineralogy, petrology, mineral deposits and ceramics. J. Micros. 95:145-164.

Anderson, W.M., and R.R. Binnie. 1961. Modal analysis of soils. Soil Sci. Soc. Am. Proc. 25:499-503.

Ashley, G.M. 1973. Impregnation of fine-grained sediments with a polyester resin: A modification of Altemuller's method. J. Sediment. Petrol. 43:298-301.

Binnig, G., and H. Rohrer. 1985. The scanning tunneling microscope. Sci. Am. 253:50-56.

Bisdom, E.B.A., H.A. van Adrichem Boogaert, G. Heintzberger, D. Schoonderbeek, and F. Thiel. 1983. Porosity measurements form analysis of mineral grains in thin sections from oil-gas reservoir rocks using Quantimet 720 and BESI. Geoderma 30:323-337.

Bouma, J., A. Jongerius, O. Boersma, A. Jager, and D. Schoonderbeek. 1977. The function of different types of macropores during saturated flow through four swelling soil horizons. Soil Sci. Soc. Am. J. 41:945-950.

Bradbury, S. 1977. Quantitative image analysis. p. 91-116. *In* G.A. Meek and H.Y. Elder (ed.) Analytical and quantitative methods in microscopy. Cambridge Univ. Press, New York.

Brewer, R. 1976. Fabric and mineral analysis of soils. Robert E. Krieger Publ. Co., Huntington, NY.

Buckingham, W.F., J.M. Spaw, and E.B. Peacock. 1988. Evaluation of waste-entrained concrete using image analysis and quantification. Nucl. Chem. Waste Manag. 8:261-268.

Bui, E.N., A.R. Mermut, and M.C.D. Santos. 1989. Microscopic and ultramicroscopic porosity of an Oxisol determined by image analysis and water retention. Soil Sci. Soc. Am. J. 53:661-665.

Bullock, P., and C.P. Murphy. 1980. Towards the quantification of soil structure. J. Micros. 120:317-328.

Bullock, P., and A.J. Thomasson. 1979. Rothamsted studies of soil structure--II. Measurement and characterization of macroporosity by image analysis and comparison with data from water retention measurements. J. Soil Sci. 30:391-413.

Buol, S.W., and F.D. Hole. 1961. Clay skin genesis in Wisconsin soils. Soil Sci. Soc. Am. Proc. 25:377-379.

Cady, J.G. 1950. Rock weathering and soil formation in the North Carolina Piedmont region. Soil Sci. Soc. Am. Proc. 15:337-352.

Cady, J.G., L.P. Wilding, and L.R. Drees. 1986. Petrographic microscopic techniques. p. 185-218. *In* A. Klute (ed.) Methods of soil analysis. Part 1. 2nd ed. Agron. Monogr. 9. SSSA, Madison, WI.

Chayes, F. 1949. A simple point counter for thin-section analysis. Am. Mineral 34:1-11.

Chayes, F. 1956. Petrographic modal analysis. John Wiley & Sons, New York.

Chayes, F., and H.W. Fairbairn. 1951. A test on the precision of thin-section analysis by point count. Am. Mineral. 36:704-712.

Crook, T. 1907. Titaniferous volcanic rocks. Geology Magazine, Decade 5,4:157-165.

Davis, B.L., and M.J. Walawender. 1982. Quantitative mineralogical analysis of granitoid rocks: a comparison of X-ray and optical techniques. Am. Mineral. 67:1135-1143.

DeHoff, R.T., and F.N. Rhines. 1968. Quantitative microscopy. McGraw-Hill Book Co., New York.

Delgado, M., and C. Dorronsoro. 1983. Image analysis. p. 71-86. *In* P. Bullock and C.P. Murphy (ed.) Soil micromorphology, Vol. 1. Techniques and applications. AB Acad. Publ., Berkhamsted, Hertz, England.

Dott, R.H. 1983. Episodic sedimentation--how normal is average? How rare if rare? Does it matter? J. Sediment. Petrol. 53:5-23.

Dravis, J.J., and D.A. Yurewicz. 1985. Enhanced carbonate petrology using florescence microcopy. J. Sediment. Petrol. 55:795-804.

Drees, L.R., and L.P. Wilding. 1983. Microradiography as a submicroscopic tool. Geoderma 30:65-76.

Drees, L.R., and L.P. Wilding. 1987. Micromorphic record and interpretations of carbonate forms in the Rolling Plains of Texas. Geoderma 40:157-175.

Dunn, T.L., R.B. Hessing, and D.L. Sandkuhl. 1985. Application of voice recognition in computer-assisted point counting. J. Sediment. Petrol. 55:602-603.

Ehrlich, R., S.K. Kennedy, S.J. Crabtree, and R.L. Cannon. 1984. Petrographic image analysis--I. Analysis of reservoir pore complexes. J. Sediment. Petrol. 54:1365-1378.

Eswaran, H. 1968. Point-count analysis as applied to soil micromorphology. Pedologie 18:238-252.

Fleet, W.F. 1926. Petrological notes on the Old Red Sandstone of the West Midlands. Geology Magazine 63:505-516.

Friedman, G.M. 1959. Identification of carbonate minerals by staining methods. J. Sediment. Petrol. 29:87-98.

Friedman, G.M. 1960. Chemical analysis of rocks with the petrographic microscope. Am Mineral. 45:69-78.

Friedman, G.M. 1971. Staining. p. 511-530. *In* R.E. Carver (ed.) Procedures in sedimentary petrology. John Wiley & Sons, New York.

Fry, W.H. 1933. Petrographic methods for soil laboratories. USDA Tech. Bull. 344. U. S. Gov. Print. Office, Washington, DC.

Galehouse, J.S. 1969. Counting grain mounts: Number percentage vs. number frequency. J. Sediment. Petrol. 39:812-815.

Galehouse, J.S. 1971. Point Counting. p. 385-407. *In* R.E. Carver (ed.) Procedures in sedimentary petrology. Wiley Interscience, New York.

Gilkes, R.J. 1994. Transmission electron micrscope analysis of soil materials. p. 177-204. *In* J.E. Amonette and LW. Zelazny (ed.) Quantitative methods in soil mineralogy. SSSA Misc. Publ. SSSA, Madison, WI.

Glagolev, A.A. 1934. Quantitative analysis with the microscope by the "point" method. Eng. Min. J. 135:399-400.

Gross, D.L., and S.R. Moran. 1970. A technique for the rapid determination of the light minerals of detrital sands. J. Sediment. Petrol. 40:759-761.

Grossman, R.B. 1964. Composite thin sections for estimation of clay-film volume. Soil Sci. Soc. Am. Proc. 28:132-133.

Guerra, J.M. 1990. Photon tunneling microscopy. Appl. Optics 29:3741-3752.

Hagni, R.D. 1966. The preparation of thin sections of fragmental materials using epoxy resin. Am. Mineral. 51:1237-1242.

Halley, R.B. 1978. Estimating pore and cement volumes in thin section. J. Sediment. Petrol. 48:642-650.

Hansma, P.K., V.B. Elings, O. Marti, and C.E. Bracker. 1988. Scanning tunneling microscopy and atomic force microscopy: Application to biology and technology. Science (Washington, DC) 242:209-216.

Harris, W.G., and L.W. Zelazny. 1985. Criteria assessment for micaceous and illitic classes in Soil Taxonomy. p. 147-160. *In* J.A. Kittrick (ed.) Mineral classification of soils. SSSA Spec. Publ. 16. SSSA, Madison, WI.

Harris, W.G., L.W. Zelazny, J.C. Baker, and D.C. Martens. 1985. Biotite kaolinization in Virginia Piedmont soils--I. Extent, profile trends, and grain morphological effects. Soil Sci. Soc. Am. J. 49:1290-1297.

Harris, W.G., J.C. Parker, and L.W. Zelazny. 1984. Effects of mica content on engineering properties of sand. Soil Sci. Soc. Am. J. 48:505-505.

Hartman, H., G. Sposito, A. Yang, S. Manne, S.A.C. Gould, and P.K. Hansma. 1990. Molecular-scale imaging of clay mineral surfaces with the atomic force microscope. Clays Clay Min. 38:337-342.

Hewitt, J.S., and A.R. Dexter. 1981. Measurement and comparison of soil structures. Appl. Math. Model. 5:2-12.

Higham, N. 1963. A very scientific gentleman: The major achievements of Henry Clifton Sorby. The MacMillan Co., New York.

Hilliard, J.E. 1968. Measurement of volume in volume. p. 45-76. *In* R. T. DeHoff and F.N. Rhines (ed.) Quantitative microscopy. McGraw-Hill Book Co., New York.

Holmes, A. 1930. Petrographic methods and calculations. Thomas Murphy & Co., London.

Houghton, H.F. 1980. Refined techniques for staining plagioclase and alkali feldspars in thin section. J. Sediment. Petrol. 50:629-631.

Hunt, W.F. 1924. An improved Wentworth recording micrometer. Am. Mineral. 9:190-193.

Hurlbut, C.S. 1939. An electric counter for thin-section analysis. Am. J. Sci. 237:253-261.

Hutchison, C.S. 1974. Laboratory handbook of petrology techniques. John Wiley & Sons, New York.

Innes, R.P., and D.J. Pluth. 1970. Thin section preparation using epoxy impregnation for petrographic and electron microprobe analysis. Soil Sci. Soc. Am. Proc. 34:483-485.

Inoue, S. 1986. Video microscopy. Plenum Press, New York.

Johannsen, A. 1919. A planimeter method for the determination of the percentage composition of rocks. J. Geol. 27:276-285.

Jones, M.P. 1977. Automatic image analysis. p. 167-199. *In* J. Zussman (ed.) Physical methods in determinative mineralogy. Acad. Press, New York.

Jones, M.P. 1987. Applied mineralogy: A quantitative approach. Graham & Trotman, London.

Jongerius, A. 1974. Recent developments in soil micromorphometry. p. 67-83. *In* G.K. Rutherford (ed.) Soil microscopy. The Limestone Press, Kingston, Ontario, Canada.

Jongerius, A., D. Schoonderbeek, A. Jager, and St. Kowalinski. 1972. Electro-optical soil porosity investigation by means of Quantimet-B equipment. Geoderma 7:177-198.

Keeler, R. 1991. Confocal microscopes. R&D Magazine 33(5):40-42.

Kubiena, W.L. 1938. Micropedology. Collegiate Press, Inc., Ames, IA.

Kuk, Y, and P.J. Silverman. 1989. Scanning tunneling microscope instrumentation. Rev. Sci. Instrum. 60:165-180.

Langford, F.F. 1962. Epoxy resin for oil immersion and heavy mineral studies. Am. Mineral. 47:1478-1480.

Larsen, E.S., and F.S. Miller. 1935. The Rosiwal method and the modal determination of rocks. Am. Mineral. 20:260-273.

Lawrence, G.P. 1977. Measurement of pore sizes in fine-textured soils: A review of existing techniques. J. Soil Sci. 28:527-540.

Lin, C., and J.W. Harbaugh. 1984. Graphic display of two- and three-dimensional Markov computer models in geology. Van Nostrand Reinhold Co., New York.

Lincoln, F.C., and H.L. Rietz. 1913. The determination of the relative volumes of the components of rocks by mensuration methods. Econ. Geol. 8:120-139.

Lindholm, R.C., and R.B. Finkelman. 1972. Calcite staining: semiquantitative determination of ferrous iron. J. Sediment. Petrol. 42:239-242.

Loewinson-Lessing, F.Y. 1954. A historical survey of petrology. Oliver & Boyd, London.

Macdonald, I.F., P. Kaufmann, and F.A.L. Dullien. 1986a. Quantitative image analysis of finite porous media--I. Development of genus and pore software. J. Micros. 144:277-296.

Macdonald, I.F., P. Kaufmann, and F.A.L. Dullien. 1986b. Quantitative image analysis of finite porous media--II. Specific genus of cubic lattice models and Berea sandstone. J. Micros. 144:279-316.

Machel, H.G. 1985. Cathodoluminescence in calcite and dolomite and its chemical interpretation. Geoscience Can. 12:139-147.

Marshall, D.J. 1988. Cathodoluminescence of geological material. Allen & Unwin, Boston.

McBratney, A.B., and C.J. Moran. 1990. A rapid method of analysis for soil micropore structure--II. Stereological model, statistical analysis and interpretation. Soil Sci. Soc. Am. J. 54:509-515.

McCaughey, W.J., and W.H. Fry. 1913. The microscopic determination of soil-forming minerals. USDA Bur. Soils Bull. 91. U. S. Gov. Print. Office, Washington, DC.

McCrone, W.C. 1991. Microscopy in the 1990's. Am. Lab. 23(6):17-26.

McKeague, J. A., R.K. Guertin, F. Page, and K.W.G. Valentine. 1978. Micromorphological evidence of illuvial clay in horizons designated Bt in the field. Can. J. Soil Sci. 58:179-186.

McKeague, J.A., R.K. Guertin, K.W.G. Valentine, J. Belisle, G.A. Bourbeau, A. Howell, W. Michalyna, L. Hopkins, F. Page, and L.M. Bresson. 1980. Estimating illuvial clay in soils by micromorphology. Soil Sci. 129:386-389.

Miedema, R., and S. Slager. 1972. Micromorphological quantification of clay illuviation. J. Soil Sci. 23:309-314.

Milfred, C.J., F.D. Hole, and J.H. Torrie. 1967. Sampling for pedographic modal analysis of an argillic horizon. Soil Sci. Soc. Am. Proc. 31:244-247.

Morris, W.J. 1985. A convenient method of acid etching. J. Sediment. Petrol. 55:600.

Murphy, C.P. 1983. Point counting pores and illuvial clay in thin section. Geoderma 31:133-150.

Murphy, C.P. 1986. Thin section preparation of soils and sediments. AB Acad. Publ., Berkhamsted, England.

Murphy, C.P., and C.F. Banfield. 1978. Pore space variability in a sub-surface horizon of two soils. J. Soil Sci. 33:156:166.

Murphy, C.P., P. Bullock, and R.H. Turner. 1977a. The measurement and characterization of voids in soil thin sections by image analysis. Part 1. Principles and techniques. J. Soil Sci. 28:498-508.

Murphy, C.P., P. Bullock, and K.J. Biswell. 1977b. The measurement and characterization of voids in soil thin sections by image analysis. Part 2. Applications. J. Soil Sci. 28:509-518.

Murphy, C.P., and R.A. Kemp. 1984. The over-estimation of clay and the under-estimation of pores in soil thin sections. J. Soil Sci. 35:481-495.

Norman, J.A.D. 1974. Improved techniques for selective staining of feldspars and other minerals using amaranth. U.S. Geol. Surv. J. Res. 2:73-79.

Pohl, D.W., U. Ch. Fischer, and U.T. Durig. 1988. Scanning near-field microscopy (SNOM). J. Micros. 152:853-861.

Poole, A.B., and A. Thomas. 1975. A staining technique for the identification of sulphates in aggregates and concretes. Mineral. Magazine 40:315-316.

Puentes, R., and L.P. Wilding. 1990. Structural restoration in Vertisols under pastures in Texas. Trans. 14th Int. Congr. Soil Sci. 7:244-249.

Puentes, R., L.P. Wilding, and L.R. Drees. 1992. Microspatial variability and sampling concepts in soil porosity studies of Vertisols. Geoderma 53:373-385.

Rebertus, R.A., and S.W. Buol. 1989. Influence of micaceous minerals on mineralogy class placement of loamy and sandy soils. Soil Sci. Soc. Am. J. 53:196-201.

Reddick, R.C., R.J. Warmack, and T.L. Ferrell. 1989. New form of scanning optical microscopy. Phy. Rev. B. 39:767-770.

Redlich, G.C. 1940. Determination of soil structure by microscopical investigation. Soil Sci. 50:3-13.

Ringrose-Voase, A.J. 1987. A scheme for the quantitative description of soil macrostructure by image analysis. J. Soil Sci. 38:343-356.

Ringrose-Voase, A.J., and P. Bullock. 1984. The automatic recognition and measurement of soil pore types by image analysis and computer programs. J. Soil Sci. 35:673-684.

Ringrose-Voase, A.J., and S. Nortcliffs. 1987. The application of stereology to the estimation of soil structural properties: A preview. p. 81-88. In N. Fedoroff et al. (ed.) Soil micromorphology. Proc. 7th Int. Working Meetings on Soil Micromorphology, Paris. Association Francaise pour l'Etude du Sol, Paris, France.

Ross, C.S. 1924. A method of preparing thin sections of friable rocks. Am. J. Sci. 7:483-485.

Ross, C.S. 1926. Methods of preparation of sedimentary materials for study. Econ. Geol. 21:454-468.

Russ, J.C., and J.C. Russ. 1984. Image processing in a general purpose microcomputer. J. Micro. 135:89-102.

Russ, J.C. 1990. Computer-assisted microscopy: The measurement and analysis of images. Plenum Press, New York.

Ruzyla, K. 1986. Characterization of pore space by quantitative image analysis. Soc. Petrol. Eng. Format. Eval. 1986:389-398.

Ruzyla, K., and D.I. Jezek. 1987. Staining method for recognition of pore space in thin and polished sections. J. Sediment. Petrol. 57:777-778.

Schafer, A., and T. Teyssen. 1987. Size, shape and orientation of grains in sands and sandstones - image analysis applied to rock thin-sections. Sediment. Geol. 52:251-271.

Scott, G.J.T., R. Webster, and S. Nortcliff. 1988a. The topology of pore structure in cracking clay soil--I. The estimation of numerical density. J. Soil Sci. 39:303-314.

Scott, G.J.T., R. Webster, and S. Nortcliff. 1988b. The topology of pore structure in cracking clay soil--II. Conductivity density and its estimation. J. Soil Sci.39:315-326.

Serra, J. 1982. Image analysis and mathematical morphology. Acad. Press, New York.

Shand, S.J. 1916. A recording micrometer for geometrical rock analysis. J. Geol. 24:394-404.

Snedecor G.W., and W.G. Cochran. 1967. Statistical methods. The Iowa State Univ. Press. Ames, IA.

Soil Survey Staff. 1975. Soil taxonomy: A basic system of soil classification for making and interpreting soil surveys. USDA-SCS Agric. Handb. 436. U.S. Gov. Print. Office, Washington, DC.

Solomon, S.T., and G.M. Walker. 1985. The application of cathodoluminescence to interpreting and diagenesis of an ancient calcrete profile. Sediment. 32:877-896.

Swanson, C.L.W., and J.B. Peterson. 1940. Differences in the microstructure of profiles on the Marshall and the Shelby silt loam. Soil Sci. Soc. Am. Proc. 5:297-303.

Swanson, C.L.W., and J.B. Peterson. 1942. The use of the micrometric and other methods for the evaluation of soil structures. Soil Sci. 53:173-185.

Thompson, E. 1930. Quantitative microscopic analysis. J. Geol. 38:193-222.

Underwood, E.E. 1970. Quantitative stereology. Addison-Wesley Publ. Co., Reading, MA.

Van der Plas, L., and A.C. Tobi. 1965. A chart for judging the reliability of point counting results. Am. J. Sci. 263:87-90.

Velbel, M.A. 1989. Weathering of hornblende to ferruginous products by a dissolution - reprecipitation mechanism: petrography and stoichiometry. Clays Clay Miner. 37:515-524.

Walter, R.J., and M.W. Berns. 1986. Digital image processing and analysis. p. 327-393. *In* S. Inoue (ed.) Video microscopy. Plenum Press, New York.

Warne, S.St.J. 1962. A quick field or laboratory staining scheme for the differentiation of the major carbonate minerals. J. Sediment. Petrol. 32:29-38.

Weibel, E.R. 1979. Stereological methods. Vol 1. Practical methods for biological morphometry. Acad. Press, New York.

Wentworth, C.K. 1923. An improved recording micrometer for rock analysis. J. Geol. 31:228-232.

Wickramasinghe, H.K. 1989. Scanned-probe microscopes. Sci. Am. 261:98-105.

Yanguas, J.E., and J.J. Dravis. 1985. Blue fluorescent dye technique for recognition of microporosity in sedimentary rocks. J. Sediment. Petrol. 55:600-602.

6 Transmission Electron Microscope Analysis of Soil Materials

R. J. Gilkes
University of Western Australia
Nedlands, Western Australia, Australia

The optical microscope provides information on the size, shape and mineral species of grains larger than about 5 μm (i.e., the silt and sand fractions) as has been discussed by Drees and Ransom (1994, see Chapter 5). If the optical microscope is used in conjunction with a laser-excited optical-emission spectrometer, scanning electron microscope (SEM), or electron probe microanalyser (EPMA), the chemical composition of the grains may be determined. Consequently, the nature of minerals that make up the sand and silt fractions of soils is well understood and the systematic characterization of these minerals is a common component of pedological and other studies. Many of the chemical and physical properties of soils reflect the properties of materials that make up the clay fraction where particles are micron-sized or smaller and cannot be investigated by optical microscopy. For example, most crystals of clay minerals in soils are 0.1- to 2-μm in size and crystals of some sesquioxides may be very small (10-100 nm). The larger clay particles can be resolved by SEM and their chemical composition determined by concurrent energy-dispersive x-ray spectrometry (EDX) although only limited crystallographic information is provided by SEM, and x-rays from adjacent grains may contribute to spectra. Frequently most of the clay fraction of soils consists of particles that are too small to be resolved and analyzed by SEM techniques; fortunately such particles are ideally suited for investigation using the transmission electron microscope (TEM). Indeed, recent advances in the design of TEM instruments have provided facilities for: (i) the determination of crystal morphology of particles as small as 10 nm, (ii) the direct resolution of 0.1-nm-size crystal-lattice spacings and structural defects using high-resolution transmission electron microscopy (HRTEM), (iii) combined scanning-transmission electron microscopy (STEM) of surfaces of submicron particles, (iv) electron diffraction (ED) to determine lattice constants and structure in single crystals, to identify constituents of mineral assemblages and to measure the relative orientations of crystals in polycrystalline materials, (v) chemical analysis of particles by analytical electron microscopy (AEM), which measures the electron-excited x-ray spectrum of the particle using EDX, and (vi) facilities to measure energies of scattered and Auger electrons. All these analyses can be performed on a single submicron particle and microscopy can be combined with a variety of pretreatments to provide further diagnostic information. Thus, the mineralogist can now use this powerful technique to investigate submicron-sized particles present in the clay fraction of soils to obtain the same wealth of information that was previously only possible for sand-size grains. Furthermore the comprehensive armory of image-analysis and statistical procedures that have been developed by optical microscopists (Drees and Ransom, 1994; see Chapter 5) can be directly employed in the interpretation of TEM images.

THE TRANSMISSION ELECTRON MICROSCOPE

The construction of a transmission electron microscope (TEM) closely resembles that of an optical microscope used in projection mode (Fig. 6-1). The interior of an electron microscope is

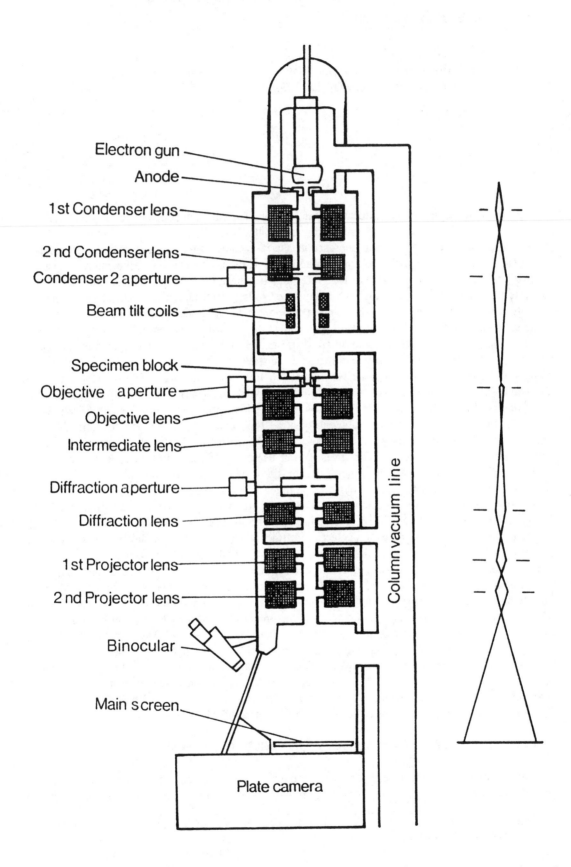

Electron gun

Anode

1st Condenser lens

2nd Condenser lens

Condenser 2 aperture

Beam tilt coils

Specimen block

Objective aperture

Objective lens

Intermediate lens

Diffraction aperture

Diffraction lens

1st Projector lens

2nd Projector lens

Binocular

Main screen

Plate camera

Column vacuum line

Fig. 6-1. Construction of a transmission electron microscope showing the major components and a ray diagram illustrating the formation of an image.

maintained under a high vacuum to minimize image degradation and specimen contamination. The sample is illuminated by a beam of electrons produced from a tungsten filament or LaB_6 crystal maintained at a negative voltage of 50 to 300 kV relative to the anode (some electron microscopes provide much higher potentials). Electrons accelerated by the potential difference acquire enough momentum to penetrate thin specimens. For equivalent specimens the transmissivity of the electron beam increases with the value of the accelerating potential. Generally, for clay-sized materials, potentials of 100 to 300 kV are most useful. The electron beam is focused and collimated by a series of condenser lenses and apertures and may be directed onto the specimen as a coherent parallel beam or, in some microscopes, as a convergent highly focused beam with a diameter as small as a few nanometers. The beam may also be tilted from its usual position perpendicular to the specimen plane to facilitate some imaging procedures.

The specimen is supported on a holder. For soil mineral investigations, specimens usually consist of either a very thin slice of resin-impregnated soil or a powder distributed over a thin C film on a metal grid. Various types of specimen holders and stages are available for specific applications that may involve specimen tilting, heating/cooling, and detection of scattered and secondary electrons and x-rays. The transmitted electron beam is imaged and magnified by a series of magnetic lenses to provide a final image on a fluorescent screen or photographic film. An important capability of an electron microscope as opposed to an optical microscope is that for crystalline materials two discrete images are formed by the objective lens and both can be magnified and imaged by the lenses lower in the microscope column. These images are of the (i) specimen which occurs in the image plane of the objective lens, and (ii) diffraction pattern of the specimen which occurs in the back focal plane of the objective lens (Fig. 6-2a). The two image types may be rotated and inverted by the lenses so that the final recorded images of the specimen and its diffraction pattern may not retain their true relative orientation. In these instances, an appropriate correction should be applied.

Many modern TEMs are equipped with an energy-dispersive x-ray detector located adjacent to the specimen so that the chemical composition of the specimen can be determined. Under favorable conditions, as little as 10^{-19} g of an element can be detected by EDX and many elements measured at concentrations as small as 3 g kg^{-1} for micrometer-sized particles.

THE INTERACTIONS OF AN ELECTRON BEAM WITH CRYSTALS

With the SEM and EPMA techniques, polished and fractured surfaces of thick samples of soil materials are usually examined. These types of specimens are too thick to be directly examined by TEM, so the following discussion is restricted to thin samples. We can define a thin sample as one where most of the incident electron beam is transmitted by the sample. For the accelerating voltages used in modern electron microscopes and the common soil minerals, a thin particle will thus, generally, be <0.1 μm in thickness. The various signals generated by the interaction of the electron beam with a thin crystalline particle are shown in Fig. 6-3 and are discussed separately below. Readers may also refer to recent publications by Malla and Komarneni (1990) and Nadeau and Tait (1987) for further discussion of the applications of TEM to soil materials.

Diffracted Electrons

A major difference between x-ray diffraction (XRD) and ED is that electrons are much more strongly ($\sim 10^4$ times) scattered by atoms than are x-rays so that a very small crystal may provide an intense ED pattern for a short exposure time. The term elastic scattering is used to indicate that the electron experiences no change in energy (wavelength) due to scattering. Electrons have a dual particle and wave nature with the electron wavelength being approximately inversely related to accelerating voltage (Gard, 1971). The theory of ED is similar to that for XRD (Klug & Alexander, 1974) and will not be repeated here. For a 100-kV accelerating voltage the electron wavelength is 0.0037 nm which is much smaller than the wavelength of x-rays (e.g., $CuK\alpha$ = 0.15

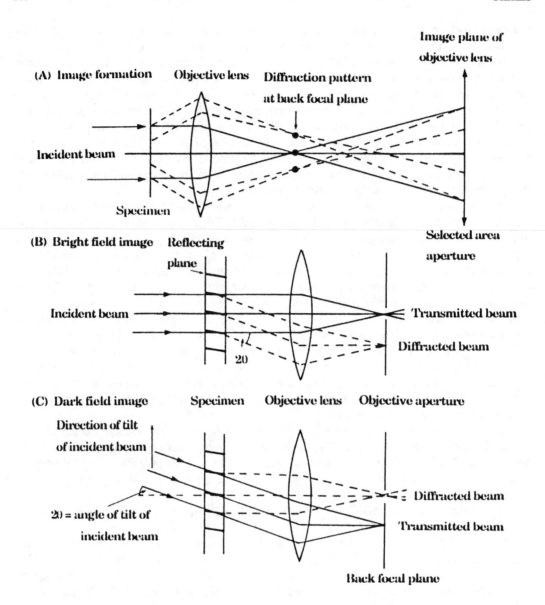

Fig. 6-2. Different types of image formation by a TEM: *A*) images of the specimen and diffraction pattern formed by the objective lens; *B*) bright-field image with diffracted beams excluded; and *C*) dark-field image with transmitted beam excluded.

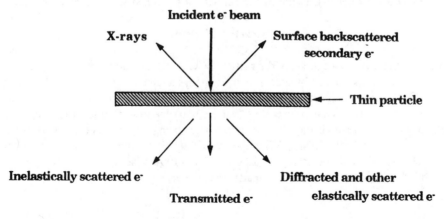

Fig. 6-3. Signals generated by the interaction of an electron beam with a crystalline particle.

nm) typically used for XRD studies of soil minerals (Bish, 1994; Jones & Malik, 1994; Hughes et al., 1994; Schulze, 1994; see Chapters 9, 10, 11, and 13). Consequently, Bragg angles (2θ) (i.e., dispersion) in ED patterns are much smaller than for the corresponding reflections in XRD patterns and are generally less than one degree. A benefit of this difference in scale is that ED patterns can be easily interpreted by employing the reciprocal-lattice concept (Andrews et al., 1968) in which the three-dimensional spatial arrangement of reflecting planes in real or crystal space can be regarded as having an associated three-dimensional reciprocal lattice of points. A point in the reciprocal lattice corresponds to a particular reflection *hkl* which falls on a line perpendicular to the reflecting plane and at a distance of $1/d(hkl)$ or $d*$ from the origin. Thus, higher orders of this reflection ($2h\ 2k\ 2l$, $3h\ 3k\ 3l$, etc.) will occur at regular intervals [$2/d\ (hkl) = 2d*$, $3/d\ (hkl) = 3d*$, etc.] on the same line. The three-dimensional arrangement of these lines generates the reciprocal lattice. This relationship is illustrated for a plane (two-dimensional) rectangular lattice in Fig. 6-4 but can be extended to all three-dimensional lattice geometries.

The Bragg equation (Klug & Alexander, 1974) that describes the relationship between the Bragg diffraction angle (2θ) and the interplanar spacing in the crystal [d(hkl)] is conventionally written as

$$n\lambda = 2d(hkl)\sin\theta \qquad [1]$$

where λ is the wavelength of the radiation (electrons) and n is an integer (the order of the reflection).

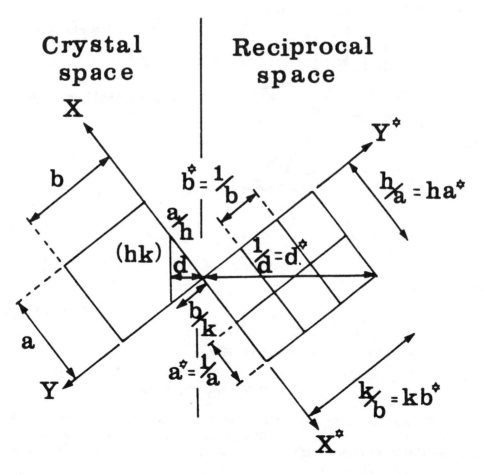

Fig. 6-4. The relationship between real or crystal space and the reciprocal lattice for a two-dimensional rectangular lattice (see text for full explanation).

If the reciprocal-lattice dimension, $d_n{}^*$, is used, the Bragg equation reduces to

$$d_n{}^* = n/d(hkl) = 2\sin\theta/\lambda \qquad [2]$$

The Bragg condition for diffraction can then be geometrically represented by the intersection of the reciprocal lattice with a sphere of radius λ^{-1} (Ewald sphere) with the center of the sphere located at a distance of λ^{-1} from the common origin of the real and reciprocal lattices and in the direction of the incident beam (Fig.6-5). From this diagram we can write

$$2\lambda^{-1}\sin\theta = 1/d_n \qquad [3]$$

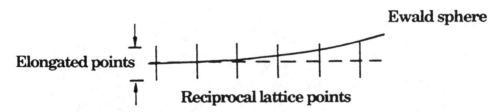

Fig. 6-5. The generation of an electron diffraction pattern represented by the intersection of the reciprocal lattice and Ewald sphere. The large radius of the Ewald sphere and the elongation of reciprocal-lattice points for thin crystals results in diffracted beams being formed for many reciprocal-lattice points.

or

$$2\sin\theta/\lambda \ = \ d_n{}^*$$

which is Eq. [2], the Bragg condition. Thus, a reciprocal-lattice point that intersects the Ewald sphere is indicative of a plane in the correct orientation for diffraction. The direction of the diffracted beam is parallel to the radius of the Ewald sphere that intercepts the reciprocal-lattice point (Point P in Fig. 6-5).

A major benefit of the small wavelength of electrons is that because the radius of the Ewald sphere (radius $= \lambda^{-1}$) is much larger than for x-rays the surface of the sphere can be regarded as being almost flat over limited regions of reciprocal space (Fig. 6-5). Since, as $\theta \rightarrow 0$, $\sin\theta \approx \theta$ (where θ is measured in radians) the Bragg relationship at small angles can be written

$$d_n{}^* \ = \ 2\sin\theta/\lambda \ \approx \ 2\theta/\lambda \qquad\qquad\qquad [4]$$

Thus, reciprocal-lattice dimensions are directly proportional to the Bragg diffraction angle. For such small diffraction angles the following approximation relates the distance (x) between a reflection and the undeviated beam on the screen or film, to the effective ED camera length (L) and the Bragg angle (2θ)

$$2\theta \ = \ x \ /L \qquad\qquad\qquad [5]$$

We can combine Eq. [4] and [5] to obtain the simple relationship

$$d^* \ = \ k^{-1}x \text{ or } d \ = \ kx^{-1} \qquad\qquad\qquad [6]$$

where k is a camera constant which incorporates L and λ. Consequently, ED patterns of single crystals that are generated when the reciprocal lattice intersects the Ewald sphere appear on a photographic plate as an almost undistorted projection of that plane of the reciprocal lattice that coincides with the surface of the Ewald sphere. The size of a reciprocal-lattice point is inversely related to crystal size (or more specifically to the size of the coherently diffracting domain in a crystal) so that the probability of a point intersecting the Ewald sphere and diffraction occurring increases for small crystals. For platy crystals of clay minerals where the electron beam is orthogonal to the (001) plane and where the crystals are very thin in the c direction, the hk reciprocal-lattice points become extended into partial or continuous rods along the c axis and all central $hk0$ rods intersect the Ewald sphere (Fig. 6-5). For this reason and due to their similar pseudohexagonal structures in the ab plane, crystals of the platy clay minerals produce similar pseudohexagonal ED patterns when positioned approximately normal to the electron beam (Fig. 6-6a,b). More complex ED patterns are produced by: (i) thicker crystals where only a few reciprocal lattice points intersect the Ewald sphere (Fig. 6-6c,d), and the crystal may need to be tilted until a suitable complete symmetrical pattern is obtained (reflections from much of the reciprocal lattice can be obtained in this way and crystal polytypes distinguished), (ii) tubular crystals such as halloysite for which reciprocal-lattice points become rings in reciprocal space which intersect the Ewald sphere to give streaks in ED patterns resembling XRD fiber-texture or rotation patterns (Klug & Alexander, 1974) (Fig. 6-6e,f), (iii) crystals that contain dislocations, strain, and other structural defects such as super- or substructures that produce extra reflections or streaks of diffuse scattering in ED patterns, and (iv) polycrystalline materials that may produce full or partial random-powder patterns depending on the degree of preferred orientation of crystallites (Fig. 6-6g,h).

Various types of ED pattern can be obtained with a modern TEM depending on the required application. The most commonly used ED procedures for the investigation of minerals in the clay fraction of soils are discussed in the paragraphs that follow.

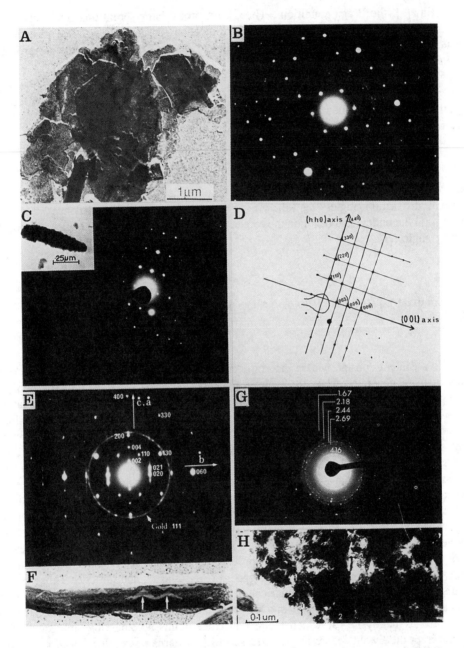

Fig. 6-6. Electron diffraction patterns and micrographs of: (**a,b**) a compound kaolinite crystal showing a pseudohexagonal arrangement of many *hk* reflections and a ring pattern due to the gold internal standard; (**c,d**) a large crandallite crystal showing that for a thick crystal only a few reflections occur and that they define a rectangular reciprocal lattice based on the (*hh*0) and (00*l*) axes; (**e,f**) a halloysite crystal and SAD pattern showing the characteristic streaking of reflections for tubular crystals. The crystal is elongated along its b-axis, reflections of the type *h*0*l*, *h*1*l*, *h*2*l*, etc. each fall on a common axis perpendicular to the *b** elongation of the tube [the crystal has been shadowed with Au to provide a measure of thickness and also to provide a powder (ring) ED pattern which acts as an internal standard]; (**g,h**) small (i) and large (ii) aggregates of soil hematite crystals in a random organization giving a powder pattern consisting of the major hematite reflections, spacings are expressed in angstrom units.

Selected-Area Electron Diffraction

An aperture located in the image plane of the objective lens is used to exclude electrons originating from regions of the specimen field other than the desired area or crystal (Fig. 6-2a). The ED pattern is then focused and contains only beams originating from the selected area. Areas as small as about 0.2 μm^2 can be selected but aberrations of the imaging system result in some contribution to the pattern from areas of the specimen adjacent to the selected area. This is usually not a significant problem for separated particles on a thin C film but can be misleading if selected-area electron diffraction (SAED) patterns are obtained from grains included in a crystalline matrix such as are provided by microtomed or ion-beam-thinned samples. The SAED patterns can be obtained by locating the selector aperture at a position off the optical axis of the microscope, but pattern quality is superior if the selected area is first translated to the center of illumination (i.e., closest to the optical axis).

Micro- and Nanoprobe Electron Diffraction

Some electron microscopes are equipped with condenser lenses that can be used to produce convergent electron beams with widths of a few nanometers. This capability provides diffraction patterns of individual crystals or regions within crystals and can be used to simultaneously provide a chemical analysis of the crystal using the EDX facility.

Convergent-Beam Electron Diffraction

When a crystal is illuminated with a finely focussed converging electron beam, 10 to 50 nm in width, the surface of the Ewald sphere is replaced by a family of surfaces. Consequently, spots in ED patterns become circular images of the diffracting region. The convergent-beam electron diffraction (CBED) pattern may also be used to determine the space group and accurate unit-cell axial ratios for crystals. For thick crystals observed with CBED, Kossel-Mellenstedt patterns occur in images. Similar lines can be observed with a parallel incident beam (Kikuchi lines) and can be interpreted to provide crystallographic information using the theory of multiple scattering of electrons (Andrews et al., 1968).

Forward-Scattered Electrons

Forward-scattered electrons contribute to both the diffraction pattern and image of the crystal as is discussed later. Some electrons lose energy by collisions with atoms within the sample (inelastic scattering) and are not brought to a sharp focus in the image but do contribute to the diffuse background of the ED pattern and image, thereby affecting image quality.

Surface-Backscattered and Secondary Electrons

Some electrons that are scattered from the surfaces of particles may experience no change in energy (elastic scattering) and the fraction of electrons scattered in this way increases with the density or average atomic number of the particle. In a scanning transmission electron microscope (STEM) a finely focused electron beam traverses the particle and the backscattered-electron image provides an indication of variations in density or average atomic number across the particle as in a conventional SEM. Resolution of surface features as small as 2 nm is possible for suitable specimens. Similarly, the secondary electrons emitted from the surface of a particle can be imaged to show the surface morphology of the particle. For thin crystals, simultaneous secondary- and transmitted-electron images may be generated by the scanning electron beam (Fig. 6-7f,g).

X-Rays

Inelastic collisions of electrons with atoms within the particle may remove electrons from the inner electron shells of the atoms. These atoms then emit characteristic x-ray spectra as outershell electrons migrate to fill the inner shell vacancies (Amonette & Sanders, 1994; see Chapter 1). In addition to the characteristic spectral lines a continuum x-ray background is produced by deceleration of some electrons within the specimen (i.e., Bremsstrahlung radiation). Thus, the x-ray emission from a particle in the electron beam will consist of the characteristic spectra of the elements within the particle superimposed on a continuous background. The EDX collection of the x-ray signal enables the composition of the particle to be determined and is the basis of the technique of analytical electron microscopy (AEM).

For a thin particle such as is normally examined by TEM, complex matrix corrections are not required for the calculation of chemical composition from the EDX data. The ratio of intensities of spectral lines for any two elements may be directly proportional or simply related to the ratio of concentrations of the two elements in the particle. We may therefore determine amounts of elements in a particle relative to a reference element and, by use of appropriate calibration constants derived from standard minerals, it is possible to determine the composition of the particle (Lorimer, 1987).

Transmitted Electrons

The electron beam that is transmitted through a particle suffers a loss of intensity due to the elastic and inelastic scattering processes described above. The image of the particle formed by the transmitted electron beam on a photographic film placed below the particle consequently appears as a shadow of the particle with variations in intensity across the shadow due to variations in the extent of scattering. The nature of the actual image obtained with an electron microscope is more complex as it is dependent on the extent to which the final image also includes contributions from scattered electrons that have been imaged by the lenses of the microscope. Unlike image formation in an optical microscope, elastically scattered beams associated with Bragg diffraction play an important role in determining image contrast and detail. The three major types of image that can be produced by a modern electron microscope, bright-field, dark-field, and high-resolution, are discussed below.

Bright-Field Image

Many of the electrons that are scattered at higher angles and particularly those scattered by diffraction can be prevented from being focussed into the image by positioning a small aperture about the transmitted beam in the focal plane of the objective lens [i.e., the objective aperture (Fig. 6-2b)]. Under these conditions, the image of the particle will appear as a relatively dark shadow upon a bright background caused by unscattered electrons. Variations in contrast across the particle reflect differences in scattering due to differences in thickness and composition and also to the extent to which diffraction of the electron beam occurs as regions of the crystal fulfil the Bragg condition for strong reflections. Crystals of clay minerals are often gently curved so that images contain curved dark bands indicating regions where the Bragg condition is fulfilled (Fig. 6-7a).

Dark-Field Image

The objective aperture can be positioned so as to exclude the transmitted beam and include one or more diffracted beams (Fig. 6-2c). The image formed from these beams shows only those parts of a crystal that are contributing to the diffracted beam(s). Holes in the crystal and the surrounding nondiffracting region do not contribute to the image and so appear dark, hence the use of the term dark-field image (Fig. 6-7b,c,d). This condition is best achieved by tilting the incident beam relative to the optic axis of the microscope by the 2θ angle corresponding to the chosen diffracted beam. In this way, the diffracted beam is imaged under optimum axial conditions. Dark-field

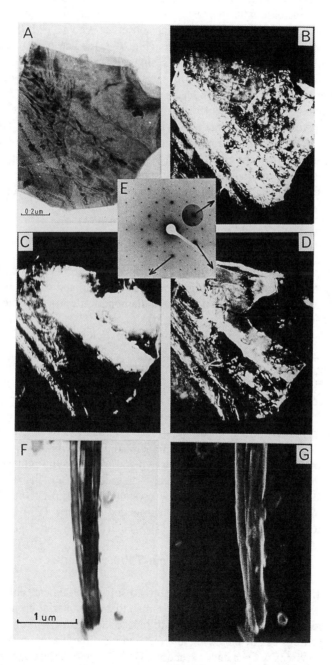

Fig. 6-7. Different types of image formed by the TEM: (*a-e*) bright-field image (*a*) of a mica particle showing dark bands due to increased elastic scattering of electrons by regions of the crystal which are at a suitable orientation for diffraction to occur (i.e., at the Bragg angle for a strong reflection). Three dark-field images (*b,c,d*) of the crystal are shown, these were formed from the indicated reflections (*e*). (*f,g*) STEM transmitted (*f*) and secondary (*g*) electron images of halloysite crystals.

images can also be obtained with a nontilted beam by moving the objective aperture to a nonaxial position to isolate the chosen reflection, but under these nonaxial conditions image quality is inferior.

High-Resolution Image

The spatial resolution of all imaging systems increases as apertures are widened to transmit beams that diverge at large angles from the optic axis. However, focussing errors, including those due to lens aberrations, increase with increasing divergence of the beam and limit the effective resolution of the system. When a crystalline particle is observed with a TEM, the diffracted beams are imaged in the focal plane of the objective lens (Fig. 6-2a). The objective aperture can be positioned to allow only the transmitted beam and a single diffracted beam to be imaged. Under these circumstances the image will contain of a set of equally spaced parallel fringes that correspond to the crystal lattice planes producing the diffracted beam (Fig. 6-14). If several diffracted beams on a single axis in the diffraction-pattern reciprocal space are used to form the image, great detail is revealed of the periodicity within the crystal structure in the corresponding direction in real or crystal space. If diffracted beams from two axes within the diffraction pattern are used to form the image, then the corresponding two-dimensional periodicity of the crystal is revealed. Various precautions are required to avoid spurious images, but lattice images of suitable crystals can be routinely obtained with modern electron microscopes at a resolution of about 0.3 nm (Cowley, 1981; Veblen, 1990).

SPECIMEN PREPARATION

The two major methods used for the preparation of specimens for TEM of soils are deposition of the dispersed clay-fraction upon a thin carbon or collodion film and cutting of thin sections of a resin-impregnated soil or dried clay-fraction with a diamond knife in a microtome.

Deposition of Dispersed Particles

This is a simple procedure but one that is capable of producing a number of artifacts to mislead the unwary microscopist. A drop of very dilute suspension (a few μg solid/mL) of a well dispersed soil clay, mineral concentrate, or crushed grain in distilled water, alcohol or acetone is deposited upon a continuous or holey C film. Unfortunately many clay and oxide minerals tend to form aggregates during this process which makes the TEM examination of single crystals difficult. Various strategies can be employed to reduce aggregation. The microscope grid with its C film can be quickly immersed in the suspension, withdrawn and excess liquid removed by shaking or with the edge of a filter paper. Some minerals resist aggregation if the pH of the suspension is raised/lowered to modify the surface charge on the crystals; the best pH is obtained by trial and error. Further information on the preparation of suspended clays for TEM examination can be found in recent reviews by Malla and Komarneni (1990) and Nadeau and Tait (1987).

Microtomed and Ion-Beam-Thinned Samples

Weathered sand-size mineral grains and undisturbed soil materials can only be examined by TEM if reduced to a thickness less than about 100 nm so that the material becomes transparent to electrons. For high-resolution TEM it is desirable to reduce the thickness of the specimen to about 20 nm. Various procedures can be employed and these mostly involve the initial impregnation of the material with a low-viscosity resin under vacuum. The thickness of the material is then reduced by ultramicrotomy with a diamond knife or by grinding/polishing to prepare a conventional petrological polished thin section. The thin slice of material is then mounted on an electron microscope and may be observed with the TEM. Commonly only small regions of the sample are thin enough for TEM examination and additional thinning by ion-milling is required. The grid is bombarded at a glancing angle with Ar or other ions accelerated by a potential of up to about 10 kV. Frequent examination of the specimen is necessary and, generally, the thinning is not uniform so that holes develop in softer materials. An advantage of ion-milling is there is little physical deformation of the sample. Further details of these procedures is provided by Paulus et al., (1975).

Supplementary procedures may be employed to assist analyses including: (i) vacuum coating of particular samples with an evaporated metal or C film at an inclined angle of incidence (e.g., 30°) chosen to give an elongated shadow that enables the measurement of particle thickness; (ii) vacuum-coating samples with Au, Pt, Si or other crystalline materials that generate simple diffraction patterns for use as internal diffraction standards; and (iii) coating the surface of a soil particle with an organic or evaporated C film that, after dissolution of the particle in hydrofluoric or other acids and subsequent shadowing, provides a thin replica of the topography of the particle surface (Sudo et al., 1981).

It is important to realize that most of the difficulties encountered in the TEM study of soil minerals are associated with specimen preparation rather than microscopy. It is very easy to introduce artifacts that may be mistakenly interpreted as original features of minerals. For example, the suspension containing particles to be deposited on a C film should not contain dissolved materials as these will be concentrated by evaporation and will crystallize on the film. Similarly, suspensions should not be stored for long periods as they become colonized by bacteria and other organisms. Samples prepared by microtomy, petrological thin sectioning and ion-beam thinning procedures are all susceptible to artifacts caused by chemical contamination, physical damage during cutting and polishing, thermal damage, contamination and differential thinning during ion-beam thinning. High-vacuum conditions combined with the high temperatures and irradiation experienced by specimens during electron microscopy from the very high electron current densities can rapidly damage and alter some minerals. Many soil minerals are hydrous and have a low thermal stability so that they are very easily damaged. Cooled specimen holders, low beam currents, environmental cells and other procedures to reduce exposure (MacKinnon, 1990b) may be used to limit damage to specimens within the electron microscope, but the microscopist should always be alert for the presence of experimental artifacts.

EXAMPLES OF TRANSMISSION ELECTRON MICROSCOPE STUDIES OF SOIL MINERALS

As discussed earlier, TEM studies of soil minerals can be conveniently subdivided into: (i) studies of the properties of individual crystals and aggregates in dispersed preparations, and (ii) studies of undisturbed mineral assemblages and fabrics in thin sections of soils and weathered mineral grains. In practice, many studies will employ both types of analysis together with supplementary optical, SEM, thermal, infrared, chemical, XRD and other analyses. Indeed, since TEM techniques examine very small amounts of material (roughly μm^3 or 10^{-12} g) it is essential that the spatial and compositional relationships linking the TEM specimen to the bulk material are well-defined at every scale of spatial and compositional resolution. The problem of obtaining a truly representative 10^{-12} gram subsample of a whole soil for TEM examination places a considerable demand on the microscopist for a critical step-by-step evaluation of the procedures and interpretation used in an investigation. There is a very human tendency for operators to select the spectacular crystal, the curious morphology, the anticipated textural relationship, or a strongly diffracting crystal for recording and publication when such materials may comprise only a minor nonrepresentative part of a mostly much less photogenic specimen. There can be few other situations in soil mineral research where the requirement for systematic investigation and self criticism is as necessary as when one sits at the controls of a high resolution AEM and explores a soil!

Investigations of Dispersed Particles

The majority of TEM studies of soil minerals and sediments carried out during the past 40 yr have been of dispersed clay fractions deposited on thin films (Gard, 1971; Sudo et al., 1981). Similarly, studies of synthesized minerals that are directed towards obtaining an understanding of analogs of soil minerals have also employed TEM measurements on dispersed samples (e.g., Fe oxides; Cornell & Giovanoli, 1988). The capacity of an AEM to provide a complete understanding of the mineralogy of a soil clay may be illustrated by a study of deferrated soil clays that consist of kaolinite with very minor amounts of quartz and anatase (Balwant Singh & Gilkes, 1992). Georgia kaolinite was also examined to provide comparative data and had a much sharper XRD 001 reflection than did the seven soil kaolinites (FWHM $=0.27°\,2\theta$ vs. a mean value of $0.40°\,2\theta$, CuKα), a more ordered structure [Hughes & Brown (1979) index of 44 vs. 6], a smaller surface area (24 vs. 52 m^2 g^{-1}), a higher dehydroxylation temperature (540 vs. 480 °C), a smaller CEC [0.4 vs. 4.9 cmol$_c$ kg^{-1}], and a smaller maximum P sorption capacity (89 vs. 560 μg g^{-1}). Simple TEM of well-dispersed samples provided micrographs from which variations in crystal morphology could be measured. As would be anticipated, Georgia kaolinite consisted of much larger crystals than the soil kaolinites (Fig. 6-8a,b).

Electron micrographs provide only a qualitative impression of particle characteristics and differences between specimens, yet micrographs can be interpreted in a systematic, quantitative manner. For example, a variety of crystal morphological properties can be determined from these electron micrographs as is illustrated in Fig. 6-8c through 6-8h. Crystal width may be readily measured for many crystals and crystal-size distribution functions calculated (Fig. 6-8c,d). In the present example, both kaolinites contained crystals of many different sizes with soil kaolinite containing a much higher proportion of small crystals and consequently having a much smaller median crystal size (0.07 vs. 0.28 μm). If the crystals are shadowed with metal prior to microscopy, their thickness can also be determined. These measures of crystal size can be used to calculate crystal volume and mass which are directly comparable with values obtained by sedimentation procedures. The measurement of crystal morphology can be extended to include measures of crystal shape such as the ratio of largest to smallest dimension of platy particles (i.e., axial ratio). In this example, the axial ratio for soil and Georgia kaolinites varied substantially (from 1-1.8 and 1-1.4, respectively) but median values were close to 1 (i.e., 1.25 and 1.17, respectively) indicating that most of the crystals were nearly equant within the (001) plane (Fig. 6-8e,f). Another major difference observed between the soil and Georgia kaolinites is that the Georgia kaolinite consists almost entirely of euhedral crystals whereas relatively few crystals of soil kaolinite have one or more straight (hk0) faces (Fig. 6-8g,h). The above measurements of morphological features can be readily carried out using the automated image-analysis procedures described in Chapter 5 (Drees & Ransom, 1994).

The intensive comparison of properties of many individual crystals can be extended to include comparisons of ED patterns, which commonly consist of a pseudohexagonal set of hk reflections for both soil and reference kaolinites. Unit-cell dimensions can be obtained from these patterns but the accuracy and precision of SAED and other ED techniques are inadequate to detect the very small differences in the a- and b-axis dimensions between soil and reference kaolinites. For the present example, XRD measurements of the spacing of the 060 reflection of soil kaolinites gave values ranging from 0.1488 to 0.1490 nm. These small differences ($\sim$0.1%) between the kaolinites are significant and systematically related to the Fe content of kaolinite (Rengasamy et al., 1975), but are much too small to be accurately determined by ED, which has a resolution of about 1% (Mackinnon, 1990a).

Single crystals of kaolinite as small as 0.05 μm in width (i.e., weighing about 10^{-16} g) provide sufficiently intense x-ray spectra in the AEM to enable the calculation of chemical analyses. Individual kaolinite crystals from a soil and from a Georgia kaolinite sample were analysed and differences in chemical composition between crystals determined. Typical EDX spectra for single crystals of Georgia and soil kaolinites are shown in Fig. 6-9. The two strongest x-ray lines are for Al and Si, which are the only cations present in ideal kaolinite. The weak K x-ray line indicates that very minor amounts of K are also present in both crystals, presumably as rare interlayers of mica (Lee et al., 1975). The samples had been saturated with Ba prior to AEM examination but, due to the low CEC of kaolinite and short counting time, the Ba x-ray lines cannot be resolved from the background in the illustrated spectra. A little Fe is present in the crystal of Georgia kaolinite and a considerable amount of Fe occurs in the crystal of soil kaolinite shown in Fig. 6-9. A significant proportion of the octahedral-cation sites in soil kaolinite seem to be occupied by Fe (which replaces Al), hence the relatively weaker Al and stronger Fe x-ray lines

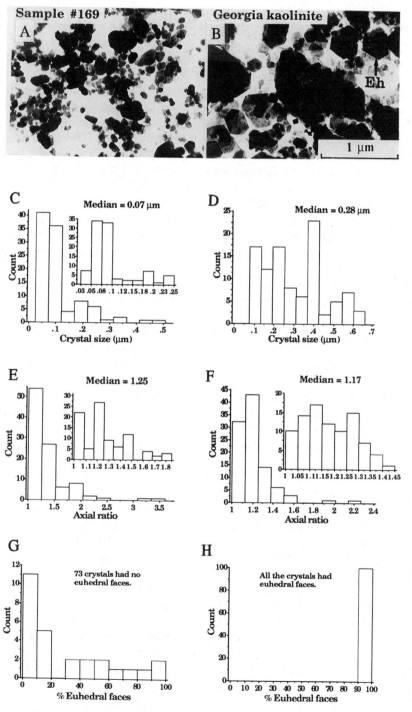

Fig. 6-8. The TEM measurement of properties of a soil kaolinite (Sample 169) and Georgia kaolinite illustrating the capacity of microscopy to provide quantitative morphological data: *a,b*) electron micrographs; *c,d*) crystal-size distributions; *e,f*) crystal-shape (axial-ratio) distributions; *g,h*) percentage of euhedral faces (Balwant Singh Gilkes, 1992).

for this crystal relative to the values for the crystal of Georgia kaolinite. When a large number of soil kaolinite crystals was analyzed by AEM a statistically highly significant inverse relationship between the contents of Fe and Al was obtained. This relationship is consistent with the presence of octahedral Fe and with analyses of bulk samples of soil kaolinites by Rengasamy et al. (1975) and other workers. However, analyses of bulk samples do not indicate the extent of variation in chemical composition among individual crystals, which can be of considerable importance in providing an understanding of soil chemical behavior. Such information on intercrystal variability is provided by AEM analysis of many individual crystals and is illustrated by the distribution function shown in Fig. 6-10 for soil and Georgia kaolinites and for a reference, well-crystallized kaolinite from a laterite pallid-zone. The variations in Si and Al contents due solely to instrumental factors must be smaller than are indicated by the standard deviation values of 7 and 8 g kg^{-1}, respectively, for the pallid-zone kaolinite. The corresponding variations for Georgia kaolinite are larger (16 and 15 g kg^{-1}) and suggest that crystals of Georgia kaolinite differ in chemical composition. Much larger relative variations for Fe and Ba contents were obtained but, as the amounts of these elements were at levels approaching the detection limit of the AEM technique, it is probable that much of this variation is of instrumental origin. The variation in Si and Al contents of individual crystals was even greater for the soil kaolinite indicating that a great diversity of kaolinite compositions can occur in a single soil. The much higher exchangeable-Ba content of the soil kaolinite is consistent with the relatively higher CEC of bulk-soil kaolinite samples. Indeed, the median value of 3 g kg^{-1} BaO is equivalent to about 4 cmol$_c$ kg^{-1}, which is the value obtained for the bulk sample of this soil kaolinite. This investigation demonstrates the capacity of AEM to analyze individual crystals in a clay fraction with the same ease that petrologists study sand-size grains. The technique has clear potential for solving many of the previously intractable problems of soil mineralogy.

In addition to the use of AEM to characterize the *major* minerals in a clay fraction as was demonstrated above, a *unique* and powerful capability of AEM is to provide chemical, structural and morphological information for clay-size particles that are only very *minor* constituents of mixtures. This capability will be illustrated by a study of an unusual clay mineral in a laterite pallid-zone (Balwant Singh & Gilkes, 1991). A smectite is an alteration product of muscovite and occurs as a very minor phase in the clay fraction of deeply weathered rocks at the Hedges gold mine, Western Australia. Chemical analyses and the *b*-axis unit-cell dimension of the smectite could not be obtained because it could not be physically separated from kaolinite and halloysite, which were the dominant constituents of the clay. The XRD basal-spacing measurements and thermal analysis results obtained on the clay fraction were consistent with the smectite mineral being beidellite but this could only be confirmed by chemical analysis of the pure beidellite. As this was clearly an impossible task using conventional procedures, the problem could only be resolved by the analysis of individual crystals of the clay minerals by TEM. Three distinct crystal morphologies are apparent in TEM micrographs of the clay (Fig. 6-11). These are quite large (1- to 2-μm) irregular flakes of an unknown clay mineral, tubular halloysite crystals, and euhedral hexagonal kaolinite crystals. The SAD of kaolinite and halloysite were typical of these minerals and SAD of the unknown mineral gave a pattern consisting of *hk*0 reflections which show the mineral to be dioctahedral. The AEM data show the halloysite and kaolinite crystals to have identical compositions close to that of ideal kaolinite but containing minor amounts of Fe. The unknown mineral has a distinctly different composition with more Fe and a much higher Si/Al ratio that resembles that of beidellite. The sample had been saturated with Ba prior to analysis so that the CEC of individual beidellite crystals could be derived from their Ba content. The average CEC value (*n* = 18) was 45 cmol$_c$ kg^{-1}, which is about half the value reported for standard beidellite (Weir & Greene-Kelly, 1962) and half that required to satisfy the calculated layer charge. The mineral also contained 47 cmol$_c$ kg^{-1} of K that was not displaced during the Ba saturation treatment and exactly satisfied the remainder of the layer charge.

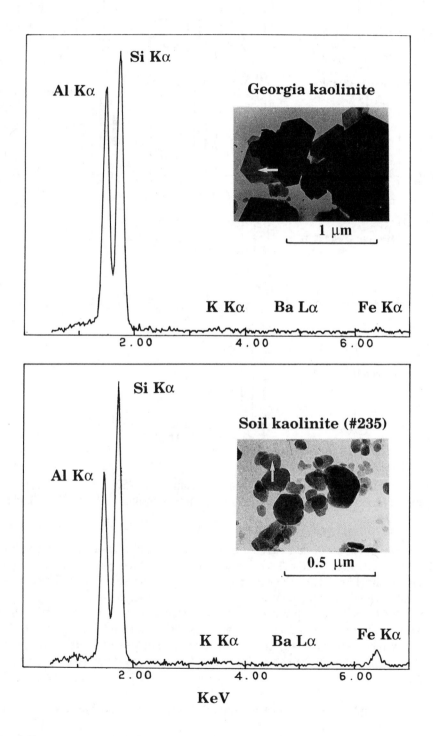

Fig. 6-9. Micrographs and EDX spectra of the indicated single crystals of soil and Georgia kaolinites obtained by AEM (Balwant Singh & Gilkes, 1992).

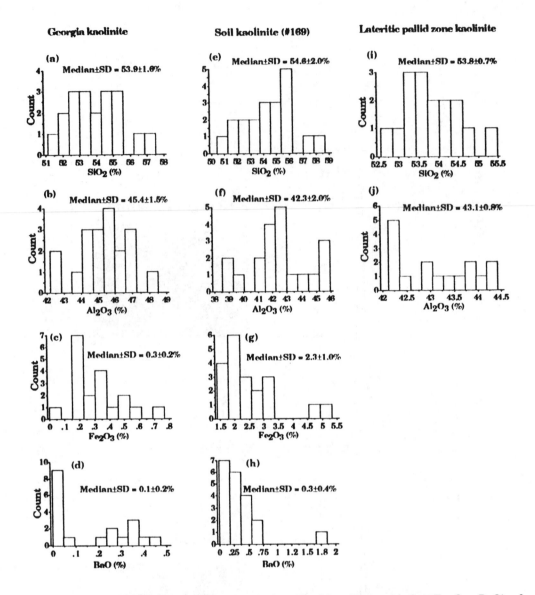

Fig. 6-10. Distribution functions for the chemical composition (SiO_2, Al_2O_3, Fe_2O_3, BaO) of individual crystals of kaolinite (Georgia, soil, pallid zone) obtained by AEM (Balwant Singh & Gilkes, 1992).

This mineral, which could be described as K-beidellite, had a quite variable composition as indicated by the wide range (0.26-0.35) of values for [Al_2O_3/(Al_2O_3 + SiO_2)] obtained by AEM of individual crystals (Fig. 6-12). The wide range of composition was not due to random analytical error as the corresponding range of values for the associated kaolinite and halloysite was much smaller (0.41-0.47), and variability due solely to instrumental factors would be even smaller. Thus, AEM not only provided a chemical analysis of the K-beidellite that could not have been obtained by any other technique but, as in the investigation of soil kaolinites, has also measured the variation in composition of the K-beidellite, something that could not have been determined by other techniques.

Studies of Undisturbed Mineral Assemblages and Fabrics

This type of investigation is generally carried out either to identify the structure of soil materials at a submicron scale (Tessier, 1984) or to determine pathways and mechanisms of mineral

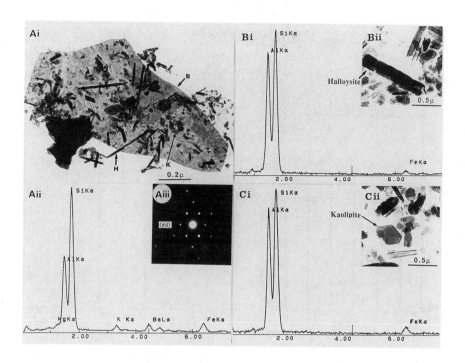

Fig. 6-11. Electron micrographs and EDX spectra for Ba-saturated beidellite (indicated as B), kaolinite (K) and halloysite (H) crystals in the clay fraction of a pallid-zone clay: (Ai) irregular flake of beidellite with attached halloysite and kaolinite crystals, (Aii) EDX spectrum of beidellite, (Aiii) SAED pattern of beidellite; (Bi,ii) tubular halloysite and EDX spectrum; (Ci,ii) euhedral hexagonal kaolinite and EDX spectrum (Balwant Singh & Gilkes, 1991).

alteration (Gilkes et al., 1986). If thin-sectioning facilities are unavailable some success may be achieved by examining the thin edges of cleaved or fractured specimens. However, this type of specimen preparation relies on the fortuitous occurrence of suitable mineral assemblages at grain boundaries and such regions may not be representative of the major phases in a sample. In order to make selected grains or the matrix sufficiently thin to be transparent to electrons, the material must be first stabilised by impregnation with resin before a thin slice is prepared for TEM investigation. The slice may be prepared by some combination of petrological thin-sectioning, microtomy using a diamond knife, and ion-beam thinning (Sudo et al., 1981; MacKinnon, 1990a). The microscopist should be alert to the specimen damage and generation of artifacts that commonly occur during these procedures.

Sectioning of partly weathered mineral grains along various axes that relate to the crystallographic axes of the parent and secondary minerals has been used to good effect by many workers to determine mechanisms of mineral weathering and diagenesis (e.g., illite-smectite associations, Jiang et al., 1990; biotite-chlorite, Eggleton & Banfield, 1985). An illustration of this procedure is provided by a study of the alteration of the mica fuchsite (Cr-muscovite) to kaolinite during the weathering of a micaceous sandstone (Balbir Singh & Gilkes, 1991). Conventional separation of the clay fraction and its deposition on a C-covered grid results in all the mica crystals lying flat on the grid. Consequently, micrographs and ED patterns of these conventionally prepared specimens only provide information on the mutual orientation of fuchsite and kaolinite within the (001) plane (Fig. 6-13). In this instance, the kaolinite crystals grew in exact parallel orientation to the mica

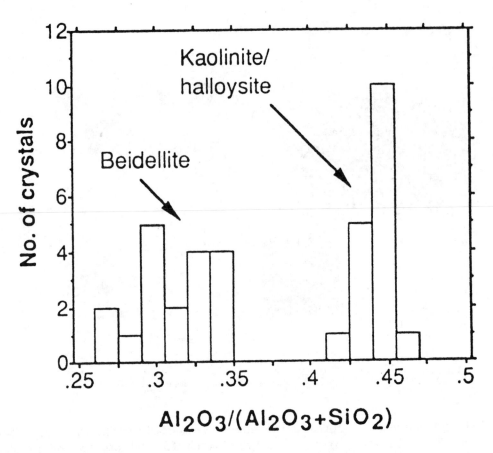

Fig. 6-12. The distribution of values of $[Al_2O_3/(Al_2O_3 + SiO_2)]$ for beidellite and associated kaolinite/halloysite crystals in a pallid-zone clay determined by AEM (Balwant Singh & Gilkes, 1991).

so that the a^* and b^* axes and hk reflections of the two minerals coincide. The unit cell dimensions of fuchsite and kaolinite were almost identical ($b = 0.899$ and 0.896 nm, respectively) so that no splitting of *hk* reflections could be detected due to the poor resolution of the SAED technique. However, analysis of ultramicrotomed sections of partly altered fuchsite crystals cut perpendicular to the basal plane revealed the organization of fuchsite and kaolinite along the c^* axis when the TEM was used in HRTEM mode (Fig. 6-14). In some regions of partly altered fuchsite grains there were packets of kaolinite (lattice-image spacing 0.7 nm) parallel to packets of fuchsite (1.0 nm) indicating that alteration is proceeding in the c^* direction by sequential replacement of whole fuchsite layers (Fig. 6-14a). Elsewhere in the crystal, one layer of fuchsite (1.0 nm) may pass laterally into one layer of kaolinite (0.7 nm) (Fig. 6-14b) or two layers of kaolinite (2 x 0.7 nm = 1.4 nm) (Fig. 6-14c). Use of the AEM technique shows that the kaolinite contains none of the K and Mg and about half of the Fe and Cr that were present in the fuchsite. This type of topotaxial and partly or wholly Al-conservative alteration of layer silicates has been reported by several workers (e.g., biotite, Banfield & Eggleton, 1988; Gilkes & Suddhiprakarn, 1979a,b).

The TEM may also be used to investigate more complex types of alteration such as the alteration of platy muscovite to tubular halloysite (Robertson & Eggleton, 1991; Balbir Singh & Gilkes, 1992a). Muscovite crystals may be replaced both by highly oriented kaolinite crystals as described for fuchsite and by long (up to 10 μm) halloysite tubes that are elongated parallel to the 001 plane of the parent muscovite (Fig. 6-15). The complex spatial relationships of these components is illustrated by the micrograph of an altered muscovite grain that has been sectioned perpendicularly to the coincident b axes of the parent muscovite and the secondary kaolinite and halloysite (Fig. 6-16). The tubes and laths in this porous region have retained a very high degree

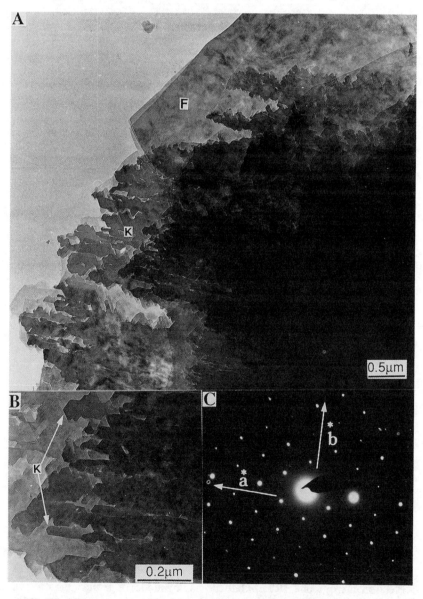

Fig. 6-13. (*a*,*b*) Electron micrograph of kaolinite (K) crystals replacing a fuchsite (F) crystal within the (001) plane of the fuchsite; (*c*) The SAED pattern was derived from one fuchsite crystal and many kaolinite crystals yet appears to be the *hk*0 net of reflections for a *single* crystal of a dioctahedral layer silicate due to the high degree of parallel alignment of the crystals (Balbir Singh & Gilkes, 1991).

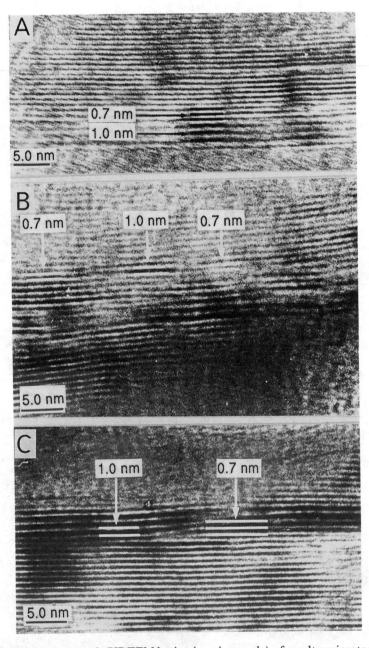

Fig. 6-14. Electron micrograph (HRTEM lattice-imaging mode) of an ultramicrotomed thin-section of partly kaolinised fuchsite cut parallel to c^* showing the replacement of fuchsite [d (001) = 1.0 nm] by kaolinite (0.7 nm). The micrographs a,b,c refer to different alteration mechanisms whereby fuchsite layers alter to kaolinite layers in directions parallel (a) and normal (b,c) to the c^* axis. Single fuchsite layers can alter to one (b) or two (c) kaolinite layers (Balbir Singh & Gilkes, 1991).

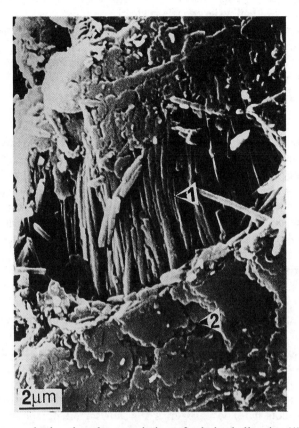

Fig. 6-15. SEM micrograph showing the association of tubular halloysite (1) and platy kaolinite (2) crystals that are aligned parallel to the cleavage of a former mica crystal (Balbir Singh & Gilkes, 1992a).

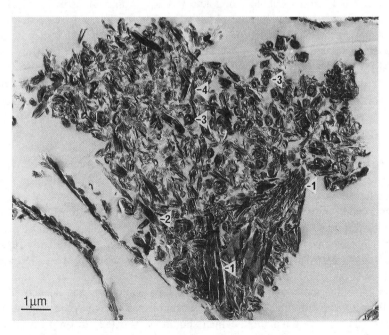

Fig. 6-16. TEM micrograph of an ultramicrotomed thin section of an altered mica crystal. The section has been cut perpendicular to the *b*-axis of the former mica crystal and also to the *b*-axis of various types of platy kaolinite crystals (1,2,4) and tubular halloysite crystals (3) which are the alteration products (Balbir Singh & Gilkes, 1992a).

of *b*-axis parallelism, which is also revealed by the aligned ED patterns (not shown). The complex cross-sectional geometry of halloysite crystals is clearly revealed in very-high-magnification TEM micrographs of ultramicrotomed sections of this material (Fig. 6-17). These properties are not apparent when halloysite tubes are examined by TEM of dispersed clay (Fig. 6-18), but such micrographs together with the transmitted and secondary-electron STEM images (Fig. 6-7f,g) indicate that the halloysite crystals are not simple, perfect tubes.

Amorphous Soil Constituents

The above examples have demonstrated the versatility of the TEM in investigations of assemblages of crystalline minerals. Many studies of mineral weathering sequences and soil mineralogy, chemistry and physics have indicated that amorphous materials are important constituents of soils. Such materials are a major constituent of allophanic soils (Wada, 1989) and other soil types may contain amorphous and poorly crystalline materials including opaline silica, protoimogolite-allophane, and ferrihydrite (Drees et al., 1989; Farmer et al., 1983; Schwertmann et al., 1982). Recognition of such materials by electron microscopy provides a particular challenge to soil mineralogists as these materials commonly have a nondescript morphology (as opposed to the distinct characteristic and often euhedral morphology of crystalline minerals), have variable chemical compositions, and may occur as thin coatings on or in intimate association with crystalline minerals.

The excellent spatial resolution of TEM combined with the capacity of electrons to provide diffraction patterns of small volumes of materials having partly ordered structures encourages the view that TEM offers the best potential for characterising poorly crystalline materials in soils. For example, amorphous SiO_2 usually adheres to soil particles and may occur in some soils as a void filling (Drees et al., 1989). Very small amounts of this SiO_2 cement may strongly influence the structural properties of soils and larger amounts can form indurated horizons which restrict the development of plant roots and shoots. Figure 6-19a shows a micrograph of amorphous silica from within a void in a silicified laterite pallid-zone (Balbir Singh & Gilkes, 1992b). The material is granular with conchoidal fracture and shows few of the straight edges which are a characteristic of crystalline minerals. The ED pattern of this material (Fig. 6-19b) consists of diffuse rings, which are indicative of an amorphous compound (Gard, 1971). Other particles in the same soil consist of kaolin crystals enclosed by amorphous silica (Fig. 6-19c). These particles may give ED patterns consisting of the usual *hk* net for basally oriented kaolinite (which may be in the form of a powder pattern if several kaolinite crystals are present), combined with a diffuse ring due to the amorphous silica (Fig. 6-19d). This intimate submicroscopic association of silica and kaolinite could not be determined by other techniques and can be further investigated by AEM measurements of the chemical composition of amorphous silica-kaolinite associations.

FUTURE DIRECTIONS

The use of high-resolution analytical transmission electron microscopy (ATEM) to investigate soil materials is in its infancy and future developments may mostly be limited by the imagination of users. The technique has application in many areas of physical and chemical soil science, plant nutrition, and soil microbiology. For example, spatial relationships in soil/root/root hair/mycorrhiza/fertilizer interfaces could be explored at a nm scale. The most pressing need is to develop reliable methods of preparation to generate specimens that are truly representative of bulk materials and which have not been altered by the preparation process. Novel methods of specimen preparation could include pretreatments with reagents so as to impose diagnostic properties on minerals and organic materials (e.g., Ba saturation has been used to provide a measure of the CEC of individual crystals and organic materials may be stained with indicator heavy metals). There is a temptation, which should be resisted, to only investigate the more spectacular and amenable samples provided by discrete weathered mineral grains rather than the much more heterogeneous matrix of soils and in particular the mineral-organic assemblages present in topsoils. Computer-assisted methods of image analysis are available which can greatly assist in

Fig. 6-17. Very high magnification TEM micrograph of a thin section of parallel halloysite crystals cut perpendicular to the tube axis (crystallographic *b* axis) (Balbir Singh & Gilkes, 1992a).

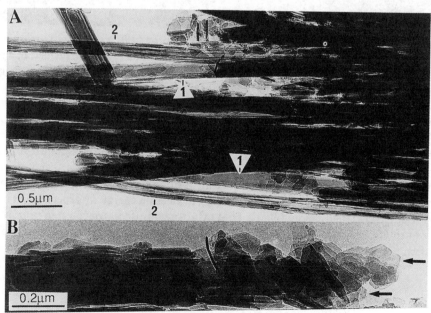

Fig. 6-18. TEM micrograph of dispersed halloysite tubes showing their complex morphology which includes terminating hexagonal plates (1 and arrowed) and tubular sections (2) (Balbir Singh & Gilkes, 1992a).

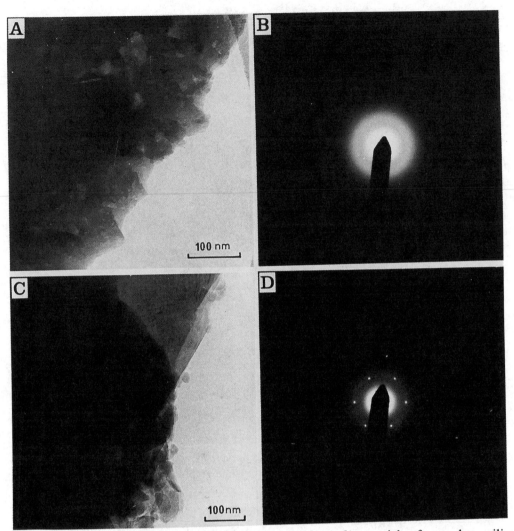

Fig. 6-19. TEM micrographs and microbeam ED patterns for (*a,b*) a particle of amorphous silica in a soil void, the ED pattern is a diffuse ring, (*c,d*) a kaolinite crystal embedded in amorphous silica, the sharp *hk*0 reflections of kaolinite are superimposed on a diffuse ring due to amorphous silica.

the interpretation of electron micrographs. The wealth of information provided by modern ATEM procedures must be related to other properties of soils. A minimum requirement for investigators should be that the materials investigated by ATEM should have a clearly established provenance and they should also be characterised by optical microscopy, SEM, EPMA and spectroscopic techniques. Finally, it must be stressed that the field of electron-optical analysis of materials is developing rapidly. Emerging TEM techniques such as electron energy loss spectrometry, electron chanelling (ALCHEMI) and energy filtered TEM (MacKinnon, 1990a) have yet to be proven useful in studies of soil materials, but are likely to find applications in many investigations of soil mineral assemblages.

REFERENCES

Amonette, J.E., and R.W. Sanders. 1994. Nondestructive techniques for bulk elemental analysis. p. 1-48. *In* J. Amonette and L. W. Zelazny (ed.) Quantitative methods in soil mineralogy. SSSA Misc. Publ. SSSA, Madison, WI.

Andrews, K.W., D.J. Dyson, and S.R. Keown. 1968. Interpretation of electron diffraction patterns. Adam Hilger Ltd., London.

Banfield, J.F., and R.A. Eggleton. 1988. Transmission electron microscope study of biotite weathering. Clays Clay Miner. 36:47-60.

Bish, D.L. 1994. Quantitative x-ray diffraction analysis of soils. p. 000-000. *In* J. Amonette and L. W. Zelazny (ed.) Quantitative methods in soil mineralogy. SSSA Spec. Publ. 35. SSSA, Madison, WI.

Cornell, R.M., and B. Giovanoli. 1988. Acid dissolution of akaganeite and lepidocrocite: The effect on crystal morphology. Clays Clay Miner. 36:385-390.

Cowley, J.M. 1981. Diffraction physics, 2nd ed. North Holland Publ., Amsterdam.

Drees, L.R., and M.D. Ransom. 1994. Light microscopic techniques in quantitative soil mineralogy. p. 000-000. *In* J. Amonette and L.W. Zelazny (ed.) Quantitative methods in soil mineralogy. SSSA Spec. Publ. 35. SSSA, Madison, WI.

Drees, L.R., L.P. Wilding, N.E. Smeck, and A.L. Senkayi. 1989. Silica in soils: quartz and disordered silica polymorphs. p. 913-974. *In* J.B. Dixon and S.W. Weed (ed.) Minerals in soil environments. SSSA., Madison, WI.

Eggleton, R.A., and J.F. Banfield. 1985. The alteration of granitic biotite to chlorite. Am. Mineral. 70:902-910.

Farmer, V.C., J.D. Russell, and B.F.L. Smith. 1983. Extraction of inorganic forms of translocated Al, Fe, and Si from a podzol B_s horizon. J. Soil Sci. 34:571-576.

Gard, J.A. (ed.). 1971. The electron-optical investigation of clays. Mineral. Soc. Monogr. 3, Alden Press, Oxford, England.

Gilkes, R.J., and A. Suddhiprakarn. 1979a. Biotite alteration in deeply weathered granite--I. Morphological, mineralogical and chemical properties. Clays Clay Miner. 17:349-360.

Gilkes, R.J., and A. Suddhiprakarn. 1979b. Biotite alteration in deeply weathered granite--II. The oriented growth of secondary minerals. Clays Clay Miner. 27:361-367.

Gilkes, R.J., R.R. Anand, and A. Suddhiprakarn. 1986. How the microfabric of soil may be influenced by the structure and chemical composition of parent minerals. p. 1093-1106. *In* Proc. Congr. Int. Soil Sci. Soc., Hamburg, 13th.

Hughes, R.E., D.M. Moore, and H.D. Glass. 1994. Qualitative and quantitative analysis of clay minerals in soils. p. 000-000. *In* J. Amonette and L.W. Zelazny (ed.) Quantitative methods in soil mineralogy. SSSA Spec. Publ. 35. SSSA, Madison, WI.

Hughes, J.C., and G. Brown. 1979. A crystallinity index for soil kaolins and its relation to parent rock, climate and soil maturity. J. Soil Sci. 30:557-563.

Jiang, W.T., D.R. Peacor, R.J. Merriman, and B. Roberts. 1990. Transmission and analytical electron microscope study of mixed-layer illite/smectite formed as an apparent replacement product of diagenetic illite. Clays Clay Miner. 38:449-468.

Jones, R.C., and H.U. Malik. 1994. Analysis of minerals in oxide-rich soils by x-ray diffraction. p. 000-000. *In* J. Amonette and L.W. Zelazny (ed.) Quantitative methods in soil mineralogy. SSSA Spec. Publ. 35. SSSA, Madison, WI.

Klug, H.P., and L.E. Alexander. 1974. X-ray diffraction procedures for polycrystalline and amorphous materials. Wiley, New York.

Lee, S.Y., M.L. Jackson, and J.L. Brown. 1975. Micaceous occlusions in kaolinite observed by ultramicrotomy and high resolution electron microscopy. Clays Clay Miner. 23:125-129.

Lorimer, G.W. 1987. Quantitative x-ray microanalysis of thin sections in the transmission electron microscope: A review. Miner. Mag. 51:49-60.

Mackinnon, I.D.R. 1990a. Introduction to electron-beam techniques. p. 1-14. *In* I.D.R. Mackinnon and F.A. Mumpton (ed.) Electron-optical methods in clay science. Clay Miner. Soc., Boulder, CO.

Mackinnon, I.D.R. 1990b. Thin-film elemental analyses for precise characterisation of minerals. p. 32-53. *In* L.M. Coyne et al. (ed.) Spectroscopic characterization of minerals and their surfaces. Am. Chem. Soc. Symp. Ser., Vol. 415. ACS, Washington, DC.

Malla, P.B., and S. Komarneni. 1990. High-resolution TEM in the study of clays and soils. Adv. Soil Sci. 12:159-186.

Nadeau, P.H., and J.M. Tait. 1987. Transmission electron microscopy. p. 210-247. *In* M. J. Wilson (ed.) A handbook of determinative methods in clay mineralogy. Blackie, London.

Paulus, M., A. Dubon, and J. Etienne. 1975. Application of ion-thinning to the study of the structure of argillaceous rocks by TEM. Clay Miner. 10:417-426.

Rengasamy, P., G.S.R. Krishna Murti, and V.A.K. Sarma. 1975. Isomorphous substitution of iron for aluminium in some soil kaolinites. Clays Clay Miner. 23:211-214.

Robertson, I.D.M., and R.A. Eggleton. 1991. Weathering of granitic muscovite to kaolinite and halloysite and of plagioclase-derived kaolinite to halloysite. Clays Clay Miner. 39:113-126.

Schulze, D.G. 1994. Differential x-ray diffraction analysis of soil minerals. p. 412-429. *In* J. Amonette and L.W. Zelazny (ed.) Quantitative methods in soil mineralogy. SSSA Misc. Publ. 35. SSSA, Madison, WI.

Schwertmann, V., D.G. Schulze, and E. Murad. 1982. Identification of ferrihydrite in soils by dissolution kinetics, differential x-ray diffraction and Mössbauer spectroscopy. Soil Sci. Soc. Am. J. 46:869-875.

Singh, Balbir, and R.J. Gilkes. 1991. Weathering of chromian-muscovite. Clays Clay Miner. 39:571-579.

Singh, Balbir, and R.J. Gilkes. 1992a. An electron-optical investigation of the alteration of kaolinite to halloysite. Clays Clay Miner. 40:212-229.

Singh, Balbir, and R.J. Gilkes. 1992b. An electron-optical investigation of aluminosilicate cements in silcretes. Clays Clay Miner. 40:707-721.

Singh, Balwant, and R.J. Gilkes. 1991. A potassium-rich beidellite from laterite pallid zone in Western Australia. Clay Miner. 26:233-244.

Singh, Balwant, and R.J. Gilkes. 1992. Properties of soil kaolinites from south-western Australia. J. Soil Sci. 43:645-667.

Sudo, T., S. Shimoda, H. Yotsumoto, and S. Aita. 1981. Electron micrographs of clay minerals. Elsevier, New York.

Tessier, D. 1984. Etude experiméntale de l'organisation des matériaux argileux. D.Sc. thesis. INRA, Paris.

Veblen, D.R. 1990. Transmission electron microscopy: scattering processes, conventional microscopy and high-resolution imaging. p. 15-40. *In* I.D.R. Mackinnon and F.A. Mumpton (ed.) Electron-optical methods in clay science. Clay Miner. Soc., Boulder, CO.

Wada, K. 1989. Allophane and imogolite. p. 1051-1087. *In* J.B. Dixon and S.B. Weed (ed.) Minerals in soil environments. SSSA., Madison, WI.

Weir, A.H., and R. Greene-Kelly. 1962. Beidellite. Am. Miner. 47:137-146.

7 Analysis of Soil Surfaces by X-Ray Photoelectron Spectroscopy

D. L. Cocke and R. K. Vempati
Lamar University
Beaumont, Texas

R. H. Loeppert
Texas A&M University
College Station, Texas

Surface characterization has been an active area of research in recent years. X-ray photoelectron spectroscopy, XPS, is one of the most commonly used surface characterization techniques because it: (i) provides quantitative surface and near surface composition; (ii) provides surface chemical information (i.e., oxidation state, chemical bonding, site occupancy, adsorbed species, surface reactivity, etc.); (iii) can be used to study all elements except for H and He atoms; (iv) can be used to study conductive, nonconductive, crystalline, and amorphous materials; (v) is generally a nondestructive technique; and (vi) requires only a very small sample size (<10 ng). The use of XPS in surface-reactivity studies of real systems is an exciting extension of the technique (Yoon & Cocke, 1987; Wright et al., 1987; Cocke et al., 1988a,b, 1990; Mehbratu et al., 1991). There are several basic texts (Shirley, 1972; Kane & Larrabee, 1974; Carlson, 1975; Czanderna, 1975; Brundle & Baker, 1977; Briggs, 1977; Muilenburg, 1979; Briggs & Seah, 1983; Eland, 1984) on XPS that the reader is encouraged to consult for greater detail. The applications of XPS to the study of biomaterials (Ratner, 1983, 1988), catalysis (Hofman, 1986), corrosion (Baer & Thomas, 1982), material science (Windawi & Ho, 1982) and polymers (Clark, 1979) have been reviewed. Applications to the study of clay minerals and soils have been limited, but interest is growing (Hochella, 1988; Perry et al., 1990). The intent of this chapter is to illustrate the application of XPS to the study of soil chemistry and soil mineralogy. For more detailed examples and further review, the reader is referred to the articles cited in the text.

THEORY

An understanding of the chemistry of the surface and of surface reactivity is the ultimate goal of the surface chemist or the soil chemist interested in surfaces and interfaces. The XPS provides information regarding the surface and its chemistry from the chemical shift, satellite structure, multiplet splitting, valence-band structure, Auger-peak shape and position, and Auger parameter. In order to understand the relevance of surface chemistry to the XPS spectrum and the information that can be obtained by this methodology, one must understand the basic processes and properties that influence the XPS spectrum.

Basic X-Ray Photoelectron Spectroscopy

The XPS is used to analyze all the elements except H and He because they have low photoionization cross sections for 1s electrons (i.e., the photon has a very low probability of ejecting an electron from these atoms). The XPS spectrum for each element is unique because the electron configuration is unique and the energies of the electrons of each element are quantized. Figure 7-1 shows a wide-scan spectrum of montmorillonite; the intensity of the photoelectron peaks are plotted on the ordinate, and the binding energy is plotted on the abscissa (Seyama & Soma, 1988). The valence-electron or bonding-electron bands occur in the region of 0 to 20 eV; however

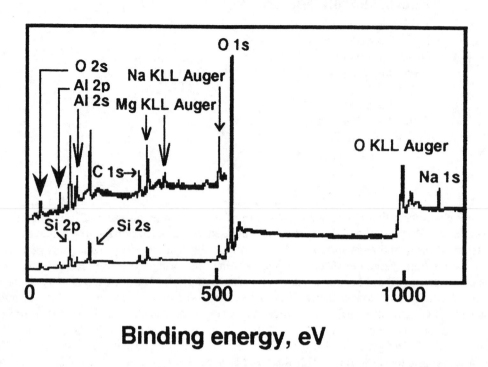

Fig. 7-1. The wide XPS scan of montmorillonite (after Seyama & Soma, 1988).

it is difficult to probe the bonding interactions at the surface of multicomponent materials because of: (i) the overlap of the many transitions, (ii) the low photoionization cross section (which results in low peak intensity), and (iii) the poor spectral resolution stemming from environmental contamination (Briggs & Riviera, 1983). Therefore, most of the bonding interactions at the surfaces of solids are studied utilizing the core electronic states which are easily accessed by XPS.

The photoelectron and Auger lines of montmorillonite indicate the presence of the following elements (with electron-shell designations in parentheses): Si(2s, 2p), C(1s), Al(2s, 2p), Mg(KLL), O(1s, 2s, KLL) and Na(1s, KLL). The scan from 0 to 1200 eV shows a series of steps, which is a characteristic feature of all x-ray spectra. A larger step is observed near a higher intensity peak, e.g., the Si(2p), Mg(KLL) Auger, O(1s), and O(KLL) peaks in Fig. 7-1. This phenomenon is due to the fact that many of the photoelectrons undergo inelastic collisions from the time of their creation until they exit the solid surface, and this effect is manifested in the form of higher background at the high binding-energy side of the peak. The major peaks, termed as no-loss peaks, are due to those electrons that escape the surface without energy loss. Since the electron mean-free-path in solids is relatively short (on the order of nm), the no-loss peaks result from electrons which orginate from atoms at or very near the surface (mainly 1.5 nm or less).

All sample surfaces contain excess C and O which are derived as contaminants from air or residual gases from the vacuum-pump oil. The use of the C(1s) peak at 285 eV for the sample-charge correction is discussed in detail in the Binding Energy section.

Usually Mg- or Al-anode x-ray sources without monochromators are used to generate the XPS spectrum. Consequently, besides the principle $K\alpha_{1,2}$ characteristic x-ray line, many other characteristic lines are present, but, of these, only the $K\alpha_3$ and $K\alpha_4$ characteristic x-ray lines are intense enough to generate small photoelectron peaks on the lower binding-energy side of the main peaks. For example, in Fig. 7-1, the less-intense peaks on the lower binding-energy side of the O(1s) peak are the $K\alpha_3$ and $K\alpha_4$ satellite peaks. These peaks do not interfere with the main XPS peak. The use of a monochromator eliminates the $K\alpha_3$ and $K\alpha_4$ satellites, but decreases the intensity of the primary peaks and the signal-to-noise ratio of the spectrum.

Binding Energy

One of the main uses of binding-energy measurements is to provide direct evidence for the presence of a given oxidation state or multiple oxidation states. The best interpretation comes from a basic understanding of the photoemission process. In this process as illustrated in Fig. 7-2, the absorption of a photon induces the ionization of a core or valence electron associated with a surface or near-surface atom. The electron that is ejected is the photoelectron that is used in XPS. The relaxation process can result in the ejection of another electron, an Auger electron, or an x-ray photon (see Chapter 1 for a discussion on the fundamentals of x-ray emission spectroscopy; Amonette & Sanders, 1994). The energetics with respect to the instrument are shown in Fig. 7-3.

To a first approximation, the photoelectron spectrum should indicate the density of occupied electronic states in the surface. The orbital binding energy, E_b, of the state the electron leaves is assumed to be the same as it was before the photon interaction, and all other electrons in the system are assumed to be in the same state as before the photoionization event. This situation is described by Koopman's theorem,

$$KE = h\nu - E_b \qquad\qquad [1]$$

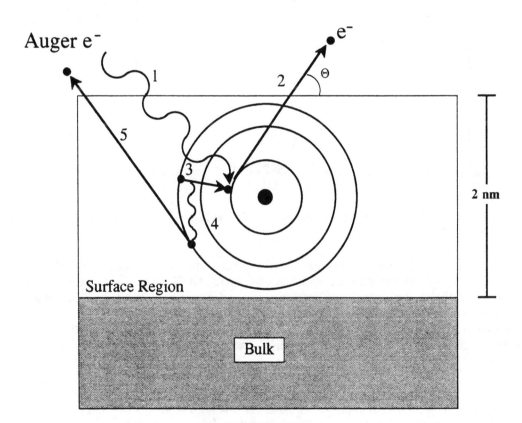

Auger Electron Process:

1) Absorption of x-rays
2) Ejection of photoelectron
3) Relaxation of higher shell electron
   into the hole
4) Energy of Relaxation
5) Ejection of Auger electron or x-ray
   photoelectron

Fig. 7-2. Illustration of photoelectron and Auger electron processes induced by x-ray bombardment.

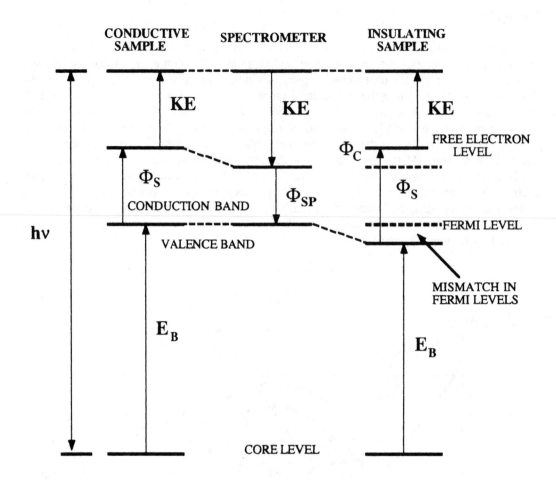

Fig. 7-3. X-ray photoelectroemission energy level diagram for conductive and insulating samples and spectrometer.

where *KE* is the kinetic energy of the photoelectron with respect to the vacuum level (i.e., in the absence of any interference such as stray magnetic fields) and $h\nu$ is the energy of the incident x-ray photon, and E_b is also known as Koopman's energy. Koopman's energy is never observed, however, because of an intraatomic relaxation shift, E_A, caused by other electrons within the atom relaxing in energy to partly screen the positive hole forming as the electron exits. This electronic structure adjustment results in more energy being available to the outgoing photoelectron,

$$KE = h\nu - (E_b - E_A) \qquad [2]$$

Because the photoemission process is rapid and equilibrium may not be reached, additional energy-balancing processes may occur in which another electron is excited to a bound state of the atom or in which it is ejected into the continuum of unbound states above the vacuum level. These processes produce satellite lines in the XPS spectrum corresponding to photoelectrons having lower kinetic energies than the primary photoelectrons.

If there is interaction with the solid, highly mobile electrons will screen the core hole, leading to an additional "extraatomic" relaxation shift, E_{ES}, and to a still larger kinetic energy for the primary photoelectron,

$$KE = h\nu - (E_b - E_A - E_{ES}) \qquad [3]$$

where E_A and E_{ES} are on the order of a few electron volts or less. These electronic energy terms can be collected into a single term, $E_B = E_b - E_A - E_{ES}$, which is the measured binding energy.

The uncertainties associated with the relaxation energies must be kept in mind when interpreting E_B values.

A small amount of work must be done on the photoelectron as it arrives at the electron multiplier to be counted. The amount of energy required to perform this work is termed Φ_{sp}, the work function of the spectrometer. Similarly, the small amount of energy required to completely remove the electron from the sample is termed Φ_s, the sample work function. Thus, as shown in Fig. 7-3, the measured binding energy for conducting samples is found by

$$E_B = h\nu - KE - \Phi_S = h\nu - KE - \Phi_{sp} \qquad [4]$$

For conducting samples, the Fermi levels (i.e., the energy at which the binding energy of the electron is zero) match that of the spectrometer, and E_B can be determined without knowledge of Φ_S, so long as Φ_{SP} is known. The E_B can be compared directly from sample to sample measured on the same instrument; however, it is still best to use internal standards on all samples if good electrical connection with the instrument is in doubt.

For insulating samples, e.g., soil and geologic materials, the sample surface will be electrically isolated from the spectrometer. Consequently, its Fermi level will float with respect to the spectrometer and E_B's measured using Eq. [4] will not be correct. A new term, Φ_C, the static surface charge, must be considered. This charge is typically positive because of the loss of photoelectrons. For insulating samples, then, Eq. [4] can be modified to

$$E_B = h\nu - KE - (\Phi_S - \Phi_C), \qquad [5]$$

in which E_B depends on the work function of the sample and on the surface charge (Fig. 7-3). These values can be obtained from the difference between the measured binding energy and the known binding energy of a standard on the sample surface. Two standards, adventitious C and metallic Au, are commonly used. The adventitious-C method (Swift, 1982) depends on the contaminant carbon that exists on most samples that have been exposed to air. The position of the adventitious C(1s) line is generally considered to be near 285 eV. The exact value used depends on the interaction of the adventitious C with the surface and peak positions derived via this method could conceivably be in error by as much as 0.6 eV. The surface-Au method (Stephenson & Binkowski, 1976; Swift et al., 1983) requires that a small amount of Au be deposited on the surface of the sample using a thin film evaporator mounted in the instrument. The Au(4f$_{7/2}$) binding energy is generally taken as 84.0 eV. Stipp and Hochella (1991) reported that for calcite the use of the near-surface Au (4f$_{7/2}$) peak was a better charge-referencing peak than the C(1s) XPS peak. Even though deposition of a small amount of Au onto the sample seems to provide the best charge-referencing peak (Hochella, 1988; Stipp & Hochella, 1991), the C(1s) adventitious peak is probably the most routinely adopted technique.

The correction of the referencing problem and of charging can lead to reliable binding energies that can be used for chemical interpretation of surface properties. The use of binding energies in the determination of the oxidation states of transition metals is illustrated in this paper using Mn. Manganese forms complex chemical species with O. A number of mixed-valence compounds occur in addition to the simple oxides of Mn(II), Mn(III), and Mn(IV). In general, the binding-energy shifts are used to identify oxidation states. Manganese oxides are distinguishable on the basis of their Mn(2p) binding energies and peak shapes (Aoki, 1976; Oku et al., 1975; Evans & Raftery, 1982; Foord et al., 1984). The results of the XPS examination of several Mn oxides of different stoichiometries (Carver et al., 1972; Oku et al., 1975; Rao et al., 1979; Evans & Raftery, 1982; Foord et al., 1984; Zhoa & Young, 1984) are compared in Fig. 7-4, where the Mn(2p) binding energy of a particular compound is plotted against the oxidation state in its oxides. Although discrepancies occur, there appears to be a systematic correlation between the Mn(2p) binding energy and the effective charge on the Mn atom, as suggested by Carver et al. (1972) and Foord et al. (1984). The determination of average oxidation states for mixed-valence compounds is difficult using the binding-energy parameter alone. Thus, neither the two oxidation states of Mn in Mn_3O_4 nor the Mn site differences in γ-Mn_2O_3 are distinguishable by XPS (Oku et al., 1975).

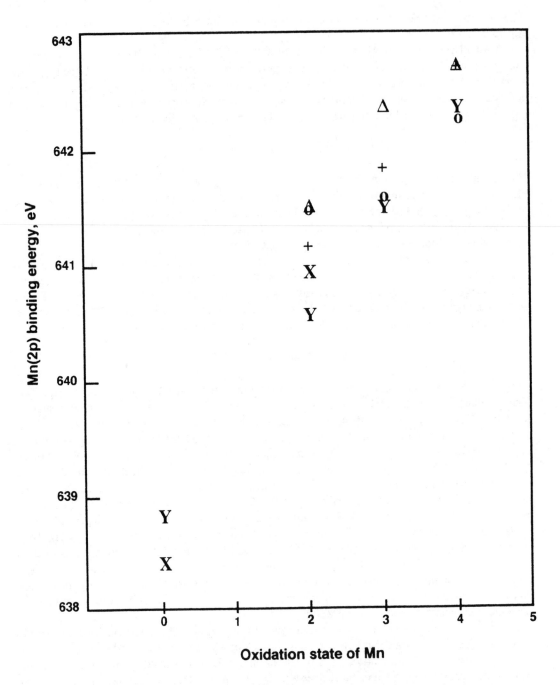

Fig. 7-4. A plot of Mn(2p) binding energy (eV) vs. Mn oxidation state from: (o) Carver, et al. (1972); (Δ) Oku et al. (1975); (X) Evan and Raftery (1982); (+) Zhao and Young (1984); and (Y) Foord et al. (1984).

Amorphous and mixed oxide systems of Mn present similar problems. If more than one oxide exists, the binding-energy shifts between them may not be large enough to provide absolute confirmation of the oxidation states. The Auger-parameter method (see below) is not suitable either, because the differences are not large enough and the Mn Auger line is rather broad. Therefore, one must resort to other means to determine the Mn oxidation states. This determination is done by analyzing the Mn(2p) or the Mn(3s) satellite structure (see section on Satellites). The satellite lines of Mn(2p) levels are explained as arising from shake-up processes, whereas those of Mn(3s) levels are thought to arise largely from multiplet splitting of the levels (Frost et al., 1974). Evans and Raftery (1982) have determined the oxidation states of manganese in MnO, Mn_3O_4, Mn_2O_3, MnO_2 and lepidolite using the Mn(2p)-O(1s) binding-energy differences, and the multiplet-splitting distance of the Mn(3s) peak.

Auger Parameter

Observed binding energies are equal to the difference in total energy of the final-state ion due to rearrangement of orbitals as a consequence of the ionization process and the initial state due to the chemical state of the compound. The chemical and final-state effects cannot be separated by measuring the changes in binding energies alone, and chemical shifts can be completely masked by the final-state effects. Thus, a measure of the relaxation energies is needed. As discussed above, the measured binding energy, E_B, is equal to the orbital energy (i.e., Koopman's energy), E_b, of an electron occupying the initial energy level minus a relaxation energy, $R = E_A + E_{ES}$, where E_A and E_{ES} are the intraatomic and extraatomic relaxation energies, respectively. During the photoionization process, intrarelaxation occurs because of the flow of electrons within the host atom; whereas, extrarelaxation occurs because of the flow of the electrons from the surrounding atoms to the host atom. Therefore,

$$E_B(i) = E_b(i) - R(i). \tag{6}$$

The difference in measured binding energy between two chemical compounds is then given by

$$\Delta E_B(i) = \Delta E_b(i) - \Delta R_{ES}(i), \tag{7}$$

where the difference in relaxation energy is essentially extra-atomic since the intra-atomic part is about the same for a given atom in different compounds.

Three energy levels (ijk) must be considered for the shift in an Auger line, because the Auger process involves three electrons. Auger electrons are produced when the atom's inner shell (i) is ionized (i.e., the atom is in an excited state because of the influence of the incident electron or photon on the atom). This excitation creates a hole in the core level, resulting in transfer of an electron from a higher atomic level (j) to fill the inner-core hole. The de-excitation process or release of excess energy results in emission of an x-ray photon, leading to ejection of another electron (k). Therefore, the nomenclature of Auger electrons involves three electrons,

$$E_k(ijk) = E_B(i) - E_B(j) - E_B(k) - E(jk) + R(jk), \tag{8}$$

where the term $E(jk)$ is the interaction energy between the two holes in the final state and $R(jk)$ is the total relaxation energy. The difference in the Auger kinetic energy between two chemical compounds would be

$$\Delta E_k(ijk) = \Delta E_B(i) - \Delta E_B(j) - \Delta E_B(k) - \Delta E(jk) + \Delta R_{ES}(jk). \tag{9}$$

The relation between the measured kinetic energy of an x-ray induced Auger line, $E_k(ijk)$, and its apparent binding energy, E_B^a, is (similar to Eq. [4])

$$E_k(ijk) = h\nu - E_B^a - \Phi_S. \tag{10}$$

Using $KE(P) = h\nu - E_B - \Phi_S$ for photoelectrons, an equation for the Auger parameter, α, has been defined by Wagner (Williams & Nason, 1974) as the difference between the kinetic energy of the most intense Auger line, $E_k(ijk)$, and the most intense photoelectron line, $KE(P)$,

$$\alpha = E_k(ijk) - KE(P), \tag{11}$$

which, in terms of the binding energies, becomes

$$\alpha = E_B - E_B^a .$$ [12]

In order to keep the Auger parameter positive and independent of photon excitation energy, it is usually presented as

$$\alpha' = \alpha + h\nu,$$ [13]

where α' is the modified Auger parameter. Thus, the Auger parameter is the difference in the kinetic energy between two lines in the same spectrum. This difference is independent of sample charging and, consequently, is important for studying soil samples, minerals and other geologic materials, many of which are insulators and can exhibit charging. In addition, the Fermi level need not be precisely determined.

Binding-energy data referenced to the Fermi level and corrected for charging and Auger kinetic-energy data can be combined into a "chemical-state" plot, which provides useful chemical information (Fig. 7-5). In these plots, the x-axis gives the binding energies of photoelectron lines, the y-axis gives the kinetic energies of the Auger lines, and the modified Auger parameter is shown as a series of diagonal lines.

The Auger parameters for many different compounds have been measured and tabulated and are known to be sensitive to the chemical environment (Wynblatt & Ku, 1977; Hultgren et al., 1973). For KLL Auger transitions, the following approximation (Wagner & Joshi, 1980; Seah, 1983) is valid:

$$\Delta\alpha = 2\Delta R_{ES}(i)$$ [14]

This relationship suggests that the change in the Auger parameter is a direct measure of changes in the extraatomic relaxation, which is related to the polarizability of the material (i.e., the greater the polarizability of a compound, the greater the Auger parameter). For conductors, the hole is completely screened and the Auger parameter is large. However, for geologic materials that are likely to be good dielectrics a slight displacement of the electron clouds of neighboring atoms occurs towards the positive hole, and in those samples that are ionic crystals, the anion and cation interactions dominate (Ho & Lewis, 1976). Hence, the Auger parameters for geologic materials and for ionic crystals fall between the values for the corresponding gaseous atom and the metal.

Spin-Orbit Splitting

Another obvious feature of the XPS spectrum is the occurrence of doublets for p, d and f electrons (Fig. 7-6). These doublets, which result from spin-orbit splitting, are useful for chemical-state interpretation. As shown in Table 7-1 for $\ell > 1$, where ℓ is the orbital angular momentum quantum number, the XPS spectrum is split into two peaks, with the peak for the higher j level occuring at the lower binding energies. The relative intensities of these peaks are 1:2, 2:3, and 3:4 for p, d, and f sublevels, respectively.

The presence of Auger peaks in the XPS spectrum is quite useful in that they can be used for chemical state differentiation when the core-level photoelectron peaks display no chemical shifts. For example, this method is the principal way to differentiate between the Cu(0) and Cu(I) oxidation states. In addition, the combined use of Auger and photoelectron peaks to yield the Auger parameter provides chemical information where final-state effects mask the chemistry (Seyama & Soma, 1988). Further details concerning the fundamentals of the XPS and Auger electron spectroscopy (AES) techniques can be found in Hochella (1988).

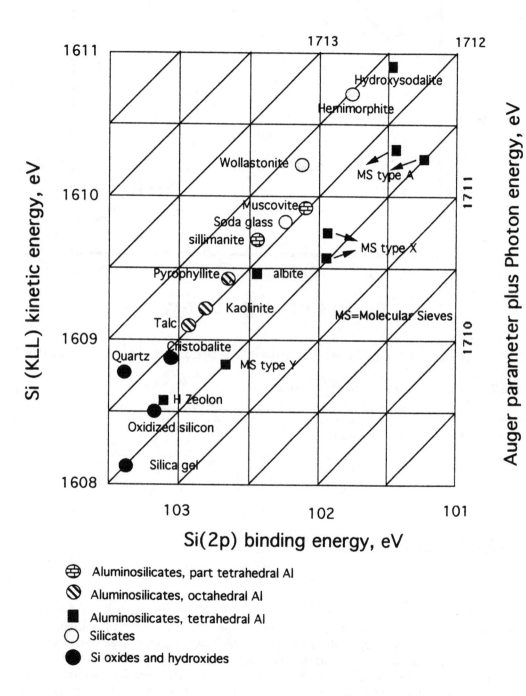

Fig. 7-5. Silicon chemical-state plot (after Wagner et al., 1982).

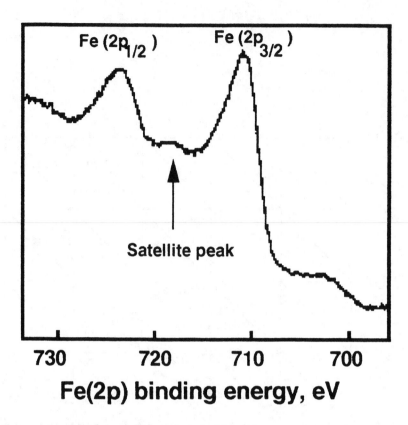

Fig. 7-6. XPS spectrum of Fe(2p) peaks in ferrihydrite (adapted from Vempati et al., 1990a).

Table 7-1. X-ray nomenclature of the spin-orbit splitting.

Subshell	ℓ† 1s	Quantum number (j-coupling)	Area ratio
s	0	1/2	--
p	1	1/2, 3/2	1:2
d	2	3/2, 5/2	2:3
f	3	5/2, 7/2	3:4

† ℓ is the orbital angular momentum quantum number.

Satellites

Satellites due to multiple-electron excitations and ligand-transfer processes can be usefully applied to materials characterization. If such an interaction occurs during a photoionization event, the photoelectron will lose some of its energy and additional peaks will be observed in the spectrum. These peaks occur at the high-binding-energy side of the main photoelectron peak and are termed shake-up or shake-off satellites depending on whether the other electrons being excited are only promoted to an excited state or to the continuum state, respectively. The transition-metal and rare-earth elements that have unpaired electrons (paramagnetic) produce the most intense satellite structures. Chromium(III) d^3, high-spin Co(II) d^7, high-spin Fe(III) d^5 (Fig. 7-6), high-spin Fe(II) d^6, Ni(II) d^8, and Cu(II) d^9 are prime examples of ions that give strong satellites.

Multiplet Splitting.   The emission of a core-level photoelectron can be interactively coupled to one or more valence electrons through a phenomenon known as multiplet splitting. In multiplet splitting the unpaired "hole" created by the photoemission in the core level interacts with the unpaired valence electrons. This phenomenon is particularly useful for studying the paramagnetic metal ions. Although the multiplet-splitting phenomenon causes broadening of the 2p photoelectron spectra of transition metal ions, it also produces a useful splitting in the 3s level as exemplified by Mn ions (Fig.7-7). The splitting distance between the Mn(3s) peaks is referred to as ΔBE and has been used to identify the oxidation state of Mn (Murray et al., 1985). A unit change in oxidation state results in about a 0.7-eV change in the multiplet splitting for the Mn(3s) peak. The multiplet splitting can help to shed light on the surface chemical changes that are manifested only weakly in the binding energy shift. These satellite-peak intensities depend not only on species, but also on the surface conditions. The MnO has shake-up peaks at about 5 eV from both the Mn($2p_{3/2}$) and Mn($2p_{1/2}$) levels (Oku et al., 1975; Hu and Rabalais, 1981). Both Mn_2O_3 and MnO_2 have detectable shake-up peaks (discussed below) with a Mn($2p_{1/2}$) peak that has a binding energy about 10 eV higher than the Mn($2p_{3/2}$) satellite peak due to the shake-up process (Wallbank et al., 1975). A similar phenomenon is observed for the Mn($2p_{3/2}$) satellite peaks of Mn_2O_3 and MnO_2. The energy separations between the Mn(2p) main peaks and the satellite peaks for Mn_2O_3 and MnO_2 are about twice that for MnO. In the first-transition series, the shake-up satellite energy suddenly increases from d^5 to d^4 (Wallbank et al., 1975). The Mn(II) in MnO has d^5 electrons while Mn(III) and Mn(IV) ions have d^4 and d^3 electrons, respectively. Thus, the analogous relationship between the number of 'd' electrons and the energy separation also occurs in the Mn ion series. However, shake-up satellites are often broad and are consequently seldom distinguishable against a high-intensity background, except in the instance of stoichiometric oxides (Allen et al., 1982).

According to Van Vleck's theorem, the Mn(3s) splitting, E(3s), can be given by:

$$E(\text{3s}) = (2S + 1)G2(\text{3s,3d})/5 \qquad [15]$$

where S is the initial state spin and $G2$(3s,3d) is the appropriate 3s-3d Slater exchange integral. Except for Mn metal [Mn(m)], the Mn(3s) splitting is approximately proportional to the initial-state total spin, $2S + 1$ (Fig. 7-7). Because the 4s electrons of Mn(m) reside in an s-p conduction band, screening effects yield a smaller 3s-3d overlap than in MnO and a smaller observed multiplet splitting (Bagus et al., 1973). The pair-correlation energy is about 1.35 eV per 3s-3d pair (Shirley, 1975). Experimentally determined values for the multiplet splitting of the Mn(3s) level at different total spin states show a relatively good agreement among researchers (Fig. 7-7).

Plasmon-loss Lines.   As it passes through the solid, the photoelectron can interact in a discrete manner with the valence and conduction bands of the solid producing one or more high-binding-energy peaks. This "plasmon loss" is proportional to the electron density of the atom, and a change in the plasmon-loss peaks can be related to changes in the initial electron densities of the material. Plasmon losses occur in conducting samples but are rare in nonconducting samples like silicate minerals (Perry et al., 1990). However, energy-loss lines have been observed in some silicate minerals at the high-binding-energy side of the photoelectron peak (15-20 eV), and these energy-loss lines are 10 eV or more in width at half peak height (see Hochella, 1988).

EXPERIMENTAL TECHNIQUES

Instrumentation

The XPS is an ultrahigh-vacuum technique (10^{-6} to 10^{-8} Pa), which places restrictions on its use in soil chemistry. The basic components of the instrumentation are shown in Fig. 7-8. The heart of the instrument is the analyzer; the one most commonly used is the hemispherical-sector type. Here, the photoelectrons are subjected to a retarding field and resolved by their different

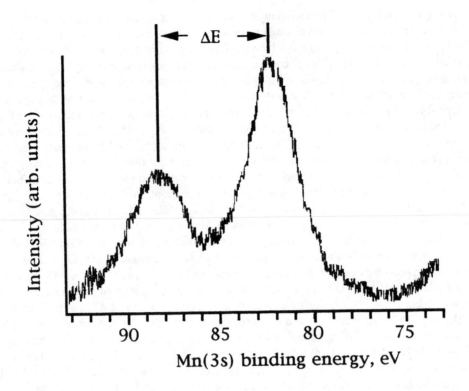

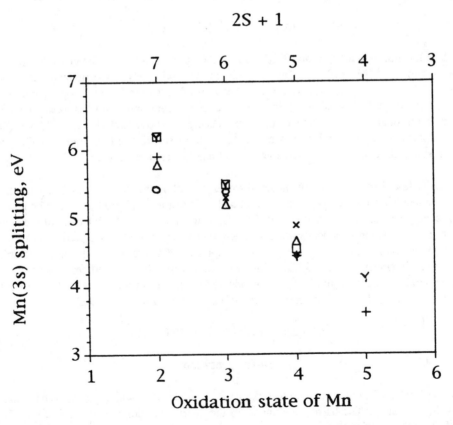

Fig. 7-7. (top) Multiplet splitting in the Mn(3s) XPS spectrum, and (bottom) the experimentally determined relationship between the splitting distance and Mn oxidation state: (o) Carver et al., 1972; (□) Wertheim et al., 1973; (△) Oku et al., 1975; (X) Evan and Raftery, 1982; (+) Zhao and Young, 1984; and (Y) Foord et al., 1984.

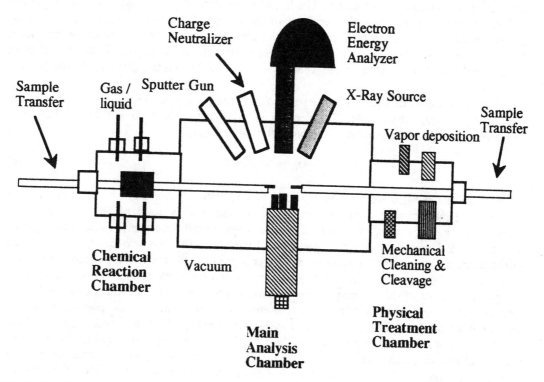

Fig. 7-8. Sketch of a typical XPS spectrophotometer.

paths between the two concentric charged hemispheres. Both single-channel and multi-channel electron detectors are used; the latter increase the speed of the relatively slow XPS experiment. The x-ray source is usually a dual-anode type, with aluminum and magnesium as the source metals. Other metals are used on a limited basis to provide alternative energies. A charge neutralizer which provides a flux of low-energy electrons is used in some instruments to help compensate for the charging of poorly conducting samples. A sputter gun is usually present to provide etching of the surface by bombardment with argon ions, thus allowing for depth profiling of samples. The sample is generally placed on an apparatus that allows it to be tilted with respect to the analyzer to provide for angle-resolved XPS studies. Since samples often require some form of in situ treatment before or during analysis, sample-treatment chambers are usually a part of the total apparatus.

Sample Handling

Sample handling before entry into the spectrometer requires that the surface be protected from extraneous contamination. This objective is achieved by avoiding contact with contaminating atmospheres or surfaces. After insertion into the spectrometer, the sample can be treated in special chambers as shown in Fig. 7-8. Here, the surface can be cleaned by mechanical means while under vacuum, using metal brushes, scrapers or sample cleavers. Reference materials such as Au can be deposited directly on the sample in vacuum, or reactants can be added via the gas phase or by deposition in inert atmospheres. The samples can be heated or cooled during chemical treatments.

Quantitative Studies

Quantitative analysis with accuracies as good as 50 g kg^{-1} are possible using XPS, but care in the method of calibration is required. In any such attempt, it is usual to regard the photoemission as a three-step process: (i) photoionization, (ii) transport of the photoelectron to the surface, and

(iii) transmission through the surface. The last of these three steps is rarely a matter of concern in XPS; however, when the kinetic energies of the exiting electrons are low, as in ultraviolet-photoelectron spectroscopy, total internal reflection can occur for electrons approaching the surface at the grazing angle. Usually with XPS, the energies are sufficiently large and internal reflection is not a problem. The next step in the XPS process is the collection and detection by the electron analyzer.

The total intensity of a photoelectron peak, N_i, is the sum of emissions, δN_i, originating from various depths within the sample. The electron intensity originating from an infinitely thin layer at a given depth can be expressed as the product of six factors,

$$\delta N_i = \Phi N A E \Theta I \qquad [16]$$

where Φ is the x-ray flux reaching the layer, N is the number of atoms of i in the layer, A is the differential photoionization cross section of the relevant energy level of species i, E is the probability of a no-loss escape of electrons from the layer, Θ is the acceptance angle of the electron analyzer; and I is the instrumental detection efficiency. The last two of these factors are instrumental in nature and depend on the kinetic energy of the photoelectron and the sample area illuminated by the x-ray source. Normally, the penetration depth of the x-rays is very large compared with that of the escaping photoelectrons, so the spatial (depth) dependence of the first term (Φ) can be neglected; however, the spatial distribution does influence the total signal through the fourth term (E), primarily because of inelastic scattering of the electrons. Thus, a large concentration of species i in several atomic layers below the surface may give less signal than a much smaller concentration of this species in the uppermost surface layer. In some instances, it is possible to gain information about the depth distribution of components by studying the angular dependence of the emission signal.

Sputtering Method

The sputtering technique provides a way to study the variability in the chemical composition of minerals with depth below the surface. The mineral surfaces are eroded by bombarding the surface with energetic ions, usually Ar^+, emitted by an ion gun. The depth of XPS analysis is achieved by calibrating the sputtering rate of the surface of a known mineral, usually a SiO_2/Si wafer of known thickness. Studies have shown that sputtering can cause damage to the mineral surface by altering the oxidation state of the surface elements and/or causing the formation of amorphous products [see Hochella (1988) and the references therein]. In recent studies, comparisons of the altered and unaltered material have been made to aid in the interpretation of the depth-profile analysis (Hellmann et al., 1990; Inskeep et al., 1991).

Inskeep et al. (1991), using the Ar-sputtering method, determined the influence of weathering on the composition of labradorite, a feldspar containing calcic and sodic lamellar phases, with respect to depth. A comparison was made between the unweathered and weathered labradorite grains which were treated at pH 3.7 for 415 d. The sputtering of the unweathered labradorite for 60 min, i.e., to a depth of approximately 100 nm, indicated that (i) Na was sputtered at a faster rate than the other elements, (ii) a slight increase in the Ca surface:bulk ratio was observed, and (iii) Al, Si and O were sputtered at approximately constant rates (Fig. 7-9). The comparison of the depth-profile analyses of the untreated and treated samples indicated that the Ca surface:bulk ratio was considerably lower for the treated sample, and the depth of Ca depletion upon acidic treatment was ~50 nm (Fig. 7-10). Similar results were obtained for Al, but the amount of depletion was less than that for Ca. In the case of Na and Si, surface enrichments were observed to a depth of 50 nm for the treated sample.

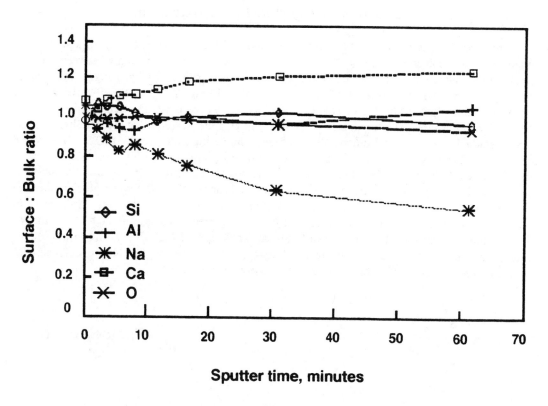

Fig. 7-9. Sputter depth profile of unweathered labradorite sample. Note preferential sputtering of Na compared to Al, Si, Ca, and O. The calibration with SiO_2/Si indicated a sputtering rate of 1.57 nm min^{-1} (after Inskeep et al., 1991).

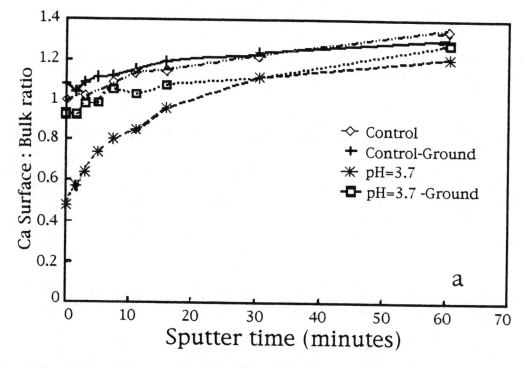

Fig. 7-10. Surface:bulk ratio of Ca versus sputter time for unweathered labradorite (control) and weathered labradorite at pH 3.7 for 415 d (after Inskeep et al., 1991).

Angle-Resolved X-Ray Photoelectron Spectroscopy

Angle-resolved XPS is a nondestructive method to study the composition of the top few nanometers of a mineral surface. The analysis depth is a function of the angle between the axis normal to the sample surface and the electron-signal collection path; hence by changing the sample-tilt angle, a variation in the depth of analysis is achieved (Fig. 7-11). The effective depth of analysis is calculated using the equation: $d = 3\lambda\cos\Theta$, where d is the effective depth of the analysis, λ is the attenuation depth (escape depth), and Θ is the sample tilt angle. Hellmann et al. (1990) studied surfaces of hydrothermally weathered albite using angle-resolved XPS. They found significantly lower Na/Si ratios in the near-surface regions (0-3.4 nm and 0-6.9 nm) than in the 0- to 7.8-nm region of the weathered or unweathered sample. The increase in the full-width-at-half-maximum (FWHM) of the O(1s) peak for the weathered sample (0- to 7.8-nm region) relative to that for the unweathered sample indicated the addition of hydrated-O environment. A decrease in the O(1s) FWHM in the shallowest region (0-3.4 nm) suggested a decrease in the total number of O environments resulting from the loss of surface Na-O and Al-O (Table 7-2). Furthermore, changes in the O/Si ratio from 8:3 for fresh albite to 2:1 for the leached sample (data not shown) suggested the formation of an amorphous-Si or polymerized-Si phase.

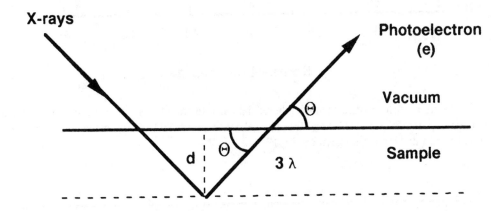

Fig. 7-11. Schematic diagram showing the effective depth of analysis and the angle of electron ejection (after Hochella, 1988).

Table 7-2. The angle-resolved XPS study of albite treated at pH 2.33 and 225 °C for 14 h (after Hellmann et al., 1990).

Sample	Angle of tilt	Depth	Na/Si atomic ratio	Al/Si atomic ratio	O(1s) FWHM
	°	nm			eV
Untreated	0	7.8	0.173	0.175	2.2
Treated	0	7.8	0.160	0.203	2.3
Treated	30	6.9	0.139	0.178	2.2
Treated	65	3.4	0.132	0.180	2.0

APPLICATIONS

In this section, the application of XPS and Auger spectroscopy to soil chemical and mineralogical problems, both qualitative and quantitative in nature, will be discussed. Uses to be discussed include those involving characterization of adsorbed species, oxidation states, chemical weathering, active sites, amorphous materials, and environmental samples.

Semi-Quantitative Analysis

X-ray photoelectron spectroscopy is not a highly accurate tool for quantitative analysis of a mineral for several reasons, i.e., the presence of C and minor phases on the mineral surface and the variable sensitivity factors of the elements under consideration. Bancroft et al. (1977), however, demonstrated that XPS is a sensitive method for analyzing surface-adsorbed ions in the nanogram per gram range. In their experiment, Ba from a $Ba(NO_3)_2$ solution was adsorbed onto a cleaved calcite surface. The results indicated a significant correlation between the intensity ratio of the Ba(3d) and Ca(2p) XPS peaks and the amount of Ba adsorbed to the calcite surface (Fig. 7-12). In another experiment with a series of clay minerals, González et al. (1988) observed a 10 to 30% deviation between surface and bulk chemical analyses (Table 7-3). The XPS has also been shown to be a valuable tool for detecting minor isomorphic substitutions, such as Fe^{2+} for Mn^{2+} in almandine, Na^+ for K^+ in orthoclase, and Na^+ for Ca^{2+} in anorthite (Seyama & Soma, 1988).

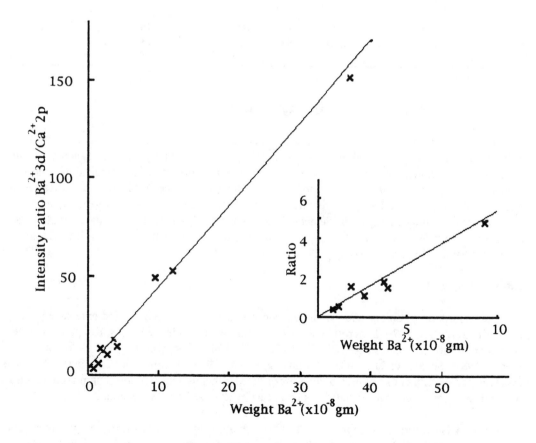

Fig. 7-12. Plot of Ba(3d)/Ca(2p) peak area ratio vs. Ba adsorbed on the calcite surface (adapted from Bancroft et al., 1977).

Table 7-3. Surface (S) and bulk (B) elemental composition of selected phyllosilicates (adapted from González et al., 1988).

Mineral		Framework Elements				Interlayer Elements		
		O	Si	Al	Mg	K	Na	Ca
					g kg^{-1}			
Pyrophyllite	(S)	653	233	113	--†	--	--	--
	(B)	669	223	107	--	--	--	--
Muscovite	(S)	656	168	138	--	30	2	--
	(B)	638	168	148	--	42	2	--
Vermiculite	(S)	652	142	79	123	--	--	18
	(B)	619	149	61	145	--	--	23
Margarite	(S)	635	110	215	--	--	7	32
	(B)	632	111	204	--	--	10	42
Talc	(S)	634	207	--	157	--	--	--
	(B)	632	211	--	157	--	--	--
Phlogopite	(S)	623	159	73	107	36	--	--
	(B)	606	144	63	142	44	--	--

† None expected.

Qualitative Analysis

Most of the elements have electron binding energies with ranges of approximately 1 to 3 eV. Shifts in binding energy provide information regarding electronic and geometric structures of the minerals and chemical processes occurring at the mineral surface, i.e., chemical weathering, oxidation or reduction, adsorption of ions, etc. The shift in the XPS peak is indicative of differences in the chemical and/or structural environment of the elements. In general, increases in the electron binding energy of an element are attributable to (i) an increase in the oxidation state and the resulting decrease in electronic population in the valence shell, and (ii) an increase in the electronegativity of the surrounding ions.

Adsorbed Species

The adsorption of anions and cations on mineral surfaces has been studied extensively using XPS (Alvarez et al., 1976; Bancroft et al., 1977; Count et al., 1973; Dillard & Koppelman, 1982; Goodman, 1986; Koppelman & Dillard, 1977; 1978; 1980; Koppelman et al., 1980; Martin & Smart, 1987; Schenk et al., 1983; Seyama & Soma, 1985; Vempati et al., 1990a,b). One disadvantage of XPS is that hydrated minerals or those minerals sensitive to vacuum are not suited for analysis. However, hydroxylated minerals can be analyzed by this technique.

Anion Adsorption. Vempati et al. (1990a, 1990b) studied the adsorption of phosphate and silicate by ferrihydrite. For the untreated ferrihydrite samples, the decomposition of the O(1s) peak indicated the presence of oxide-type O at 530.1 eV and OH at 531.8 eV. After phosphate treatment a decrease in surface-OH concentration was observed (Table 7-4), which was in agreement with the ligand-exchange model of Hingston et al. (1968) in which phosphate ions adsorb by replacing the surface-OH ions. Furthermore, the presence of the P XPS peak at 133.3 eV indicated that HPO_4^{2-} was the predominant species adsorbed on the ferrihydrite surface (Vempati et al., 1990a). Adsorption of silicate ions on the ferrihydrite surface was also studied by XPS. For ferrihydrite samples treated with ≤ 37.5 g Si kg^{-1}, only a broad peak at 100.9 eV was observed, indicating the presence of monomeric silicate or small units of polymerized silica linked at only one or two corners of the O-Si-O tetrahedra. For ferrihydrite samples treated with ≥ 37.5 g Si kg^{-1}, the Si(2p) XPS peak was decomposed into two peaks, at approximately 101.6 and 103.8 eV (Fig. 7-13). The latter peak occurred at a position similar to that observed for Si-rich minerals. For these silica-rich samples, the O(1s) peak was decomposed into three peaks: 530.4 (oxide-O group), 531.9 (OH-group) and 533.7 (Si-O group) (Fig. 7-14). The presence of the 103.8- and 533.7-eV XPS peaks confirmed the precipitation of a Si-rich mineral in the samples treated with ≥ 75 g Si kg^{-1} (Vempati et al., 1990b).

Cation adsorption. The adsorption of cations to mineral surfaces has been studied extensively using XPS (Bancroft et al., 1977; Count et al., 1973; Goodman, 1986; Koppelman & Dillard, 1977, 1978, 1980; Koppelman et al., 1980; Dillard & Koppelman, 1982; Schenk et al.,1983). Seyama and Soma (1988) studied the surface atomic composition of montmorillonite treated with selected cations and the bonding states of these exchangeable cations. The atomic concentrations of the major elements in the samples were calculated relative to that of the Al atom (Table 7-5). The results indicated that the surface compositions were consistent (within 10%)

Table 7-4. Results of XPS analyses of untreated and phosphate-treated ferrihydrite samples (adapted from Vempati et al., 1990a).

Treatment	C	Fe	O	OH	P
	------------------------------------- Atomic % -------------------------------------				
Untreated	36.2	9.2	33.8	20.8	--
P-treated	9.5	9.5	56.9	7.8	2.7

Table 7-5. Surface atomic ratios with respect to Al of the major elements present in montmorillonite saturated with different cations (Seyama & Soma, 1988).

Element	Saturating cation			
	Na	Mg	Ca	Cd
Al	1.00	1.00	1.00	1.00
Si	2.17	2.13	2.24	2.25
Mg	0.21	0.34†	0.22	0.23
Na	0.26	--‡	--	--
Ca	--	--	0.11	--
Cd	--	--	--	0.11

† 0.34 (total Mg) = 0.22 (structural Mg) + 0.12 (adsorbed Mg).
‡ -- = None expected.

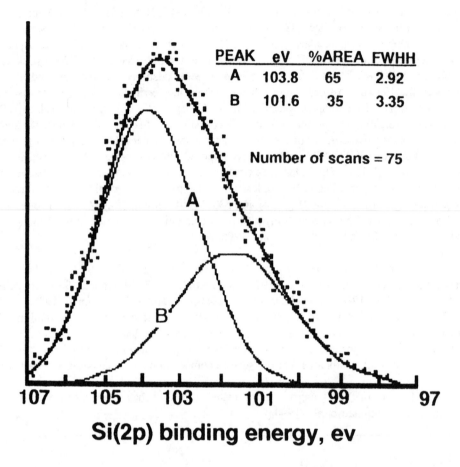

PEAK	eV	%AREA	FWHH
A	103.8	65	2.92
B	101.6	35	3.35

Number of scans = 75

Si(2p) binding energy, ev

Fig. 7-13. Decomposition of the Si(2p) x-ray photoelectron spectroscopy peak of ferrihydrite treated with ≥ 75 g Si kg^{-1}. The dots surrounding the composite peak represent the original spectrum (adapted from Vempati et al., 1990b).

regardless of the saturating cation. The atomic ratio with respect to Al of the adsorbed divalent cations (0.11) was roughly one-half that of the adsorbed monovalent cation (0.26), in approximate agreement with the cation exchange capacity.

The bonding state of the divalent cations was determined using a chemical-state plot, which incorporates the XPS binding-energy and the Auger kinetic-energy values. The extra-atomic relaxation energy difference, ΔR_s, was calculated by {[XPS binding energy (sample) + Auger kinetic energy(sample)] - [XPS binding energy (standard) + Auger kinetic energy (standard)]} (Table 7-6). Figure 7-15 shows the chemical-state plot for Mg; the exchangeable Mg^{2+} plotted between $MgCl_2$ and MgF_2 whereas the nonexchangeable ion plotted near MgO and other minerals containing structural Mg. The ΔR_s of the nonexchangeable Mg ion is greater than that of exchangeable ion (Table 7-6), indicating that the density of electrons surrounding the nonexchangeable Mg is greater.

Oxidation States

Stucki et al. (1976) were the first to use XPS to study the chemical reduction of structural Fe in nontronites. The nontronite was reduced using hydrazine or dithionite and the surface Fe^{3+}/Fe^{2+} ratio as measured by relative intensities of the Fe-XPS peaks was compared with the bulk Fe^{3+}/Fe^{2+} ratio measured by colorimetric techniques. The binding energies for Fe^{3+} and Fe^{2+} XPS Fe(2p) peaks are 711.8 and 708.6 eV, respectively. The surface ratio of Fe^{3+}/Fe^{2+} measured by XPS was in agreement with the bulk ratio obtained by colorimetric methods (Table 7-7). The oxidation states of other elements (e.g., Mn, Co, Pb, Ti, Cu, Ni, and S) have been measured by

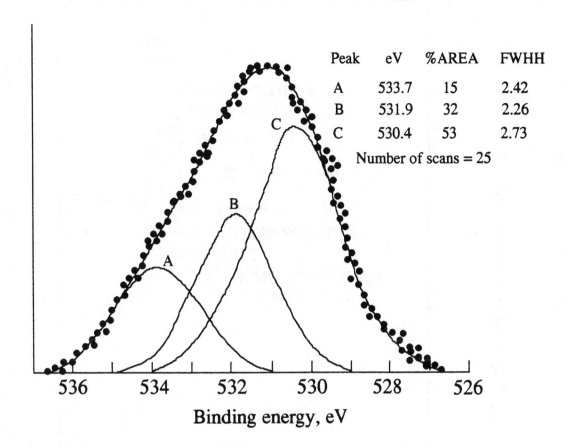

Peak	eV	%AREA	FWHH
A	533.7	15	2.42
B	531.9	32	2.26
C	530.4	53	2.73

Number of scans = 25

Binding energy, eV

Fig. 7-14. Decomposition of the O(1s) x-ray photoelectron spectroscopy peak of ferrihydrite treated with ≥ 75 g Si per kilogram. The dot surrounding the composite peak represents the original spectrum (Vempati et al., 1990b).

Table 7-6. Mg(1s) binding energies, Mg $KL_{23}L_{23}$ Auger kinetic energies and ΔR_S values of Mg compounds (Seyama & Soma, 1988).

Compounds	Mg(1s)	Mg $KL_{23}L_{23}$	ΔR_S †
	---------------- eV ----------------		
MgF_2	1306.5	1176.8	0.0
Exchangeable	1305.3	1179.0	1.0
Nonexchangeable	1303.8	1181.0	1.5
$MgCl_2 \cdot 6H_2O$	1304.8	1180.2	1.7
MgO	1303.9	1181.3	1.9
$MgBr_2 \cdot 6H_2O$	1305.3	1180.7	2.7

†$\{BE[Mg(1s)] + KE[Mg(KL_{23}L_{23})]\}MgX - \{[BE[Mg(1s)] + KE[Mg(KL_{23}L_{23})]\}MgF_2$.

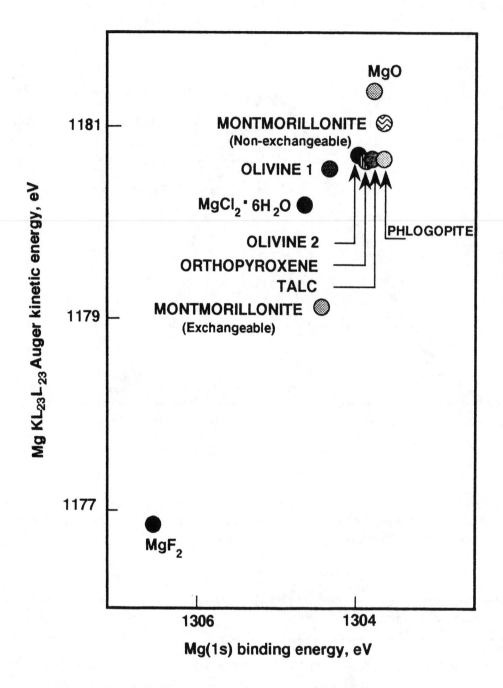

Fig. 7-15. Magnesium chemical-state plot (adapted from Seyama & Soma, 1988).

XPS (Craig et al., 1974; Dillard et al., 1981, 1984; Dillard & Schenck, 1986; Hyland & Bancroft, 1989; Koppelman & Dillard, 1977; Murray & Dillard, 1979; Murray et al., 1985; Myhra et al., 1988; Perry, 1986; Schenck et al., 1983).

Crowther et al. (1983) studied the adsorption of Co^{2+} on the birnessite (γ-MnO_2) surface at different pH values. At pH 4 to 7, the Co^{2+} was oxidized to Co^{3+} whereas, at pH 8 to 10, Co^{2+} was removed from solution and presumably adsorbed to the birnessite surface. The absence of a satellite peak and the ΔE [obtained from subtracting $Co(2p_{1/2})$ and $Co(2p_{3/2})$ XPS binding-energy values] of 15 eV for Co^{2+}-treated birnessite samples at pH 4 to 7 indicated the presence of Co^{3+}. At pH 8 to 10, on the other hand, the presence of a satellite peak and the ΔE value of ~16 eV suggested the probable precipitation of $Co(OH)_2$ compounds (Table 7-8). The Mn(3s) splitting distance of 4.7 for birnessite indicated that the oxidation state of Mn was 4+. For a sample treated

Table 7-7. The Fe^{3+} and Fe^{2+} ratios as measured by colorimetry and relative intensity of the Fe(2p) XPS peaks in unaltered nontronite and in nontronite treated with hydrazine or dithionite (Stucki et al., 1976).

Treatment	Fe(III)/Fe(II)	
	Colorimetry	XPS
Unaltered	13.1	--†
Hydrazine-treated	6.4	6.0
Dithionite-treated	2.2	2.5

† XPS peak of Fe^{2+} could not be resolved.

at pH 6.5, either under aerobic or anaerobic conditions, the oxidation states of the Co and Mn were (III) and (IV), respectively, suggesting that O was not responsible for the oxidation of Co^{2+} to Co^{3+} in solutions at pH values of 4 to 7. When the birnessite sample was repeatedly washed with Co^{2+} solution in an anaerobic environment, however, Co^{2+} was oxidized to Co^{3+} and the Mn(3s) splitting distance increased to 5.1 eV, suggesting that a portion of Mn(IV) was reduced to Mn(II). Furthermore, the XPS-Mn($2p_{3/2}$) binding-energy value of 642.2 eV suggested the formation of a Mn_2O_3-like surface phase.

Chemical Weathering

The XPS has been extensively used to investigate chemical weathering processes occurring at mineral surfaces (Crovisier et al., 1983; Eggleston et al., 1989; Hellman et al., 1990; Holdren & Berner, 1979; Hochella et al., 1988; Hochella & Brown, 1988; Inskeep et al., 1991; Muir et al., 1989; Petrovic et al., 1976; Schott et al., 1981; Schott & Berner, 1983; Schott & Petite, 1987; Thomassin et al., 1977; White et al., 1986; Zing & Hercules, 1978). These studies provide clues to the stability and decomposition of minerals in terrestrial environments, and also to the release of plant nutrients and toxic elements to the rhizosphere and groundwater. Prior to the development of XPS, such information was obtained primarily by investigations of the kinetics of dissolution of minerals. From kinetic studies, it had been suggested that chemical weathering may produce uniformly altered layers which are thousands of micrometers in thickness, thus limiting the diffusion of surface ions. However, XPS studies have shown that mineral dissolution occurs at high-energy sites at the mineral/solution interface (surface nonhomogeneities, irregularities and defects), and that the altered layer (protective layer) which forms is only on the order of 1 to 10 nm in thickness (Petrovic et al., 1976; Berner & Holdren, 1977; Holdren & Berner, 1979).

As discussed previously, Inskeep et al. (1991) studied the weathering of labradorite in a batch study. The samples were chemically weathered at pH 3.7 and 4.1 in 0.01 M LiCl in the presence and absence of Na-citrate and oxalate buffers. The suspensions were shaken slowly, and the solutions periodically changed to prevent supersaturation of the secondary phases. Under the above conditions, the surface was depleted of Ca and Al in the range of 30 to 50% and 10 to 15%, respectively. And, the surface was enriched by Na and Si in the range of 10 to 15% and 13 to 25%, respectively (Table 7-9). The surface depletion of Ca and the enrichment of Na measured by XPS were in agreement with the results of the solution analyses. Analyses by scanning electron microscopy showed that the weathered labradorite grains contained nonuniform surface alterations (etch pits), and transmission electron microscopy analyses indicated that the calcic lamellae weathered more readily than the sodic lamellae. Therefore, the XPS surface composition changes

Table 7-8. The XPS analysis of Co(II) adsorption on birnessite at different pH values and reference data for several Mn and Co oxides (Crowther et al., 1983).

pH	Co(2p$_{3/2}$)	ΔE†	Mn(2p$_{3/2}$)	Mn(3s) peak splitting
	-- eV --			
4	780.4	15.1	642.1	4.7
6	780.4	15.0	642.2	4.8
7	780.4	15.0	642.2	4.8
8	781.0	16.0	642.2	4.8
10	780.8	16.0	642.2	4.7
6.5‡	780.3	15.1	642.2	4.7
6.5§	780.3	15.0	642.2	4.7
6.5¶	780.3	16.0	642.2	5.1
Reference Values				
γ-MnO$_2$	--	--	642.2	4.7
MnO	--	--	640.6	5.8
α-Mn$_2$O$_3$	--	--	641.9	5.2
Mn$_3$O$_4$	--	--	642.2	5.3
CoOOH	780.2	15.1	--	--
Co(OH)$_2$	781.0	15.9	--	--

† Co(2p$_{1/2}$)-Co(2p$_{3/2}$)
‡ aerobic.
§ anaerobic.
¶ anaerobic, repeated washing.

for the unweathered and weathered samples were attributed to the preferential weathering occurring at the fracture zones or dislocations that intersect the grain surfaces. Since the lateral resolution of XPS in this study was 2000 μm, and the sodic and calcic lamella were 5- and 15 μm in diameter, respectively, it was not possible to assess the composition of the individual lamallae. The difference in the depth composition between the unaltered and altered labradorite was attributed to the preferential dissolution of the calcic phase, and H$^+$ attack at the low activation-energy sites, i.e., fracture zones, twin boundaries and dislocations. Hence, Inskeep et al. concluded that XPS was not sensitive enough to accurately analyze the surface composition of weathered minerals having exsolution lamellae or intimately intergrown phases because of the relatively large beam diameter.

Acid Sites

Many industrial and biochemical reactions occur at the Lewis and Brønsted acid sites at mineral surfaces. Clay mineral surfaces contain these acidic sites that may catalyze the transformations of organic compounds. Generally, Lewis and Brønsted acid sites are evaluated using infrared spectroscopy, but XPS can also be used to identify and quantify these sites (Borade et al., 1990; Defosse et al., 1978). Borade et al. (1990) probed the acidic sites in zeolites using pyridine. The pyridine molecules were sorbed onto the zeolites, and the samples then evacuated at 10^{-3} Pa for 16 h. Examination of the N(1s) XPS peak at ~400 eV corresponding to the pyridine chemisorbed on the zeolite showed that it could be decomposed into three peaks at 398.7, 399.8 and 401.6 eV.

Table 7-9. Surface:bulk concentration ratios for major elements in unweathered and weathered labradorite (Inskeep et al., 1991).

Treatment	Si	Al	Na	Ca
Untreated	1.00	1.00	1.00	1.00
pH 3.7, 25 °C	1.21	0.86	1.15	0.56
pH 4.1, 40 °C	1.19	0.99	1.25	0.71
pH 4.1, 40 °C†	1.22	0.90	0.97	0.64
pH 4.1, 40 °C‡	1.21	0.91	1.06	0.64

† weathered in 100 μM Na-oxalate.
‡ weathered in 100 μM Na-citrate.

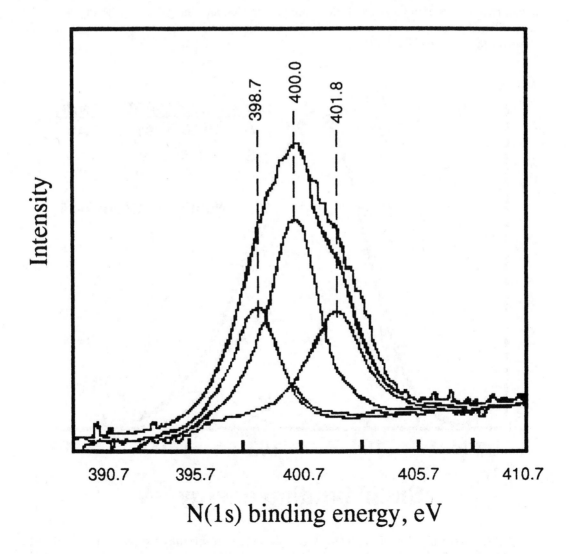

Fig. 7-16. N(1s) XPS spectrum of pyridine chemisorbed on zeolite (after Borade et al., 1990).

These peaks correspond to a Lewis acid site, and weak and strong Brønsted acid sites, respectively (Fig. 7-16). The relative ratio of Brønsted to Lewis acid sites calculated by XPS was in agreement with the ratio obtained by infrared spectroscopy.

Study of Amorphous Materials

One of the advantages of XPS is that both crystalline and amorphous materials can be analyzed (Seyama & Soma, 1987; Vempati et al., 1990a,b). Vempati et al. (1990a,b) studied the surface composition, reactivity and Si-binding state of coprecipitated synthetic Si-ferrihydrites with Si/Fe molar ratios ranging from 0 to 1. The O(1s) peak could be decomposed into oxide-type O at 530.1 eV and hydroxyl-O at 531.8 eV in the Si-ferrihydrite samples. Russell (1979), using infrared spectrometry, suggested that ferrihydrite contains surface-OH as well as inner- and/or structural-OH. The XPS spectra indicated a decrease in OH content (Table 7-2) upon phosphate treatment, thus confirming the presence of surface OH in the original spectrum. The Fe(2p) XPS peaks were observed at 711 and 724 eV, the typical Fe-oxide peaks. The observation of an additional weak XPS peak at 719 eV, ~8 eV above the main Fe(2p) peak (711 eV), suggests that the Fe is in a paramagnetic state (See Fig. 7-6). For the Si-ferrihydrite samples with Si/Fe molar ratios ≤ 0.05, a broad Si peak was observed at 100.7 eV, which was likely due either to unlinked-SiO_4 tetrahedra bound to the ferrihydrite surface or to the presence of small units of polymerized silica linked at one or two corners of the SiO_4 tetrahedra. For the samples containing Si/Fe molar ratios ≥ 0.10, an additional band was observed at 102.8 eV, which was attributed to the presence of Si in the ferrihydrite structure (Fig. 7-17).

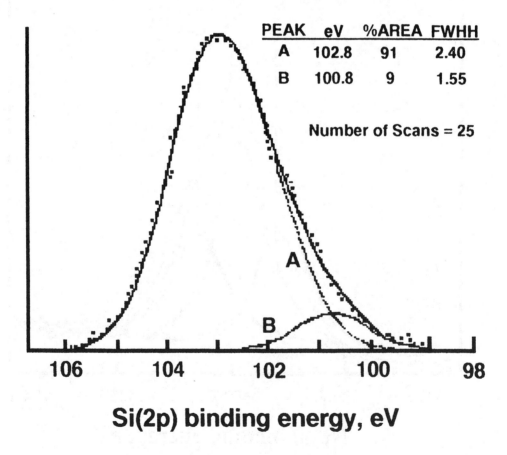

PEAK	eV	%AREA	FWHH
A	102.8	91	2.40
B	100.8	9	1.55

Number of Scans = 25

Si(2p) binding energy, eV

Fig. 7-17. Decomposition of the Si(2p) x-ray photoelectron spectroscopy peak of the 1.0 Si/Fe molar ratio ferrihydrite sample. The dots surrounding the composite peak represent the original spectrum (adapted from Vempati et al., 1990b).

Environmental Samples

The XPS has also been used as a tool in the area of environmental/aerosol chemistry (Cofer et al., 1987; Craig et al., 1974; Dillard et al., 1979; Koppelman & Dillard, 1975; Mossotti et al., 1987; Wightman, 1982). Soma et al. (1985) studied the surface composition of three reference sediment materials, a pond sediment (Japan NIES CRM no. 2), a river sediment (US NBS SRM 1645) and an estuarine sediment (US NBS SRM 1646). The surface concentration of Cr was three times higher than that of Si in the river sediment; also, the fractionation of the river sediment (Table 7-10) revealed that most of the Cr was present in the fine-size fraction. Even the coarse fraction, which was separated by ultrasonification, contained a higher Cr/Si ratio in the surface than in the bulk sample, indicating that the surface was coated with a Cr-rich compound. The oxidation state of the Cr-rich compound was Cr^{3+} as revealed by the $Cr(2p_{3/2})$ XPS peak at 577.2 eV. The $S(2s)$ XPS peaks at 168.5 and 163.5 eV indicated that the oxidation states of S were +6 and -2, due to the presence of sulfate and metal sulfides, respectively. The occurrence of the $Fe(2p_{3/2})$ peak at 711.2 eV and the $N(1s)$ peak at 399.8 eV indicated the presence of Fe^{3+} and an amino-group, respectively.

Table 7-10. Surface Cr/Si atom ratios for the fractionated river sediment, NBS 1645. (Soma et al., 1985).

Particle-size fraction	Cr/Si atomic ratio
44-105μm	3.60
105-180μm	2.10
105-180μm†	0.86
Bulk‡	0.067

† ultrasonically washed.
‡ bulk sediment.

FUTURE DEVELOPMENTS

Tremendous improvements have been made in XPS instrumentation in recent years. Currently, it is possible to map a region at one photoelectron energy at a time with a spatial resolution of 10 μm. The energy resolution of the instrument has also improved considerably. For example, with a monochromatic source, the FWHM of the Ag(3d) line is about 0.4 eV. We would expect both the spatial and energy resolution of XPS instrumentation to continue to improve. The development of improved electron analyzers and detectors is expected to result in a decrease in XPS analysis time and thus allow analysis of samples that are sensitive to electron-beam damage (Hochella, 1988). Last, it may be possible to extend the XPS technique to the study of vapor/solid and liquid/solid interfaces. The development of liquid-phase XPS, now in its infancy (Barr, 1991; Siegbahn, 1990), would greatly enhance our ability to characterize soil minerals and the processes that occur at their surfaces.

SUMMARY

Surfaces of minerals play a pivotal role in governing the physical and chemical properties of the soil environment. In recent years, several surface-sensitive techniques have been developed, but XPS is becoming more popular among geochemists, mineralogists and soil scientists since it provides both quantitative and qualitative information. For soil scientists, XPS is a valuable tool to study soil chemical properties and processes, e.g. chemical weathering, oxidation-reduction reactions, adsorbed species, chemical bonding states, amorphous materials, etc. The possibility exists for continued advancements through the application of XPS and Auger spectroscopy.

ACKNOWLEDGMENTS

The contribution of DLC was partially supported by the Gulf Hazardous Research Center, the Texas Advanced Technological Research Program of the Texas Higher Education Coordinating Board and the Robert A. Welch Foundation. The portion contributed by RKV was written when he was a National Research Council-National Academy of Sciences postdoctoral research associate at the National Aeronautics and Space Administration-Johnson Space Center, Houston, Texas. The contribution of RHL was partly supported by a grant from the Texas Advanced Technological Research Program of the Texas Higher Education Coordinating Board. The helpful comments of J. E. Amonette and two anonymous reviewers are gratefully acknowledged.

REFERENCES

Allen, G. C., P. M. Tucker, and R. K. Wild. 1982. Characterization of iron/oxygen surface reactions by x-ray photoelectron spectroscopy. Philos. Mag. B46:411-421.

Alvarez, R. C., C. S. Fadley, J. A. Silva, and G. Uehara. 1976. A study of silicate adsorption on gibbsite (Al(OH)₃ by x-ray photoelectron spectroscopy (XPS). Soil Sci. Soc. Am. J. 40: 615-617.

Amonette, J.E., and R.W. Sanders. 1994. Nondestructive techniques for bulk elemental analysis. p. 1-48. In J. Amonette and L.W. Zelazny (ed.) Quantitative methods in soil mineralogy. SSSA Misc. Publ. SSSA, Madison, WI.

Aoki, A. 1976. Photoelectron spectroscopic studies on ZnS:MnF₂. Jpn. J. Appl. Phys. 15:305-311.

Baer, D. R, and M. T. Thomas. 1982. Use of surface analytical techniques to examine metal corrosion problems. Am. Chem. Soc. Symp. Ser. 199:251-282.

Bagus, P. S., A. J. Freeman, and F. Sasaki. 1973. Production of new multiplet structure in photoemission experiments. Phys. Rev. Lett. 30:850-852.

Bancroft, G. M., J. R. Brown, and W. S. Fyfe. 1977. Calibration studies for quantitative x-ray photoelectron spectroscopy of ions. Anal. Chem. 49:1044-1048.

Barr, T. L. 1991. Advances in the application of x-ray photoelectron spectroscopy (ESCA)--2. New methods. Crit. Rev. Anal. Chem. 22:229-325.

Berner, R. A., and G. R. Holdren, Jr. 1977. Mechanism of feldspar weathering: Some observational evidence. Geology 5:369-372.

Borade, R. B., A. Adnot, and S. Kaliaguine. 1990. Characterization of acid sites in pentasil zeolites by X-ray photoelectron spectroscopy. J. Catal. 126:26-30.

Briggs, D. 1977. Handbook of x-ray and ultraviolet spectroscopy. Heydon and Son, London.

Briggs, D., and M. P. Seah. 1983. Practical surface analysis. Wiley, New York.

Briggs, D., and J. C. Riviera. 1983. Spectral interpretation. p. 87-139. In D. Briggs and M. P. Seah, (ed.) Practical surface analyses by Auger and x-ray photoelectron spectroscopy. Wiley, New York.

Brundle, C. R., and A. D. Baker. 1977. Electron spectroscopy: Theory, techniques, and applications. Acad. Press, New York.

Carlson, T. A. 1975. Photoelectron and Auger spectroscopy. Plenum Press, New York.

Carver, J. C., G. K. Schweitzer, and T. A. Carlson. 1972. X-ray photoelectron spectroscopy to study bonding in Cr, Mn, Fe and Cu compounds. J. Chem. Phys. 57:973-982.

Clark, D. T. 1979. Structure, bonding and reactivity of polymer surfaces studied by means of ESCA. p. 1-51. In R. Vanselow (ed.) Chemistry and physics of solid surfaces, Vol. 2. CRC Press, Boca Raton, FL.

Cocke, D. L. 1986. Heterogenous catalyst of amorphous materials. J. Metals 1986(Feb.):70-74.

Cocke, D. L., M. S. Owens, and R. B. Wright. 1988a. Surface oxidation and reduction of zirconium and nickel compounds examined by XPS. Appl. Surf. Sci. 31:341-346.

Cocke, D. L., M. S. Owens and R. B. Wright. 1988b. The preparation by intermetallic compound oxidation of supported Ni and TiO₂ overlayers. Langmuir 4:1311-1318.

Cocke, D. L., T. R. Hess, T. Mehbratu, D. E. Mencer, Jr., and D. G. Naugle. 1990. The surface reactivity of Ti-Cu and Ti-Al alloys and the ion chemistry of their oxide overlayers. Solid State Ionics 43:119-131.

Cofer, W. R., G. G. Lala, and J. Wightman. 1987. Analysis of mid-tropospheric space shuttle exhausted aluminum oxide particles. Atmos. Environ. 21:1187-1196.

Count,, M. E., J. S. C. Jen, and J. P. Wightman. 1973. An electron spectroscopy for chemical analysis study of lead adsorbed on montmorillonite. J. Phys. Chem. 77:1924-1926.

Craig, N. L., A. B. Harker, and T. Novakov. 1974. Determination of the chemical states of sulfur in ambient pollution aerosols by x-ray photoelectron spectroscopy. Atmos. Environ. 8:15-21.

Crovisier, J. L., J. H. Thomassin, T. Juteau, J. P. Eberhart, J. C. Touray, and P. Baillif. 1983. Experimental seawater-basaltic glass interaction at 50 °C: Study of early developed phases by electron microscopy and x-ray photoelectron spectrometry. Geochim. Cosmochim. Acta 47:377-387.

Crowther, D. L., J. G. Dillard, and J. W. Murray. 1983. The mechanism of Co(II) oxidation on synthetic birnessite. Geochim. Cosmo. Acta 47:1399-1403.

Czanderna, A. W. 1975. Methods of surface analysis. Elsevier, New York.

Defosse, C., P. Canesson, P. G. Rouxhet, and B. Delmon. 1978. Surface characterization of silica-aluminas by photoelectron spectroscopy. J. Catal. 51:269-277.

Dillard, J. G., R. D. Seals, and J. P. Wightman. 1979. Electron spectroscopy for chemical analysis (ESCA) study of aluminum-containing atmospheric particles. Atmos. Environ. 14:129-135.

Dillard, J. G., and M. H. Koppelman. 1982. X-ray photoelectron spectroscopic (XPS) surface characterization of cobalt on the surface of kaolinite. J. Colloid Interface Sci. 87:46-55.

Dillard, J. G., D. L. Crowther, and S. E. Calvert. 1984. X-ray photoelectron spectroscopic study of ferromanganese nodules: Chemical speciation for selected transition metals. Geochim. Cosmochim. Acta 48:1565-1569.

Dillard, J. G., and C. V. Schenck. 1986. Interaction of Co(II) and Co(III) complexes on synthetic birnessite: Surface characterization. Am. Chem. Soc. Symp. Ser. 323:503-522.

Dillard, J. G., M. H. Koppelman, D. L. Crowther, C. V. Shenck, J. W. Murray, and L. Balistrieri. 1981. X-ray photoelectron spectroscopy (XPS) studies on the chemical nature of metal ions adsorbed on clays and minerals. p.227-240. In P. H. Tewari (ed.) Adsorption from aqueous solutions. Plenum Press, New York.

Eggleston, C. M., M. F. Hochella, Jr., and G. A. Parks. 1989. Sample preparation and aging effects on the dissolution rate and surface composition of diopside. Geochim. Cosmochim. Acta 50:2481-2497.

Eland, J. H. D. 1984. Photoelectron spectroscopy. Butterworths, Boston.

Evans, S., and E. Raftery. 1982. Determination of the oxidation state of manganese in lepidolite by x-ray photoelectron spectroscopy. Clay Miner. 17:477-481.

Foord, J. S., R. B. Jackman, and G. C. Allen. 1984. An x-ray photoelectron spectroscopic investigation of the oxidation state of manganese. Philos. Mag. A49:657-663.

Frost, D. C., C. A. McDowell, and I. S. Woolsey. 1974. X-ray photoelectron spectra of Co compounds. Mol. Phys. 27:1473-1489.

Gonzàlez, A. R., J. P. Espinós, G. Munuera, J. Sanz, and J. M. Serratosa. 1988. Bonding-state characterization of constituent elements in phyllosillicate minerals by XPS and NMR. J. Phys. Chem. 92:3471-3476.

Goodman, B. A. 1986. Adsorption of metal ions and complexes on aluminosilicate minerals. Am. Chem. Soc. Symp. Ser. 323:342-361.

Hellmann R., C. M. Eggleston, M. F. Hochella, Jr., and D. A. Crerar. 1990. The formation of leached layers on albite surfaces during dissolution under hydrothermal conditions. Geochim. Cosmochim. Acta 54:1267-1281.

Hingston, F. J., R. J. Atkinson, A. M. Posner, and J. P. Quirk. 1968. Specific adsorption of anions on goethite. Trans. Int. Congr. Soil Sci., 9th. 1:669-678.

Ho, P. S., and J. E. Lewis. 1976. Deconvolution method for composition profiling by Auger sputtering techniques. Surf. Sci. 65:335-348.

Hochella, M. F. Jr. 1988. Auger electron and x-ray photoelectron spectroscopies. p. 573-637. In F. C. Hawthorne (ed.) Spectroscopic methods in mineralogy and geology, Reviews in mineralogy, Vol. 18, Miner. Soc. Am., Washington, DC.

Hochella, M. F. Jr., and G. E. Brown. 1988. Aspects of silicate surface structure analysis using X-ray photoelectron spectroscopy (XPS). Geochim. Cosmochim. Acta 52:1641-1648.

Hochella, M. F. Jr., H. B. Ponader, A. M. Turner, and D. W. Harris. 1988. The complexity of mineral dissolution as viewed by high resolution scanning microscopy: Labradorite under hydrothermal conditions. Geochim. Cosmochim. Acta 52:385-394.

Hofman, S. 1986. Practical surface analysis: State of the art and recent developments in AES, XPS, ISS and SIMS. Surf. Interface Anal. 9:3-20.

Holdren, G. R., Jr., and R. A. Berner. 1979. Mechanism of feldspar weathering - I. Experimental studies. Geochim. Cosmochim. Acta 43:1161-1171.

Hu, H. K., and J. W. Rabalais. 1981. Chemisorption and the initial stage of oxidation of Mn. Surf. Sci. 107:376-390.

Hultgren, R., P. A. Desai, D. T. Hawkins, M. Gleiser, and K. K. Kelly. 1973. Selected values of the thermodynamic properties of metals. Am. Soc. Metals, Metals Park, OH.

Hyland, M. H., and G. M. Bancroft. 1989. Palladium sorption and reduction of sulfide mineral surfaces: An XPS and AES study. Geochim. Cosmochim. Acta 54:117-130.

Inskeep, W. P., E. A. Nater, D. S. Vandervoort, P. R. Bloom, and M. S. Erich. 1991. Characterization of laboratory weathered Labradorite surfaces using x-ray photoelectron spectroscopy and transmission electron microscopy. Geochim. Cosmochim. Acta 55:787-801.

Kane, P. F., and G. B. Larrabee. 1974. Characterization of solid surfaces. Plenum, New York.

Koppelman, M. H., and J. G. Dillard. 1975. An ESCA study of sorbed metal ions on clay minerals. Am. Chem. Soc. Symp. Ser. 18:186-201.

Koppelman, M. H., and J. G. Dillard. 1977. A study of the adsorption of Ni(II) and Cu(II) by clay minerals. Clays Clay Miner. 25:457-462.

Koppelman, M. H., and J. G. Dillard. 1978. An x-ray photoelectron spectroscopic (XPS) study of cobalt adsorbed on the clay mineral chlorite. J. Colloid Interface Sci. 66:345-351.

Koppelman, M. K., and J. G. Dillard. 1980. Adsorption of $Cr(NH_3)_6^{3+}$ and $Cr(en)_3^{3+}$ on clay minerals and the characterization of chromium by x-ray photoelectron spectroscopy. Clays Clay Miner. 28:211-216.

Koppelman, M. H., A. B. Emerson, and J. G. Dillard. 1980. Adsorbed Cr(III) on chlorite, illite and kaolinite: An x-ray photoelectron spectroscopic study. Clays Clay Miner. 28:119-124.

Martin, R. R., and R. St. C. Smart. 1987. X-ray photoelectron studies of anions adsorption on goethite. Soil Sci. Soc. Am. J. 51:54-56.

Mehbratu, T., T. R. Hess, D. E. Mencer, Jr., K. G. Balke, D. G. Naugle, and D. L. Cocke. 1991. Photoelectron spectroscopic investigations of the surface reactivity of crystalline and amorphous Ti-Cu alloys. Mater. Sci. Eng. A134:1041-1042.

Mossotti, V. G., J. R. Lindsay, and M. F. Hochella, Jr. 1987. Alteration of limestone surfaces in an acid rain environment by gas-solid reaction mechanisms. Mater. Perf. 26:47-52.

Muilenburg, G. E. 1979. Handbook of X-ray photoelectron spectroscopy. Perkin-Elmer Corp., USA.

Muir, I. J., G. M. Bancroft, and H. W. Nesbitt. 1989. Characteristics of altered labradorite surfaces by SIMS and XPS. Geochim. Cosmochim. Acta 53:1235-1241.

Murray, J. W., and J. G. Dillard. 1979. The oxidation of cobalt(II) adsorbed on manganese dioxide. Geochim. Cosmochim. Acta 43:781-787.

Murray, J. W., J. G. Dillard, R. Giovanoli, H. Moers, and W. Stumm. 1985. Oxidation of Mn(II): Initial mineralogy oxidation state and ageing. Geochim. Cosmochim. Acta 49:463-470.

Myhra, S., T. J. White, S. E. Kesson, and J. C. Riviere. 1988. X-ray photoelectron spectroscopy for the direct identification of titanium in hollandites. Am. Miner. 73:161-167.

Oku, M., K. Hirokawa, and S. Ikeda. 1975. X-ray photoelectron spectroscopy of manganese-oxygen systems. J. Electron Spectrosc. Relat. Phenom. 7:465-473.

Perry, D. L. 1986. Applications of surface techniques to chemical bonding studies of minerals. Am. Chem. Soc. Symp. Ser. 323:389-402.

Perry, D. L., J. A. Taylor, and C. D. Wagner. 1990. X-ray-induced photoelectron and Auger spectroscopy. p. 45-86. In D. L. Perry (ed.) Instrumental surface analysis of geologic materials. VCH, New York.

Petrovic, R. R., R. A. Berner, and M. B. Goldhaber. 1976. Rate control in dissolution of alkali feldspards--I. Study of residual feldspar grains by x-ray photoelectron spectroscopy. Geochim. Cosmochim. Acta 40:537-548.

Rao, C. N., D. D. Sarma, S. Vasudevan, and M. S. Hegde. 1979. Study of transition metal oxides by photoelectron spectroscopy. Proc. R. Soc. London 367:239-252.

Ratner, B. D. 1983. Study of biomaterials by electron spectroscopy for chemical analysis. Ann. Biomed. Eng. 11:313-336.

Ratner, B. D. (ed.) 1988. Surface characterization of biomaterials. Elsevier, New York.

Russell, J. D. 1979. Infrared spectroscopy of ferrihydrite: Evidence for the presence of structural hydroxyl groups. Clay Miner. 14:109-114.

Schenck, C. V., J. G. Dillard, and J. W. Murray. 1983. Surface analysis and the adsorption of Co(II) on goethite. J. Colloid Interface Sci. 95:398-409.

Schott, J., and R. A. Berner. 1983. X-ray photoelectron studies of the mechanism of iron silicate dissolution during weathering. Geochim. Cosmochim. Acta 47:2233-2240.

Schott, J. R. and J. C. Petite. 1987. New evidence for the mechanisms of dissolution of silicate minerals. p. 293-315. *In* W. Stumm (ed.) Aquatic surface chemistry. Wiley Intersci., New York.

Schott, J., R. A. Berner, and E. L. Sjöberg. 1981. Mechanism of pyroxenes and amphiboles weathering--I. Experimental studies of iron-free minerals. Geochim. Cosmochim. Acta 45:2123-2135.

Seah, M. P. 1983. A review of quantitative Auger electron microscopy. Scan. Electron Microsc. 1983(2):521-536.

Seyama, H., and M. Soma. 1985. Bonding-state characterization of the constituent elements of silicate minerals by X-ray photoelectron spectroscopy. J. Chem. Soc., Faraday Trans. 1. 81:485-495.

Seyama, H., and M. Soma. 1987. Fe 2p spectra of silicate minerals. J. Electron Spectrosc. Rel. Phenom. 42:97-101.

Seyama, H., and M. Soma. 1988. Applications of x-ray photoelectron spectroscopy to the study of silicate minerals. Res. Rep. no. 11, Natl. Inst. Environ. Stud., Japan.

Shirley, D. A. 1972. Electron spectroscopy. North-Holland, Amsterdam.

Shirley, D. A. 1975. Hyperfine interaction and ESCA data. Phys. Scr. 11:117-120.

Siegbahn, K. 1990. From x-ray to electron spectroscopy and new trends. J. Electron Spectr. Rel. Phenom. 51:11-36.

Soma, M., H. Seyama, and K. Oksmoto. 1985. Characterization of sediment reference materials by x-ray photoelectron spectroscopy. Talanta 32:177-181.

Stephenson, D. A., and N. J. Binkowski. 1976. X-ray photoelectron spectroscopy of silica in theory and experiment. J. Non-Cryst. Solids 22:339-421.

Stipp, S. L. and M. F. Hochella. 1991. Structure and bonding environments at the calcite surface as observed with X-ray photoelectron spectrosocpy (XPS) and low energy electron diffraction (LEED). Geochem. Cosmochim. Acta 55:1723-1736.

Stucki, J. W., C. B. Roth, and W. E. Baithinger. 1976. Analysis of iron-bearing clay minerals by electron spectroscopy for chemical analysis (ESCA). Clays Clay Miner. 24:289-292.

Swift, P. 1982. Adventitious carbon-the panacea for energy referencing? Surf. Interface Anal. 4:47-51.

Swift, P., D. Shuttleworth, and M. P. Seah. 1983. Static charge referencing techniques. p. 437-444. *In* D. Briggs and M. P. Seah (ed.) Pratical surface analysis by Auger and x-ray photoelectron spectroscopy. Wiley, New York.

Thomassin, J. H., J. Goni, J. C. Touray, and P. Baillif. 1977. An XPS study of the dissolution kinetics of chrysotile in 0.1N oxalic acid at different temperature. Phys. Chem. Min. 1:385-398.

Vempati, R. K., R. H. Loeppert, and D. L. Cocke. 1990a. Mineralogy and reactivity of amorphous Si-ferrihydrite. Solid State Ionics 38:53-61.

Vempati, R. K., R. H. Loeppert, D. C. Dufner, and D. L. Cocke. 1990b. X-ray photoelectron spectroscopy to differentiate silicon-bonding state in amorphous iron oxides. Soil Sci. Soc. Am. J. 54:695-698.

Wagner, C. D. and A. J. Joshi. 1980. The Auger parameter, its utility and advantages: A review. J. Electron Spectrosc. Relat. Phenom. 47:283-313.

Wagner, C. D., D. E. Passoja, H. F. Hillery, T. G. Kinisky, H. A. Six, W. T. Jansen, and J. A. Taylor. 1982. Auger and photoelectron line energy relationships in aluminum-oxygen and silicon-oxygen compounds. J. Vac. Sci. Technol. 21:933-944.

Wallbank, B., I. G. Main, and C. E. Johnson. 1974. 2p and 2s shake up satellites in solid compounds of 3d ions. J. Electron Spectrosc. Relat. Phenom. 5:259-266.

Wertheim, G. K., S. Hufner, and H. J. Guggenheim. 1973. Systematics of core electron exchange splitting in 3d group transition metal compounds. Phys. Rev. B7:556-558.

White, A. F., L. V. Benson, and A. Yee. 1986. Chemical weathering of the May 18, 1980, Mount St. Helens ash fall and the effect on the Iron Creek Watershed, Washington. p.351-375. *In* S. M. Colman and D. P. Dethier (ed.) Rates of chemical weathering of rocks and minerals. Acad. Press, Orlando, FL.

Wightman, J. P. 1982. XPS analysis of Mount St. Helens ash. Colloids Surf. 4:401-406.

Windawi, H., and F. F. L. Ho. 1982. Applied electron spectroscopy for chemical analysis. John Wiley, New York.

William, F. L., and D. Nason. 1974. Binary alloy surface compositons from bulk alloy thermodynamic data. Surf. Sci. 45:377-408.

Wright, R. B., J. G. Jolly, M. S. Owens, and D. L. Cocke. 1987. Catalytic hydrogenation activity of a Zr-Ni intermatallic alloy. J. Vac. Sci. Technol. A5:586-589.

Wynblatt, P., and R. C. Ku. 1977. Surface segration in alloys. p. 115-136. *In* W. C. Johnson and J. M. Blakely (ed.) Interfacial segregation. Am. Soc. Metals, Metals Park, OH.

Yoon, C., and D. L. Cocke. 1987. Potential of amorphous materials as catalyst. J. Non-Crystall. Solids 79:217-245.

Zing, D. S., and D. M. Hercules. 1978. Electron spectroscopy for chemical analysis studies of lead sulfide oxidation. J. Phys. Chem. 82:1992-1995.

Zhao, L. Z., and V. Young. 1984. XPS studies of carbon-supported films formed by the resistive deposition of manganese. J. Electron Spectrosc. Relat. Phenom. 34:45-54.

8   Preconcentration Techniques in Soil Mineralogical Analyses

D. A. Laird
USDA-ARS-MWA
Ames, Iowa

R. H. Dowdy
University of Minnesota
St. Paul, Minnesota

The first step in most soil mineralogical investigations is separation of various particle-size fractions from samples of soil materials. Size fractionation not only separates soil materials by size, but also tends to segregate different mineral constituents into different fractions. Thus, size fractionation is a technique for preconcentrating soil minerals. Other preconcentration techniques separate soil materials by density, magnetic susceptibility, electrophoretic mobility, dielectric constant, particle morphology, or surface properties.

Various mineral-separation techniques are commonly used as research tools in the geological sciences and for ore processing in the mining industry (Muller, 1977). Most of these techniques are limited to separations of sand-size materials. Indeed, Muller (1977) recommended that the <53-μm material ". . . be discarded at the outset. . . ." With soil mineralogical investigations, the fine materials, and especially the clay fraction, cannot be ignored because clay minerals dominate the physical and chemical properties of most soils. The physical properties of soil particles are generally related to their mineralogy. However, due to weathering effects and small variations in elemental composition, different particles of the same mineral species may have substantially different properties. Ranges in physical properties for different mineral species often overlap making separation of monomineralic specimens difficult if not impossible. With clay-size materials, particle-particle interactions further limit the extent to which mineral species can be separated.

The intent of this chapter is to provide theoretical bases, critical evaluations, and specific examples of various preconcentration techniques used in soil mineralogical investigations. Step-by-step instructions are not provided, but where possible, references giving such instructions have been cited. Special emphasis has been placed on techniques for separating clay mineral species commonly found in soils.

PARTICLE-SIZE TECHNIQUES

As a general rule, soils contain a continuum of particle sizes from boulders to submicron clays. The particle-size distribution of any given soil, however, reflects the processes and environment under which that soil formed. Soils formed in loess may contain only silt and clay-size particles, whereas soils derived from talus may be dominated by boulders. Associated with particle-size variations are systematic variations in mineralogy.

Most primary minerals are unstable in soil environments, and therefore are transformed by weathering processes into secondary minerals. Weathering rates for primary minerals depend on both the relative stability and the amount of exposed surface area of the mineral particles (Marshall, 1977). Surface area increases with decreasing particle size, and hence, chemical-weathering rates tend to increase with decreasing particle size. Relatively unstable primary minerals, such as olivine and Ca-plagioclase, are almost invariably depleted from the clay fraction, even in young soils, whereas relatively stable primary minerals, such as quartz and K-feldspars, are commonly found in the coarse-clay fraction of young soils. Due to particle-size effects, even relatively stable primary minerals are usually depleted from fine-clay fractions of young soils, and all primary minerals are depleted from both coarse- and fine-clay fractions of highly weathered soils.

Recognition of the relationships between mineralogy and particle size led Marshall (1931) and Truog et al. (1936) to suggest fractionation of soils at 2 μm for separation of clays (<2 μm), and further fractionation of clays at 0.2 μm for segregation of coarse (0.2-2 μm) from fine (<0.2 μm) clays. Jackson (1985) recommended fractionation of soils into coarse sand (>100 μm), very fine sand (50-100 μm), coarse silt (20-50 μm), medium silt (5-20 μm), fine silt (2-5 μm), coarse clay (0.2-2 μm), medium clay (0.08-0.2 μm), and fine clay (<0.08 μm). The fractionation scheme one chooses, however, should be tailored to the properties of the soils being investigated and the nature of the question being addressed. Procedures for particle-size fractionation of soils have been described by Jackson (1985), and Whittig and Allardice (1986).

Pretreatments and Dispersion

Dispersion is key to achieving particle-size separations. A wide variety of chemical and physical treatments has been employed to enhance dispersion. Chemical treatments are used to remove flocculating and aggregate-cementing agents, such as carbonates, organic matter, oxides and oxyhydroxides of Fe and Al, amorphous silica, soluble salts, and multivalent inorganic cations, and to neutralize positively charged sites on mineral surfaces. Physical treatments are used to apply shear forces to aggregates in order to separate individual mineral grains and tactoids.

Soluble salts are easily removed from soil materials by washing with distilled water (Kunze & Dixon, 1986). Alkaline-earth carbonates are removed by slow dissolution in a sodium acetate--acetic acid (pH = 5) buffer solution (Grossman & Millet, 1961; Rabenhorst & Wilding, 1984; Jackson, 1985; Kunze & Dixon, 1986). Strong mineral acids, such as HCl, should be avoided as they may seriously degrade soil constituents, especially clays.

Oxidation with H_2O_2 is the most commonly used method for removal of organic matter from soil materials (Jackson, 1985; Kunze & Dixon, 1986). Soil samples must be acidified prior to treatment with H_2O_2 because H_2O_2 decomposes in alkaline environments. Jackson (1985) recommends that oxidation with H_2O_2 be conducted in a sodium acetate buffer (pH = 5) solution. Although effective for removing organic matter from soil materials, the use of H_2O_2 has three disadvantages. First, acidity released during the oxidation of organic matter by H_2O_2 may damage clay minerals (Douglas & Fiessinger, 1971; van Langeveld et al., 1978). Second, under certain circumstances, the use of H_2O_2 may promote formation of calcium oxalate (Martin, 1954). And third, H_2O_2 is a strong oxidizing agent and must be handled with extreme care.

Organic matter can be removed from soil materials by oxidation with sodium hypochlorite (Hughes et al., 1994; see Chapter 11; Moore & Reynolds, 1989, p. 185-186) or sodium hypobromite (Troell, 1976; Bourget & Tanner, 1953), and these oxidants are effective under both alkaline and acidic conditions. Lavkulich and Wiens (1970) compared the H_2O_2 and sodium hypochlorite techniques and concluded that sodium hypochlorite removed more organic matter from soil materials with less destruction of oxide minerals than H_2O_2. Moore and Reynolds (1989) recommended the sodium hypochlorite technique as ". . . cheaper, quicker, and safer . . ." than the H_2O_2 technique. Van Langeveld et al. (1978) recommended oxidation with bromine in a sodium bicarbonate solution (40 g L^{-1}; Mitchel & Smith, 1974) for removal of organic matter from clays.

In the past, a number of techniques have been used for completely removing all iron oxides and oxyhydroxides from soil materials (Aguilera & Jackson, 1953). Currently, the only widely accepted technique is the dithionite-citrate-bicarbonate (DCB) method (Mehra & Jackson, 1960; Jackson, 1985; Kunze & Dixon, 1986). Removal of Fe oxides and oxyhydroxides is often necessary to achieve dispersion of soil clays, especially with Oxisols and Ultisols. Some soil minerals, however, may be altered by DCB treatments (Douglas, 1967; Ghabru et al., 1990a,b). Silicon and Al are commonly present in DCB extracts, and are thought to be derived from partial dissolution of amorphous and poorly crystalline aluminosilicates (Dahlgren, 1994, see Chapter 14; Kunze & Dixon, 1986). Furthermore, poorly organized, and hence more labile, hydroxy-interlayer materials are often removed from chloritic intergrades by DCB treatments (Harward et al., 1962; Brewster, 1980; Ghabru et al., 1990a).

Theoretical considerations pertaining to dispersion and flocculation of clays have been reviewed by Van Olphen (1987). In practice, after aggregate-cementing agents have been removed, dispersion is best enhanced by: (i) saturating the exchange complex of clays with small, strongly hydrated, monovalent cations; (ii) reducing the electrolyte concentration in solution to as low a level as possible; (iii) neutralizing any positively charged sites on clay surfaces; and (iv) applying shear forces to physically separate aggregates into individual mineral grains and tactoids.

Saturating the exchange complex of clays with Li or Na and reducing the electrolyte concentration in solution leads to expansion of the diffuse double layers between clay particles, and hence facilitates dispersion. Positively charged sites in phyllosilicates arise largely from exposed octahedral cations on lateral edges of clay platelets. Such sites are attracted to negatively charged sites on other particles, causing flocculation. Positively charged sites can be neutralized by raising the solution pH above the zero-point-of-charge for those sites. Positively charged sites can also be neutralized by chemisorption of organic anions (Van Olphen, 1987). Pyrophosphates and metaphosphates are commonly used as dispersants for particle-size analyses. Use of such dispersants, however, is not recommended for mineralogical analyses as these agents may alter some clays (Omueti & Lavkulich, 1988). Soil clays can often be dispersed in distilled water, and when difficulties do arise, dilute Na_2CO_3 solutions (pH = 9.5) are very effective at facilitating dispersion (Whittig & Allardice, 1986; Jackson, 1985), especially when combined with ultrasonic treatments.

Application of shear forces to soil materials is essential to achieving thorough sample dispersions. Shaking samples for long periods of time in a reciprocal shaker has been the traditional method of applying shear forces for particle-size separations. Ultrasonic dispersion using a probe-type sonicator is also an effective method of dispersing soil materials (Edwards & Bremner, 1964, 1967; Genrich & Bremner, 1972). Busacca et al., (1984) advocate use of cuphorn sonicators for dispersing soils to avoid contamination of samples with Ti from the sonicator probe tip. With ultrasonic treatments, dispersion is often achieved without the prior removal of aggregate-cementing agents or the use of chemical dispersants. Ultrasonic treatments, however, may fragment fragile clay-size materials and physically abrade large primary-mineral particles into clay-size particles.

Whereas the selection of pretreatments is important for qualitative mineralogical analyses (Harward & Theisen, 1962), it becomes critical for quantitative mineralogical analyses. Pretreatments can cause partial or even complete dissolution of some mineral phases, and others may be transformed from one phase to another. Furthermore, selective dispersion (Harris et al., 1987; Ali et al., 1987; Velasco-Molina et al., 1971) may bias samples for one mineral species relative to another. Indeed, problems pertaining to how one obtains a "representative sample" for quantitative soil-clay mineralogy may prove as difficult as the development of accurate techniques for quantification.

Stokes' Law

The optimum technique for particle-size separations varies with the size of particles being separated. Separations of particles 50 μm or larger are generally conducted by dry or wet sieving. The use of water during sieving facilitates the transport of fines through the sieve, enhances slaking of aggregates, and reduces dust. Dry sieving, however, is easier and faster for fractionation of coarse materials and may be required for samples containing soluble constituents. Sedimentation under the influence of gravity is used for separation of particles between 50 and 2 μm, and sedimentation under the influence of a centrifugal force is used for separations of particles smaller than 2 μm.

Stokes' law provides the theoretical basis for separation techniques based on differences in settling velocities of particles suspended in liquids. In general, the terminal settling velocity (V_t) of a spherical particle with a given diameter (d) in a liquid medium, is determined by a balance between the buoyancy (F_b), drag (F_d), and gravitational (F_g) forces acting on the particle (Fig.

8-1A). When the particle is initially suspended it is acted upon by F_g, which results from the mutual attraction between the particle mass and that of the earth, and by F_b, which results from the displacement of the liquid medium by the particle. The gravitational force is proportional to the density (ρ_s) of the particle, the volume $(\pi d^3/6)$ of the particle, and the acceleration due to gravity (g):

$$F_g = \rho_s \pi d^3 g/6. \qquad [1]$$

The buoyancy force is proportional to the density of the liquid medium (ρ_l), the volume of displaced liquid $(\pi d^3/6)$, and g:

$$F_b = \rho_l \pi d^3 g/6. \qquad [2]$$

The net force acting on the particle initially, then is

$$F_N = F_g - F_b = (\rho_s - \rho_l)\pi d^3 g/6. \qquad [3]$$

As the particle accelerates in the liquid in response to F_N, a third force acts on the particle in opposition to F_N. This force, termed the drag force, F_d, results from the resistance of the liquid to passage of the particle through it. F_d increases with the particle velocity until it is equivalent to F_N. At this point, the velocity of the particle has reached a maximum (i.e., V_t), the net force on the particle is nil, and

$$F_d = 3\pi d \eta V_t, \qquad [4]$$

where η is the viscosity of the liquid medium. Since $F_d = F_g - F_b$ once the particle has reached V_t, Eq. [3] and [4] can be equated and rearranged to yield

$$V_t = d^2 g (\rho_s - \rho_l)/18\eta. \qquad [5]$$

Integration of Eq. [5] yields the time (t) required for a particle to fall a given distance (h).

$$t = 18\eta h/d^2 g (\rho_s - \rho_l). \qquad [6]$$

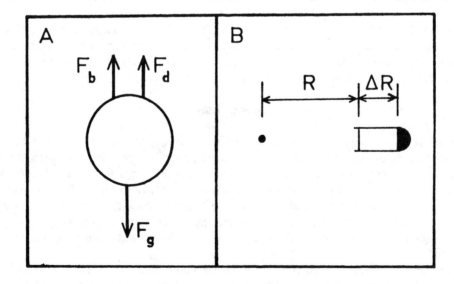

Fig. 8-1. (*A*) Relationship between buoyancy (F_b), drag (F_d), and gravitational (F_g) forces acting on a particle; (*B*) distance from the center of rotation (R) and settling distance (ΔR) of particles in a centrifuge.

Stokes' law is modified for sedimentation of particles within a centrifuge, such that g is replaced by the acceleration ($a = u^2/R$) of the particle within the centrifuge, where u is the linear velocity of the particle and R is the distance of the particle from the center of rotation (Fig. 8-1B). Since

$$u = 2\pi RN, \tag{7}$$

where N is the rotational frequency of the centrifuge in hertz, the instantaneous settling velocity (V_i) of a particle in a centrifuge is

$$V_i = d^2 2\pi^2 RN^2 (\rho_s - \rho_l)/9\eta . \tag{8}$$

The time required for a particle to settle a distance ΔR is given, after integration of Eq. [8], by

$$t = 9\eta [\ln(R+\Delta R)-\ln(R)]/[d^2 2\pi^2 N^2 (\rho_s - \rho_l)]. \tag{9}$$

For all the preceding equations, η is expressed in 10^{-2} pascal seconds (i.e., decipoise); R, ΔR, d, and h in meters; ρ_s and ρ_l in kilograms per cubic meter; V_t, V_i, and microns in meters per second; grams in meters per seconds squared; t in seconds; and N in hertz. Jackson (1985) discusses the application of Stokes' law to angle-head and supercentrifuges.

The size of particles separated by sedimentation is expressed in terms of their equivalent spherical diameter (esd) in recognition that numerous factors affect the settling velocity of particles in addition to particle size. Particle densities are generally assumed to equal that of quartz (2.65 Mg m^{-3}). Iron-bearing minerals, however, may be substantially more dense than quartz while smectites are less dense. The assumption that all particles are spherical is obviously untrue for phyllosilicates. Because of their platy morphology, settling velocities of phyllosilicates are slower than spherical particles of the same particle volume. Application of Stokes' law also assumes that particles settle laminarly in an infinitely dilute suspension. Turbulence, thermal convection, and particle-particle interactions all influence the effective settling velocity of particles. Because of these factors, esds are generally less than diameters estimated by electrically induced birefringence or electron microscopy (Oakley & Jennings, 1982).

Size Fractionation of Soil Clays

Soil-clay mineralogical investigations invariably involve separation and analysis of the clay (<2 μm) fraction. Separation and analyses of size fractions finer than 2 μm are uncommon, although these investigations can yield significant amounts of information. In temperate-region soils, quartz typically dominates the coarse clay (0.2- to 2-μm fraction), but is rarely identified in fractions finer than 0.2 μm. Kaolinite and illite are concentrated in the medium clay (0.08- to 0.2-μm fraction), and smectite is the dominant mineral phase in the fine clay (<0.08 μm) fraction. Plagioclase, K-feldspar, and other primary minerals can often be detected by x-ray diffraction (XRD) analysis of coarse-clay samples, whereas XRD peaks for these minerals are usually difficult to distinguish from background noise in XRD patterns of whole-clay (<2.0 μm) samples.

The nature, and hence size, of a smectite "particle" is ill-defined, because individual smectite platelets are grouped in loose associations known as tactoids and both the extent to which individual tactoids are dispersed and the average number of platelets per tactoid depend on the chemical and physical pretreatments to which a sample is subjected. Theoretically, smectites can be completely delaminated so that only individual platelets or elementary particles (Nadeau et al., 1984) are present in suspension. Under these circumstances the smectite would be entirely within the very fine-clay (<0.02 μm) fraction but, due to incomplete dispersion some smectite is invariably present in all size fractions separated from smectite-containing soils.

Numerous researchers (Wilson, 1987; Badraoui et al., 1987; Laird et al., 1988; Badraoui & Bloom, 1990) have reported evidence for interstratification of illitic layers in soil smectites. The smectitic and illitic phases are generally considered inseparable, although Laird et al. (1991) were able to distinguish the elemental compositions of smectitic and illitic phases in a sample of interstratified smectite-illite from a typical midwestern agricultural soil. To do so, they separated six fine-clay fractions (<0.020, <0.026, <0.036, <0.045, <0.060, and <0.090 μm) from a bulk sample of H_2O_2-DCB-treated soil clay. Elemental analyses of the fine-clay fractions revealed strong linear correlations between the concentrations of nonexchangeable K and other elements in the samples (Fig. 8-2). Based on the linearity of the relationships between the concentrations of K and the other elements present in the samples and on the lack of XRD evidence for other mineral phases (Fig. 8-3), it was assumed that the fine-clay samples were mixtures with varying proportions of homogeneous smectitic (K-free) and illitic (K-bearing) phases. A nonlinear elemental-mass-balance optimization procedure was then used to estimate the elemental compositions and the relative proportions of the two phases in the fine-clay fractions. The nonlinear analysis indicated that the smectitic phase consisted of dioctahedral, Fe-rich montmorillonite particles with a layer charge of 0.482 per formula unit, 47% of which was located in the tetrahedral sheet. The illitic phase was estimated to consist of dioctahedral, elementary-illite particles (Nadeau et al., 1984) with a layer charge of 0.473 per formula unit, 87% of which was located in the tetrahedral sheet.

DENSITY TECHNIQUES

Density-separation techniques have been used with geologic materials since the nineteenth century, and with soil materials throughout the twentieth century (Mitchell, 1975). Muller (1977) reviewed density-separation techniques commonly used in geological sciences. These techniques were developed primarily for separation of sand-size materials and are not generally applicable for separation of clays. Early density separations of soil materials were also restricted to sand fractions. Volk (1933), one of the first researchers to attempt density separation of clay-size soil materials, used density separation in demonstrating that added K was ". . . fixed largely in the form of muscovite in the colloidal fraction. . . ." Despite their long history of use, density-separation techniques are infrequently used with soil materials.

Density of Soil Minerals

Densities of commonly occurring soil minerals are listed in Table 8-1. Although ideal elemental compositions and unit-cell dimensions prescribe unique densities for each mineral species, ranges in density are commonly observed. These ranges reflect structural defects, variations in elemental compositions, inclusions of other mineral phases, and weathering effects. Densities of expanding phyllosilicates are particularly variable, as they are affected by the extent of interlayer hydration and the nature of the saturating cations. Density ranges for many soil minerals overlap (Table 8-1), and this phenomenon sets fundamental limits on the extent to which soil minerals can be separated by density fractionation alone.

Density-Fractionation Techniques

There are two basic types of density-separation techniques, sink-or-float and density- gradient. Sink-or-float separations are performed using heavy liquids with uniform densities. Particles either sink or float in the heavy liquid, depending on their density relative to that of the liquid. Density-gradient separations are performed using two miscible liquids with different densities that are mixed in varying proportions within a gradient tube, such that the heavier liquid is concentrated at the bottom of the tube and the lighter liquid is concentrated at the top. Ideally, the density of the liquid in a gradient tube increases linearly with depth. A particle introduced at the top of a linear density-gradient will settle under gravitational or centrifugal force until it reaches its isopycnic point (the depth in a density gradient at which densities of the particle and liquid are equal).

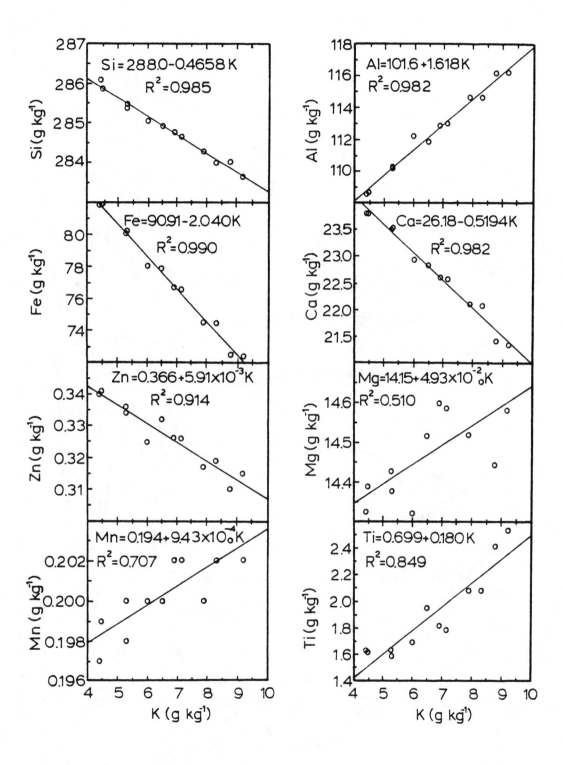

Fig. 8-2. Relationship between K content and Si, Al, Fe, Ca, Zn, Mg, Mn, or Ti content in various fine-clay fractions separated from an Ap-horizon sample of Webster soil (adapted from Laird et al., 1991).

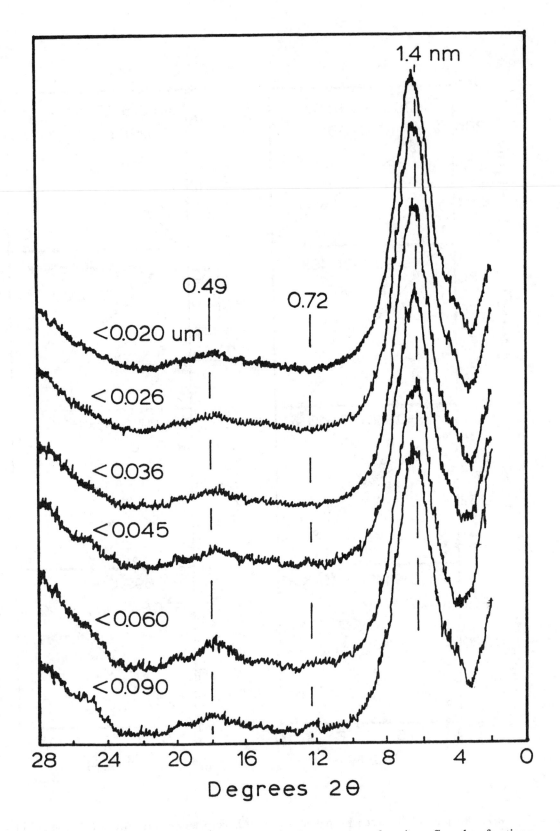

Fig. 8-3. X-ray (CuKα radiation) powder diffraction patterns of various fine-clay fractions separated from an Ap-horizon sample of Webster soil (adapted from Laird et al., 1991).

Table 8-1. Densities of common soil minerals [compiled from Frye (1974), Hurlbut (1971), and Jackson (1985)].

Mineral	Density	Mineral	Density
	Mg m^{-3}		Mg m^{-3}
Oxides and hydroxides		Primary silicates	
Hematite	5.26	Quartz group	
Magnetite	5.18	Quartz	2.65
Goethite	3.3-4.4	Cristobalite	2.32
Ilmenite	4.7	Opal	2.0-2.3
Rutile	4.25	Feldspar group	
Anatase	3.90	Orthoclase	2.57
Gibbsite	2.2-2.5	Microcline	2.56
Phosphates and sulphates		Plagioclase group	
Apatite	3.1-3.4	Albite	2.62
Gypsum	2.32	Anorthite	2.76
Carbonates		Zeolite group	
Calcite	2.71	Analcime	2.27
Magnesite	3.0-3.2	Chabazite	2.10
Dolomite	2.85	Amphibole group	
Siderite	3.96	Tremolite	3.0-3.3
Silicate clays		Hornblende	3.0-3.3
Clay mica group		Pyroxene group	
Illite	2.2-2.8	Enstatite	2.2-3.6
Glauconite	2.4-3.0	Diopside	3.2-3.3
Celadonite	2.6-3.0	Augite	3.2-3.4
Smectite group	1.8-2.4	Tourmaline group	3.0-3.3
Vermiculite group	2.0-2.6	Olivine group	3.3-4.4
Kaolinite group		Garnet group	3.5-4.3
Kaolinite	2.2-2.7	Zircon group	4.6-4.7
Halloysite	2.0-2.6	Beryl group	2.7-2.8
Serpentine group		Mica group	
Antigorite	2.6	Muscovite	2.7-2.9
Chrysotile	2.55	Biotite	2.8-3.2
Talc	2.6-2.9	Phlogopite	2.7-2.9
Pyrophyllite	2.8	Lepidolite	2.8-2.9
Imogolite	1.7-2.8	Margarite	3.0-3.1
Allophane	1.8-2.8	Chlorite group	2.6-3.3

The sink-or-float technique was used for early density separations of soil samples (Volk 1933; Pearson & Truog, 1938; Loughnan, 1957) and is still commonly used for separations of sand-size materials (Mitchell, 1975; Muller, 1977; Mermut et al., 1986; Huang, 1989). Density-gradient techniques were originally developed for use with biological materials (Brakke, 1951; Low & Richards, 1952; Lammers, 1967) and later adapted for use with soil materials (Halma, 1969a,b; Bonner et al., 1970; Francis et al., 1970). The principal advantage of the density-gradient technique over the sink-or-float technique is that numerous density fractions can be separated simultaneously from one sample. Several density fractions, however, can be separated from one sample using the sink-or-float technique. To do so, a sample is refractionated several times and each time the density of the heavy liquid is decreased by dilution with an appropriate solvent (Pearson & Truog, 1937; Loughnan, 1957).

The movement of particles in heavy liquids is governed by the same forces that govern movement of particles in water, and therefore is described by Stokes' law. Application of Stokes' law for sink-or-float separations requires knowledge of the density and viscosity of the heavy liquid, but the equations are the same as for particle-size separations using water (i.e., Eq. [1]-[9]). By contrast, equations describing particle movement in density gradients are complicated by variations in density and viscosity with depth (Essington et al., 1985). In theory, an infinite amount of time is required for complete particle separation in a density gradient, because the terminal settling-velocity of a particle approaches zero at its isopycnic point. In practice, 6 to 18 h of centrifugation in an ultracentrifuge are sufficient for isopycnic banding of clay samples (Oster & Yamamoto, 1963; Francis et al., 1972; Essington et al., 1985).

Three different techniques have been developed for forming linear density-gradients. Beavers (1961) and Halma (1969a) described systems that rely on two pyramid-shaped containers, inverted with respect to each other, and gravity to supply light and heavy liquids in varying proportions to a mixing chamber and then to a gradient tube. Oster and Yamamoto (1963) described both the two-syringe and the two-tank methods of forming linear density-gradients. For the two-syringe method, programmed cams drive syringes filled with heavy and light liquids, so that the liquids are delivered to a mixing chamber in varying proportions. For the two-tank method, a peristaltic pump delivers the heavy liquid from one tank to a second tank that initially contains the light liquid. At the same time, liquid is pumped from the second tank to a gradient tube at twice the rate that it is being pumped from the first tank to the second. If the second tank is kept thoroughly mixed during the pumping process a linear density-gradient will be formed in the gradient tube.

Pearson and Truog (1937) and Jackson (1985) discussed criteria for selecting heavy liquids for density separations of soil materials. Tetrabromoethane (TBE) and bromoform, the most commonly used heavy liquids, are largely inert with respect to soil minerals. Brantley et al. (1974), however, reported that significant amounts of heavy metals are extracted from soils and sediments by TBE. Bromoform is less viscous but more volatile then TBE, although neither viscosity nor volatility generally limits the use of these heavy liquids. Both bromoform and TBE are recoverable (Pearson & Truog, 1937; Jackson, 1985). The most serious disadvantage to the use of bromoform and TBE is that they are extremely poisonous and even relatively mild exposure may cause serious liver damage and/or cancer. Such heavy liquids should be handled only by trained personnel under conditions of adequate ventilation. The densities of bromoform and TBE (2.89 and 2.97 Mg m^{-3}, respectively) are high enough to fractionate most of the silicate minerals found in soils but are too low to fractionate some accessory minerals. Diiodomethane (3.33 Mg m^{-3}) may be used for fractionations involving accessory minerals (Muller, 1977).

Recently, aqueous solutions of $ZnCl_2$, $ZnBr_2$, and $3Na_2WO_4 \cdot 9WO_3 \cdot H_2O$ (sodium polytungstate) have been used as heavy liquids for density fractionations of sand-size mineral grains. Sodium polytungstate (Plewinsky & Kamps, 1984) is an inorganic salt that can be mixed with distilled water to prepare solutions having densities between 1 and 3.1 Mg m^{-3}. Aqueous zinc bromide solutions can be prepared with densities between 1 and 2.6 Mg m^{-3}. These inorganic salts have two distinct advantages over bromoform and TBE for density fractionations of soil materials: (i) dispersion of soil materials is generally easier in aqueous solutions than nonpolar organic solvents; and (ii) they are considerably less toxic than bromoform and TBE. Sodium polytungstate solutions have not been evaluated as mediums for density fractionation of clays.

The density of heavy liquids can be measured with a hydrometer, a specific-gravity balance, or pycnometrically (Jackson, 1985). These methods require a large amount of heavy liquid and substantial effort. The refractive index, on the other hand, is linearly related to density, and can be determined in seconds on a single drop of liquid using a refractometer (Phillips, 1971, p. 62-65). Volk (1933) used the refractive index-density relationship to determine densities of heavy liquids, and this is still the most convenient method. The effects of temperature and surfactant concentration must be included in refractive index-density calibrations.

Several techniques are used for recovery of density fractions after separations are complete. The suction technique is widely used for density-gradient separations. With the suction technique, suspended particles and heavy liquid are removed from the top of a gradient column, layer by layer, using a pipette (Mattigod & Ervin, 1983). Use of separatory funnels and decantation are common sample-recovery techniques for sink-or-float separations. Matelski (1951) suggested freezing the heavy liquid after separations and then thawing just the upper portion to facilitate recovery of the light fraction by decantation.

Entrainment of minerals, air bubbles, and agglomerations are problems often encountered during density separations of soil materials. Entrainment of minerals results from overloading the separation vessel, and can be avoided by separating smaller samples. Air bubbles adhering to particles may float particles that otherwise would sink. Problems of this nature can be avoided by degassing heavy liquids and prewetting samples prior to separations (Batson & Truog, 1939; Jackson, 1985). Agglomeration of particles is a particularly serious problem for separations involving clay-size materials. To minimize agglomeration, samples have been dispersed ultrasonically and treated with surfactants and various inorganic cations (Batson & Truog, 1939; Woolson & Axley, 1969; Halma, 1969b; and Bonner et al., 1970; Francis et al., 1972). The best approach appears to be a combination of ultrasonic dispersion and the use of an appropriate surfactant.

Halma (1969b) and Bonner et al. (1970) investigated numerous compounds for use as surfactants to facilitate dispersion of clay in nonpolar solvents such as bromoform and TBE. Halma (1969b) recommended the use of the cationic surfactants dodecyltrimethylammonium bromide and dodecylpyridinium bromide. Bonner et al. (1970), by contrast, found both cationic and anionic surfactants ineffective and suggested the use of polyvinylpyrrolidone (PVP), an organic polymer that is both nonionic and water-soluble (Francis, 1973). Since 1970 most researchers have used PVP as a dispersant for density separations involving clays, although the controversy over the efficacy of cationic surfactants has yet to be resolved.

The use of PVP as a dispersant for density separations could affect the results of subsequent investigations in which density fractions are used because PVP is retained on clay surfaces (Bonner et al., 1970; Francis et al., 1970). Francis (1973) found that washing with ethanol removed most, but not all, of the PVP from clay samples. Jaynes and Bigham (1986), on the other hand, reported complete removal of PVP from clay surfaces by boiling samples in a $1M$ Na-acetate(pH 5)-30% H_2O_2 solution.

Francis et al. (1976) introduced the use of large-scale zonal rotors in density-gradient separations. The principal advantage in the use of zonal rotors is an increase in the size of samples that can be separated in a single operation (from 0.25-15 g). Since their introduction, large-scale zonal rotors have seen increased use (Bergman et al., 1979; Jaynes & Bigham, 1986; Jaynes et al., 1989).

Density Separation of Soil Minerals

Environmental concerns have been a major impetus behind the use of density separations with soil materials in recent years. Olson and Skogerboe (1975) used density-gradient centrifugation in combination with magnetic-separation techniques to concentrate minerals containing lead in soils contaminated by aerosols from auto exhaust. Francis et al. (1976) fractionated both soils and sediments to determine associations between various heavy metals and minerals. Bergman et al. (1979) established that most of the heavy metals in sewage sludge were associated with low-density organic-rich fractions. Various combinations of particle-size, density-gradient, and magnetic-separation techniques have been used to fractionate fly ash (Mattigod, 1982; Mattigod & Ervin, 1983) and soils contaminated with zinc mining wastes (Mattigod et al., 1986).

Other researchers (Francis et al., 1970; Francis & Tamura, 1972; Jaynes & Bigham, 1986; Jaynes et al., 1989) fractionated soil clays to improve mineralogical characterizations. Francis and Tamura (1972) used density-gradient centrifugation to fractionate E-horizon clay (<2 μm) from a sample of Dodge (Typic Hapludalf) silt loam. The clay separated into nine distinct bands with substantial mineralogical differences (Fig. 8-4). Jaynes and Bigham (1986), using a large-scale zonal rotor and a linear TBE-ethanol density-gradient, fractionated fine-clays (<0.2 μm) from

247

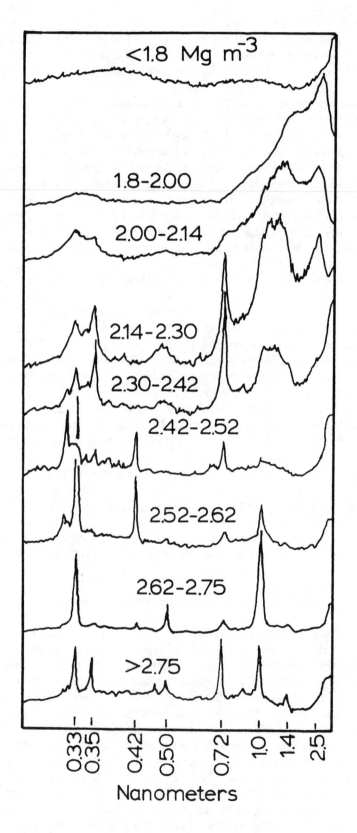

Fig. 8-4. X-ray powder diffraction patterns of various density fractions separated from an E-horizon sample of Dodge soil (adapted from Francis & Tamura, 1972).

three Ohio soils. The XRD analyses (Fig. 8-5) indicated that smectite was the only mineral present in significant quantities in the 1.8 to 2.1 Mg m^{-3} fraction, that the 2.1 to 2.3 Mg m^{-3} fraction was enriched in kaolinite, vermiculite, and mica, and that the >2.4 Mg m^{-3} fractions were enriched with hematite, goethite, and lepidocrocite. More recently, Jaynes et al. (1989) identified an interstratified 1:1-2:1 phyllosilicate in soil material collected near a loess-paleosol contact from a polygenetic soil in southern Ohio. The interstratified mineral was concentrated in the 2.03 Mg m^{-3} fraction and could not have been identified without density fractionation.

MAGNETIC TECHNIQUES

The main classes of magnetic behavior are listed in Table 8-2 along with descriptions of the magnetic susceptibility and origin of magnetism for each class. Most soil minerals are paramagnetic, but a few Fe-bearing minerals, such as magnetite (Fe_3O_4), maghemite (γ-Fe_2O_3), and pyrrhotite (Fe_7S_8), are ferrimagnetic. Others such as hematite (α-Fe_2O_3), goethite (α-FeOOH), lepidocrocite (γ-FeOOH), ilmenite ($FeTiO_3$), and pyrolusite (MnO_2) are antiferromagnetic. Ferrimagnetic minerals are readily magnetized to saturation by an applied magnetic field, and therefore, the extent of magnetization of these minerals is only weakly dependent on the strength of the applied magnetic field. Paramagnetic and antiferromagnetic minerals, by contrast, are only weakly magnetized, but the extent of magnetization of these minerals increases with the strength of an applied magnetic field (McBride, 1986). The magnetic susceptibility of paramagnetic minerals depends on their elemental composition. Iron, Ni, and Co are the only strongly magnetic elements but at least 56 other elements are considered weakly magnetic (Kolm et al., 1975). For paramagnetic soil minerals, magnetic susceptibility is primarily related to Fe content.

Ferrimagnetic minerals can be easily separated from soil materials with a hand magnet. Magnetic separation of paramagnetic and antiferromagnetic minerals, by contrast, requires the use of powerful electromagnets capable of producing magnetic-field strengths greater than 1 T. Muller (1977) has reviewed various techniques for dry magnetic separation of sand-size minerals, many of which involve use of the Franz Isodynamic Separator[1] (S.G. Franz Co., Inc., Trenton, NJ).

Magnetic Separation of Fine Materials

Berry and Jorgensen (1969) were among the first researchers to fractionate fine-grained mineral samples with an electromagnet. They constructed two plexiglas separator cells for use with a Franz Isodynamic Separator. Their system relied on magnetic trapping of paramagnetic particles during gravitational settling. The technique was slow, relatively inefficient, and inapplicable for samples with particles <2 μm. Despite the limitations, relatively pure chlorite samples were separated from the 2 to 4 μm fraction of a Norwegian quick-clay using this technique.

Relatively efficient magnetic fractionation of fine-grained materials became possible with the development of high-gradient magnetic separation (HGMS). This technique was developed for use in the kaolin industry to remove impurities (mostly mica and various Fe-, Al-, and Ti-oxyhydroxides) from kaolin ores (Iannicelli, 1976). Although not widely used outside the kaolin industry, other potential industrial applications for HGMS include purification of pigments and glass sands, processing of low-grade Fe ores, and even water purification (Kolm et al., 1975). Within the scientific community, the first use of HGMS for fractionation of soil minerals was reported in 1979 (Schulze & Dixon, 1979). Since then, HGMS has been increasingly used as a means of preconcentrating soil minerals for subsequent identification, characterization, and quantification (Senkayi et al., 1981; Hughes, 1982; Hughes & La Mare, 1982; Russell et al., 1984; Righi & Jadault, 1988; Ghabru et al., 1988; Santos et al., 1989; Ghabru et al., 1990a,b; Weed & Brown, 1990; Sayin et al., 1990).

---

[1] Trade names and company names are included for the benefit of the reader and do not imply any endorsement or preferential treatment of the product listed by the USDA or the Minnesota Agricultural Experiment Station.

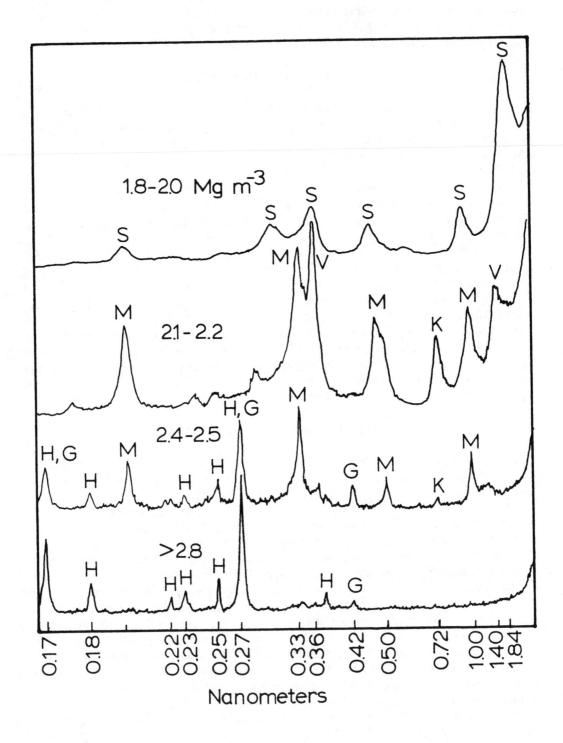

Fig. 8-5. X-ray (CuKα radiation) powder diffraction patterns of various density fractions separated from a 2BC-horizon sample of an Upshur soil; S=smectite, M=clay mica, V=vermiculite, K=kaolinite, G=goethite, and H=hematite (adapted from Jaynes & Bigham, 1986).

Table 8-2. Classes of magnetic behavior, magnetic susceptibility, and origin of magnetism in materials.

Class	Magnetic susceptibility	Origin of magnetism
Ferromagnetic	Strong Positive	Alignment of spins for unpaired electrons of atoms within domains.
Ferrimagnetic	Strong Positive	Incomplete alignment of spins for unpaired electrons of atoms within domains.
Antiferromagnetic	Weak Positive	Antiparallel alignment of spins for unpaired electrons of atoms within domains.
Paramagnetic	Weak Positive	Alignment of spins for unpaired electrons of individual atoms.
Diamagnetic	Very weak Negative	Alignment of polarized electron orbitals of individual atoms.

High-gradient magnetic separations are performed by pumping a suspension of dispersed particles through a separator that is placed between the poles of an electromagnet. The separator consists of several plexiglas tubes connected in series and filled with a ferromagnetic collector. After the suspension has passed through the separator, the magnetic field is turned off and the separator is flushed with water to recover particles held on surfaces of the collector. Fractions with different magnetic susceptibilities may be recovered by cycling the same sample through the separator several times while increasing the strength of the applied magnetic field with each cycle.

A gradient in the strength of an applied magnetic field must exist within a particle for a net magnetic force to act on that particle (Fig. 8-6). With HGMS, magnetic-field gradients are achieved by placing a ferromagnetic collector within the field of a strong electromagnet. Stainless steel wool is commonly used as a collector because steel wool strands have sharp edges which propagate strong magnetic field gradients and a large surface area for trapping and retaining particles (Schulze & Dixon, 1979).

The magnetic force acting on a particle during HGMS is opposed by the hydrodynamic-drag force of the flowing suspension acting on the particle. Both of these forces decrease with particle radius (r). The magnetic force decreases with r^3 and the hydrodynamic-drag force decreases with r^2. Therefore, particles smaller than a critical size will not be magnetically trapped because the hydrodynamic-drag force exceeds the magnetic force acting on those particle (Schulze & Dixon, 1979). In practice, HGMS is effective for fractionating silt (2-50 μm) and coarse-clay (0.2-2 μm) samples, but relatively ineffective for fractionating fine-clay (<0.2 μm) samples.

Liquid magnetic separation (LMS) is a technique for magnetic fractionation of sand-size materials. The LMS is similar to HGMS, except that a steel rod is used as a ferromagnetic collector and sand-size materials pass through the separator via gravitational settling (Ghabru et al., 1987).

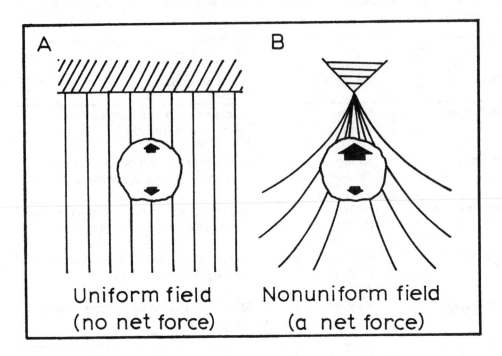

Fig. 8-6. Effect of (*A*) uniform and (*B*) nonuniform applied magnetic fields on the net magnetic force acting on a particle (adapted from Schulze & Dixon, 1979).

Magnetic Separation of Soil Minerals

The HGMS was first used with soil clays (Schulze & Dixon, 1979; Hughes, 1982; Hughes & La Mare, 1982) to facilitate identification and characterization of Fe-oxyhydroxide minerals. Schulze and Dixon (1979) separated the 0.2 to 2 μm and the <0.2-μm size fractions of six soils into magnetic fractions (material retained on the collector at 1.6 T) and tailings (material not retained at 1.6 T). Goethite, hematite, lepidocrocite, chlorite, and anatase were concentrated in magnetic fractions. Quartz was concentrated in tailings, while mica, vermiculite, kaolinite, and smectite were detected in both magnetic fractions and tailings. The Fe-oxyhydroxide minerals were readily identifiable by x-ray powder diffraction analysis of the magnetic fractions. Hughes (1982) and Hughes and La Mare (1982) used HGMS to fractionate clays from various tropical soils. Total carbon (C) and acid-ammonium-oxalate extractable Fe (Fe$_{ox}$) and Al (Al$_{ox}$) were determined for samples of the whole clay, magnetic fractions, and tailings. Relationships between Fe$_{ox}$, Al$_{ox}$, and C indicated an association between organic C and Fe-oxyhydroxide minerals in these soils. Santos et al. (1989) reached a similar conclusion for analyses involving HGMS of soil microaggregates from Brazilian Oxisols.

Russell et al. (1984) introduced two modifications to the HGMS method of Schulze and Dixon (1979): (i) a variable-speed peristaltic pump to recirculate the suspensions; and (ii) a voltage regulator to control the strength of the applied magnetic field. These modifications permitted several fractions with different magnetic susceptibilities to be separated from the same sample.

Righi and Jadault (1988) fractionated three soil clays with HGMS, including one obtained from the C horizon of a Dystrochrept developed on a chlorite-rich, mica schist (Fig. 8-7). Vermiculite, mica, chlorite, kaolinite, and quartz were identified in the whole-clay fraction. By contrast, a subfraction of this material retained on the collector at 0.05 T was nearly pure chlorite. Furthermore, smectite was sufficiently concentrated in materials retained on the collector at 0.75

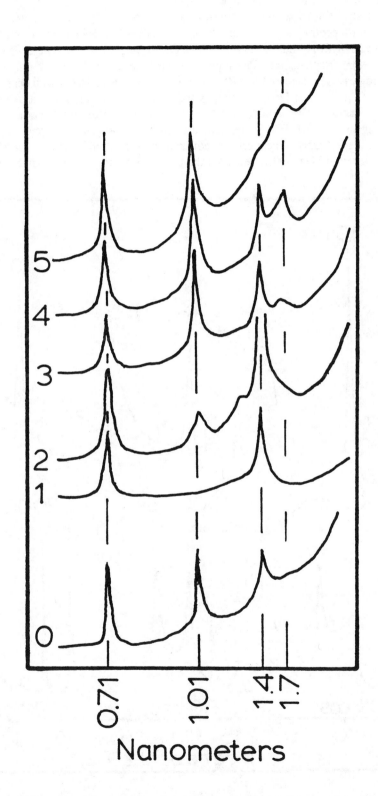

Fig. 8-7. X-ray (CoKα radiation) powder diffraction patterns of (0) the original bulk clay, and fractions obtained by HGMS at (1) <0.05 T, (2) 0.05 to 0.20 T, (3) 0.20 to 0.75 T, (4) 0.75 to 1.25 T, and (5) >1.75 T from a C-horizon sample of a Dystrochrept developed in chlorite-rich mica-schist in Nepal (adapted from Righi & Jadault, 1988).

and 1.25 T and for material passing the collector at 1.75 T to be identified by XRD, whereas smectite was undetected in the whole clay by XRD.

Ghabru et al. (1988) fractionated coarse-clay (0.2- to 2-μm) samples obtained from the Bt horizon of a north-central Saskatchewan Boralf. Mica, mixed-layer mica-vermiculite, kaolinite, quartz, feldspars, hydroxy-interlayer-vermiculite (HIV), and amphiboles were all identified in the coarse clay before fractionation. Prominence of XRD peaks for HIV (1.4 nm) and amphiboles (0.84 nm) progressively increased in XRD patterns of fractions collected with decreasing magnetic field strength (Fig. 8-8). Indeed, amphiboles and HIV were the only mineral phases identified in the <0.20-T fraction. The <0.20-T fraction was treated with dithionite-citrate-bicarbonate and fractionated again at 0.20 T (Fig. 8-9) to separate nearly monomineralic amphibole (<0.20 T) and HIV (>0.20 T) specimens. These results indicated an association between Fe oxides and HIV in soil clays. Ghabru et al. (1990b) later used the citrate-dithionite treatment of Tamura (1956) to

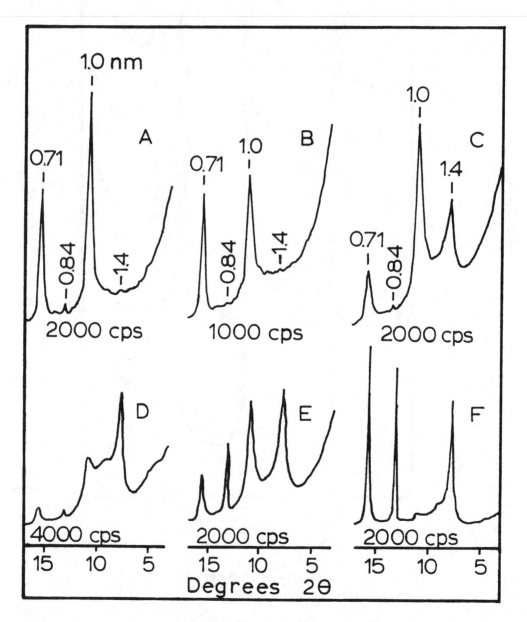

Fig. 8-8. X-ray (FeKα radiation) powder diffraction patterns of (A) the coarse clay, and fractions of the coarse clay obtained by HGMS at (B) >1.38 T, (C) 0.68 to 1.38 T, (D) 0.48 to 0.68 T, (E) 0.20 to 0.48 T, and (F) <0.20 T from a Bt-horizon sample of a north-central Saskatchewan Boralf (adapted from Ghabru et al., 1988).

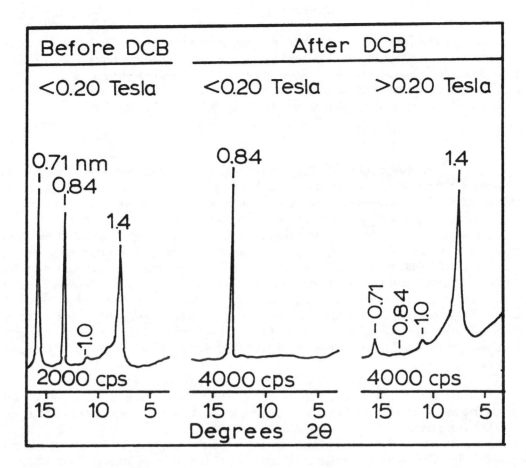

Fig. 8-9. X-ray powder (FeKα radiation) diffraction patterns of the <0.20-T HGMS fraction separated from the coarse-clay fraction of a Bt-horizon sample of a north-central Saskatchewan Boralf before and after removal of DCB-extractable iron (adapted from Ghabru et al., 1988).

extract hydroxy-interlayer material from the monomineralic HIV specimen and determined that ~95% of the interlayer cations were Fe.

Weed and Brown (1990) used HGMS to fractionate clays from Ultisols of the North Carolina Coastal Plain. Fractions retained by the collector at 0.77 T appeared to be pure HIV by XRD analysis. Differential thermal analyses, however, revealed that the samples contained small amounts of kaolinite. Based on structural interpretations of chemical analyses, Weed and Brown (1990) suggested a phengitic muscovite or celadonite precursor for the HIV. In other recent studies HGMS has been used to evaluate associations between DCB-extractable iron and clay minerals (Ghabru et al., 1990a) and to determine phosphate sorption-desorption characteristics of magnetically separated soil fractions (Sayin et al., 1990).

ELECTROPHORETIC TECHNIQUES

Electrophoresis has been in use since the beginning of the twentieth century, and is a very important technique for analysis of biological materials. With proteins and other biocolloids, electrophoresis is routinely used for separation of specimens, purity testing, and characterization of surface properties. The use of electrophoresis with inorganic materials is considerably less prevalent. Electrophoresis, however, has been used with inorganic materials to characterize surface properties, evaluate interactions, and separate both crystalline and noncrystalline phases.

Smoluchowski Equation

Electrophoresis is the motion of charged particles under an electric field in a liquid medium. At a minimum, an electrophoretic cell consists of two electrodes immersed in a liquid medium. When an electrical potential is applied across the electrodes, negatively charged particles within the cell move toward the anode and positively charged particles move toward the cathode. The electrophoretic mobility (μ) of a particle is given by the Smoluchowski equation,

$$\mu = \zeta \epsilon_0 \epsilon_r (\eta)^{-1} \qquad [10]$$

where ζ is the electrostatic potential at the plane of shear, ϵ_0, the permittivity of a vacuum, ϵ_r, the relative permittivity of the medium, and η, the effective viscosity (Adamson, 1982; Harsh et al., 1988). Within the Smoluchowski equation μ is expressed in meters squared per volt second but μ is conveniently measured in micrometers per second volt centimeter.

Zeta potential (ζ) has been used to estimate the potential on the outer surface of the Stern layer, but the theoretical relationship between ζ and surface potential is uncertain (Low, 1981). This uncertainty stems from the fact that the shear plane is not necessarily coincident with the solid-liquid phase boundary. Due to frictional forces, liquid close to particle surfaces tends to move with the particles, and ζ is strongly influenced by counter ions entrained with that liquid. Indeed, charge reversals at the plane of shear are easily achieved by adding polyvalent inorganic cations (Beavers & Marshall, 1950; Park & Lewis, 1969; Drever, 1969), cationic surfactants (Horikawa, 1975), and hydroxy-aluminum polycations (Oades, 1984; Bottero et al., 1988) to clay suspensions.

The effective viscosity in the Smoluchowski equation includes both hydrodynamic and electrostatic retardation forces acting on particles in an electrophoretic field. Counter ions beyond the shear plane, but within the diffuse double-layer, move in a direction opposite to that of the particle. These counter ions entrain liquid, and hence, increase the hydrodynamic drag acting on the particle over that for the same particle moving in a static solution. In addition, due to the opposite relative motions of particles and counter ions, there is an asymmetric distribution of counter-ion charge with respect to particle charge. This asymmetry in charge distribution leads to electrostatic drag on the mobility of both the particles and the counter ions.

Much of the electrical current in an electrophoretic cell is carried by H^+ and OH^- that are produced by electrohydrolysis of water at the electrodes. Protons move toward the cathode where they are reduced to H_2 gas:

$$H_3O^+ + 2e^- = H_2(g) + OH^- \qquad [11]$$

Hydroxyls move toward the anode where they are oxidized to O_2 gas:

$$2OH^- = O_2(g) + 4e^- + 2H^+ \qquad [12]$$

Because of electrohydrolysis and the opposite relative motion of the H^+ and OH^-, a pH gradient develops within an electrophoretic cell. The liquid medium is generally buffered to decrease the size of the pH gradient.

Electrophoretic Techniques

Early moving-boundary electrophoretic techniques were based on the movement of particles in free suspensions (Longsworth, 1959). These techniques were used primarily for determining ζ and not for phase separations. Resolution with moving-boundary techniques was generally limited due to the effects of gravitational settling and thermal convection.

Zone-electrophoresis techniques were first developed in the 1950s. With zone electrophoresis, particles move within, or on top of, a supporting medium that is saturated with a pH buffer solution. The supporting medium minimizes thermal convection and gravitational settling effects. Some supporting media, such as filter paper, cellulose acetate, and weak agar gels have little influence on the electrophoretic movement of particles. Other media, such as starch and acrylamide gels, exert a sieving action, and thus separate particles by both electrophoretic mobility and particle size.

Isoelectric focusing is an electrophoretic technique that uses a pH gradient to facilitate separations. With isoelectric focusing the liquid medium contains various ampholines. Under the influence of an electric field the ampholines are distributed across an electrophoretic cell in relation to their isoelectric points, establishing a controlled pH gradient within the cell. Under these conditions, particles move in response to a potential gradient until they reach a position within the cell having a pH equal to their isoelectric point. Isoelectric focusing is an effective and commonly used technique for separation of biocolloids. This technique might have some utility for fractionation of variable-charge clays but would not be effective for separating permanent-charge clays because of their low isoelectric points.

Paper electrochromatography (Wunderly, 1959; McNeal & Young, 1963) uses a hanging curtain, a vertically oriented piece of filter paper, as a supporting medium. Buffer is uniformly supplied across the top of the curtain via capillarity and descends under gravity to a series of drip points along the bottom of the curtain. Particle suspensions are introduced using a motorized syringe at a selected location near the top of the curtain. Under these conditions and with an electrical-potential gradient running horizontally across the curtain, particle fractions with different electrophoretic mobilities may be collected at the various drip points.

Continuous-particle electrophoresis (Strickler et al., 1966; Strickler, 1967) was a significant advance over paper electrochromatography. With this technique, separations are made within a curtain (5 cm wide, and 30 cm long) of flowing buffer, which is constrained between cell plates 1.5 mm apart. Fractions with different electrophoretic mobilities are collected from a series of exit tubes at the bottom of the cell (Fig. 8-10). This technique allows rapid continuous fractionation of particles in free suspensions. Thermal convection is not a problem with continuous-particle electrophoresis because cell plates can be water cooled and the curtain buffer is generally replaced before it heats up. Deleterious effects due to gravitational settling are minimal because settling is generally slower than the flow of buffer within the curtain.

Electrophoretic Separation of Soil Minerals

Early work with reference clays (Beavers & Marshall, 1950; Beavers & Larson, 1953), using moving-boundary techniques, established that different mineral species had different electrophoretic mobilities. Beavers and Larson (1953), however, were unable to distinguish mobility differences for various mineral species in soil clays.

The first electrophoretic separations of multimineralic clay specimens were performed using paper electrochromatography. Using this technique, McNeal and Young (1963) successfully separated smectite from prepared mixtures of reference clays containing kaolinite and vermiculite. They were unable, however, to separate kaolinite from vermiculite.

Several researchers (Drever, 1969; Park & Lewis, 1969; Arshad et al., 1971; Dunning et al., 1982) used continuous-particle electrophoresis for fractionation of mineral samples. Drever (1969) successfully separated mixtures of reference clays containing kaolinite and montmorillonite, but had more difficulty separating mineral species present in a shallow-water marine sediment (Fig. 8-11). The separations were run in 0.0004 M Na_2CO_3 (pH = 9.0) solutions containing 50 mL ethylene glycol L^{-1}. Attempts to alter shear-plane potentials of the clays by addition of polyvalent cations (Ba^{2+} and La^{3+}), cetyltrimethylammonium chloride, and sodium polymetaphosphate failed to improve separations because flocculation occurred before electrophoretic mobilities were altered sufficiently to improve the separations.

Arshad et al. (1971) used continuous-particle electrophoresis for fractionation of fine (<0.08 μm) soil-clay samples. The separations were run in 0.001 M Na_2CO_3 solutions (adjusted to pH 4.5 with HCl) at 6°C with a 40 V cm^{-1} potential gradient. Dilute clay suspensions (5 g L^{-1}) were introduced at 0.1 mL min^{-1} and the curtain flow was maintained at 15 mL min^{-1}. Under these conditions five fractions were collected and analyzed by XRD (Fig. 8-12). Fraction 1 contained

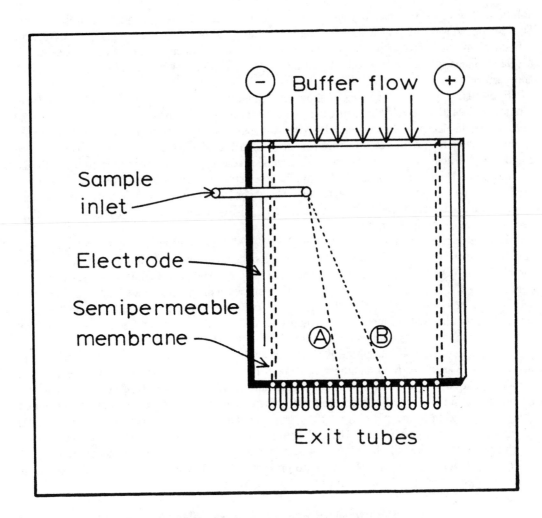

Fig. 8-10. Design of continuous-particle electrophoresis cell; lines A and B show paths followed by particles with different electrophoretic mobilities.

x-ray amorphous material rich in Al, Fe, and P, and low in Si. Other fractions contained various amounts of amorphous material and roughly equal amounts of smectite and illite. No evidence for separation of the smectitic and illitic phases was observed.

In general, separation of reference-clay mixtures by electrophoresis have been more successful than separations of naturally occurring multimineralic soil and sediment clays. One reason for this difference is the effects of the various organic and noncrystalline inorganic constituents in soils and sediments on electrophoretic mobilities. Cavallaro and McBride (1984) demonstrated that removal of organic matter by sodium hypochlorite oxidation from acidic soil clays greatly increased the positive electrophoretic mobility of the clays, whereas removal of Fe-oxyhydroxides with DCB treatments led to negative mobilities (Fig. 8-13). Other interactions leading to charge reversals have been demonstrated for clay systems containing imogolite (Horikawa, 1976) and polycations of Al and Fe (Oades, 1984). The presence of these materials in soil- and sediment-clay samples tends to mask differences in electrophoretic mobilities between clay species.

Another major disadvantage of electrophoresis is a tendency for exchange cations to be stripped from particle surfaces and replaced by H^+ during electrophoretic separations. Hydrogen saturation promotes hydrolysis of structural oxygens and the release of Al and other structural cations. As a result of electrophoresis, the chemical and physical properties of clays can be altered to the extent

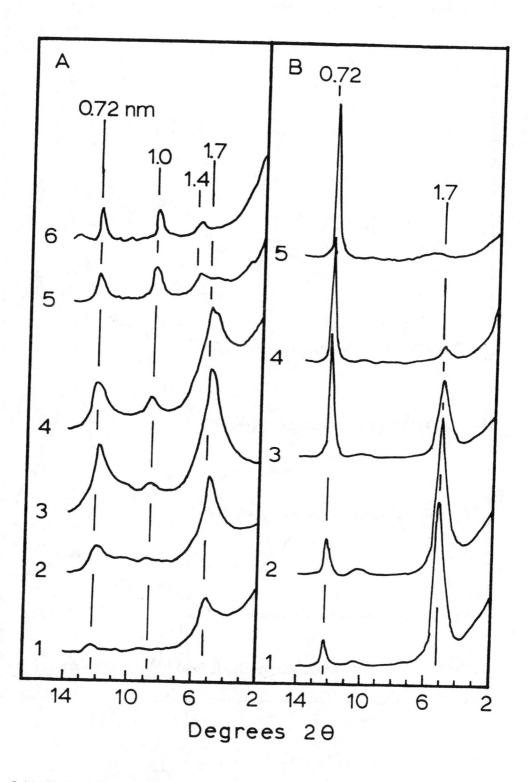

Fig. 8-11. X-ray (CuKα radiation) powder diffraction patterns of various electrophoretic fractions separated from (A) a shallow-water marine sediment, and (B) a mixture of kaolinite and montmorillonite (adapted from Drever, 1969).

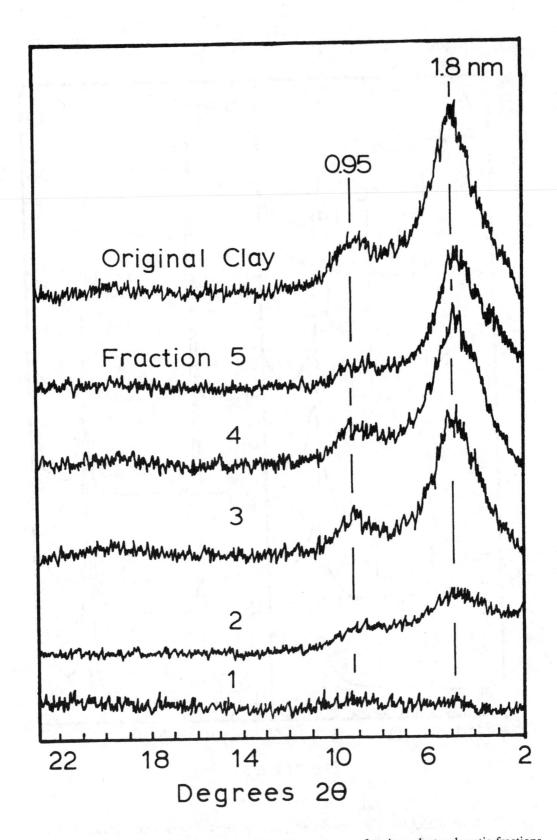

Fig. 8-12. X-ray (CuKα radiation) powder diffraction patterns of various electrophoretic fractions separated from a fine (<0.08-μm) soil-clay sample (adapted from Arshad et al., 1971).

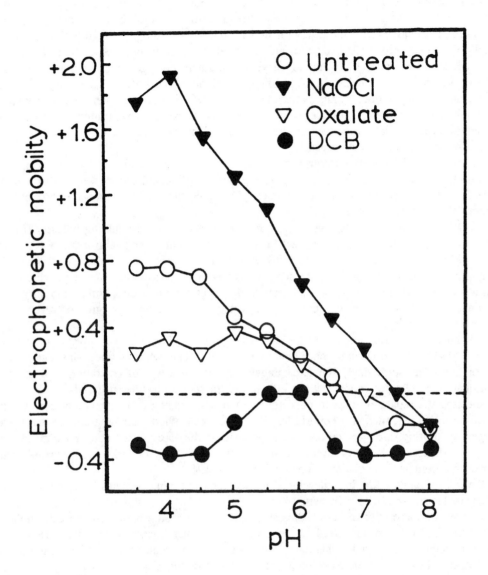

Fig. 8-13. Electrophoretic mobility (μm s^{-1} V^{-1} cm^{-1}) of untreated and treated soil clay (adapted from Cavallaro & McBride, 1984).

that the separates are not representative of the mineral phases in the original sample. For some purposes, limited hydrolysis of mineral particles may be acceptable, but this process inherently limits the utility of electrophoresis for physical separation of clay species. As hydrolysis proceeds, clays become increasingly Al saturated.   The Al saturation causes clays to flocculate, and flocculation inhibits physical separation.   Increasing the concentration of a buffer or neutral electrolyte in the liquid medium reduces hydrolysis, but increases the tendency of clays to flocculate.

Since the early 1970s interest in the use of electrophoresis as a separation technique for soil clay minerals has declined. Instead, research with electrophoresis has focused on charge properties of clay surfaces (Low, 1981; Pashley 1985; Cavallaro & McBride, 1984), interactions between hydroxy-aluminum polycations and clay surfaces (Oades, 1984; Bottero et al., 1988; Harsh et al., 1988), and interactions between imogolite and other clay minerals (Horikawa, 1976).

OTHER METHODS OF MINERAL SEPARATION

Abudelgawad et al. (1985) evaluated three techniques for separating palygorskite from reference samples containing smectite and sepiolite impurities. For all three methods, the samples (<0.2-μm fractions) were saturated with dodecylammonium ions. With the first technique, samples were dispersed in distilled water and separated by centrifugation-decantation. Because the impurities adsorbed substantially more dodecylammonium than palygorskite, the impurities tended to flocculate and settle out under centrifugation; whereas the palygorskite remained suspended. The second and third separation techniques were based on water:decanol and water:nitrobenzene phase partitioning, respectively. All three techniques resulted in some concentration of palygorskite. Abudelgawad et al. (1985) recommended the differential-dispersion technique because it was simple and did not require hazardous reagents.

Maynard et al. (1969) described a differential-dispersion technique for removing TiO_2 impurities from kaolin ores. To achieve these separations, dispersing agents were added in excess of the amount required for maximum dispersion. The kaolin suspension remained stable, but TiO_2 minerals flocculated and settled out.

Flotation is widely used for separating ore from waste within the mining industry (Muller, 1977; Adamson, 1982) and has been used for beneficiation within the kaolin industry (Greene & Duke, 1962). Horikawa (1975) demonstrated that flotation can be used to separate allophane and imogolite from other clay species. The separations were most successful when clays were treated with anionic surfactants, and pH was adjusted to the point of zero electrophoretic mobility. This technique merits further research, as it may be widely applicable for separation of variable- and permanent-charge clays.

Asymmetric vibrators are flat plates which are tilted slightly and then given lateral pulses (Muller, 1977). Round mineral grains tend to roll off the plate, whereas flat grains tend to move horizontally. This surprisingly simple technique is an excellent way of separating sand-size mica from quartz and feldspar grains, but has little utility for separating finer materials.

Electrostatic separators depend on differences in the extent to which mineral grains retain electrostatic charge (Mitchell, 1975; Muller, 1977). In one method, mineral grains are electrically charged by passing through a corona-discharge field as they are poured onto a grounded metal cylinder. Mineral grains that lose their charge quickly fall off the cylinder and grains that retain their charge cling to the cylinder (Woodley & Duffell, 1964).

Several techniques have been developed for separation of mineral grains based on differences in their dielectric properties (Mitchell, 1975; McEuen 1964). For these methods, mineral grains are suspended in a solution of known dielectric, and two electrodes with a potential difference of several hundred volts are immersed in the suspension. Mineral grains with a dielectric constant greater than the solution tend to migrate to the space between the electrodes. Mineral grains with a dielectric constant less than the solution are repelled from this region.

Bush et al. (1966) proposed a centrifugal method for separating swelling clays. The samples were pretreated with silicone and dispersed in distilled water. The clay suspensions were placed above a heavy liquid and centrifuged. To break the water-heavy liquid interface, the samples were stirred violently during centrifugation. Bush et al. (1966) reported quantitative recovery of swelling clays by this means.

Finally, monomineralic samples may be obtained by hand-picking particles with the desired mineralogy from soil materials using a pair of tweezers or a suction pick and a binocular microscope (Muller, 1977). Hand picking proceeds rapidly with coarse sands but may be quite tedious for medium sands and nearly impossible for fine sands. Even with coarse sands, hand-picking is a viable separation technique only when small amounts of sample are required. For studies in which separation of monomineralic samples is critical, however, hand-picking may be the only way of obtaining monomineralic samples with confidence. Preconcentration of a desired mineral species by particle-size-, density-, and/or magnetic-fractionation reduces the amount of time and effort required during hand-picking (Ghabru et al., 1989).

CONCLUSIONS

Preconcentration techniques are commonly used as pretreatments for soil mineralogical analyses to facilitate the identification, characterization, and quantification of the various mineral phases in soil materials. Ideally, preconcentration techniques separate monomineralic samples for each mineral species present in a soil sample. This ideal, however, cannot be achieved with present technology and is unlikely to be achieved for many years.

The most effective techniques for separating soil minerals are generally those performed under conditions that maximize the dispersive forces between particles, and hence, enhance separations based on the properties of the individual particles. Difficulties with mineral separations are due to the heterogeneous nature of soil minerals and to particle-particle interactions. Particles of the same mineral species may differ from one another in their elemental composition, size, density, morphology, magnetic susceptibility, dielectric properties, and surface properties. Because the various preconcentration techniques separate mineral particles based on differences in these properties, such heterogeneity places fundamental limits on the extent to which mineral species can be separated.

Three types of information can be obtained by analyses of subfractions of soil materials that generally cannot be obtained by analyses of bulk samples. First, minerals present in trace quantities but undetectable in the bulk samples often are detected in one or more of the subfractions. Second, physical and chemical properties of bulk-soil materials can be at least partly partitioned among the individual mineral constituents. And third, variations in properties of individual mineral species can sometimes be determined.

Particle-size fractionation is by far the most commonly used preconcentration technique in soil mineralogical analyses. Size fractionation is generally sufficient for most investigations involving qualitative characterization and even semiquantitative analyses of soil minerals. Other preconcentration techniques, such as density, magnetic, and electrophoretic separations, are used infrequently, and are almost always used in combination with particle-size fractionation. Use of these techniques, however, can provide additional mineralogical information not obtainable by particle-size fractionation alone. Density and magnetic separations are particularly useful for concentrating accessory minerals.

In the future, some improvements in preconcentration techniques can be anticipated. However, fractionations based on particle size, density, and electrophoretic mobility are mature separation techniques, and advances in these methods are likely to be incremental. High-gradient magnetic separation, by contrast, has been little used for soil mineral separations and recent developments indicate room for substantial improvement in this technique. The most significant advances, however, are likely to be in the area of data analysis. The wide availability of digital computers and the development of sophisticated numerical optimization techniques [e.g., the Marquardt (Bevington, 1969) and simplex (Sneddon, 1990) algorithms] has made it possible to obtain information about single phases in polymineralic samples that is equivalent to that obtained from direct analysis of monomineralic samples.

REFERENCES

Abudelgawad, G., B.E. Viani, and J.B. Dixon. 1985. Palygorskite separation from dodecylammonium-treated clays. Clays Clay Miner. 33:438-442.

Adamson, A.W. 1982. Physical chemistry of surfaces. 3rd ed. John Wiley & Sons, New York.

Aguilera, N.H., and M.L. Jackson. 1953. Iron oxide removal from soils and clays. Soil Sci. Soc. Am. Proc. 17:359-364.

Ali, O.M., M. Yousaf, and J.D. Rhoades. 1987. Effect of exchangeable cation and electrolyte concentration on mineralogy of clay dispersed from aggregates. Soil Sci. Soc. Am. J. 51:896-900.

Arshad, M.A., R.J. St. Arnaud, and P.M. Huang. 1971. Characterization of electrophoretic separates of soil clays. Soil Sci. 112:46-52.

Badraoui, M., and P.R. Bloom. 1990. Iron-rich high-charge beidellite in Vertisols and Mollisols of the High Chaouia region of Morocco. Soil Sci. Soc. Am. J. 54:267-274.

Badraoui, M., P.R. Bloom, and R.H. Rust. 1987. Occurrence of high charge beidellite in a Vertic Haplaquoll of northwestern Minnesota. Soil Sci. Soc. Am. J. 51:813-818.

Batson, D.M., and E. Truog. 1939. Further improvements in the mineralogical subdivision of fine clay by means of heavy liquid specific gravity separations. Soil Sci. Soc. Am. Proc. 4:104-105.

Beavers, A.H. 1961. Preparation of sensitive linear density gradients. Soil Sci. Soc. Am. Proc. 25:357-359.

Beavers, A.H. and B.L. Larson. 1953. Electrophoresis of clays by the schlieren moving boundary procedure. Soil Sci. Soc. Am. Proc. 17:22-26.

Beavers, A.H., and C.E. Marshall. 1950. The cataphoresis of clay minerals and factors affecting their separation. Soil Sci. Soc. Am. Proc. 15:142-145.

Bergman, S., C.J. Ritter, E.E. Zamierowski, and C.R. Cothern. 1979. The use of zonal centrifugation in delineating trace element distributions in sewage sludges from the Dayton, Ohio, area. J. Environ. Qual. 8:416-422.

Berry, R., and P. Jorgensen. 1969. Separation of illite and chlorite in clays by electromagnetic techniques. Clay Miner. 8:201-212.

Bevington, P.R. 1969. Data reduction and error analysis for the physical sciences. McGraw-Hill, New York.

Bonner, W.P., T. Tamura, C.W. Francis, and J.W. Amburgey, Jr. 1970. Zonal centrifugation - a tool for environmental studies. Environ. Sci. Technol. 4:821-825.

Bottero, J.Y., M. Bruant, and J.M. Cases. 1988. Interactions between hydroxy-aluminum species and homoionic Na- and Ca-montmorillonite particles, as manifested by ζ potential, suspension stability and x-ray diffraction. Clay Miner. 23:213-224.

Bourget, S.J., and C.B. Tanner. 1953. Removal of organic matter with sodium hypobromite for particle-size analysis of soils. Can. J. Agric. Sci. 33:579-585.

Brakke, M.K. 1951. Density gradient centrifugation: A new separation technique. J. Am. Chem. Soc. 73:1847-1848.

Brantley, J.N., J.P. Breillatt, Jr., F.S. Brinkley, C.W. Francis, B.A. Halshall, R. Levy, and S.G. Rush. 1974. Zonal centrifugation: Applied aspects in elucidating chemical and biological forms, distribution and availability of heavy metals in the environment. p. 195-221. *In* W. Fulkerson et al. (ed.) Ecology and analysis of trace contaminants. Oak Ridge Natl. Lab., Oak Ridge, TN.

Brewster, G.R. 1980. Effect of chemical pretreatments on x-ray powder diffraction characteristics of clay minerals derived from volcanic ash. Clays Clay Miner. 28:303-310.

Busacca, A.J., J.R. Aniku, and M.J. Singer. 1984. Dispersion of soils by an ultrasonic method that eliminates probe contact. Soil Sci. Soc. Am. J. 48:1125-1129.

Bush, D.C., R.E. Jenkins, and S.B. McCaleb. 1966. Separation of swelling clay minerals by a centrifugal method. Clays Clay Miner. 14:407-418.

Cavallaro, N., and M.B. McBride. 1984. Effect of selective dissolution on charge and surface properties of an acid soil clay. Clays Clay Miner. 32:283-290.

Dahlgren, R.A. 1994. Quantification of allophane and imogolite. p. 430-452. *In* J. Amonette and L.W. Zelazny (ed.) Quantitative methods in soil mineralogy. SSSA Misc. Publ. SSSA, Madison, WI.

Douglas, L.A. 1967. Sodium-citrate-dithionite-induced alteration of biotite. Soil Sci. 103:191-195.

Douglas, L.A., and F. Fiessinger. 1971. Degradation of clay minerals by H_2O_2 treatments to oxidize organic matter. Clays Clay Miner. 19:67-68.

Drever, J.I. 1969. The separation of clay minerals by continuous particle electrophoresis. Am. Miner. 54:937-942.

Dunning, J.D., B.J. Herren, R.W. Tipps, and R.S. Snyder. 1982. Fractionation of mineral species by electrophoresis. J. Geophys. Res. 87:10781-10788.

Edwards, A.P., and J.M. Bremner. 1964. Use of sonic vibration for separation of soil particles. Can. J. Soil Sci. 44:366.

Edwards, A.P., and J.M. Bremner. 1967. Dispersion of soil particles by sonic vibration. J. Soil Sci. 18:47-63.

Essington, M.E., S.V. Mattigod, and J.O. Ervin. 1985. Particle sedimentation rates in the linear density gradient. Soil Sci. Soc. Am. J. 49:767-771.

Francis, C.W. 1973. Adsorption of polyvinylpyrrolidone on reference clay minerals. Soil Sci. 115:40-54.

Francis, C.W., W.P. Bonner, and T. Tamura. 1972. An evaluation of zonal centrifugation as a research tool in soil science--I. Methodology. Soil Sci. Soc. Am. Proc. 36:366-372.

Francis, C.W., F.S. Brinkley, and E.A. Bondietti. 1976. Large-scale zonal rotors in soil science. Soil Sci. Soc. Am. J. 40:785-792.

Francis, C.W., and T. Tamura. 1972. An evaluation of zonal centrifugation as a research tool in soil science--II. Characterization of soil clays. Soil Sci. Soc. Am. Proc. 36:372-376.

Francis, C.W., T. Tamura, W.P. Bonner, and J. W. Amburgey, Jr. 1970. Separation of clay minerals and soil clays using isopycnic zonal centrifugation. Soil Sci. Soc. Am. Proc. 34:351-353.

Frye, K. 1974. Modern mineralogy. Prentice-Hall, Inc., Englewood Cliffs, NJ.

Genrich, G.A., and J.M. Bremner. 1972. A reevaluation of the ultrasonic-vibration method of dispersing soils. Soil Sci. Soc. Am. Proc. 36:944-947.

Ghabru, S.K., R.J. St. Arnaud, and A.R. Mermut. 1987. Liquid magnetic separation of iron-bearing minerals from sand fractions of soils. Can. J. Soil Sci. 67:561-569.

Ghabru, S.K., R.J. St. Arnaud, and A.R. Mermut. 1988. Use of high gradient magnetic separation in detailed clay mineral studies. Can. J. Soil Sci. 68:645-655.

Ghabru, S.K., R.J. St. Arnaud, and A.R. Mermut. 1990a. Association of DCB-extractable iron with minerals in coarse soil clays. Soil Sci. 149:112-120.

Ghabru, S.K., A.R. Mermut, and R.J. St. Arnaud. 1989. Characterization of garnets in a Typic Cryoboralf (Gray Luvisol) from Saskatchewan, Canada. Soil Sci. Soc. Am. J. 53:575-582.

Ghabru, S.K., A.R. Mermut, and R.J. St. Arnaud. 1990b. Isolation and characterization of an iron-rich chlorite-like mineral from soil clays. Soil Sci. Soc. Am. J. 54:281-287.

Greene, E.W., and J.B. Duke. 1962. Selective froth flotation of ultrafine minerals or slimes. Mining Eng. 14:51-55.

Grossman, R.B., and J.C. Millet. 1961. Carbonate removal from soils by a modification of the acetate buffer method. Soil Sci. Soc. Am. Proc. 25:325-326.

Halma, G. 1969a. A simple and rapid method to obtain a linear density gradient. Clay Miner. 8:47-57. .

Halma, G. 1969b. The separation of clay mineral fractions with linear heavy liquid density gradient columns. Clay Miner. 8:59-69.

Harsh, J.B., H.E. Doner, and D.W. Fuerstenau. 1988. Electrophoretic mobility of hydroxy-aluminum and sodium-hectorite in aqueous solutions. Soil Sci. Soc. Am. J. 52:1589-1592.

Harris, W.G., V.W. Carlisle, and K.C.J. Van Rees. 1987. Pedon zonation of hydroxy-interlayered minerals in Udic Haplaquods. Soil Sci. Soc. Am. J. 51:1367-1372.

Harward, M.E., and A.A. Theisen. 1962. Problems in clay mineral identification by x-ray diffraction. Soil Sci. Soc. Am. Proc. 26:335-341.

Harward, M.E., A.A. Theisen, and D.D. Evans. 1962. Effect of iron removal and dispersion methods on clay mineral identification by x-ray diffraction. Soil Sci. Soc. Am. Proc. 26:535-541.

Horikawa, Y. 1975. Effects of ionic surface active agents on the floatability and electrophoretic mobility of selected clays. Clay Sci. 4:281-290.

Horikawa, Y. 1976. Electrophoretic mobility of binary mixtures of imogolite and some other clay minerals in aqueous suspensions. Clay Sci. 5:43-50.

Huang, P.M. 1989. Feldspars, olivines, pyroxenes, and amphiboles. p. 975-1050. *In* J.B. Dixon and S.B. Weed (ed.) Minerals in soil environments. 2nd ed. SSSA., Madison, WI.

Hughes, J.C. 1982. High gradient magnetic separation of some soil clays from Nigeria, Brazil and Colombia--I. The inter-relationships of iron and aluminum extracted by acid ammonium oxalate and carbon. J. Soil Sci. 33:509-519.

Hughes, J.C., and P.H. La Mare. 1982. High gradient magnetic separation of some soil clays from Nigeria, Brazil and Colombia--II. Phosphate adsorption characteristics, the interrelationships with iron, aluminum, and carbon, and comparison with whole soil data. J. Soil Sci. 33:521-533.

Hughes, R.E., D.M. Moore, and H.D. Glass. 1994. Qualitative and quantitative analysis of clay minerals in soils. p. 330-359. *In* J. Amonette and L.W. Zelazny (ed.) Quantitative methods in soil mineralogy. SSSA Misc. Publ. SSSA, Madison, WI.

Hurlbut, C.S. Jr. 1971. Dana's manual of mineralogy. 18th ed. John Wiley & Sons, Inc., New York.

Iannicelli, J. 1976. High extraction magnetic filtration of kaolin clay. Clays Clay Miner. 24:64-68.

Jackson, M.L. 1985. Soil chemical analysis - Advanced Course. 2nd ed. M.L. Jackson, Madison, WI.

Jaynes, W.F., and J.M. Bigham. 1986. Concentration of iron oxides from soil clays by density gradient centrifugation. Soil Sci. Soc. Am. J. 50:1633-1639.

Jaynes, W.F., J.M. Bigham, N.E. Smeck, and M.J. Shipitalo. 1989. Interstratified 1:1-2:1 mineral formation in a polygenetic soil from southern Ohio. Soil Sci. Soc. Am. J. 53:1888-1894.

Kolm, H., J. Oberteuffer, and D. Kelland. 1975. High-gradient magnetic separation. Sci. Am. 233:46-54.

Kunze, G.W., and J.B. Dixon. 1986. Pretreatment for mineralogical analysis. p. 91-100. *In* A. Klute (ed.) Methods of soil analysis. Part 1. 2nd ed. Agron. Monogr. 9. SSSA, Madison, WI.

Laird, D.A., T.E. Fenton, and A.D. Scott. 1988. Layer charge of smectites in an Argialboll-Argiaquoll sequence. Soil Sci. Soc. Am. J. 52:463-467.

Laird, D.A., P. Barak, E.A. Nater, and R.H. Dowdy. 1991. Chemistry of smectitic and illitic phases in interstratified soil smectite. Soil Sci. Soc. Am. J. 55:1499-1504.

Lammers, W.T. 1967. Separation of suspended and colloidal particles from natural water. Environ. Sci. Technol. 1:52-57.

Lavkulich, L.M., and J.H. Wiens. 1970. Comparison of organic matter destruction by hydrogen peroxide and sodium hypochlorite and its effects on selected mineral constituents. Soil Sci. Soc. Am. Proc. 34:755-758.

Longsworth, L.G. 1959. Moving boundary electrophoresis - Theory. p. 91-136. *In* M. Bier (ed.) Electrophoresis theory, methods, and applications. Acad. Press, New York.

Loughnan, F.C. 1957. A technique for the isolation of montmorillonite and halloysite. Am. Miner. 42:393-397.

Low, P.F. 1981. The swelling of clay--III. Dissociation of exchangeable cations. Soil Sci. Soc. Am. J. 45:1074-1078.

Low, B.W., and F.M. Richards. 1952. The use of the gradient tube for the determination of crystal densities. J. Am. Chem. Soc. 74:1660-1666.

Marshall, C.E. 1931. Studies in the degree of dispersion of the clays--I. Notes on the technique and accuracy of mechanical analysis using the centrifuge. J. Soc. Chem. Ind. 50T:444-450.

Marshall, C.E. 1977. The physical chemistry and mineralogy of soils. John Wiley & Sons, New York.

Martin, R.T. 1954. Calcium oxalate formation in soil from hydrogen peroxide treatment. Soil Sci. 77:143-145.

Matelski, R.P. 1951. Separation of minerals by subdividing solidified bromoform after centrifugation. Soil Sci. 71:269-272.

Mattigod, S.V. 1982. Characterization of fly ash particles. Scan. Elect. Microsc. 2:611-617.

Mattigod, S.V., and J.O. Ervin. 1983. Scheme for density separation and identification of compound forms in size-fractionated fly ash. Fuel 62:927-931.

Mattigod, S.V., A.L. Page, and I. Thornton. 1986. Identification of some trace metal minerals in mine-waste contaminated soil. Soil Sci. Soc. Am. J. 50:254-258.

Maynard, R.N., N. Millman, and J. Iannicelli. 1969. A method for removing titanium dioxide impurities from kaolin. Clays Clay Miner. 17:59-62.

McBride, M.C. 1986. Magnetic methods. p. 91-100. *In* A. Klute (ed.) Methods of soil analysis. Part 1. 2nd ed. Agron. Monogr. 9. SSSA, Madison, WI.

McEuen, R.B. 1964. Dielectrophoretic behavior of clay minerals--I. Dielectrophoretic separation of clay mixtures. Clays Clay Miner. 12:549-556.

McNeal, B.L., and J.L. Young. 1963. Paper electrochromatography of clay minerals. Nature (London)197:1132.

Mehra, O.P., and M.L. Jackson. 1960. Iron oxide removal from soils and clays by a dithionite-citrate system buffered with sodium bicarbonate. Clays Clay Miner. 7:317-327.

Mermut, A.R., K. Ghebre-Egziabhier, and R.J. St. Arnaud. 1986. Quantitative evaluation of feldspar weathering in two Boralfs (Gray Luvisols) from Saskatchewan. Soil Sci. Soc. Am. J. 50:1072-1079.

Mitchel, B.D. and B.F.L. Smith. 1974. The removal of organic matter from soil extracts by bromine oxidation. J. Soil Sci. 25:239-241.

Mitchell, W.A. 1975. Heavy minerals. p. 449-480. *In* J.E. Gieseking (ed.) Soil components. Vol. 2. Inorganic components. Springer-Verlag, New York.

Moore, D.M., and R.C. Reynolds, Jr. 1989. X-ray diffraction and the identification and analysis of clay minerals. Oxford Univ. Press, New York.

Muller, L.D. 1977. Laboratory methods of mineral separation. p. 1-34. *In* J. Zussman (ed.) Physical methods in determinative mineralogy. Acad. Press, New York.

Nadeau, P.H., J.M. Tait, W.J. McHardy, and M.J. Wilson. 1984. Interstratified XRD characteristics of physical mixtures of elementary clay particles. Clay Miner. 19:67-76.

Oades, J.M. 1984. Interactions of polycations of aluminum and iron with clays. Clays Clay Miner. 32:49-57.

Oakley, D.M., and B.R. Jennings. 1982. Clay particle sizing by electrically-induced birefringence. Clay Miner. 17:313-325.

Olson, K.W., and R. K. Skogerboe. 1975. Identification of soil lead compounds from automotive sources. Environ. Sci. Technol. 9:227-230.

Omueti, J.A.I., and L.M. Lavkulich. 1988. Identification of clay minerals in soil: The effect of sodium-pyrophosphate. Soil Sci. Soc. Am. J. 52:285-287.

Oster, G., and M. Yamamoto. 1963. Density gradient techniques. Chem. Rev. 63:257-268.

Park, R.G., and G.C. Lewis. 1969. Electrophoretic separation and fractionation of clay mixtures. Am. Miner. 54:1473-1476.

Pashley, R.M. 1985. Electromobility of mica particles dispersed in aqueous solutions. Clays Clay Miner. 33:193-199.

Pearson, R.W., and E. Truog. 1937. Procedure for the mineralogical subdivision of soil separates by means of heavy liquid specific gravity separations. Soil Sci. Soc. Am. Proc. 2:109-114.

Pearson, R.W., and E. Truog. 1938. Further results on the mineralogical subdivision of soil separates by means of heavy liquid specific gravity separations. Soil Sci. Soc. Am. Proc. 3:20-25.

Phillips, W.R. 1971. Mineral optics. W.H. Freeman & Company, San Francisco, CA.

Plewinsky, B., and R. Kamps. 1984. Sodium metatungstate, new medium for binary and ternary density gradient centrifugation. Makromol. Chem. 185:1429-1439.

Rabenhorst, M.C., and L.P. Wilding. 1984. Rapid method to obtain carbonate-free residues from limestone and petrocalcic materials. Soil Sci. Soc. Am. J. 48:216-219.

Righi, D., and P. Jadault. 1988. Improving soil clay minerals studies by high-gradient magnetic separation. Clay Miner. 23:225-232.

Russell, J.D., A. Birnie, and A.R. Fraser. 1984. High-gradient magnetic separation (HGMS) in soil clay mineral studies. Clay Miner. 19:771-778.

Santos, M.C.D., A.R. Mermut, and M.R. Ribeiro. 1989. Submicroscopy of clay microaggregates in an Oxisol from Pernambuco, Brazil. Soil Sci. Soc. Am. J. 53:1895-1901.

Sayin, M., A.R. Mermut, and H. Tiessen. 1990. Phosphate sorption-desorption characteristics by magnetically separated soil fractions. Soil Sci. Soc. Am. J. 54:1298-1304.

Schulze, D.G., and J.B. Dixon. 1979. High gradient magnetic separation of iron oxides and other magnetic minerals from soil clays. Soil Sci. Soc. Am. J. 43:793-799.

Senkayi, A.l., J.B. Dixon, and L.R. Hossner. 1981. Transformation of chlorite to smectite through regularly interstratified intermediates. Soil Sci. Soc. Am. J. 45:650-656.

Sneddon, J. 1990. Simplex optimization in atomic spectroscopy. Spectroscopy 5(7):33-36.

Strickler, A., A. Kaplan, and E. Vigh. 1966. Continuous microfractionation of particle mixtures by electrophoresis. Microchem. J. 10:529-544.

Strickler, A. 1967. Continuous particle electrophoresis: A new analytical and preparative capability. Separ. Sci. 2:335-355.

Tamura, T. 1956. Weathering of mixed-layered clays in soils. Clays Clay Miner. 3:413-422.

Troell, E. 1976. The use of sodium hypobromite for the oxidation of organic matter in the mechanical analysis of soils. J. Agric. Sci. (Cambridge) 21:476-483.

Truog, E., J.R. Taylor, Jr., R.W. Simonson, and M.E. Weeks. 1936. Mechanical and mineralogical subdivision of the clay separate of soils. Soil Sci. Soc. Am. Proc. 1:175-179.

Van Langeveld, A.D., S.J. van der Gaast, and D. Eisma. 1978. A comparison of the effectiveness of eight methods for the removal of organic matter from clay. Clays Clay Miner. 26:361-364.

Van Olphen, H. 1987. Dispersion and flocculation. p. 203-236. In A.C.D. Newman (ed.) Chemistry of clays and clay minerals. John Wiley & Sons, New York.

Velasco-Molina, H.A., A.R. Swoboda, and C.L. Godfrey. 1971. Dispersion of soils of different mineralogy in relation to sodium adsorption ratio and electrolytic concentration. Soil Sci. 111:282-287.

Volk, N.J. 1933. Formation of muscovite in soils, and refinements in specific gravity separations. Am. J. Sci. 26:114-126.

Weed, S.B., and L.H. Bowen. 1990. High-gradient magnetic concentration of chlorite and hydroxy-interlayer minerals in soil clays. Soil Sci. Soc. Am. J. 54:274-280.

Whittig, L.D., and W.R. Allardice. 1986. X-ray diffraction techniques. p. 331-362. In A. Klute (ed.) Methods of soil analysis. Part 1. 2nd ed. Agron. Monogr. 9. SSSA, Madison, WI.

Wilson, M.J. 1987. Soil smectites and related interstratified minerals: Recent developments. p. 167-173. In L.G. Schultz et al. (ed.) Proc. Int. Clay Conf., Denver, Colorado. 1985. Clay Miner. Soc., Bloomington, IN.

Woodley, D.J.A., and C.H. Duffel. 1964. A high-tension disc separator. Mining Mag. 111:313-318.

Woolson, E.A., and J.H. Axley. 1969. Clay separation and identification by a density gradient procedure. Soil Sci. Soc. Am. Proc. 33:46-48.

Wunderly, C. 1959. Paper electrophoresis. p. 179-224. In M. Bier (ed.) Electrophoresis theory, methods, and applications. Acad. Press, New York.

9   Quantitative X-Ray Diffraction Analysis of Soils

D.L. Bish
Los Alamos National Laboratory
Los Alamos, New Mexico

The inorganic fraction of most soils consists of a mixture of fine-grained, often poorly crystalline minerals, frequently with one or more amorphous phases. Because of the fine-grained nature of soils, x-ray powder diffraction methods have traditionally been applied in their study. Although chemical, optical, and spectroscopic methods are often applied, for many soils, x-ray diffraction provides the most convenient and unambiguous method for determining mineralogical composition. The first studies of soils by x-ray diffraction methods were in the early 1930s by Hendricks and Fry (1930) and Kelley et al. (1931). Since then, x-ray diffraction has been established as probably the most important technique applied to the study of soil mineralogy.

Qualitative identification of soil components by x-ray powder diffraction has been described by numerous authors, and the methodology for distinguishing between the variety of minerals possible in a soil, particularly clay minerals, is well documented (see, for example, the summaries by Brown & Brindley, 1980; Whittig & Allardice, 1986; Moore & Reynolds, 1989). In contrast, the methods for quantitative mineralogical analysis of soils are less well defined, in part because of the difficulties in quantitatively analyzing poorly crystalline mixtures. Brindley (1980), Moore and Reynolds (1989), and Pevear and Mumpton (1989) presented excellent detailed treatments of quantitative analysis of clay minerals, and the principles described in their papers are directly applicable to quantitative analysis of soils. The fundamentals of quantitative analysis will be briefly reviewed here, with emphasis on a few proven methods and several new and novel techniques.

Specific examples of quantitative phase analysis by XRD are not numerous in the literature, but several successful applications have been published. Brindley (1980) provided an excellent summary of quantitative analysis of clay minerals using x-ray powder diffraction, and the methods and caveats described therein can be equally well applied to any mineral system. Quite often, quantitative analyses have been directed at determining the amounts of only one or two phases, such as the analyses of Parker (1978) for analcime in pumice. Carter et al. (1987) used either an internal standard or the measurement of the mass absorption coefficient (to correct for absorption effects) to determine the amounts of quartz and cristobalite in bentonites. Their results were excellent, with lower limits of detection (LLD) of 0.01 g kg^{-1} for quartz and 0.03 g kg^{-1} for cristobalite. They estimated the lower limit of quantification as a level 10 times the LLD, and absolute errors at the 95% confidence level were less than ± 3.5 g kg^{-1} for both phases. These results for quartz and cristobalite show that much of the conventional wisdom regarding precision and accuracy of analyses and limits of detection is incorrect; with proper methodology, high-precision quantitative analyses can be obtained, with very low detection limits. In another analysis for only two phases, Boski and Herbillon (1988) determined the amounts of hematite and goethite in bauxites using a combination of XRD and heating methods. Comprehensive whole-rock analyses analogous to modal analyses were performed by Davis and Walawender (1982) who compared XRD reference-intensity-ratio analyses with optical results. They obtained good agreement between the two methods for most rocks analyzed, demonstrating that it is possible to obtain accurate modal analyses quickly by XRD. Pawloski (1985) applied the Chung (1974b) external-standard method to volcanic tuffs from the Nevada Test Site, and she obtained reproducible results for the fine-grained rocks. Maniar and Cooke (1987) performed modal analysis of natural and artificial granitic rocks using the quartz present in all samples as an internal standard. They obtained rapid modal analyses with mean standard deviations of 70 g kg^{-1}. Bish and Chipera (1988) reviewed a number of the problems inherent in quantitative analysis of complex mixtures, using results for tuffaceous rocks as examples. Clearly, although quantitative analysis by XRD has not been widely used to obtain modal analysis of soils and rocks, the methods have the potential for quickly and accurately providing useful mineralogical information.

SAMPLE PREPARATION

As with many analytical methods, the preparation of samples for x-ray powder diffraction analysis is one of the most critical steps, and this is particularly so for quantitative analysis of soils. Quantitative results require accurate and precise (i.e., reproducible) intensities, and minimizing systematic sample-related effects is probably more important than accounting for instrumental systematic errors in the collection of accurate and precise diffraction data. Due to the nature of soil materials and differences among modern x-ray diffractometers, several sample preparation problems are commonly encountered. Some of these concern the preparation of the powdered material and others arise during mounting of the powders in a suitable holder.

Particle-Size Requirements

Accurate x-ray diffraction intensities from a powder sample require that the grain (i.e., crystallite) size be small, at least 10 μm and preferably smaller. For intrinsically fine-grained soils, a particle produced by grinding may consist of an aggregate of crystals that are much smaller. For coarse-grained minerals in soils that are ground in the laboratory, grain size is usually equal to crystal size. The problems encountered with coarse grain size, including extinction, particle statistics, and microabsorption, are discussed at length in Bish and Reynolds (1989). Extinction is an important process in thick and nearly perfect crystals, but it is not a common problem with the kinds of fine powders normally dealt with in powder diffraction techniques and will not be discussed here. Particle statistics and microabsorption will be briefly discussed because of their importance in quantitative analysis.

Particle Statistics

Particle statistics can be understood by examining two extremes of crystallite size and distribution. Consider, for example, two megascopic crystals of different minerals that are oriented at random on a substrate and analyzed in a flat-specimen diffractometer. Under most circumstances, no measurable diffraction will occur from these crystals because no lattice planes from either are oriented suitably for diffraction. At the other extreme, if the crystals are finely ground so that a very large number ($> 10^6$) of particles are randomly oriented, then many lattice planes are suitably aligned to produce diffraction peaks as the diffractometer detector scans over the appropriate angles of 2θ. Between the extremes of these two examples, the relative intensities of the diffraction lines from each mineral depart from the theoretical intensity distribution because, by chance, some orientations are missing, and/or some are over-represented because of the finite number of crystals and hence the finite number of possible orientations. This condition is easily detected by a lack of repeatability of the powder diffraction pattern and is well illustrated by quartz data presented by Klug and Alexander (1974, p. 365-368). They showed that reproducible intensities could only be obtained from quartz powders with Cu-Kα radiation when size fractions < 15 μm were used. Clearly, < 325-mesh (44-μm) powders are not sufficiently fine for anything but qualitative measurements. The 5- to 15-μm data quoted by Klug and Alexander (1974) are a good compromise between reliability and particle size, and suggest that 10-μm powders are suitable for applications where a few percent error can be tolerated in the intensities from intermediate-absorbing silicate minerals irradiated by Cu-Kα radiation. It is important to remember, however, that although good results can be obtained using 5- to 15-μm powders of typical silicate minerals, the required sizes for powders of more highly absorbing materials, such as Fe oxides, are much smaller.

Bish and Reynolds (1989) showed, with a few assumptions, that typical sample and instrumental parameters yield approximately 23 000 000 10-μm quartz crystallites ($\mu^* = 91$ cm^{-1}) bathed in an x-ray beam. Clearly, doubling the grain size from 10 μm to 20 μm increases the volume of individual crystallites eightfold, but *decreases* the total number of "active" crystals by the same factor. In addition, use of a different radiation or a different material with a significantly larger linear absorption coefficient can cause a similarly large decrease in the number of "active" crystallites because of the shallower penetration depth. For example, the use of Cr-Kα radiation increases the absorption of quartz by about threefold, thus decreasing by a similar factor the number of crystallites that contribute to the diffraction pattern. Use of the same instrumental and sample parameters with hematite (Fe$_2$O$_3$, $\mu^* = 1151$ cm^{-1}) yields approximately 1 940 000 10-μm hematite crystallites bathed in the Cu-Kα x-ray beam. Simply changing from quartz to hematite decreased the number of "active" crystallites by more than an order of magnitude. Clearly the most practical method of achieving good particle statistics is to reduce the particle size of the powder as much as possible, consistent with present-day grinding technology and minimal crystal-structure damage. Although particle statistics is not as serious a problem with soil samples as it is with more coarsely grained materials, it is important always to consider its importance in quantitative analysis.

Microabsorption

Microabsorption is an important effect that causes two or more phases in a mixture to contribute diffraction intensities that are related to both their relative proportions and their mass absorption coefficients (see Brindley, 1945; de Wolff, 1947; Wilchinsky, 1951; Gonzalez, 1987; Hermann & Ermrich, 1987). The mass absorption coefficient (μ) is defined as the linear absorption coefficient (μ^*) divided by the density (ρ). When microabsorption is significant, corrections are required if the results of the quantitative analysis are to be accurate. Microabsorption occurs in coarse powders if the constituents have very different mass absorption coefficients, and it causes underestimation of the highly absorbing constituents. The effect diminishes with decreases in particle size and the powder mixture behaves ideally if the particle size is less than a critical value that is dictated by the phase with the highest absorption coefficient. Absorption coefficients depend on x-ray wavelength as well as chemical composition, so a powder suitable for quantitative analysis with one wavelength may not be suitable for another. For soil mineralogists, the most serious manifestations of microabsorption arise from the use of Cu-Kα radiation with minerals rich in Fe. Results given by Bish and Reynolds (1989) show that accurate quantitative analysis cannot be made for Fe-rich phases in common rocks using Cu-Kα radiation when grain sizes exceed 10 μm. Iron and Co radiation are preferred because their Kα lines lie to the low-μ side of the Fe absorption edge and, consequently, the mass absorption contrasts are greatly reduced. Microabsorption is seldom a problem for clay minerals because of their intrinsically small particle size. Calculations performed by R. C. Reynolds, Jr. (personal communication) indicate that for kaolinite-glauconite and kaolinite-chlorite mixtures, grain thicknesses of 1 μm produce errors of only a few percentages in randomly oriented mixtures. With oriented micaceous or clay-mineral aggregates microabsorption becomes markedly 2θ-dependent. The effect is enhanced at low diffraction angles because the apparent thickness of the grains is greater than at high diffraction angles.

Sample Grinding

Effects on Analysis

Probably the most common fault in sample preparation for quantitative x-ray diffraction analysis is the use of samples that are insufficiently ground, and in fact, many authors incorrectly recommend the use of powder that has passed through a 325-mesh sieve (<44 μm). As discussed by several authors (see Bish & Reynolds, 1989, for a summary) materials of this size are much too coarse for most quantitative powder-diffraction experiments using Cu-Kα radiation for a variety of reasons, including extinction, particle statistics, and microabsorption effects. Using powder that is too coarse can give rise to inaccurate and imprecise intensities, and spurious or unusually intense and sharp reflections will often be obtained from material that is poorly ground. Fortunately, many soil minerals occur naturally with crystallite sizes smaller than is usually required for quantitative analysis.

The overgrinding of materials, particularly those that are very soft (such as many clays), can present other problems. However, overgrinding is seldom a problem when appropriate grinding procedures are followed, particularly if samples are ground under a liquid. If overgrinding is encountered, the method of grinding may be changed or the ground sample may be periodically sieved during grinding to remove material of sufficiently small size (being sure to recombine the complete sample after grinding multicomponent mixtures). The eventual effect of grinding a material excessively is to broaden some (anisotropic grinding effects) or all reflections and to produce small amounts of amorphous surface layers (see Nakamura et al., 1989). Materials will usually begin to show some particle-size related broadening when individual crystallites are smaller than about 100 nm. Strain-related broadening, however, can be evident in some materials having crystallites much larger than this. It is a common misconception that x-ray reflection intensities are a function of peak breadth. Although the peak height decreases and the breadth or full-width-at-half-maximum (FWHM) of observed reflections increases with decreasing crystallite size, the *integrated* intensities remain the same. Thus it is imperative that integrated intensities be used when performing conventional quantitative analyses using individual peak intensities. It is also important to realize that the position of a broadened reflection may be shifted from the position of the unbroadened reflection, particularly for low-angle reflections, due to variations in the Lorentz-polarization factor and possibly the structure factor across the broad reflection (Reynolds, 1968; Ross, 1968). Although not common with soil samples, numerous authors have shown that overgrinding can also induce phase transformations and solid-state reactions. Jenkins et al. (1986) summarized some of the reactions that have been observed, including the transformation of calcite to aragonite, of wurtzite to sphalerite, of kaolinite to mullite, and of vaterite to calcite. Such reactions seldom if ever occur if correct grinding procedures are used.

A final aspect of sample grinding that may be important in some multicomponent samples is differentiation of particle sizes due to differential grinding. For example, a soil containing quartz, feldspars, mica, and smectite will probably yield a distribution of particle sizes after grinding, the softer materials producing smaller particles than the hard materials. This size distribution results both from the greater resistance of the hard minerals to grinding by the grinding device and from the grinding of softer materials by the harder materials in the sample. In addition, platy materials such as micas are poorly ground by some methods. This effect can be minimized by sieving during grinding and recombining the completely ground sample, or it can be ignored in many cases as long as the harder materials are sufficiently ground and the softer materials are not overground. In most cases, wet grinding of such mixtures produces appropriate particle sizes of the hard minerals while not overgrinding the softer, generally finer-grained materials.

Methods of Grinding and Disaggregation

There are numerous methods and equipment available to reduce the particle sizes of different types of samples, ranging from a simple hand mortar and pestle or percussion mortar to automatic grinding machines that may cost as much as $10 000. As mentioned above, there are several things to consider when grinding samples, such as not adversely affecting the materials being ground (e.g., by overgrinding or contaminating the sample) and obtaining a relatively homogeneous particle size. Although most methods of grinding introduce low levels of contamination that may interfere with chemical measurements, contamination is, in general, not a problem for diffraction measurements. Commercially available hand mortars and pestles made of agate, mullite, or corundum are acceptable for general grinding. On the other hand, porcelain or iron mortars are generally too soft for grinding mineral samples and can introduce considerable contamination. Grinders made of tungsten carbide are usually excellent for reducing particle sizes, although some tungsten carbide contamination can occur in powders after grinding silicate samples in a shatterbox. One of the most important requirements in grinding samples is the use of a lubricating and cooling liquid such as alcohol, acetone, or water. Lengthy grinding of a liquid-saturated sample will result in the desired reduction in particle size with little or no structural damage to the sample. Several automatic grinders, such as ball mills or shatterboxes, that do not easily accommodate liquids, often produce significant amounts of damage to the sample. The most significant drawback to hand grinding in a conventional mortar and pestle is the difficulty in obtaining sufficiently small particle sizes. The automatic "mortar and pestle" made by Retsch and marketed in the USA by Brinkmann gives very good results with hard silicates and oxides. This unit consists of a mortar with a cylindrical cavity and a cylindrical pestle with a step on the bottom and a cam ground on the side. During operation, both the mortar and pestle rotate, usually at different speeds. Mortars are available in agate and tungsten carbide (and others). This device can routinely produce powders with average particle sizes of ~ 3 μm, as measured by a commercial particle-size analyzer, after grinding under acetone for 12 min. Contrary to reports in the mineralogical literature on the preparation of <10-μm powders, the author has not been able to prepare homogeneous powders of this particle size using a hand mortar and pestle without considerable effort. Typical literature reports of the production of particle sizes below 10 μm do not specify if and how particle sizes were measured, and thus the quoted particle sizes are suspect. Another device that is able to produce very fine particle sizes easily is the McCrone micronizing mill, yielding material averaging <10 μm.

Sieving is an effective method for isolating particles of the appropriate size but should not be used with mixtures unless all of the mixture is passed through the sieve. However, it is difficult manually to sieve powders much finer than 400 mesh (38 μm), although automatic devices are available (e.g., sonic sifter) that perform well down to micrometer sizes if static charge and agglomeration are limited.

Some materials, notably micaceous minerals, are not amenable to grinding by conventional methods. Numerous authors have obtained good results using a metal file on such materials, and S. W. Bailey (personal communication) has determined that filing micas and chlorites produces much less layer-stacking disorder than attempting to grind these minerals by conventional methods. The common laboratory or kitchen blender (Waring) has also been used to reduce the particle size of micaceous materials, although the final particle size achieved is usually relatively large. It is possible that the newer high-speed laboratory homogenizers, such as those marketed by Brinkmann, may prove superior to conventional blenders at reducing the particle size of platy or fibrous materials. Good results can be obtained with fibrous materials such as chrysotile using a device commonly used to grind biologic materials known as a Thomas-Wiley mill. As with the laboratory blender, this device produces material of relatively large particle size, but it is effective at cutting fibrous materials into a powder that can be ground further by other methods.

Ultrasonic vibration is a method often used to disaggregate loosely consolidated materials such as many soils. Both ultrasonic baths and high-power probes are available. The high-power ultrasonic probes are quite effective at disaggregating even slightly consolidated soils and are particularly useful for suspending clay minerals. However, these devices can heat the suspension, an effect that may be undesirable with some samples. Cuphorn sonifiers eliminate many of the problems associated with ultrasonic probes. The ultrasonic baths are relatively ineffective at disaggregating all but poorly consolidated soils due to their low power.

Sample Mounting

The factors to be considered when mounting powdered samples for quantitative analysis by x-ray diffraction have been thoroughly reviewed (see Bish & Reynolds, 1989), and only the most important points will be outlined here. Perhaps the most common mounting error is the use of sample holders with sample areas of insufficient size. Because the irradiated area increases with decreasing diffraction angle, 2θ, it is important to ensure that the x-ray beam is fully within the sample at the lowest angle of interest (see Parrish et al., 1966; Reynolds, 1989a, p. 10-12). This is particularly important in examinations of soils containing layer silicates, such as chlorites, micas, and smectites, that have low-angle reflections. It is important to know accurately the size and shape of the irradiated area at several angles and with a variety of divergence slits for the diffractometer used. This information can be determined using a fluorescent screen in place of a sample or it can be calculated by simple trigonometry using the appropriate values for the goniometer radius and the angular divergence of the divergence slit (Reynolds, 1989a). If the sample length is too short to contain the beam at the lowest angle of interest, intensities from reflections occurring at angles where the beam is not fully within the sample will be relatively weak compared with higher-angle reflections where the x-ray beam is within the sample. This effect is particularly important during quantitative analysis, when accurate intensities are required. It is also beneficial to ensure that the width of the sample is sufficient to contain the beam completely. Using sample widths that are too small can produce diffraction from the sample holder and give rise to spurious diffraction peaks or increased background in addition to altering relative intensities from the sample.

When mounting powdered soil samples for diffraction analysis, the sample must be thick enough so that essentially all of the incident x-ray beam interacts with the sample and does not pass through it. The minimum thickness depends on sample packing density and the linear absorption coefficients of the components, but typical sample mounts at least several hundred μm thick satisfy the requirement of infinite thickness (see Reynolds, 1989a, p. 12-15).

Sample Packing

Attention to sample packing is important because diffraction occurs from a volume of sample, not just the surface. Loose sample packing can give rise to transparency effects in addition to those related intrinsically to the sample, i.e., the sample may not be infinitely thick. Systematic errors related to sample packing and thickness are relatively unimportant for materials having high linear absorption coefficients, such as hematite (Fe_2O_3) with Cu-Kα radiation, since the incident x-ray beam penetrates little more than a single layer of crystallites on the surface of the sample. However, as discussed above, materials with high absorption coefficients have other inherent problems.

Sample Surface

Ideally the surface of a mount should be flat, with no roughness or curvature, and not tilted in any direction. Any roughness or curvature has the potential to produce systematic deviations in the positions and breadths of observed reflections related to sample-height-displacement and flat-specimen errors (Klug & Alexander, 1974, p. 302). Unless specifically desired, for example in a study of preferred orientation, sample tilting should be avoided because it changes the 2:1 angular relationship between the receiving slit and the sample surface and can give rise to systematic errors in intensity and peak breadth (de Wolff et al., 1959).

Randomly Oriented Powder Mounts

One of the most serious problems in quantitative analysis arises from the tendency of most sample particles to orient preferentially. Numerous methods have been proposed to minimize preferred orientation and these are summarized in several publications (e.g., Klug & Alexander, 1974, p. 364-376; Smith & Barrett, 1979; Bish & Reynolds, 1989). Procedures for minimizing preferred orientation or producing a randomly oriented powder mount for diffractometry include side- or back-packing a powder into a cavity mount, mixing a powder sample with a filler material (usually amorphous) (e.g., Calvert et al., 1983), mixing a powder with a viscous material, dispersion in a binder (e.g., Bloss et al., 1967), and more exotic methods such as spray drying (e.g., Florke & Saalfeld, 1955; Jonas & Kuykendall, 1966; Hughes & Bohor, 1970; Smith et al., 1979a,b), tubular aerosol suspension (Davis, 1986), or liquid-phase spherical agglomeration (LPSA) (Calvert & Sirianni, 1980). Most of these procedures are relatively simple and produce sample mounts with varying degrees of randomness. One of the most important prerequisites for preparing an orientation-free sample mount is the use of material of sufficiently fine particle size. Materials with good cleavage will orient much more effectively when in the size range greater than about 30 μm than they will when ground to <5 μm. Some of the more involved methods such as spray drying and LPSA require special equipment and are not well-documented in the mineralogical or soil literature. For clay-rich soil samples, the spray-drying method probably holds the greatest potential for minimizing preferred orientation. However, in the absence of a spray-drying apparatus, it is often acceptable to ensure that the degree of preferred orientation in all samples to be analyzed is the same as that in standard samples, thereby implicitly correcting for preferred-orientation effects (see Bish & Chipera, 1988; Reynolds, 1989a, p. 8). This method of using standards close to or identical to the samples to be analyzed has the potential to correct implicitly for a variety of problems, including microabsorption, preferred orientation, compositional variations, and even particle statistics. These and other methods of sample preparation for clay-containing samples have been evaluated by Gibbs (1965), and workers in clay mineralogy will profit by a study of that work. There is still a finite number of laboratories that continue to use sample preparation methods that are entirely unsuitable for quantitative analysis, yet their results are reported with confidence. Demonstrations of precision do not guarantee accuracy.

In addition to sample preparation methods, several instrumental techniques are reported to minimize preferred orientation. These include the use of a sample spinner and slightly rocking the sample (by ~1°) about the θ axis of the diffractometer. The former improves particle statistics but does little to eliminate preferred orientation of crystallites parallel to the sample surface (Parrish & Huang, 1983). The latter will probably be of little help in eliminating all but small degrees of preferred orientation. In addition, de Wolff et al. (1959) commented that the rocking method, primarily used to improve particle statistics, must be used with care because it changes the 2:1 angular relationship between the receiving slit and the specimen surface. Even small changes in the 2:1 relationship will cause line broadening and a reduction in peak intensity. Systematic intensity errors may result because this effect increases with decreasing 2θ. Although it is usually preferable to minimize orientation during sample preparation, preferred orientation can be qualitatively evaluated using a conventional θ-2θ diffractometer by rocking the sample about the θ axis of the diffractometer (e.g., Yukino & Uno, 1986). Because this is a very specialized area of powder diffraction, more closely related to texture studies of materials, measurement of orientation will not be addressed in this chapter.

CONVENTIONAL QUANTITATIVE ANALYSIS

Most conventional methods of quantitative analysis today are based on the familiar equation given in Klug and Alexander (1974, Eq. 7-10) that relates the intensity of a reflection from a phase in a mixture (I_i) to its concentration and density and to the mass absorption coefficient of the unknown mixture:

$$I_i = K_i \frac{X_i}{\rho_i \mu_m} ,$$

[1]

where X_i is the weight fraction of component i, K_i is a constant depending on the diffractometer and component i, ρ_i is the density of component i, and μ_m is the mass absorption coefficient of the mixture. The traditional problem in quantitative analysis is to eliminate μ_m and determine K_i, usually by "flushing out" the matrix factors (Chung, 1974a).

Reference Intensity Ratio Method

One of the easiest means to remove the effects of matrix factors is to use the Reference Intensity Ratio (RIR) concept. A general RIR (K in Eq.. [1]) can be defined using any internal standard and relative intensities (e.g., Snyder & Bish, 1989), but a more convenient definition has evolved that requires the use of integrated intensities and corundum (α-Al_2O_3) as the internal standard. This definition, often called "I over I corundum" or I/I_c, was first proposed by Visser and de Wolff (1964). The I/I_c is the ratio of the integrated intensity of the strongest line of a given phase to that of the strongest line of corundum for a 1:1 mixture by weight. When I/I_c values are used with a known amount of corundum internal standard, the mass absorption coefficients of the pure samples and of the unknown mixture cancel and the following equation can be derived (Chung, 1974a; Snyder & Bish, 1989)

$$X_i = \frac{X_c}{k_i} \frac{I_i}{I_c} ,$$

[2]

where k_i is the I/I_c value for phase i, X_c is the amount of added corundum internal standard, I_i is the intensity diffracted by phase i in the mixture, and I_c is the intensity diffracted by the strongest line of corundum in the mixture. This equation can be easily modified for the use of I/I_c values for other than the strongest lines or for the use of RIR values for any internal standard (Snyder & Bish, 1989). This method of quantitative analysis can provide accurate results if multiple lines are used, if the RIR values have been determined on standard materials as close as possible to the components of the unknown mixture, and if all analyses are done on identical or similarly equipped diffractometers. Because the method does not constrain the sum of components to 100%, it is the preferred method for use with samples potentially containing amorphous components. In ideal circumstances, amounts of amorphous components can be determined by difference from 100%. Significant other benefits of this method are that it is capable of implicitly correcting for some serious systematic errors. The use of multiple lines of appropriate standards can partly compensate for preferred orientation effects. Also, extinction effects can be compensated for if a given standard is of approximately the same crystallite size as that phase in the mixture. In addition, the use of standard mixtures similar in composition to the unknowns can partly correct for microabsorption effects that are particularly important with mixtures of Fe- or Mn-rich minerals and silicate minerals when using Cu-Kα radiation.

Although often only a single RIR ratio is measured, for improved accuracy, it is preferable to use multiple sample mounts and to measure several lines from the sample (so that the RIR value is not only sample-specific but depends on $hk\ell$). This will allow assessment of reproducibility and of the effects of sample properties such as preferred orientation. Whenever RIR values are determined, a diffractometer and sample mount should be used to provide constant irradiated *volume* to below the lowest angle measured, i.e., the sample mount must be sufficiently long and θ-compensating slits should not be used unless intensities are corrected to constant-volume conditions.

A convenient modification of the internal-standard method was first derived by Chung (1974b). If all phases are identified with known RIR values, then we can additionally constrain the sum of all phases in the mixture to equal 1.0 (100%). With this added constraint, Chung (1974b) showed that Eq. [2] reduces to

$$X_i = \left[\frac{k_i}{I_i} \sum_{i=1}^{n} \frac{I_i}{k_i} \right]^{-1},$$

[3]

where n is the number of phases in the mixture. This method has the advantage that it is fast, easy, and relatively accurate if RIR values are measured as recommended above. However, errors in analyses for individual components are cumulative, and unidentified or amorphous components invalidate this method, although the ratios of the crystalline components will be correct.

Absorption-Diffraction Method

An additional quantitative analysis method can be derived from Eq. [1] by writing this equation for phase α in an unknown mixture and for pure phase α,

$$\frac{I_{i\alpha}}{I_{i\alpha}^{0}} = \frac{\mu_\alpha}{\mu_m} X_\alpha,$$

[4]

where $I_{i\alpha}$ is the intensity of line i for phase α in the mixture, $I_{i\,\alpha}^{0}$ is the intensity of line i in pure phase α, and μ_α is the mass absorption coefficient of pure phase α. This equation is the basis for the absorption-diffraction method which can be applied in several special situations (Leroux et al., 1953). If the chemical composition of the pure phase and of the mixture are known, X_α can be determined simply by measuring $I_{i\alpha}$ and $I_{i\,\alpha}^{0}$ if care is taken to avoid sample-mounting differences and instrumental drift. A comparable procedure can be used if the mass absorption coefficients can be experimentally determined, for example, through x-ray transmission measurements with known sample loadings or through measurement of Compton-scattered radiation. Unfortunately the experimental determination of mass absorption coefficients is usually difficult and error prone. Davis and Johnson (1989), Reynolds (1989a), and Chipera and Bish (1991) described methods for determining the mass absorption coefficient of samples. If samples are being examined for which μ_α and μ_m are identical, e.g., mixtures of polymorphs such as cristobalite and quartz, Eq. [4] becomes a simple ratio of intensities. The same simplification holds when examining lightly loaded filters on which individual particles do not shade each other. The absorption-diffraction method may also be used to conduct quantitative analyses on two-component mixtures with appropriate modification of Eq. [4] (Snyder & Bish, 1989).

Method of Known Additions

Several other quantitative analysis procedures have been described that may be useful in the analysis of soil materials. The method of known additions (Brindley, 1980; Snyder & Bish, 1989) is useful for determining the amount of one phase in a mixture. The method involves adding a known amount of the phase to be determined to the unknown mixture and measuring the intensities from this phase before and after the addition. By using several values of addition, the original value of the unknown phase can be determined by extrapolating to zero addition, thereby eliminating a correction for matrix effects. This method may also be performed using ratios of the intensity of a reflection from the unknown to the intensity of a given reflection from any other phase in the mixture as a function of addition.

Standards

Probably the most important factor to consider when standardizing is how well the standards match the minerals in a sample. Samples should be as free as possible from preferred orientation, extinction, and microabsorption effects, or these effects should at least be constant from standard to unknown. The problem of choosing representative standards has been emphasized often (e.g., Brindley, 1980), and factors other than simply the mineralogic identity are particularly important in the analysis of soils. These factors include chemical composition, variations in degree of preferred orientation, stacking and/or structural disorder, and the presence and amount of interstratification. Often, appropriate standards for many of the phases in a given soil sample seem to be unavailable. The variation of some or all of these complicating factors can be compensated for somewhat by: (i) determining approximate chemical compositions of the phases in a mixture and using appropriate observed or calculated standard data; (ii) using purified standards separated from the mixtures to be analyzed; (iii) determining the amount of any interstratification present in the clay minerals; (iv) using reflections that are relatively insensitive to chemical variations (e.g., Reynolds, 1989a); and (v) correcting or compensating for differences in preferred orientation. Reynolds (1989a) outlined a number of these potential problems and recommended procedures to minimize their effects on quantitative analyses. Any reader beginning quantitative analyses of soil materials should consult his chapter and should be aware that the factors discussed here also apply to amorphous materials and to many common soil oxide minerals. Although an examination of the literature on quantitative analysis of soils and clay mineral mixtures suggests that many of these problems are virtually insurmountable, the information in Reynolds (1989a) provides a more optimistic view. If the variations between samples and standards are understood, they can be accounted for relatively accurately.

Intensity Measurement

When measuring RIR values or when obtaining *any* intensity data on individual reflections for quantitative analysis *integrated intensities* should be used. The use of integrated intensities improves accuracy and precision and compensates for the effects of increasing peak width due to structural disorder and decreasing crystallite size. As Reynolds (1989a) emphasized, poor crystallinity generally causes changes in peak heights but not in integrated peak intensities. Obtaining integrated intensities for use with quantitative analyses is more involved than simply measuring relative peak heights from a diffraction pattern, and several methods exist for doing so. For manual diffraction systems providing graphical output on a strip-chart recorder, either cut-and-weigh or planimeter methods may be used. Either of these methods can provide precise peak areas, but as with any method of this type, the choice of background is crucial. The tendency with the manual methods is to choose a background that is too high, thereby providing an integrated intensity that is too low. Automated diffraction systems, on the other hand, yield digital intensity information and allow the use of computerized first- and second-derivative peak-finding routines as well as graphical-integration and profile-fitting (profile-refinement) routines. Most modern first- and second-derivative routines work quite well with diffraction patterns containing peaks of consistent peak widths and shapes. When patterns typical of soils are analyzed, however, these routines frequently yield inaccurate integrated intensities. Graphical-integration methods are usually interactive computer-based routines that perform a simple integration between two points on a diffraction trace, using a background defined by the user. As with the manual methods, the choice of background in graphical-integration methods is equally important. Profile-fitting methods fit observed peaks with a calculated profile based on a given mathematical function, and these methods are useful for separating the contributions from two or more reflections to a single broadened peak. Profile fitting has the advantage that the parameters describing the background and the peaks in question are usually refined by a least-squares process, thus minimizing the errors induced by choosing inaccurate starting values. Profile-fitting is often referred to in the literature as "deconvolution," but it in no way resembles and is mathematically distinct from deconvolution. The term deconvolution should not be used in connection with separation or decomposition of overlapping reflections [see Chapter 10, Jones & Malik (1994); and Jones (1989), for excellent summaries of profile refinement applied to soil minerals]. Chipera and Bish (1988) and Jones (1989) outlined some of the pitfalls that may be encountered when performing profile refinement of clay mineral diffraction patterns. It is important to realize that several different solutions to the problem of decomposing a cluster of overlapping peaks can be obtained, depending on the assumed number of peaks, starting peak heights, peak widths, and peak profiles.

Amorphous Components

Amorphous components occur in many soils and in some instances comprise the majority of the inorganic portion of the soil. Thus, although amorphous components are not usually quantified by x-ray powder diffraction, it is often useful to be able to determine, at least semiquantitatively, the amount of amorphous material in a soil sample. Because amorphous materials present problems different than those encountered with crystalline minerals, their analysis deserves special mention. Quantitative analysis of amorphous components in a mixture has been accomplished by essentially two different methods. The first uses conventional internal- or external-standard methods together with RIR values for the amorphous material. Both RIR determination and quantitative analysis involve accurate measurement of the integrated intensity under the broad, often asymmetric amorphous scattering hump. This method of amorphous component analysis provides poor detection limits, as it is often easy to miss as much as 250 g kg^{-1} of an amorphous material in an otherwise crystalline sample (e.g., Bish & Chipera, 1988). In addition, because accurate measurement of the integrated intensity of the amorphous scattering hump is difficult, partly due to errors in background determination, the errors associated with this method of quantitative amorphous phase analysis are large, often 100% relative.

The second common method of quantitative analysis of amorphous components is to use the internal standard method, determining the total crystalline component directly and determining the total amorphous component by difference from 100%. If the internal standard analysis is done carefully, this method can provide better detection limits and precision than the above method, although the errors on the amount of amorphous material will be correlated with those for the total crystalline material.

Davis and Johnson (1982) described an unusual method for determining the amounts of individual crystalline components plus the amounts of *two different* amorphous components. If the total amount of crystalline material is known, for example from an internal-standard analysis or from optical determinations, and the mass absorption coefficients are known for the total sample, for the crystalline material, and for both of the amorphous components, the amounts of two individual amorphous components can be calculated. As with other methods of amorphous phase analysis, the errors associated with the amorphous components are large, often approaching 100% relative.

SIMULTANEOUS LINEAR EQUATIONS METHODS

Although this discussion is concerned with quantitative analysis by x-ray powder diffraction, it is worthwhile to mention associated methods that allow coupling of diffraction data with other quantitative information, such as chemical analyses, cation-exchange capacities, or thermogravimetric data. The use of additional information on a sample can provide more accurate quantitative analyses and better estimates of precision, without necessarily increasing the number of unknowns. The original and general formulation of this method was described by Copeland and Bragg (1958), and one of the first applications in soil mineralogy (McNeal & Sansoterra, 1964) used linear programming by constraining each component property to a fixed value. Hussey (1972) used a modification of the McNeal and Sansoterra method and considered the weight loss from 300 to 950 °C, the Ca/Mg cation-exchange capacity (CEC), and K/NH$_4$ CEC as observed properties. His method differed from that used by McNeal and Sansoterra (1964) in that it allowed a range of component properties rather than assigning a single, fixed value, i.e., he allowed for variations in the compositions of component phases. Johnson et al. (1985) further described Hussey's method and evaluated the different ways of performing analyses and constraining observations. Their analyses usually required that the number of individual properties (i.e., analyses for individual elements, weight losses, CEC's) equal the number of components analyzed and that the minimum number of samples must equal the number of properties measured. In a typical analysis, they set limits on the allowed ranges of component properties, e.g., by defining the maximum chemical variability a particular phase can exhibit. Using this method, they calculated component proportions and component property values for soil clays and sediments. This is potentially a very valuable method because it can provide not only quantitative phase analyses but information on various properties of the individual components. Such property information is difficult if not impossible to obtain in other ways for many fine-grained materials.

Several analogous chemical mass-balance methods have been presented, and methods similar to Hussey's have been presented by Hodgson and Dudeney (1984), Braun (1986), and Slaughter (1989). Braun's method considers compositional variations in the minerals in a mixture together with uncertainties in the x-ray diffraction data, an important improvement. Braun presented an example in which chemical information alone was used to estimate mineral abundances in a mixture of montmorillonite, plagioclase, quartz, and opal-cristobalite. As expected, he determined the amounts of montmorillonite and feldspar but only the sum of quartz and opal-cristobalite since chemical data were the only observations used. The use of x-ray diffraction data would have facilitated a complete phase analysis. These examples serve to illustrate the utility of the linear programming methods, and these techniques should prove particularly valuable in soil systems. The complete presentation of the mathematical theory behind these methods is beyond the scope of this paper, and the interested reader is referred to the above papers, particularly Braun (1986) and Slaughter (1989).

FULL-PATTERN FITTING METHODS

With the availability of automated x-ray powder diffractometers, digital diffraction data are now routinely available on computers and can be analyzed by a variety of numerical techniques. Complete digital diffraction patterns provide the opportunity to perform quantitative phase analysis using all data in a given pattern rather than considering only one or a few of the strongest reflections. As the name implies, full-pattern methods involve fitting the entire diffraction pattern, often including the background, with a synthetic diffraction pattern. This synthetic diffraction pattern can either be calculated from crystal structure data (Bish & Howard, 1986; Hill & Howard, 1987; Bish & Howard, 1988; O'Connor & Raven, 1988) or can be produced from a combination of observed standard diffraction patterns (Brown & Wood, 1985; Smith et al., 1987).

The use of complete diffraction patterns in quantitative phase analysis has several significant advantages over conventional methods that often use only a few isolated reflections or groups of reflections. In addition to eliminating the troublesome requirement of extracting reproducible intensities from complex patterns, the traditional problems of line overlap, primary extinction, and preferred orientation are minimized by using complete diffraction patterns. However, the full-pattern methods share some of the problems that affect conventional quantitative analyses, chief among them being the need to synthesize or calculate a standard pattern that embodies the same chemistry, degree of structural or stacking disorder, and degree of interstratification as the observed pattern.

Two different full-pattern fitting methods have evolved separately and each method has its own advantages and disadvantages. Both methods require the use of a computer and digital diffraction data, and data analysis is more involved than conventional quantitative analysis methods. However, the full-pattern fitting methods have the potential for providing considerable information on samples. In ideal cases, the methods can provide quantitative phase results approaching the precision and accuracy of x-ray fluorescence elemental analyses. Because the two methods are significantly different in approach, they will be treated separately here.

The Rietveld Method

Quantitative phase analysis using calculated patterns is a natural outgrowth of the Rietveld method (Rietveld, 1969), originally developed as a method of refining crystal structures using neutron powder diffraction data. Refinement is conducted by minimizing the sum of the weighted, squared differences between observed and calculated intensities *at every step* in a digital powder pattern. The Rietveld method requires a knowledge of the approximate crystal structure of all phases of interest (not necessarily all phases present) in a mixture, a trivial requirement for most phases today. However, because the method assumes three-dimensional Bragg diffraction, diffraction effects arising from layer-stacking disorder (e.g., two-dimensional diffraction) are not modeled. Therefore, diffraction patterns from disordered or amorphous materials can only be approximated. The input data to a refinement are similar to those required to calculate a diffraction pattern, namely, space-group symmetry, atomic positions, site occupancies, and lattice parameters. In a typical refinement, individual scale factors (related to the weight percentages of each phase) and profile, background, and lattice parameters are varied. In favorable cases, the atomic positions and site occupancies of major phases in a mixture can also be successfully varied, and information on the causes of peak broadening can be determined.

Theory

The methodology involved in quantitative analysis using the Rietveld method is analogous to conventional Rietveld refinement (see Post & Bish, 1989; Snyder & Bish, 1989), and the quantitative analysis theory is identical to that implemented in most conventional quantitative analyses (e.g., Klug & Alexander, 1974; Cullity, 1978). The method consists of fitting the complete experimental diffraction pattern with calculated profiles and backgrounds, and obtaining quantitative phase information from the scale factors for each phase in a mixture.

The integrated intensity of x-rays diffracted by a randomly-oriented infinitely-thick polycrystalline sample in flat-plate geometry utilizing a diffracted-beam monochromator can be written for a particular reflection as

$$I_{hkl} = \left\{ \frac{I_o A \lambda^3}{32\pi r} \left(\frac{\mu_o}{4\pi}\right)^2 \left(\frac{e^4}{m^2}\right) \left(\frac{1}{2\mu^*}\right) \left(\frac{1}{V^2}\right) \left[|F|^2 \rho \, \frac{1+\cos^2 2\theta \cos^2 2\theta_m}{\sin^2\theta\cos\theta} \right] e^{-2M} \right\}_{hkl} , \quad [5]$$

where the subscript *hkl* denotes the dependence of particular terms on the Bragg reflection *hkl*. The term $2\theta_m$ refers to the diffraction angle of the diffracted-beam monochromator crystal, μ^* is the linear absorption coefficient, and the definition of the other terms follows those of Cullity (1978). The constant (K) and variable (R_{hkl}) parameters in Eq. [5] can be separated by defining

$$K = \left(\frac{I_o A \lambda^3}{32\pi r}\right) \left(\frac{\mu_o}{4\pi}\right)^2 \left(\frac{e^4}{m^2}\right) \qquad\qquad [6]$$

and

$$R_{hkl} = \left(\frac{1}{V^2}\right) \left[|F|^2 \rho \left(\frac{1+\cos^2 2\theta \cos^2 2\theta_m}{\sin^2\theta\cos\theta} \right) e^{-2M} \right]_{hkl} . \qquad [7]$$

Eq. [5] can now be written in terms of Eq. [6] and [7] as

$$I_{hkl} = K \left(\frac{1}{2\mu^*} \right) R_{hkl} \cdot$$ [8]

In a mixture, the intensity of the $hk\ell$ reflection from the α phase is given as

$$I_{\alpha,hkl} = C_\alpha K \left(\frac{1}{2\mu_m^*} \right) R_{\alpha,hkl} ,$$ [9]

where C_α is the volume fraction of the α phase and μ_m^* is the linear absorption coefficient of the mixture. In terms of weight fraction, Eq. [9] can be rewritten as

$$I_{\alpha,hkl} = \left(\frac{w_\alpha}{\rho_\alpha} \right) K \left(\frac{\rho_m}{2\mu_m^*} \right) R_{\alpha,hkl} ,$$ [10]

where ρ_m and ρ_α are the densities of the mixture and of pure phase α, respectively.

In Rietveld refinement, the quantity minimized is

$$R = \sum_i w_i \left| y_i(o) - y_i(c) \right|^2 ,$$ [11]

where $y_i(o)$ is the observed intensity, $y_i(c)$ the calculated intensity, and w_i is a weighting factor for the ith step in the data. Therefore, it is more appropriate for Rietveld quantitative analysis to consider the intensity at a given 2θ step rather than for a given reflection. The intensity at a given step is determined by summing the contributions from background and all neighboring Bragg reflections as

$$y_i(c) = S \sum_k \left(p_k L_k \left| F_k \right|^2 G(\Delta\theta_{ik}) P_k \right) + y_{ib}(c) ,$$ [12]

where S is the scale factor, p_k is the multiplicity factor for the kth Bragg reflection, L_k is the combined Lorentz-polarization factor, F_k is the structure factor, θ_k is the Bragg angle for the kth reflection, $G(\Delta\theta_{ik})$ is the reflection-profile function, P_k is the preferred-orientation function, and $y_{ib}(c)$ is the background (Wiles & Young, 1981). The Rietveld scale factor, S, includes all of the constant terms in Eq. [5] and for x-rays can be written

$$S = \frac{K}{V^2 \mu^*} ,$$ [13]

where V is the unit-cell volume and μ^* is the linear absorption coefficient for the sample. For a multiphase mixture, Eq. [12] can be rewritten summing over the p phases in a mixture (e.g., Hill & Howard, 1987) as

$$y_i(c) = \sum_p S_p \sum_k \left(p_{kp} L_{kp} \left| F_{kp} \right|^2 G(\Delta\theta_{ikp}) P_{kp} \right) + y_{ib}(c) .$$ [14]

The scale factor for each phase can now be written

$$S_\alpha = \frac{C_\alpha K}{V_\alpha^2 \mu_m^*} \quad , \qquad \text{[15]}$$

where C_α is the volume fraction of the α phase and $\mu_m{}^*$ is the linear absorption coefficient of the mixture. Recasting Eq. [15] in terms of weight fractions and the mass absorption coefficient of the mixture, μ_m, gives

$$S_\alpha = \frac{W_\alpha K}{(\rho_\alpha V_\alpha^2 \mu_m)} \quad , \qquad \text{[16]}$$

where W_α is the weight fraction of phase α, and ρ_α and V_α are the density and unit-cell volume, respectively, of phase α. Alternatively, the unit-cell volume can be incorporated into the variable, phase-specific parameters as outlined by Bish and Howard (1988).

Therefore, in a Rietveld analysis of a multicomponent mixture, the scale factors contain the desired weight-fraction information. However, the value of K and the sample mass absorption coefficient cannot easily be determined so that an analysis of an unknown sample is usually performed by constraining the sum of the weight fractions of the phases considered to unity. Thus, for a two-phase mixture,

$$W_\alpha = \frac{W_\alpha}{(W_\alpha + W_\beta)} \quad . \qquad \text{[17]}$$

Eq. [17] can be solved for the weight fractions of the α and β phases to yield an expression for the weight fraction of phase α in terms of the scale-factor information determined in the Rietveld analysis

$$W_\alpha = \frac{S_\alpha \rho_\alpha V_\alpha^2}{(S_\alpha \rho_\alpha V_\alpha^2 + S_\beta \rho_\beta V_\beta^2)} \quad . \qquad \text{[18]}$$

In general, the weight fraction for the ith component in a mixture of n phases can be obtained from

$$W_i = \frac{S_i \rho_i V_i^2}{\sum\limits_{j=1}^{n} S_j \rho_j V_j^2} \quad . \qquad \text{[19]}$$

This method is exactly analogous to the adiabatic principle of Chung (1974b) in which reference intensity ratios are measured prior to analysis. Instead of *measuring* reference intensity ratios to put all intensities on an absolute scale, the Rietveld method *calculates* absolute intensities.

A second method of Rietveld quantitative analysis requires that a known weight fraction of a crystalline internal standard be added to the unknown mixture. The internal standard can be any well-crystallized material that is readily available in pure form. Bish and Howard (1988) used Si, but corundum appears to be preferable (1.0-μm metallurgical-grade corundum). If W_α is known, then by rearranging Eq. [16] an additional parameter (C) can be evaluated from the internal standard:

$$C = \frac{S_\alpha \rho_\alpha V_\alpha^2}{W_\alpha} = \frac{K}{\mu_m} \quad . \tag{20}$$

This parameter can then be used to determine the weight fractions for other phases in the sample. For example, the weight fraction for the β phase is determined by

$$W_\beta = \frac{S_\beta \rho_\beta V_\beta^2}{C} \quad . \tag{21}$$

where S_β is a refined parameter, ρ_β can be calculated from the composition and unit-cell volume (V_β) of phase β, and C is determined using an internal standard and Eq. [20]. Therefore, the weight fraction of phase β (W_β) can be easily determined. This second method does not constrain the sum of the weight fractions, as does the first method, and it is analogous to the internal-standard method described above. The total weight fraction of any amorphous components can also be determined with this method if the amorphous profile can be fitted with the Rietveld background polynomial. The difference between the sum of the weight fractions of the crystalline components and unity is the total weight fraction of the amorphous components. This variation of the Rietveld method of quantitative analysis is particularly valuable for soils as it can treat samples containing amorphous phases and it can provide additional information on the nature of peak broadening for the major components (e.g., Bish & Ebinger, 1989; Ebinger & Bish, 1990).

O'Connor and Raven (1988) used this method in slightly modified form, instead choosing: (i) to determine the constant parameter, K, from a single sample; (ii) to use the refined cell parameters and cell contents to evaluate density and volume; and (iii) to calculate mass absorption coefficients using the known compositions of the two phases in their single sample. In light of their analysis of a 50:50 quartz/corundum mixture, in which they concluded that their quartz contained 180 g kg[-1] amorphous component, it appears that some pitfalls may exist with this approach, primarily because of difficulties in accurately determining K.

Applications

The Rietveld method of analysis provides numerous advantages over conventional quantitative analysis methods for mixtures of crystalline phases. Since the method uses a pattern-fitting algorithm, all lines for each phase are explicitly considered, and even severely overlapped lines are not problematic. The use of all lines in a pattern minimizes the uncertainty in the derived weight fractions and the effects of nonlinear detection systems. The effects of primary extinction are also minimized, as all reflections from each phase are used in the analysis rather than just the strongest ones. Because each phase in a mixture must be explicitly included in the analysis, failure to consider a phase will yield obvious differences between the observed and calculated diffraction patterns. The Rietveld quantitative analysis method can also be used in the traditional manner for refining the structural parameters of each phase in the mixture. Thus, information obtained by this method can include atom positions, site occupancies, and precise lattice parameters, in addition to information on the nature of peak broadening. This feature allows one to calculate tailor-made standards to match the chemistry of the phases in the mixture to be analyzed. Some soils may not be amenable to quantitative analysis using this method because of their poorly crystalline nature and the occurrence of amorphous phases, although the Rietveld method has been successfully applied to goethite samples with crystallite sizes of ~10 nm (Bish & Ebinger, 1989; Ebinger & Bish, 1990). However, the method, at present, does *not* include the ability to model any type of random interstratification or layer-stacking disorder leading to two-dimensional diffraction effects. In addition, available analytical profile functions do not model the asymmetric peaks often resulting from these diffraction effects. These limitations will probably affect the application of the Rietveld method to soils containing significant amounts of disordered clay minerals or silica phases such as opal-CT.

Preferred orientation of crystallites, a problem inherent in conventional quantitative analyses, is much less of a problem with full-pattern methods. The Rietveld method uses all classes of reflections and orientation effects tend to cancel. In addition, the method presents the opportunity to correct for preferred orientation using the March function (Dollase, 1986). It is important to note that older preferred-orientation corrections do not perform properly with multi-phase refinements as none is normalized to unit integral. Because the March function is normalized to unit integral, changes in preferred-orientation correction are not reflected in changes in scale factor, and the correction can be used to improve quantitative analyses.

Few applications of Rietveld quantitative analyses have been published, particularly for natural samples, although the power and versatility of the method suggest that it will be widely applied to soil and geologic samples in the future. Hill and Howard (1987), Bish and Chipera (1988), Bish and Howard (1988), and Bish and Post (1988) have presented examples of quantitative analysis of multicomponent mixtures, and the method has been successfully applied to mixtures containing up to 10 components.

Bish and Post (1988) presented a thorough discussion of results on natural samples illustrating the power of the Rietveld quantitative analysis method. They applied the method to a variety of samples, including synthetic 50:50 mixtures, natural samples containing two volcanic feldspars, a mixture of aragonite and two distinct calcites, and a standard granite (U.S. Geological Survey, G-1). Their data were obtained using Cu-Kα x-rays on a conventional automated powder diffractometer. Results for the binary mixtures were excellent with the exception of hematite-corundum due to apparent microabsorption problems. Results for the 50:50 quartz-corundum mixture, an example of relatively simple, well-crystallized materials, gave a composition of 498 g quartz per kilogram and 502 g corundum per kilogram and cell parameters of $a = 0.49110(1)$ and $c = 0.54021(4)$ nm for quartz and $a = 0.47565(1)$ and $c = 1.2984(1)$ nm for corundum. These cell parameters are of high precision and compare well with literature values for quartz and corundum. The plot of the observed and calculated patterns ($R_{wp} = 18.8\%$) for this mixture is shown in Fig. 9-1. The R_{wp}, the weighted profile residual, is generally formulated as follows

$$R_{wp} = \left\{ \frac{\sum_i w_i [y_i(o) - y_i(c)]^2}{\sum_i w_i y_i(o)^2} \right\}^{0.5} \qquad [22]$$

where w_i is the weight assigned the intensity at each step and $y_i(o)$ and $y_i(c)$ are the observed and calculated intensities, respectively, at the ith step. The R_{wp} value for a Rietveld refinement is usually much larger than R values for conventional refinements using single-crystal x-ray or neutron diffraction data.

The 50:50 clinoptilolite/corundum mixture illustrates one of the most significant advantages of the method in analyzing mixtures of complex materials, the ability to explicitly accommodate severely overlapping reflections. Figure 9-2 is the plot of the observed and calculated diffraction patterns ($R_{wp} = 14.9\%$) for this mixture, illustrating the very large number of reflections (>300) used in the analysis. This analysis yielded 501 g clinoptilolite per kilogram and 499 g corundum per kilogram--excellent results, particularly when the severely overlapped and complex nature of the clinoptilolite pattern is considered. Cell parameters for clinoptilolite were $a = 1.76377(1)$, $b = 1.7962(1)$, $c = 0.74002(1)$ nm, and $\beta = 116.221(1)^\circ$, and those for corundum were $a = 0.47584(1)$ and $c = 1.2990(1)$ nm. Figure 9-3 is an expansion of the region between 21.0 and 25.0 °2θ illustrating the significant peak overlap in this region. The most intense doublet in this region is composed of six independent reflections, a situation that would make indexing this pattern or measuring individual integrated intensities difficult even when applying techniques such as profile refinement (Chipera & Bish, 1988; Howard & Preston, 1989; Jones, 1989). This overlap would preclude highly precise and accurate unit-cell refinement by conventional methods, but the Rietveld method readily yielded very precise cell parameters.

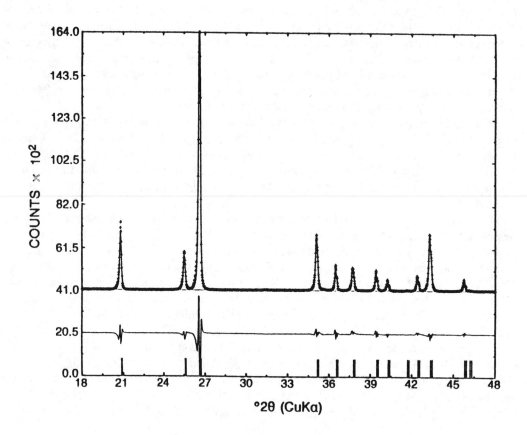

Fig. 9-1. Observed (pluses) and calculated (solid line) diffraction patterns for a 50:50 (by wt.) quartz/corundum mixture. The lower curve shows the difference between observed and calculated patterns, and vertical marks at the bottom of the plot indicate the positions of allowed Cu-Kα_1 and -Kα_2 reflections. The observed and calculated patterns have been displaced upwards for clarity.

The 50:50 biotite/corundum mixture is an example of the effects of preferred orientation on Rietveld quantitative analyses that can be expected whenever platy materials such as layer silicates are examined. Figure 9-4 is the plot of the observed and calculated data, illustrating the very poor fit obtained in this analysis due to preferred orientation (R_{wp} = 32.2%). In spite of the poor fit and the preferred orientation, quantitative results were acceptable (475 g biotite kg^{-1} and 525 g corundum kg^{-1}). Cell parameters for biotite were a = 0.5349(2), b = 0.9248(3), c = 1.0208(1) nm, and β = 100.25°, and those for corundum were a = 0.47594(1) and c = 1.2995(1) nm. This analysis highlights one of the strengths of the Rietveld method and points out a major weakness of conventional methods that rely upon ratios of only a few reflections. Because all reflections in the diffraction pattern from each phase are considered, preferred-orientation effects tend to cancel out.

The synthetic binary mixture of 50:50 hematite/corundum illustrates a potential problem with the Rietveld method when analyzing mixtures in which one or more phases has a linear absorption coefficient significantly greater than the average for the sample (e.g., samples composed of Fe- or Mn-oxides together with typical silicate minerals). Figure 9-5 shows the very good agreement between observed and calculated patterns for this mixture (R_{wp} = 21.4%). However, in spite of this agreement, the quantitative results, 440 g hematite per kilogram and 560 g corundum per kilogram, were relatively poor. The inferior results for this mixture are apparently due to microabsorption, yielding low predicted concentrations for the high linear absorption coefficient material, hematite. As noted above, this problem can be minimized by grinding samples to very

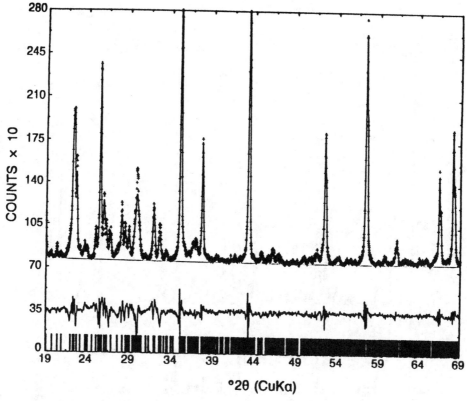

Fig. 9-2. Observed and calculated diffraction patterns for a 50:50 clinoptilolite/corundum mixture (plot conventions as in Fig. 9-1).

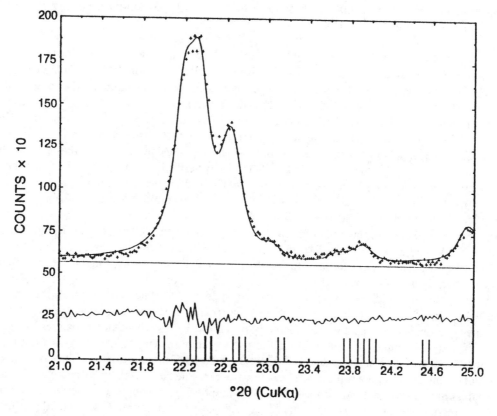

Fig. 9-3. Expansion of the 21.0 to 25.0 °2θ region of Fig. 9-2.

286   **BISH**

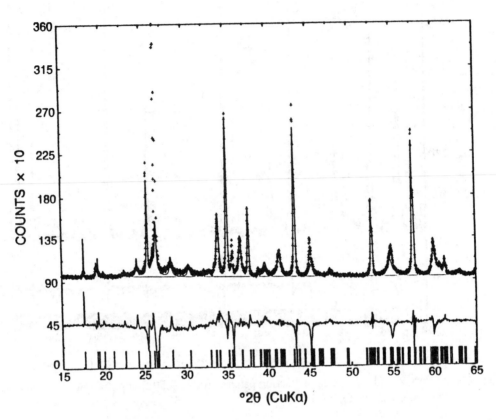

Fig. 9-4. Observed and calculated diffraction patterns for a 50:50 biotite/corundum mixture (plot conventions as in Fig. 9-1).

fine particle sizes or by choosing a different radiation. Refinements with data obtained on <38-μm natural hematite yielded even worse results as expected (200 g hematite kg^{-1} and 800 g corundum kg^{-1}), but grinding the sample to <1 μm or using Fe-Kα radiation improved the results. Two separate 50:50 mixtures of both natural and synthetic hematite with corundum yielded results of 500 g hematite per kilogram and 500 g corundum per kilogram when using Fe-Kα radiation. These mixtures are good examples of a case in which traditional methods using at most a few peaks from each phase will implicitly correct for the problem during standardization, therefore potentially yielding results superior to those obtained with the Rietveld method. In spite of the microabsorption problem, precise cell parameters were still obtained [hematite: $a = 0.50352(1)$, $c = 1.37469(2)$ nm; corundum: $a = 0.47589(1)$, $c = 1.29911(3)$ nm]. Because the Rietveld method provides an opportunity to calculate the linear absorption coefficient for every phase in a mixture, it is possible to correct for microabsorption problems assuming an average crystallite shape and size (see Klug & Alexander, 1974, p. 541-542). Bish and Howard (1988) emphasized the importance of microabsorption in Rietveld quantitative analyses, and Taylor (1991) and Taylor and Matulis (1991) recently implemented the Brindley (1945) particle-absorption-contrast factors in the Rietveld method. The only additional parameter required to correct for microabsorption is the effective particle radius for each phase. They obtained quantitative analyses with significantly improved accuracy, proving the importance of microabsorption in Rietveld quantitative analyses.

Natural soil samples are more difficult to use in demonstrating the accuracy of the Rietveld method of quantitative analysis due to difficulties in determining the true quantities of phases in mixtures, but a few examples illustrate the versatility of the method in analyzing complex mixtures. For the samples examined by Bish and Post (1988), optical point counts were used as an estimate of the true phase contents. The analysis of the U.S. Geological Survey G-1 standard granite

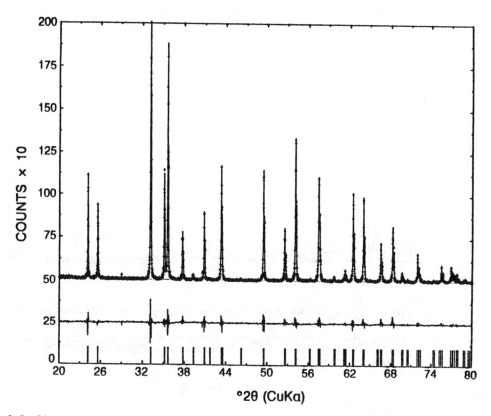

Fig. 9-5. Observed and calculated diffraction patterns for a 50:50 hematite/corundum mixture (plot conventions as in Fig. 9-1).

provided a rigorous test of the method and involved analysis for four complex phases. The results of the Rietveld analysis for quartz, albite, microcline, and biotite (Fig. 9-6) agreed with the modes of Chafes (1951) within one standard deviation and also provided precise cell parameters for all of the major phases in the rock.

The Rietveld method should prove valuable in analysis of primarily crystalline soils, particularly if detailed information on the individual components is desired. The analyses of Bish and Ebinger (1989) and Ebinger and Bish (1990) of synthetic Mn- and Cr-goethites demonstrate that useful results can be obtained from materials with average crystallite sizes close to 10 nm. Figure 9-7 illustrates the results obtained from a mixture of synthetic Fe-endmember hematite and goethite. Analyses of these samples allowed simultaneous determination of the factors causing peak broadening in both hematite and goethite, in addition to yielding precise unit-cell parameters for both phases. Peaks in the diffraction pattern of the Fe-endmember materials were broadened primarily as a result of small crystallite effects, whereas the goethite reflections for the Mn-substituted materials were significantly strain-broadened due to Jahn-Teller distortions in the octahedra induced by Mn^{+3} substituting for Fe^{+3}. In contrast, results for the Cr-substituted goethites showed that peak broadening in these samples was due primarily to small crystallite size rather than strain. This information on individual components would be difficult if not impossible to obtain using conventional single-peak methods and is considerably different from simple analyses using the Scherrer equation. For example, conventional analyses would lead to the conclusion that the crystallite sizes for the Mn-substituted goethites are significantly smaller than they actually are.

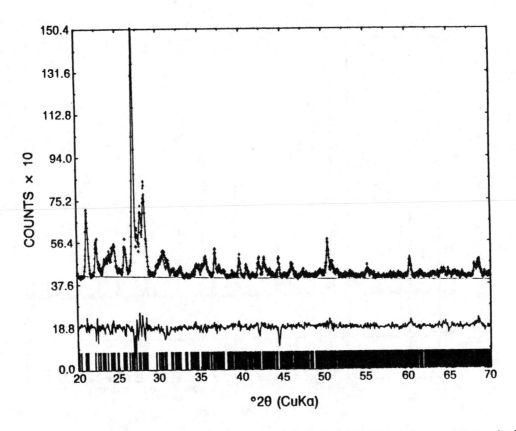

Fig. 9-6. Observed and calculated diffraction patterns for the U.S. Geological Survey G-1 standard granite containing quartz, albite, microcline, and biotite (plot conventions as in Fig. 9-1).

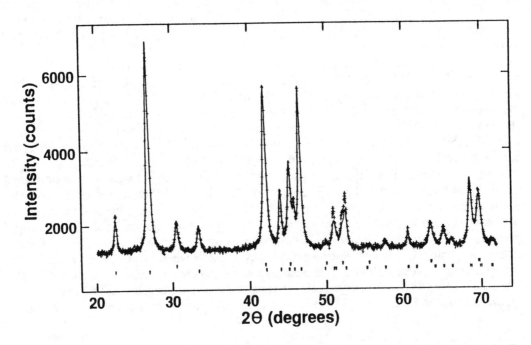

Fig. 9-7. Observed (pluses) and calculated (solid line) diffraction patterns for a synthetic Fe endmember mixture of hematite and goethite. Lower and upper vertical marks represent all possible Fe-Kα_1 and -Kα_2 reflection positions for goethite and hematite, respectively.

Results recently obtained in our laboratory on several bauxite specimens attest to the utility of this method with oxide-rich soils (Jones & Malik, 1994; see Chapter 10). Samples containing variable proportions of hematite, goethite, magnetite, gibbsite, boehmite, anatase, rutile, quartz, and kaolinite were analyzed using Fe-Kα radiation data. The results appeared excellent (e.g., Fig. 9-8, R_{wp} = 16.3, R_{exp} = 9.7%) and yielded considerable information on the minerals in the samples. The sample illustrated in Fig. 9-8 contained 119 g hematite per kilogram, 95 g goethite per kilogram, 627 g gibbsite per kilogram, 126 g boehmite per kilogram, 4 g rutile per kilogram, 15 g anatase per kilogram, and 14 g kaolinite per kilogram. Lattice parameters obtained from the analyses were used with published determinative curves to further characterize the minerals in the sample (e.g., to determine the amount of Al substitution in the goethite). In addition, as with the above goethite samples, most of the goethite reflections in the bauxite samples were anisotropically broadened, and a significant portion of the broadening was due to strain rather than finite crystallite size. Although the Rietveld method cannot explicitly model the two-dimensional diffraction effects common with phyllosilicate clay minerals, it is obviously applicable to oxide-rich soil samples and can provide considerable valuable information about these soils that is difficult to obtain with other methods.

Rietveld quantitative analysis is still evolving today as evidenced by the recent incorporation of microabsorption corrections into a Rietveld program. Future innovations may include the ability to consider amorphous materials explicitly, i.e., to calculate the radial distribution function for a given model. Incorporation of the ability to consider two-dimensional diffraction effects would considerably enhance the power of this method and extend its applicability to the analysis of soils containing abundant quantities of phyllosilicate clay minerals.

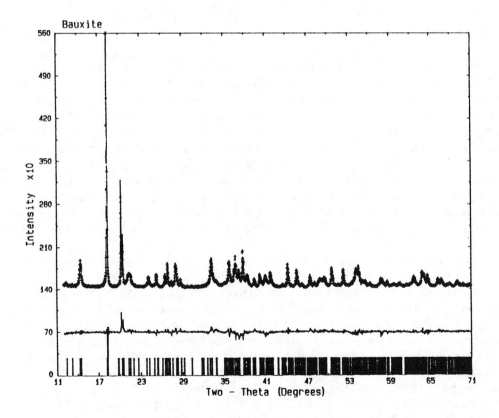

Fig. 9-8. Observed (pluses) and calculated (solid line) diffraction patterns for a natural bauxite sample containing hematite, goethite, gibbsite, boehmite, rutile, anatase, and kaolinite (plot conventions as in Fig. 9-1).

Observed Patterns

Quantitative phase analysis using complete observed diffraction data differs from the Rietveld quantitative analysis method in that it involves fitting a synthetic diffraction pattern, constructed from appropriate *observed* standard diffraction patterns, to an observed diffraction pattern. The concepts employed in this method were first described by Smith (1984) and Brown and Wood (1985) and have been described in detail by Smith et al. (1987); they are included here in brief form only.

The analysis procedure first involves measurement of standard data on pure materials using fixed instrumental conditions, preferably identical to the conditions under which unknown samples will be analyzed. These standard patterns may be processed to remove background and any artifacts, and the data may be smoothed. In some cases, reflections from unwanted phases may be removed from the standard pattern. For standard materials that are unavailable in appropriate form, diffraction patterns can either be simulated from Powder Diffraction File data and the program SIMUL (see Smith, 1989) or calculated using the program POWD10 (Smith et al., 1982). The next step in standardization is analogous to procedures used in conventional RIR quantitative analyses. Data are collected on each of the standard samples that has been mixed with a known amount of corundum in order to determine the I/I_c RIR (see above). These patterns yield the basic calibration data used in the quantitative analyses. The final step involves collection of data for the unknown samples to be analyzed, using conditions identical to those used in obtaining the standard data. Background and artifacts are also removed from unknown-sample data.

The I/I_c values are obtained in the same manner that quantitative analyses of unknown samples are performed, first assigning the standard material an I/I_c value of 1.0. The least-squares method involves minimizing the expression

$$\delta(2\theta) = I_{unk}(2\theta) - \sum_p W_p k_p I_p(2\theta) \quad ,$$

[23]

where $I_{unk}(2\theta)$ and $I_p(2\theta)$ are the diffraction intensities at each 2θ interval for the unknown and each of the standard phases, p, respectively. The W_p is the weight fraction for phase p and k_p is the RIR. For analysis of RIRs, the standard phase is assigned a ratio of 1.0 and the correct RIR is determined from the resultant weight fractions. For example (Smith et al., 1987), a standard mixture of 500 g ZnO per kilogram and 500 g corundum per kilogram yielded results of 578 g ZnO per kilogram and 422 g corundum per kilogram. The RIR for ZnO is therefore 578/422 = 1.37. It is important to note that the RIRs derived using this procedure are equivalent to peak-height RIRs rather than integrated-intensity RIRs because the minimization is conducted on a step-by-step basis. The RIR value determined in this way is valid for subsequent quantitative analysis of mixtures of variable phase composition, but it is important that the RIR be determined on a standard as similar as possible (e.g., crystallite size, chemical composition) to the unknown material.

Just as with most other quantitative analysis methods, the phases present in the mixture must be determined before performing quantitative analysis. Using the selected standard patterns, the weighted sum of these patterns is least-squares fit to the pattern of the unknown. Weight fractions are obtained from the I/I_c values, and results are typically reported on the basis of a material that is 100% crystalline (i.e., the sum of all phases is constrained to total to 100%). This analysis is analogous to the external-standard or adiabatic method of Chung (1974b). One of the components of the sample may be given a fixed weight-fraction if an internal standard has been added. This procedure, which is analogous to the internal-standard or matrix-flushing method of Chung (1974a), does not constrain the total weight fraction. Thus, it can indicate the presence of amorphous components or unidentified phases and should be ideal for many soil systems. Unlike the Rietveld method of quantitative analysis, the observed-pattern method does not explicitly vary the cell parameters of the phases in the simulated pattern. A routine has been incorporated that allows movement of the unknown pattern along the 2θ scale to give better fits, but Smith et al. (1987) commented that quantitative results are only slightly affected by errors in lattice parameters.

Results reported by Smith et al. (1987) for several mixtures are encouraging. As their most rigorous test, they analyzed a six-component mixture containing equal weight fractions (i.e., 167 g kg^{-1}) of fluorite, calcite, corundum, portlandite, halite, and quartz. Absolute standard deviations were below 20 g kg^{-1} for all phases, with relative errors less than 12%. Experience with this method has shown that it is relatively insensitive to preferred-orientation effects because the results for each phase constitute the average of many reflections, just as in the Rietveld method. Preferred-orientation and microabsorption effects are further minimized by using standards that are comparable in these respects to the unknowns. In fact, many of the factors important in standardization with conventional methods are equally important with this type of full-pattern fitting. The method should also be insensitive to several other factors such as extinction and peak overlap because the method uses the full diffraction pattern. Perhaps the most significant advantage of this method is that it does not rely on calculating the three-dimensional diffraction pattern of each component of a mixture. Clay minerals and related layer-structure materials are unique in that they often exhibit two-dimensional diffraction effects (see Reynolds, 1989b) that cannot be approximated using three-dimensional diffraction patterns. However, in theory, this full-pattern fitting method can treat materials of unknown structure, clay minerals, and even amorphous solids. Smith et al. (1986) specifically described the application of this method to both qualitative and quantitative analysis of clay minerals. Another advantage of this method is that it does not require a thorough understanding of the theory behind calculated diffraction patterns nor does it require a knowledge of the crystal structures of each of the phases present in a mixture.

Smith and Howard (1990) compared quantitative analyses obtained using the observed-pattern method with those obtained by the Rietveld method. Five mixtures simulating sedimentary rocks and containing varying amounts of quartz, calcite, dolomite, siderite, albite, microcline, pyrite, muscovite, and kaolinite were analyzed. The time required for standardization of the observed-pattern method was lengthy, whereas with the Rietveld method more time was spent in refining the data. Although the Rietveld method cannot explicitly accommodate the two-dimensional diffraction effects that occur with minerals such as kaolinite, the accuracies obtained with the two methods were comparable and usually within 30 g kg^{-1} of the true value.

Full-pattern methods of quantitative analysis (i.e., the Rietveld and observed-pattern methods) have several significant advantages over the more conventional methods, which employ at most a few of the strongest reflections from each phase in a mixture. Because these methods use the full observed diffraction patterns, problems in obtaining reproducible intensities and in dealing with severely overlapped profiles are eliminated. Ambiguities in determining background position during intensity measurement and other troublesome effects, such as preferred orientation and extinction, are minimized. In addition to determination of the amounts of crystalline phases present, the Rietveld method yields high-precision cell parameters for each phase in a mixture and has the potential for providing information on the structures and causes of peak broadening for individual crystalline components. The observed-pattern method is particularly well-suited to the study of soils because poorly crystalline minerals, amorphous phases, and clay minerals exhibiting two-dimensional diffraction effects can be explicitly accommodated. These full-pattern methods, when coupled with techniques using other observations on a given sample (e.g., chemical composition) have the potential for providing quantitative analyses of soil samples of unusually high quality.

CONCLUSIONS AND FUTURE DEVELOPMENTS

X-ray diffraction analysis will probably remain the method of choice for quantitative phase analysis of soils because of its ability to distinguish easily between various crystalline materials. Amounts of amorphous components in soils are often measured by chemical-dissolution techniques, but XRD methods should also be considered in analysis of such materials because the effects of chemical-dissolution techniques on poorly ordered soil minerals can then be ignored. Although there is a tendency to apply the most advanced methods to an analysis problem, it is obvious that judicious application of conventional quantitative XRD methods to soil systems, coupled with ancillary techniques such as simultaneous linear equations methods, will yield consistent, high-quality results. In fact, there are many problems for which conventional methods will probably yield results superior to those obtained with advanced, full-pattern methods because the full-pattern methods may not completely correct for problems with microabsorption, preferred orientation, two-dimensional diffraction, and amorphous phases.

In ideal situations, with mixtures containing only crystalline materials for which microabsorption, preferred orientation, and two-dimensional diffraction are not problems, the full-pattern methods presently can provide quantitative analyses approaching the precision and accuracy of x-ray fluorescence chemical analyses. In particular, the Rietveld method can provide not only quantitative information on amounts of phases but it can also yield precise and accurate unit-cell parameters, information on the crystal chemistry of individual phases (e.g., site occupancies), and information on the nature of broadening of individual reflections (i.e., strain and/or crystallite-size broadening). Full-pattern methods using observed patterns can presently be applied to soil systems containing clay minerals exhibiting two-dimensional diffraction effects, provided the nature of these effects is similar between standards and samples.

Future developments will likely concentrate in several major areas. First, our ability to measure accurate integrated intensities from complex, overlapped patterns is dramatically improving. Profile-fitting methods, with or without structural constraints, may soon be applied in extensions of conventional quantitative-analysis methodology, involving analysis of mixtures using many reflections from each phase. Second, continued improvements in analytical descriptions of observed profile shapes will probably occur, facilitating improved extraction of profile-related sample information. Finally, incorporation of good analytical descriptions of microabsorption and two-dimensional diffraction effects into Rietveld methodology will significantly improve the applicability of the method to soil systems. It is unlikely that future improvements in quantitative multicomponent analysis of soil systems using diffraction methods will involve significant use of synchrotron or neutron sources due to the restricted availability of such sources. However, both of these sources will play a role in providing an increasingly better picture of the structures of soil minerals.

ACKNOWLEDGMENTS

I am grateful to two anonymous reviewers and to S. Chipera for their useful comments which significantly improved this manuscript. I have also benefitted from discussions with S. Chipera, R. C. Reynolds, Jr., and R. Snyder.

REFERENCES

Bish, D. L., and S. J. Chipera. 1988. Problems and solutions in quantitative analysis of complex mixtures by x-ray powder diffraction. Adv. X-ray Anal. 31:295-308.

Bish, D. L., and M. H. Ebinger. 1989. Rietveld refinement of synthetic goethite and Mn-substituted goethite. Clay Miner. Soc. Abstr. 1989:18.

Bish, D. L., and S. A. Howard. 1986. Quantitative analysis via the Rietveld method. Workshop on Quantitative X-ray Diffraction Analysis. 23-24 June 1986, Natl. Bur. Standards, Gaithersburg, MD.

Bish, D. L., and S. A. Howard. 1988. Quantitative phase analysis using the Rietveld method. J. Appl. Cryst. 21:86-91.

Bish, D. L., and J. E. Post. 1988. Quantitative analysis of geological materials using X-ray powder diffraction data and the Rietveld refinement method. Geol. Soc. Am. Abstr. Progr. 20:A223.

Bish, D. L., and R. C. Reynolds, Jr. 1989. Sample preparation for x-ray diffraction. p. 73-99. In D. L. Bish and J. E. Post (ed.) Modern powder diffraction. Reviews in mineralogy. Vol. 20. Miner. Soc. Am., Washington, DC.

Bloss, F. D., G. Frenzel, and P. D. Robinson. 1967. Reducing orientation in diffractometer samples. Am. Miner. 52:1243-1247.

Boski, T., and A. J. Herbillon. 1988. Quantitative determination of hematite and goethite in lateritic bauxites by thermodifferential x-ray powder diffraction. Clays Clay Miner. 36:176-180.

Braun, G. E. 1986. Quantitative analysis of mineral mixtures using linear programming. Clays Clay Miner. 34:330-337.

Brindley, G. W. 1945. The effect of grain or particle size on x-ray reflections from mixed powders and alloys considered in relation to the quantitative determination of crystalline substances by x-ray methods. Philos. Mag. 36:347-369.

Brindley, G. W. 1980. Quantitative x-ray mineral analysis of clays. p. 411-438. *In* G. W. Brindley and G. Brown (ed.) Crystal structures of clay minerals and their x-ray identification. Monogr. 5, Miner. Soc., London.

Brown, G., and G. W. Brindley. 1980. X-ray diffraction procedures for clay mineral identification. p. 305-360. *In* G. W. Brindley and G. Brown (ed.) Crystal structures of clay minerals and their x-ray identification. Monogr. 5, Miner. Soc., London.

Brown, G., and I. G. Wood. 1985. Estimation of iron oxides in soil clays by profile refinement combined with differential X-ray diffraction. Clay Miner. 20:15-27.

Calvert, L. D., and A. F. Sirianni. 1980. A technique for controlling preferred orientation in powder diffraction samples. J. Appl. Cryst. 13:462.

Calvert, L. D., A. F. Sirianni, G. J. Gainsford, and C. R. Hubbard. 1983. A comparison of methods for reducing preferred orientation. Adv. X-ray Anal. 26:105-110.

Carter, J. R., M. T. Hatcher, and L. Di Carlo. 1987. Quantitative analysis of quartz and cristobalite in bentonite clay based products by x-ray diffraction. Anal. Chem. 59:513-519.

Chayes, F. 1951. Modal analyses of the granite and diabase test rocks. U. S. Geol. Surv. Bull. 980:59-68.

Chipera, S. J., and D. L. Bish. 1988. Pitfalls in profile refinement of clay mineral diffraction patterns. Clay Min. Soc. Abstr. 1988:39.

Chipera, S. J., and D. L. Bish. 1991. Measurement of mass absorption coefficients using Compton-scattered Cu radiation in x-ray diffraction analysis. Adv. X-ray Anal. 34:325-335.

Chung, F. H. 1974a. Quantitative interpretation of x-ray diffraction patterns of mixtures. I. Matrix-flushing method of quantitative multicomponent analysis. J. Appl. Cryst. 7:519-525.

Chung, F. H. 1974b. Quantitative interpretation of x-ray diffraction patterns of mixtures. II. Adiabatic principle of x-ray diffraction analysis of mixtures. J. Appl. Cryst. 7:526-531.

Copeland, L. E., and R. H. Bragg. 1958. Quantitative x-ray diffraction analysis. Anal. Chem. 30:196-206.

Cullity, B. D. 1978. Elements of x-ray diffraction. Addison-Wesley, Reading, MA.

Davis, B. L. 1986. A tubular aerosol suspension chamber for the preparation of powder samples for x-ray diffraction analysis. Powd. Diff. 1:240-243.

Davis, B. L., and L. R. Johnson. 1982. Sample preparation and methodology for x-ray quantitative analysis of thin aerosol layers deposited on glass fiber and membrane filters. Adv. X-ray Anal. 25:295-300.

Davis, B. L., and L. R. Johnson. 1989. Quantitative mineral analysis by x-ray transmission and x-ray diffraction. p. 104-117. *In* D. R. Pevear and F. A. Mumpton (ed.) Quantitative mineral analysis of clays. CMS Workshop Lectures, Vol. 1. Clay Miner. Soc., Evergreen, CO.

Davis, B. L., and M. J. Walawender. 1982. Quantitative mineralogical analysis of granitoid rocks: a comparison of x-ray and optical techniques. Am. Miner. 67:1135-1143.

de Wolff, P. M. 1947. A theory of x-ray absorption in mixed powders. Physica 13:62-78.

de Wolff, P. M., J. M. Taylor, and W. Parrish. 1959. Experimental study of effect of crystallite size statistics on x-ray diffractometer intensities. J. Appl. Phys. 30:63-69.

Dollase, W. A. 1986. Correction of intensities for preferred orientation in powder diffractometry: Application of the March model. J. Appl. Cryst. 19:267-272.

Ebinger, M. H., and D. L. Bish. 1990. Rietveld refinement of goethite and Cr-goethite from X-ray powder data. Clay Miner. Soc. Abstr. 1990:46.

Florke, O. W., and Saalfeld, H. 1955. Ein verfahren zur herstellung texturfreier Rontgen-Pulverpraparate. Z. Krist. 106:460-466.

Gibbs, R. J. 1965. Error due to segregation in quantitative clay mineral x-ray diffraction mounting techniques. Am. Miner. 50:741-751.

Gonzalez, C. R. 1987. General theory of the effect of granularity in powder x-ray diffraction. Acta Crystallogr. A43:769-774.

Hendricks, S. B., and W. H. Fry. 1930. The results of x-ray and microscopical examinations of soil colloids. Soil Sci. 29:547-580.

Hermann, H., and M. Ermrich. 1987. Microabsorption of x-ray intensity in randomly packed powder specimens. Acta Crystallogr. A43:401-405.

Hill, R. J., and C. J. Howard. 1987. Quantitative phase analysis from neutron powder diffraction data using the Rietveld method. J. Appl. Cryst. 20:467-474.

Hodgson, M., and A. W. L. Dudeney. 1984. Estimation of clay proportions in mixtures by x-ray diffraction and computerized chemical mass balance. Clays Clay Miner. 32:19-28.

Howard, S. A. and K. D. Preston. 1989. Profile fitting of powder diffraction patterns. p. 217-275. *In* D. L. Bish and J. E. Post (ed.) Modern powder diffraction. Reviews in Mineralogy. Vol. 20. Miner. Soc. Am., Washington, DC.

Hughes, R. and B. Bohor. 1970. Random clay powders prepared by spray drying. Am. Miner. 55:1780-1786.

Hussey, G. A., Jr. 1972. Use of simultaneous linear equations program for quantitative clay analysis and the study of mineral alterations during weathering. Ph.D. diss. Pennsylvania State Univ., University Park.

Jenkins, R., T. G. Fawcett, D. K. Smith, J. W. Visser, M. C. Morris, and L. K. Frevel. 1986. JCPDS-International Centre for Diffraction Data sample preparation methods in x-ray powder diffraction. Powd. Diff. 1:51-63.

Johnson, L. J., C. H. Chu, and G. A. Hussey. 1985. Quantitative clay mineral analysis using simultaneous linear equations. Clays Clay Miner. 33:107-117.

Jonas, E. C., and J. R. Kuykendall. 1966. Preparation of montmorillonites for random powder diffraction. Clay Miner. 6:232-235.

Jones, R. C. 1989. A computer technique for x-ray diffraction curve fitting/peak decomposition. p. 52-101. *In* D. R. Pevear and F. A. Mumpton (ed.) Quantitative mineral analysis of clays. CMS Workshop Lectures. Vol. 1. Clay Miner. Soc., Evergreen, CO.

Jones, R.C., and H.U. Malik. 1994. Analysis of minerals in oxide-rich soils by x-ray diffraction. 1994. p. 296-320. *In* J. Amonette and L.W. Zelazny (ed.) Quantitative methods in soil mineralogy. SSSA Misc. Publ. SSSA, Madison, WI.

Kelley, W. P., W. H. Dore, and S. M. Brown. 1931. The nature of the base exchange material of bentonite, soils, and zeolites, as revealed by chemical investigations and x-ray analysis. Soil Sci. 31:25-55.

Klug, H. P., and L. E. Alexander. 1974. X-ray diffraction procedures for polycrystalline and amorphous materials. Wiley, New York.

Leroux, J., D. H. Lennox, and K. Kay. 1953. Applications of x-ray diffraction analysis in the environmental field. Anal. Chem. 25:740-748.

Maniar, P. D., and G. A. Cooke. 1987. Modal analyses of granitoids by quantitative x-ray diffraction. Am. Miner. 72:433-437.

McNeal, B. L., and T. Sansoterra. 1964. Mineralogical examination of arid-land soils. Soil Sci. 97:367-375.

Moore, D., and R. C. Reynolds, Jr. 1989. X-ray diffraction and the identification and analysis of clay minerals. Oxford Univ. Press, Oxford, England.

Nakamura, T., K. Sameshima, K. Okunaga, Y. Sugiura, and J. Sato. 1989. Determination of amorphous phase in quartz powder by x-ray powder diffractometry. Powd. Diff. 4:9-13.

O'Connor, B. H., and M. D. Raven. 1988. Application of the Rietveld refinement procedure in assaying powdered mixtures. Powd. Diff. 3:2-6.

Parker, R. J. 1978. Quantitative determination of analcime in pumice samples by x-ray diffraction. Miner. Mag. 42:103-106.

Parrish, W., and T. C. Huang. 1983. Accuracy and precision of intensities in x-ray polycrystalline diffraction. Adv. X-ray Anal. 26:35-44.

Parrish, W., M. Mack, and J. Taylor. 1966. Determination of apertures in the focusing plane of x-ray powder diffractometers. J. Sci. Instrum. 43:623-628.

Pawloski, G. A. 1985. Quantitative determination of mineral content of geological samples by x-ray diffraction. Am. Miner. 70:663-667.

Pevear, D. R., and F. A. Mumpton (ed.). 1989. Quantitative mineral analysis of clays. CMS Workshop Lectures. Vol. 1. Clay Miner. Soc., Evergreen, CO.

Post, J. E., and D. L. Bish. 1989. Rietveld refinement of crystal structures using powder x-ray diffraction data. p. 277-308. *In* D. L. Bish and J. E. Post (ed.) Modern powder diffraction. Reviews in Mineralogy. Vol. 20. Miner. Soc. Am., Washington, DC.

Reynolds, R. C., Jr. 1968. Effect of particle size on apparent lattice spacings. Acta Crystallogr. A24:319-320.

Reynolds, R. C., Jr. 1989a. Principles and techniques of quantitative analysis of clay minerals by x-ray powder diffraction. p. 4-36. *In* D. R. Pevear and F. A. Mumpton (ed.) Quantitative mineral analysis of clays. CMS Workshop Lectures. Vol. 1. Clay Miner. Soc., Evergreen, CO.

Reynolds, R. C., Jr. 1989b. Diffraction by small and disordered crystals. p. 145-181. *In* D. L. Bish and J. E. Post (ed.) Modern powder diffraction. Reviews in Mineralogy. Vol. 20. Miner. Soc. Am., Washington, DC.

Rietveld, H. M. 1969. A profile refinement method for nuclear and magnetic structures. J. Appl. Cryst. 2:65-71.

Ross, M. 1968. X-ray diffraction effects by non-ideal crystals of biotite, muscovite, montmorillonite, mixed-layer clays, graphite, and periclase. Z. Krist. 126:80-97.

Slaughter, M. 1989. Quantitative determination of clays and other minerals in rocks. p. 120-151. *In* D. R. Pevear and F. A. Mumpton (ed.) Quantitative mineral analysis of clays. CMS Workshop Lectures. Vol. 1. Clay Miner. Soc., Evergreen, CO.

Smith, D. K. 1984. Quantitative phase analysis using the whole x-ray powder diffraction pattern. Acta Cryst. A40(Suppl.):C364.

Smith, D. K. 1989. Computer analysis of diffraction data. p. 183-216. *In* D. L. Bish and J. E. Post (ed.) Modern powder diffraction. Reviews in Mineralogy. Vol. 20. Miner. Soc. Am., Washington, DC.

Smith, D. K., and C. S. Barrett. 1979. Special handling problems in x-ray diffractometry. Adv. X-ray Anal. 22:1-12.

Smith, D. K., and S. A. Howard. 1990. Comparisons of the GMQUANT and Rietveld methods for quantitative x-ray powder analysis. Ann. Denver Conf. Applic. X-ray Anal. Abstr. 1990/1991.

Smith, D. K., G. G. Johnson, Jr., and C. O. Ruud. 1986. Clay mineral analysis by automated powder diffraction analysis using the whole diffraction pattern. Adv. X-ray Anal. 29:217-224.

Smith, D. K., G. G. Johnson, Jr., A. Scheible, A. M. Wims, J. L. Johnson, and G. Ullmann, G. 1987. Quantitative x-ray powder diffraction method using the full diffraction pattern. Powd. Diff. 2:73-77.

Smith, D. K., M. C. Nichols, and M. E. Zolensky. 1982. POWD10. A FORTRAN IV program for calculating x-ray powder diffraction patterns - Version 10. Pennsylvania State Univ., University Park.

Smith, S. T., R. L. Snyder, and W. E. Brownell. 1979a. Minimization of preferred orientation in powders by spray drying. Adv. X-ray Anal. 22:77-87.

Smith, S. T., R. L. Snyder, and W. E. Brownell. 1979b. Quantitative phase analysis of Devonian shales by computer controlled x-ray diffraction of spray dried samples. Adv. X-ray Anal. 22:181-191.

Snyder, R. L., and D. L. Bish. 1989. Quantitative analysis. p. 101-144. *In* D. L. Bish and J. E. Post (ed.) Modern powder diffraction. Reviews in Mineralogy. Vol. 20. Miner. Soc. Am., Washington, DC.

Taylor, J. C. 1991. Computer programs for standardless quantitative analysis of minerals using the full powder diffraction profile. Powd. Diff. 6:2-9.

Taylor, J. C., and C. E. Matulis. 1991. Absorption contrast effects in the quantitative XRD analysis of powders by full multiphase profile refinement. J. Appl. Cryst. 24:14-17.

Visser, J. W., and P. M. de Wolff. 1964. Absolute intensities. Rep. 641.109. Technische Physische Dienst, Delft, the Netherlands.

Whittig, L. D., and W. R. Allardice. 1986. X-ray diffraction techniques. p. 331-375. *In* A. Klute (ed.) Methods of soil analysis, Part 1, 2nd ed. ASA and SSSA, Madison, WI.

Wilchinsky, Z. W. 1951. Effect of crystal, grain, and particle size on x-ray power diffracted from powders. Acta Crystallogr. 4:1-9.

Wiles, D. B., and R. A. Young. 1981. A new computer program for Rietveld analysis of x-ray powder diffraction patterns. J. Appl. Cryst. 14:149-151.

Yukino, K., and R. Uno. 1986. "$\in$-scanning"--A method of evaluating the dimensional and orientational distribution of crystallites by x-ray powder diffractometer. Jpn. J. Appl. Phys. 25:661-666.

10 Analysis of Minerals in Oxide-Rich Soils by X-Ray Diffraction

R. C. Jones
University of Hawaii at Manoa
Honolulu, Hawaii

H. U. Malik
Masa Fujioka and Associates
Honolulu, Hawaii

When encountering oxide-rich soils for the first time, most of us, for the lack experience with such curiosities, instinctively follow the same procedures for x-ray diffraction (XRD) sample preparation as are commonly used for silica-rich soils. Furthermore, a mistake that the newcomer to oxide-rich soils often makes is in the interpretation of the XRD patterns--they primarily direct their attention to the phyllosilicates (if any). The oxide-rich soils taken as examples in this chapter will be specific to Hawaii, Puerto Rico, the Philippines, and Indonesia. More specifically, soils considered in this chapter are characterized by their high oxide, oxyhydroxide, and hydroxide contents of Al, Fe, Ti, and Mn and their relatively low silica contents. In some instances phyllosilicates are poorly crystallized and, in extreme instances, absent altogether.

Perhaps the most popularly held conception of Hawaiian soils is that they contain large quantities of x-ray amorphous materials. For some volcanic-ash soils this notion is true, but in the majority of instances, x-ray amorphous materials comprise only a very minor fraction of the soils. In mature soils of Hawaii, primary minerals such as plagioclase feldspars, pyroxenes, and olivines that were originally in the basaltic parent material have been totally lost to weathering. Therefore, only trace amounts of primary minerals such as quartz and mica occur in surface soils, and these were transported to Hawaii as tropospheric dust. The amount of eolian minerals that have been deposited are a function of rainfall, basin accumulation, and erosion. Finding quartz at depth is nearly always an indicator of buried surface horizons. There are some soils in Hawaii whose development was influenced by hydrothermal activity, in which case authigenic quartz is usually present.

When working with oxide-rich soils, the analyst must be sure of the size fraction being analyzed, an accomplishment that is not always easy to achieve. Aggregate separation and dispersion are always a challenge--very few soils respond to dispersion treatments the same way. With the exception of gibbsite and quartz, and in some instances kaolinite, XRD peaks are usually very broad, often asymmetric and shifted on the 2θ scale from the published JCPDS positions because of crystal lattice substitutions. Care must be taken not to confuse hydrated halloysite with illite. Because of interstratification, illite, smectite, and halloysite may be difficult to distinguish from one another.

Relatively fresh volcanic ash will produce XRD patterns that have no distinct peaks, suggesting that the sample contains nothing but volcanic glass. When fractionated with great care, however, smectite, imogolite, or even allophane may be found, depending on the age and weathering pressures on the ash. A "trap" that we have fallen into is assuming that once we have seen an XRD pattern of one high-oxide soil we have seen them all. Nothing could be farther from the truth! Chemical and physical properties of high-oxide soils are highly variable, but with a cursory inspection of the soil's XRD patterns they contain essentially the same minerals, i.e., gibbsite, goethite, anatase, rutile, ilmenite, hematite, maghemite, magnetite, and maybe a little boehmite and lepidocrocite. The most common phyllosilicates found in oxide-rich soils of Hawaii are kaolinite and halloysite (USDA-SCS, 1976; Bates, 1962).

We have found that the absolute abundance of a given mineral is not as important in terms of chemical and physical properties of soils as the physical properties of each mineral. That is, the diffracting-domain size (a measure of the specific surface capable of reacting, for example, with phosphate) (Jones, 1981) and the degree of crystallinity are often more important than a particular mineral's abundance.

The methods described in this chapter are primarily addressed toward acquiring a maximum amount of physical information about each mineral in a complex mixture by XRD. For selected soils, the Rietveld XRD refinement technique (Rietveld, 1969) in conjunction with XRD curve fitting/peak decomposition (Jones 1989) has shown that there is a good correlation with phosphate sorption ($r^2 = 0.86$, $n = 12$). The XRD sample preparation and data analysis techniques for oxide-rich soils that are presently used in Hawaii as well as future prospects for improvements will be addressed in this chapter.

SAMPLE PREPARATION

Dispersion

Under warm, humid conditions of prolonged leaching, oxide-rich soils tend toward a state of minimum surface potential. The potential on the surface of variable charge, low-activity clays is

$$\Phi_o = (kT/e)\ln(H^+/H_o^+) \qquad [1]$$

where Φ_o is the surface potential, k is the Boltzmann constant, T is the absolute temperature, e is the electron charge, H^+ is the hydrogen ion concentration, and H_o^+ is the hydrogen ion concentration when Φ_o is zero (Uehara & Gillman, 1981). The tendency of H^+ to approach H_o^+ was termed isoelectric weathering by Mattson (1932). Under certain conditions, H_o^+ corresponds to H^+ at the isoelectric point, which is also related to the point of zero net charge (PZNC)(Sposito, 1984). Mattson observed that as a soil weathers, the pH of the soil approaches the PZNC. From Eq. [1], we see that as the values of H^+ and H_o^+ approach one another, the surface potential, Φ_o approaches zero, a condition where the number of negative charges on the particle surfaces exactly equals the number of positive charges. For this reason, many of the oxide-rich soils in tropical regions are highly aggregated.

This phenomenon can also be understood in terms of surface charge density. Sposito (1984, p. 36) gives

$$\sigma_{in} = F(q_+ - q_-)/S \qquad [2]$$

where σ_{in} is the intrinsic surface charge density (C m^{-2}), F is the Faraday constant (C mol^{-1}), S is the specific surface area (m^2 g^{-1}), and q_+ and q_- are the amounts of adsorbed cationic and anionic charges (mol g^{-1}), respectively. If q_+ and q_- are equal, then σ_{in} = zero, the soil is at the PZNC, and it is well aggregated. Although highly weathered soils may approach their PZNC they are rarely *at* their PZNC.

In order to analyze the clay fraction it must first be separated from the silt and sand fractions. Separation involves dispersion, and for dispersion to occur a high intrinsic charge must be present on the surface to cause interparticle repulsion. From Eq. [1] and [2], it is apparent that mechanical agitation will do nothing to change the charge balance on the particle surfaces. (Usually as soon as mechanical agitation ceases the suspension flocculates.)

These equations show, however, that solution pH levels significantly above or below the PZNC will result in a high surface potential (either positive or negative) and, consequently a high intrinsic surface charge density. Clearly, then, dispersion of oxide-rich soils can be initiated by adjusting the solution pH so that it is far removed from the PZNC.

Figure 10-1 shows two generalized curves that represent the changes of σ_{in} with changes of pH. Curve A represents σ_{in} for a soil with a low PZNC. Curve B represents σ_{in} for a soil with a high PZNC. The question is, should the pH of a suspension be increased or decreased to achieve

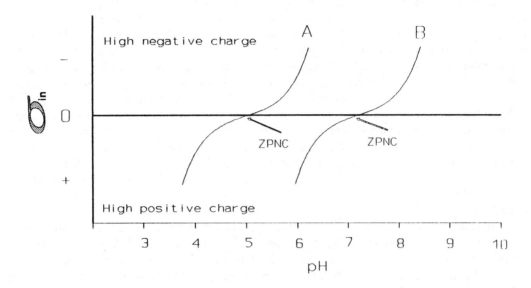

Fig. 10-1. Relationship between intrinsic surface charge density and solution pH for two hypothetical soils having different PZNCs.

the best dispersion? Experience has shown that for some soils there is no clear answer. Both raising and lowering the pH may have to be tried to achieve the best dispersion. Figure 10-1 is given as a guide to the direction that the pH might be changed. In the instance of a soil with a low PZNC (Curve *A*), dispersion is best achieved by increasing the pH which will impart a net negative charge on the particle surfaces. The stronger the net negative charge the more stable the suspension will be because of the strong interparticle repulsion. Likewise, if a soil is basic (Curve *B*), dispersion may be best achieved by lowering the pH so a net positive charge will be produced on the particle surfaces.

Wada et al. (1986) found that four Andepts from Maui were best dispersed by first treating the soils with H_2O_2. The suspensions were then repeatedly sonicated at 15-minute intervals and adjusted to pH 4.0 with 0.1 *M* HCl. For two Eutrandepts from the Island of Hawaii, Wada et al. (1990) dispersed the soils as outlined above except that they were adjusted to pH 10.0 with 0.1 *M* NaOH. One of Wada's coauthors, H. Ikawa, (personal communication) noted that the soils collected on the islands of Maui and Hawaii were similar in terms of their pH. Therefore, in the instance of the two groups of soils cited above, dispersion was best achieved by trying both high and low pH values and then employing the method that worked best.

Two of the most common techniques to break up aggregates are ultrasonic vibrations (Genrich & Bremner, 1972a,b; Busacca et al., 1984) and rapid shaking in a vortex-type mixer mill (El-Swaify, 1980). We have found that long sonication or vortex shaking times warm the suspension to a point where hydrated halloysite loses interlayer water, thereby rendering the identification of dehydrated halloysite in the presence of kaolinite impossible. We routinely sonicate or shake the suspension for only a minute or two and then cool the sample before agitation is repeated.

The method we find most successful is to increase the pH by slowly adding 0.1 *M* LiOH until dispersion occurs. The LiOH is used instead of NaOH because the clay fraction can then be analyzed for Na. With LiOH the dispersion pH rarely exceeds 8.5. Electron microscope examination of soils before and after increasing pH to very high values (i.e., pH > 10.0), shows that in many instances the high-pH samples are shrouded in gelatinous materials. This is direct evidence for dissolution of constituents, such as poorly crystalline silica, when the pH is raised too high. Furthermore, we have found that if the pH is gradually increased during mechanical agitation, a stable suspension can be achieved at a lower pH and with minimal dissolution of the mineral constituents.

Increasing suspension pH with electrolytes such as LiOH and NaOH may also initiate flocculation regardless of the net negative charge imparted to the colloid surfaces. As the electrolyte concentration in the bulk solution increases, the diffuse double layer is compressed and particles tend to come close to one another and thus initiate flocculation (van Olphen, 1977). Furthermore, van Olphen notes that the double layer (for this discussion, a negatively charged solid surface and cations in the immediate proximity of the surface) is increasingly compressed as the concentration and valence of the cations are increased. This phenomenon was demonstrated in the author's laboratory with the Sitiung soil (Typic Kandiudult) from Indonesia. The soil did not disperse until its pH was increased to 8.0 with LiOH. Then with additional LiOH to increase the pH to about 8.5, the soil again flocculated. A stable suspension was once again achieved by centrifuge-washing the soil with deionized water to remove the excess cation. Washing with deionized water decreased the counter ion concentration, thereby allowing the thickness of the diffuse double layer to expand and maintain particle-to-particle distances sufficient to prevent flocculation.

If only the phyllosilicates are of interest, then the removal of free Fe (Mehra & Jackson, 1960) is often an effective method of dispersing the soil. In the process of free-Fe removal the phyllosilicates become Na saturated and the pH is increased to 7.3 (Jackson et al., 1986, p. 119), which in the instance of most oxide-rich, acid soils is adequate to initiate dispersion. If the soil still does not disperse after the excess salts are centrifuge-washed from the sample, the pH can be increased by adding NaOH. Using a dispersing agent such as sodium hexametaphosphate is not recommended because phosphate may decompose some oxides and oxyhydroxides, gibbsite, in particular (Jackson, 1958, p. 75).

In summary, the dispersion methods that we have found are best suited for oxide-rich soils are to: (i) increase the pH of an acid soil, (ii) decrease the pH of a neutral to basic soil (in some instances, both increasing and decreasing the pH might be tried), (iii) use sonic vibrations or vortex-shaking for short periods to break up aggregates, and (iv) centrifuge-wash with deionized water to decrease counter-ion concentrations, increase the thickness of the electrical double layer, and thus decrease particle-to-particle coalescence.

Clay Subfractions

There is a difference of opinion concerning the size divisions of clay subfractions. Whittig and Allardice (1986) list the subfractions as coarse (2-0.2 μm), medium (0.2-0.08 μm), and fine (<0.08 μm) clay. The Soil Conservation Service (USDA-SCS, 1984) lists the subfractions as coarse (2-0.2 μm) and fine (<0.2 μm) clay. Regardless of which size-fraction convention is used, there is almost always a clear separation of particle sizes and densities at the bottoms of centrifuge tubes. Whittig and Allardice noted that smectite tends to concentrate in the finer (upper layer) clay fractions. We found that smectite could not be detected by XRD in the finest clay fraction of a kaolinitic Vertisol from the islands of Lanai and Molokai, Hawaii. When centrifuging the kaolinitic Vertisol at high g-forces, three subfractions were clearly separated into bands. The upper band consisted of a transparent gel, in which only dehydrated halloysite could be identified by XRD [confirmed by transmission electron microscopy (TEM)]. The lowest band contained the Fe-bearing minerals, smectite, illite and interstratified illite/smectite. The intermediate band contained barely detectable amounts of illite and interstratified illite/smectite, but no smectite was detected. Thus, unlike the soils referred to by Whittig and Allardice, smectite was detected in the coarsest fraction at the bottom of the centrifuge tube, rather than in the gel at the top of the centrifuge tube. Additional studies revealed that the fine-clay subfraction did contain smectite, but the "crystallites" were so thin that diffraction maxima were not observed. Therefore, in order to be detectable by XRD, each mineral must have crystallites that contain an adequate number of repeat layers or unit cells to produce detectable XRD maxima. As seen with the kaolinitic Vertisol one can't always predict the subfraction in which a given mineral is likely to occur.

When all three subfractions were mixed together, the bottom layer that contained detectable smectite was so diluted by the materials in the two upper layers, that no smectite was detected. These data show that there is a decided advantage to analyzing the clay subfractions separately.

However, choosing when to analyze the subfractions rather than the whole clay fraction depends upon the kind of information sought. In most instances, the whole clay fraction will yield the desired kind of information.

High-Iron Soils

We have found that soils collected in Hawaii, Puerto Rico, the Philippines, and Indonesia contain high concentrations of Fe-bearing minerals. Goethite has been found to be primarily responsible for phosphate fixation (Atkinson et al.,1974; Parfitt et al., 1975; Jones, 1981). Therefore, for tropical soils, the analysis and quantification of the Fe-bearing minerals is often more important than the analysis of the phyllosilicates. Under most circumstances, the removal of "free" Fe from tropical soils is not advisable *unless* a comparison is to be made between the XRD patterns before and after free-Fe removal such as by differential XRD (Schulze, 1981, 1994; see Chapter 13).

The problem of mass absorption is not as serious as one might think if clay-size particles are analyzed, because as the particle size is reduced so is microabsorption, i.e., the length of the x-ray path through the particles is reduced (Reynolds, 1989; Bish & Reynolds, 1989). Alternatives for high mass-absorption samples are to use either a high intensity Cu-target x-ray tube, rated at 2 kW or higher, or a Co-target tube. A Co tube must be operated at low power levels, 1 kW or less, although the mass absorption coefficient of Fe is much lower for CoKα than for CuKα radiation, 57.3 vs. 317 cm^2 g^{-1} (Bracewell & Veigele, 1971). We have found that a high-intensity Cu tube produces the same results as a standard Co tube when samples containing up to 500 g Fe-bearing minerals per kilogram are analyzed.

Tropical soils typically contain three categories of Fe: (i) Fe that is incorporated into the crystal structure of phyllosilicates, (ii) the so-called "free" Fe which is generally considered to be all Fe that is not incorporated in phyllosilicates, and (iii) noncrystalline or poorly crystalline Fe-bearing minerals. Normally, the noncrystalline Fe-bearing minerals are lumped into the "free" Fe category. "Free" Fe can be effectively removed by one or more treatments with the sodium citrate-bicarbonate-dithionite (CBD) technique of Mehra and Jackson (1960) (also see Kunze & Dixon, 1986). The acid-ammonium-oxalate method of Schwertmann (1973) can be employed to remove poorly crystalline and noncrystalline Fe from soils. A quantitative measure of the amount of noncrystalline Fe that is removed can be determined by weight difference (Hodges & Zelazny, 1980). Because most tropical soils contain much higher concentrations of Fe-bearing minerals than temperate soils, Fe removal will substantially concentrate the non-Fe-bearing minerals relative to the original soil. The resulting XRD patterns will show considerable improvements in the peak-integrated intensities which may be mistakenly construed as evidence for the mineral concentrations of the original soil. Quantitative values for the non-Fe-bearing minerals can be estimated only after taking into account the concentrations of the Fe-bearing minerals before Fe removal.

Sample Pretreatments

Sample pretreatments and types of mounts used for XRD analysis depend on the kind of information that is needed. For example, the analyst may want to semiquantitatively determine all the *detectable* minerals in a sample. In many instances, however, the analyst may be interested only in the oxide content of the sample or in the analysis of only phyllosilicates. There may also be instances where the analyst is interested only in specific minerals such as quartz, mica, and other minerals that were deposited on the soil surface as tropospheric dust.

The reader is referred to Kunze and Dixon (1986) for the customary procedures of sample pretreatments for mineralogical analyses. What we will present here are "exceptions" to the procedures outlined by Kunze and Dixon that we find are better suited for the XRD analysis of oxide-rich soils from selected areas of the Tropics. Most XRD analyses of oxide-rich soils are intended to disclose the "total" mineralogical content of the sample, i.e., all the minerals that are XRD-detectable. For this type of XRD analysis, a minimum of pretreatments is strongly suggested. The minerals identified in a typical "total" XRD analysis of an agriculturally important oxide-rich soil (e.g., a typical Hawaiian Oxisol) might include kaolinite, halloysite, gibbsite, goethite, anatase, rutile, ilmenite, hematite, magnetite, and maghemite. Note that a typical Hawaiian Oxisol, as opposed to Oxisols from Puerto Rico, the Philippines, and Indonesia, contains no quartz below the surface horizon. The mineralogy of most Ultisols is similar to Oxisols with the exception that the mineral concentration ratios may differ. The surface horizons of many Hawaiian soils contain tropospheric quartz and, to a lesser extent, mica (illite) which can be used to identify buried soils. Many Hawaiian soils contain relatively high concentrations of Mn that are not always XRD-detectable, a fact that introduces an important point. Many soil constituents are not XRD-detectable, even though they are, for all practical purposes, crystalline.

For the suite of minerals listed above, the following consequences will result if the customary pretreatments for silica-rich soils were to be conducted before XRD analysis.

Removal of Organic Matter

The majority of oxide-rich soils found in the areas addressed in this chapter, contain relatively low concentrations of organic matter. With the high temperatures and relative humidity encountered in most of these areas, biological activity is intense and organic matter rapidly decomposes. In upland regions where temperatures are lower, biological decomposition is slower, allowing a greater accumulation of organic matter than on the lower slopes. Nevertheless, organic matter removal is seldom required for the XRD analysis of oxide-rich tropical soils. When the need for organic matter removal is indicated by the appearance of the sample, poor dispersion, or high background on the XRD pattern, then a treatment with 30% H_2O_2 is recommended. Kunze and Dixon (1986) point out that in order for H_2O_2 to efficiently oxidize organic matter the soil should be buffered to a pH of about five. With very few exceptions, high-oxide soils have an acid reaction and do not require additional acidification.

Removal of Manganese-Bearing Minerals

The H_2O_2 is useful for removing Mn minerals, allowing the comparison of XRD patterns with and without poorly crystalline Mn minerals (Jackson, 1979). In addition, Mn removal substantially decreases the mass absorption coefficient for $CuK\alpha$ radiation and results in an improvement in the limit of detection of the remaining minerals. Therefore, treating the sample with H_2O_2 prior to XRD analysis may be of greater value for the dissolution of Mn minerals than for the removal of organic matter.

Soluble Salt Removal

We have found that soluble salt removal is not necessary for the majority of high-oxide soils. With the possible exception of soils that have been irrigated with brackish water or lie in basins, the high rainfall in the Tropics precludes salt accumulations. Dispersion is nearly always a problem with oxide-rich soils. Occasionally, flocculation is caused by an accumulation of salt, in which case dispersion is aided by three or more deionized-water washings each followed by centrifuging the suspension and decanting the supernatant. Centrifuging should be done in 250-mL bottles at a relative centrifugal force of 10,000 x g. The time required for the soil to settle out of suspension depends on its degree of dispersion. Soils washed for the first time may centrifuge out within 5 min. Highly dispersed soils may have to be transferred to 50-mL centrifuge tubes and spun at 28,000 x g for as much as 1 h. Salt accumulations are not the only reason why a soil may not disperse (see the section of this chapter on dispersion).

Free Iron Removal

The primary interest for the majority of soils from Hawaii and elsewhere that we analyzed by XRD is to obtain a better understanding of their chemical behavior (fertility). The Fe-bearing minerals, particularly goethite and ferrihydrite, have a strong influence on phosphate sorption, and to a lesser extent, on sulfate sorption. If free Fe is removed, in most instances the most important information to be gained from XRD analyses may be lost. As is true for all pretreatments, free Fe removal is entirely dependent upon the type of information required from the analysis. In most instances, oxide-rich soils are difficult to disperse until free Fe has been removed. The CBD free-Fe-removal method of Mehra and Jackson (1960) is very effective for goethite, lepidocrocite, and poorly crystalline Fe minerals. We have found that three or more treatments may be necessary for the complete removal of the more resistant Fe-minerals such as hematite and maghemite. Ilmenite and magnetite are not substantially affected by the CBD treatment.

SPECIMEN MOUNTS

The specimen mounts described here are those routinely used in Hawaii for the XRD analyses of oxide-rich soils. When working with particle densities that range from 2.4 g cm^{-3} for gibbsite to 5.2 g cm^{-3} for hematite, it is important to avoid particle-density separations in the specimen such as might occur with some of the centrifuge or suction methods of specimen preparation. In order to produce diffraction patterns of oxide-rich soils that faithfully reflect the relative concentrations of the minerals present, sample pretreatments and preparation of specimen mounts must not result in particle-density separations. The mounts discussed below offer a minimum of particle-density separation.

Bulk-Powder Cavity Mount

If sufficient sample is available, the bulk-powder cavity mount is the easiest and quickest preparation for semiquantitative XRD work. A cavity mount can give a fair amount of semiquantitative information if the crystallites in the specimen are oriented in an approximately random fashion. Oxide-rich soils contain variable amounts of x-ray-amorphous materials. Since the sample in a cavity mount is infinitely thick, the broad background scatter band "amorphous hump" that appears on the pattern is caused solely by the noncrystalline materials in the sample, not the mount. Therefore, the "amorphous hump" is a true indicator of the relative amounts of x-ray-amorphous material in the sample. Semiquantitative estimation of the amount of x-ray-amorphous material in a powder sample is discussed by Jones (1989).

For semiquantitative work with cavity mounts, all samples whose patterns are to be compared must be packed with the same amount of pressure. Standards and unknowns must be prepared with the same sample density and degree of orientation in order to achieve even rough semiquantitative estimates. We have found that most oxide-rich soils are not as subject to preferential orientation as soils containing high percentages of phyllosilicates. Although there is a small amount of preferential orientation of gibbsite and kaolinite, the sample surface in a cavity mount is normally jumbled by the oxides present.

Repeated trials have demonstrated that in order to achieve precision, cavity-mounted samples must be repacked and reanalyzed at least four times. The results (peak integrated intensities) must then be averaged. If there are wide variations between replicates, the sample must be repacked and reanalyzed until at least three replicates agree to within reasonable limits (to be established by the analyst). Such a procedure guarantees precision only, not accuracy. Errors that occur because of packing-density differences and/or orientation differences must be consistent between standards and unknowns. The only way that sample-preparation errors can be "standardized" is to repeat the entire sample-preparation process as many times as it takes to establish "typical" results. "Typical" results may not be accurate results, but will be reproducible, at least among three or more replicates.

We have found that very large disparities in integrated intensities (areas under the peaks) often occur when two people prepare the same sample, even though the same techniques are used. Therefore, the same person should prepare all standards *and* unknowns so there will be a "standard error" between samples.

Uncertainties occur when comparing XRD patterns of unknowns with standards because soil minerals seldom have diffracting domains that are the same size as the standards. Differences in mean diffracting-domain size, morphology, and lattice strains as well as dilution by x-ray-amorphous materials often preclude a direct comparison between soil minerals and standards. As is very often the case, soil goethite produces a different appearing pattern than the "standard" goethite from the stock room. Peak positions may shifted because of Al substitution for Fe. Peak widths will most likely differ between the "standard" and unknown samples because of differences in morphologies. For example, goethite from the island of Kauai consists of needles that average less than 30 nm in diameter by about 300 nm long; therefore, the mineral exhibits intense anisotropic peak broadening because of its acicular (needlelike) morphology. Peak widths vary according to the hkl direction of the diffracting planes in the mineral. Those planes having nonzero l Miller indices are generally narrower than peaks with $l = 0$.

A disadvantage of cavity mounts is that they are not suited for heat treatments. Therefore, the same surface cannot be analyzed because the sample must be repacked after heating and may then have a different bulk density and particle orientation.

Paste on Glass Slide

The paste method of Theisen and Harward (1962) produces a specimen surface with a fair degree of preferential orientation. The technique involves troweling a moist paste onto a glass slide with many strokes of a second glass slide until the desired thickness is achieved. Bish and Reynolds (1989) note that there is considerable disorder with pastes troweled on glass slides, but in our experience the degree of order is far superior to the surface of a cavity mount.

The primary disadvantage of a glass-slide mount is that pastes are often very thin, especially with respect to the high diffraction angles. Therefore, an "amorphous hump" usually appears in the diffraction pattern that is associated with the glass in the slide rather than x-ray-amorphous material in the specimen. Furthermore, if the specimen is not infinitely thick at the diffraction angles of the minerals of interest, peak intensities will be proportionately decreased (Reynolds, 1989).

Paste on Molybdenum Slide

Molybdenum-metal slides are microcrystalline and as such produce a flat background. The Mo slides are rugged, inexpensive, and can be reused. Pastes are applied to Mo slides in the same manner as glass slides, specifically the method described by Theisen and Harward (1962). The primary advantage of Mo slides over glass slides is that the "amorphous hump" (if any) is produced by the specimen, not the slide. Furthermore, an advantage of Mo slides over the cavity mount holders is that they can be used for solvating the sample and mild heat treatments. The Mo slides produce peaks at 40.5, 58.6, and 73.7 $°2\theta$ (CuKα) and 47.3, 69.2, and 88.1 $°2\theta$ (CoKα). There is a very weak peak at 52.5 $°2\theta$ (CuKα) on a clean Mo slide. The 52.5 $°2\theta$ peak is not noticeable when analyzing a paste. Peaks produced by Mo slides are usually not a problem because most of the useful information collected from oxide-rich soils lies in the range of 2 to 38 $°2\theta$ (CuKα).

Two of several possible sources of molybdenum sheet metal are:

1. Alfa Products, division of Johnson Matthey
   P.O. Box 8247
   Ward Hill, MA 01835-0747
   A 2.5-mm thick sheet, 100 by 100 mm is listed as stock number 10047.

2. Strem Chemicals, Inc.
   P.O. Box 108
   Newburyport, MA 01950
   A 2.5-mm thick sheet, 100 by 100 mm is listed as stock number 42-0080.

Both of these products are listed as 99.9 + % molybdenum. We have used both 1.0-mm and 2.5-mm thick Mo sheets and have found that the 2.5-mm sheets are preferable because the flatness of the slides is much truer after cutting and is much easier to maintain. The sheets can be cut to approximately the dimensions of petrographic glass slides using a band saw equipped with a blade suitable for cutting medium to soft metals. Each slide must be ground to true flatness on a plate-glass surface by using 600-mesh silicon-carbide grit or finer. A highly polished slide surface is not necessary inasmuch as a dull finish minimizes sample peeling if hydrated gels are present. The slides can be inscribed with an identifying mark on one end. Because of mechanical and heat stress, Mo slides should be periodically checked for flatness by polishing with a fine grinding compound on a plate-glass surface. Warped slides will be obvious if, after a few strokes on the glass surface, the dull finish is not uniform over the entire slide.

The primary advantage of Mo slides over zero-background slides is cost. *One* zero-background slide costs about $100.00 (Al cavity mount plus installed quartz plate plus shipping and tax, etc.). One 100- by 100-mm sheet of 2.5-mm thick Mo costs about $90.00 and *eight* slides can be cut from one 100- by 100-mm sheet. Therefore, each Mo slide costs less than $12.00 and they are virtually unbreakable.

A disadvantage of Mo slides is that they will oxidize if heated at high temperatures, thus producing additional peaks. However, if oxidation occurs, the slides can be repolished.

Paste on Zero-Background Mounts

Zero-background mounts are slices of single crystals such as quartz that are cut at an angle such that no reflecting planes are oriented parallel to the slide surface--hence no peaks are produced by the slide. As is the case for Mo slides, a zero-background mount produces a flat background. Any background on the diffraction pattern is, therefore, produced by the sample, not the slide. A paste is applied to a zero-background mount exactly as it is to a glass or Mo slide.

The primary disadvantage of zero-background mounts are their expense. Zero-background mounts for a Philips diffractometer are about $100.00 each. The second disadvantage is that the mounts are relatively fragile and can be easily broken. Samples on zero-background mounts can be solvated but not heated.

Powder on Zero-Background Mounts

The quartz plates in zero-background mounts are slightly recessed in their aluminum frames so the sample surface will not project above the focusing circle of the goniometer. Therefore, zero-background mounts are well-suited for dry powder samples. All samples, with the possible exception of those with very low mass absorption coefficients, are infinitely thick on zero-background mounts. Samples with appreciable amounts of x-ray-amorphous materials will display the characteristic "amorphous hump" without any contribution from the mount. A bonus advantage is that in instances where only a small amount of sample is available, too little to fill a standard cavity mount, the available sample may be adequate to fill a zero-background mount.

DATA COLLECTION

This section is included for those who are producing XRD patterns with instruments capable of step scanning. We assume, perhaps mistakenly, that those laboratories that collect XRD patterns by step scanning should have no trouble analyzing oxide-rich soils. For those laboratories equipped with diffractometers capable of collecting data only by a strip-chart recorder, scanning at 2 °2Θ min^{-1} may be too fast, in which case the discussion below may also apply.

Integration Interval

Integration interval is the time (in seconds) that the goniometer pauses to collect counts at each 2Θ step. Many students, as well as colleagues, hurriedly try to scan a 3000-step pattern as quickly as possible. With such impatience, very few counts are collected at each step. Total x-ray counts at such low levels follow a Poisson distribution (Bevington, 1969) as a function of time. That is, if the count rate is low, the counting time must be proportionately long in order for meaningful crystallographic information to be seen above the random background produced by the x-ray tube and scattered by the sample. The detector and the electronics also contribute to the background. Very often the minerals of particular interest are scarcely detectable above background.

The following experiment may help convince impatient analysts of the necessity for long integration intervals. First, set the integration interval to a very short time, say, 0.5 s. Then repeatedly, without moving the goniometer, collect and record the counts per each integration interval. Continue collecting counts 50 to 100 times. After a sufficiently large number of integration intervals have been counted, sort the counts according to their magnitudes. The integration intervals that contain the same number of counts are called *bins*. Each bin will then be identified by the number of counts and the number of times that count occurred. For example, a bin may represent a count of 10 that occurred 15 times. Plot the results as follows. On the x - axis, plot the bins according to the number of counts they represent. On the y -axis, plot the frequency of repeat counts in each bin.

The graph produced from the data collected from the above experiment will contain an asymmetric curve that is skewed toward the bins with the fewest counts; a Poisson distribution. If the experiment were to be repeated with, perhaps, a 10-s integration interval, a graph of the results would approach a Gaussian distribution which will be approximately symmetrical about the mean of the counts collected during the integration interval.

In analytical terms, assume the mean of a Poisson distribution is 10 counts. Liebhafsky et al. (1972, p. 334) show that the standard counting error is the square root of the mean number of counts collected during a given integration interval. Therefore, for a mean of 10 counts, the standard counting error is 3.16 or (3.16/10) x 100 $\simeq$ 32% error. With a mean of 100 counts per integration interval, the standard counting error is 10 or (10/100) x 100 = 10% error. On a peak where 1000 counts or more are collected, the counting error is 3.2% or less.

Because many of us are not convinced unless we experiment for ourselves, we suggest that you may not want to accept the analytical approach for determining the optimum integration interval. Experimentation will show that the standard counting error like the standard deviation is the full width at half maximum (FWHM) of a Gaussian distribution. You will also find by experiment that as the mean number of counts per integration interval increases, the FWHM of the Gaussian distribution will decrease. A decrease in the standard counting error means an increase in the confidence level of the data. Setting the integration interval is a trade-off between data quality and the time required to collect the data. We have found that we can collect reasonably good data from high-Fe Oxisols with a Co tube operated at 1 kW by counting for 10 s at each step. For 3000 steps the total run time is about 10 h. The lowest background is about 55 counts in 10 s. Samples with 300 g Fe per kilogram or less can be analyzed with a high intensity, 2.2-kW Cu tube. For these samples, the integration interval is set from 2 to 4 s.

Step Size

In addition to unreasonably short integration times there are those who attempt to accelerate the process by using large step sizes. Large step sizes, in the order of 0.1 $^\circ 2\theta$ produce useable data if all of the peaks are very broad and if the pattern is not intended for Rietveld refinement or curve fitting. If the pattern contains sharp peaks, e.g., gibbsite, quartz, etc., then a step size of 0.1 $^\circ 2\theta$ is too large. With too few steps, peaks could appear shifted from their true position or may be cut off at the top so that their true intensity is not evident. In any case, if the pattern is to be refined the minimum step size must be 0.025 $^\circ 2\theta$. For accuracy in curve fitting of very sharp peaks, a step size of 0.005 $^\circ 2\theta$ may be necessary. The more steps there are across a peak the more significance the chi square will have for the least-squares minimization during Rietveld refinement or curve fitting.

Theta-Compensating and Fixed Divergence Slits

All intensities listed in the powder-diffraction database maintained by the Joint Centre for Diffraction Studies (JCPDS) apply to a fixed divergence slit. With a fixed divergence slit a constant sample volume is irradiated by the x-ray beam. As the sample is rotated to higher angles, the surface area of the sample exposed to the beam decreases as a function of the sine of the angle. Also, as the sample is rotated, the beam penetrates more deeply. As a result of an increase in beam penetration coupled to a decrease in sample area exposure, a constant sample volume is exposed to the x-ray beam, assuming, of course, that the sample is infinitely thick with respect to the x-ray beam.

Theta-compensating divergence slits open to larger apertures as the diffraction angle increases, maintaining exposure of the same surface area to the x-ray beam at all angles. When properly adjusted, a sample length of 1.3 cm is exposed to the beam at all angles above $2 \, ^\circ 2\theta$. As the diffraction angle increases the volume of sample exposed to the x-ray beam also increases, resulting in more intense peaks at the higher diffraction angles than are obtained with a fixed divergence slit.

There are two primary reasons why the use of theta-compensating divergence slits are in dispute. First, the intensities obtained from a theta-compensating slit do not match the intensities given in the JCPDS database. Most search/match software will not work with data collected by use of a theta-compensating slit. Second, the intensities obtained with theta-compensating slits between 2 and $20 \, ^\circ 2\theta$ are lower than those obtained with a fixed slit. Thus, with the theta-compensating divergence slit, detection limits for most clay minerals are not as good as with the fixed-divergence-slit arrangement. At angles above $20 \, ^\circ 2\theta$, however, the opposite is true.

We find that XRD data can be collected from oxide-rich soils by using a theta-compensating slit that cannot otherwise be obtained with a fixed-divergence-slit system. The majority of minerals that influence the chemical and physical properties of high-oxide soils are the oxides, oxyhydroxides, and hydroxides of Al, Fe, Mn, and Ti. Very poorly crystallized minerals such as ferrihydrite are also important. Without exception these minerals can be detected and characterized by use of a theta-compensating slit whereas with a fixed divergence slit they may not be discernible above the background. There is a definite advantage to using fixed divergence slits for the analyses of phyllosilicates because a much larger sample area can be exposed to the x-ray beam than with theta-compensating slits.

The relationship between data collected with a theta-compensating divergence slit and that collected with a fixed divergence slit (i.e., constant-volume data) is:

Theta-Compensating   Constant-Volume
$(I_{TC} / \sin\Theta)$ ---------------------------------> (normalize)
(normalize) <-------------------------------- $(I_{CV} \sin\Theta)$

To convert theta-compensating intensities to constant-volume intensities, divide by the sine of each peak position ($2\Theta/2$) and then normalize the values on the basis of 100 for the most intense peak. To convert constant-volume intensities as listed in the JCPDS files to theta-compensating intensities, multiply by the sine of each peak position ($2\Theta/2$) and then normalize the values on the basis of 100 for the most intense peak. We have confirmed these relationships with synthetic goethite analyzed by using both a fixed and a theta-compensating divergence slit. The true peak intensities above background were then found by curve fitting. The maximum difference found between the intensities collected by the two methods was less than 3%.

Corrections for Specimen-Surface Displacement

When the surface of the specimen does not lie exactly on the focusing circle of the goniometer the locations of diffraction peaks are shifted slightly. These frequently encountered peak shifts are a function of the sample-surface displacement, S in centimeters from the goniometer focusing circle, and the diffraction angle, Θ in radians, for each peak in the pattern. The relationship given by Parrish and Wilson (1959) is

$$\Delta 2\theta = 2S \cos\theta / R \qquad [3]$$

where $\Delta 2\Theta$ is the peak shift at angle Θ in radians, and R is the goniometer radius in centimeters. The peak shift value can be converted from radians to degrees by multiplying by $(180/\pi)$. A BASIC computer program is included in this section as an aid to understanding and correcting peak positions that are shifted as a result of sample-surface displacements (Fig. 10-2).

The surface of a blank slide or a cavity mount is exactly tangent to the focusing circle if the goniometer is properly aligned. Therefore, the surface of a sample mounted on a slide lies *above* the focusing circle. The magnitude of $\Delta 2\Theta$ is dependent on the thickness of the sample. On the other hand, the surface of a properly prepared specimen in a cavity mount lies on the focusing circle, and, therefore, no peak shift occurs because of specimen-surface displacements. Normally, if the sample mounted on a slide is only as thick as it needs to be in order to be infinitely thick for the 2Θ range of interest then the magnitude of the peak shift will be only in the thousandths of degrees 2Θ---too small to be noticed on strip-chart recordings. Modern diffractometers, however, are capable of recording peak positions to within three decimal places. For exacting work, peak shifts due to sample displacements are discernible and can confuse the interpretation of the pattern. For example, goethite peak positions are useful for estimating the mole fraction of Al substituting for Fe in the lattice (Schulze, 1984). Unless the sample thickness on a slide is very thin, or the thickness is known so a correction can be made, the estimates of aluminum substitution for Fe will not be accurate.

A common suggestion for the elimination of peak shifts due to specimen-surface displacements is to cover the entire slide with the sample. If specimens could be prepared perfectly flat with a uniform thickness, then the surface of the sample would be approximately tangent to the focusing circle. However, this practice is absolutely *NOT* recommended because of the likelihood that the slide may be tilted in the goniometer, destroying the two-to-one alignment of the specimen. Furthermore, covering the entire slide with sample contaminates the sample holder causing unpredictable peak shifts and intensity losses.

MINERAL IDENTIFICATION

Halloysite and Kaolinite

Both mica and hydrated halloysite produce a peak that is close to 1.0 nm. In Hawaii, halloysite is often found in regions of high rainfall where tropospherically deposited mica is also found. Therefore, the primary objective is to distinguish between the two minerals. Because hydrated halloysite loses water at a relatively low temperature, a heat treatment of 60 °C for approximately 15 min will collapse the c-axis spacing of the mineral from 1.0 nm to about 0.73 nm. At this temperature, little or no change in the other mineral constituents occurs and the specimen may be used for other purposes including solvation.

In making quantitative estimates of the amount of halloysite or kaolinite (a distinction which must be confirmed by TEM), the area and shape of the 02ℓ peak, as well as the 00ℓ peaks, should be measured. The gibbsite 110 and 200 peaks and the goethite 110 peak lie on the high-angle tail of the halloysite 02ℓ peak. A heat treatment at 350 °C will destroy gibbsite and goethite and, consequently, a clear profile of the high-angle tail of the halloysite 02ℓ peak can be measured. We have found that the destruction of gibbsite and goethite is also useful for quantifying the concentration of spinels at a d-spacing of about 0.485 nm (gibbsite 002 peak) and hematite at a d-spacing of about 0.270 nm (goethite 130 peak).

On occasion, the $hk\ell$ reflections of halloysite and kaolinite as well as various 00ℓ orders will interfere with a clear observation of the oxide peaks. In these instances, heating the specimens at 540 °C for 2 h will destroy the halloysite and kaolinite, thereby eliminating the interference. Higher temperatures may be used if high-temperature glass is used for sample mounting. Common petrographic slides will warp at these temperatures.

```
10 ' BASIC PROGRAM TO FIND THE PEAK-SHIFT ERROR CAUSED BY SAMPLE
20 ' DISPLACEMENT ABOVE OR BELOW THE GONIOMETER FOCUSING CIRCLE.
30 '
40 ' R = GONIOMETER RADIUS - CHANGE TO APPROPRIATE VALUE AS NECESSARY.
50 ' MM = SAMPLE DISPLACEMENT IN MILLIMETERS.
60 ' S = SAMPLE DISPLACEMENT IN CENTIMETERS.
70 ' TWOTH = DEGREES 2-THETA.
80 ' THETA = DEGREES THETA.
90 ' RTHETA = THETA IN RADIANS.
100 ' DELTA = DELTA 2-THETA IN RADIANS.
110 ' D2THETA = DELTA 2-THETA IN DEGREES.
120 ' E2THETA = SUM OF PEAK POSITION + DELTA 2-THETA (DEGREES).
130 ' F = FORMAT STRING
140 '
150 DEFSTR F
160 F = "##.###"
170 PI = 3.1415926#
180 R = 17.4 ' GONIOMETER RADIUS (cm) FOR A PHILIPS DIFFRACTOMETER.
190 '
200 CLS : PRINT : PRINT : PRINT : PRINT
210 PRINT TAB(10)"ENTER THE SAMPLE DISPLACEMENT IN MILLIMETERS."
220 PRINT : PRINT TAB(10)"ENTER A POSITIVE VALUE IF THE SAMPLE SURFACE"
230 PRINT TAB(10)"IS ABOVE THE GONIOMETER AXIS OF ROTATION. FOR EX-
AMPLE,"
240 PRINT TAB(10)"A THICK SAMPLE MOUNTED ON A GLASS SLIDE."
250 PRINT : PRINT TAB(10)"ENTER A NEGATIVE VALUE IF THE SAMPLE SURFACE"
260 PRINT TAB(10)"IS BELOW THE GONIOMETER AXIS OF ROTATION. FOR
EXAMPLE,"
270 PRINT TAB(10)"IF A CAVITY MOUNT IS NOT COMPLETELY FULL."
280 INPUT MM
290 S = MM / 10 ' CONVERT MILLIMETERS TO CENTIMETERS.
300 '
310 CLS : PRINT : PRINT : PRINT : PRINT
320 PRINT TAB(10)"ENTER THE PEAK POSITION IN DEGREES 2-THETA"
330 PRINT TAB(10)"WHERE THE PEAK SHIFT IS TO BE MEASURED."
340 INPUT TWOTH
350 THETA = TWOTH / 2 ' FIND DEGREES THETA.
360 RTHETA = THETA * (PI / 180) ' CONVERT DEGREES TO RADIANS.
370 DELTA = 2 * S * COS(RTHETA) / R ' THE UNITS OF DELTA ARE IN RADIANS.
380 D2THETA = DELTA * (180 / PI) ' CONVERT RADIANS TO DEGREES.
390 E2THETA = TWOTH + D2THETA ' SUM PEAK POSITION & DELTA 2-THETA.
400 '
410 CLS : PRINT : PRINT : PRINT : PRINT
420 PRINT TAB(10)"PEAK POSITION (DEGREES 2-THETA)     = ";
430 PRINT USING F;TWOTH
440 PRINT TAB(10)"SAMPLE DISPLACEMENT                = ";
450 PRINT USING F;MM;
460 PRINT " mm"
470 PRINT TAB(10)"PEAK SHIFT (DELTA 2-THETA)         = ";
480 PRINT USING F;D2THETA
490 PRINT TAB(10)"PEAK POSITION + PEAK SHIFT         = ";
500 PRINT USING F; E2THETA
510 '
520 END
```

Fig. 10-2. BASIC source code for correction of specimen-surface displacement from goniometer
    focusing circle.

Illitic Minerals

Many oxide-rich soils developed on basaltic parent materials contain extremely low concentrations of potassium (K). With very few exceptions, Hawaiian volcanics contain virtually no K feldspars. The illitic materials found in Hawaii have been dated and found to be at least 100 times older than the volcanics of the islands (Dymond et al., 1974) and recent investigations (e.g., Betzer et al., 1988) have confirmed that the illite found has been carried in by tropospheric winds. Because illite is the major source of K in many Hawaiian soils, its detection by XRD is of interest. In most instances, the illite yields poor XRD patterns due to poor stacking order among the aluminosilicate layers. The pattern can be improved by saturating the specimen with K and heating for at least 2 h (overnight is preferred) at 110 °C. The advantage of this treatment is that a Mo slide can be used for the mount. In some instances, illite can be detected in samples receiving the heat treatment without the K-saturation step.

Interstratified Phyllosilicates

For interstratified clay minerals, diffraction maxima may not occur at the d-spacing expected for the endmembers. According to Mering's principles (Sawhney, 1989) a single, broad, asymmetric diffraction maximum will appear somewhere between the peak positions of the two end members. For example, in the absence of discrete illite and smectite, interstratified illite/smectite will produce a diffraction maximum between the d-values of 1.6 nm for smectite and 1.0 nm for illite. The exact position will depend on the concentration ratios of the two endmembers. If there is a greater abundance of illite than smectite, the interstratified phase will produce a peak between 1.3 nm (the expected d-value for equal amounts of illite and smectite) and 1.0 nm (the d-value for pure illite).

The best way to interpret XRD patterns of interstratified phases is by use of Reynolds' NEWMOD computer program (1985). With the program, "guesses" are entered for the phase ratios, the type of ordering, and the minimum and maximum number of repeat aluminosilicate layers per diffracting-domain. The program then calculates the diffraction profile for the parameters that were entered and displays a plot of the results. After several iterations, a calculated pattern similar to the experimental pattern can be produced and the analyst can be confident that the input parameters are similar to those of the actual sample.

Short-Range Order Materials

X-Ray Amorphous Materials

Oxide-rich soils developed on volcanic ash typically contain a high proportion of x-ray amorphous materials. The typic Hydrudands found on the Hamakua Coast on the island of Hawaii contain over 500 g x-ray amorphous materials per kilogram. The effects of high concentrations of x-ray amorphous materials are threefold. First, the amorphous materials "dilute" the crystalline minerals and absorb (scatter) x-rays. The diffraction intensities of the crystalline minerals in field moist [as much as 3000 g water per kilogram oven-dry soil (300% water!); SCS-USDA, 1976] samples are so diminished that in some instances the entire sample appears to be x-ray amorphous. When the sample is allowed to air dry, the proportion of hydrated amorphous gel is reduced due to water loss, and the intensity of crystalline mineral peaks such as gibbsite increase. For years there was an unpublished notion among the early investigators that as Hydrudands air dried, gibbsite was spontaneously crystallized from the soil solution. Secondly, x-ray amorphous materials scatter x-radiation in wide bands on the 2Θ scale. Glass scatters x-radiation in slightly different bands than does alumina gel. Organic matter and water scatter x-radiation in yet other 2Θ bands. Regardless of the type of x-ray amorphous material, most scattered x-radiation occurs in the range of about 20° through at least 50° 2θ for CuKα. When strip-chart recorders are used, the high background intensity from some amorphous materials may require that the scale factor be reduced to a sensitivity level where peaks produced by crystalline minerals may be difficult to distinguish above the background. Finally, x-ray amorphous materials can interfere with the orientation of tabular-shaped minerals, and produce peak intensities that are not representative of their concentrations.

Allophane

We prefer the definition of allophane as specified by van Olphen (1971), "Allophanes are members of a series which are hydrous aluminum silicates of widely varying chemical composition, characterized by short-range order, by the presence of Si-O-Al bonds [as determined by infared spectroscopy], and by a differential-thermal-analysis curve displaying a low-temperature endotherm and a high-temperature exotherm with no intermediate endotherm." van Olphen also specified that allophanes cannot be end members of the aluminosilicate series. In other words, silica or alumina gels do not qualify as allophane. Since the deliberations of the U.S.-Japan Seminar on Amorphous Clay Materials in Fukuoka, Japan, 17-19 September 1969, several researchers have observed XRD peaks or, more accurately stated, bands produced by materials believed to be allophane (Okada et al., 1975; Brown, 1980; van der Gaast et al., 1985; Parfitt, 1990).

Regardless of the definition used for allophane (van Olphen, 1971; Parfitt, 1990; Dahlgren, 1994; see Chapter 14), the material has not been found in XRD-detectable concentrations in Hawaiian surface soils. Early workers (Tamura et al., 1953; Kanehiro & Whittig, 1961; Bates, 1962; Moberly, 1963; Lai & Swindale, 1969) were convinced that Hawaiian soils were rich in allophane because of their apparent x-ray amorphous characteristics. There is a preponderance of x-ray amorphous materials in many Hawaiian soils, although, by accepted definitions, the materials do not qualify as allophane. In most instances, the x-ray amorphous material in Hawaiian Hydrudands consists of Al and Fe gels (Wada & Wada, 1976). Weathering pressures in many areas of the Tropics are responsible for silica depletion in the surface soils. The prime example of silica depletion in Hawaii is the soils on the Hamakua coast of the island of Hawaii. With all due respect to Parfitt et al. (1988), we are of the same opinion as Wada and Wada who identified very little allophane or allophane-like materials in the Hamakua-coast soils, only a preponderance of x-ray amorphous Fe- and Al-oxide gels. We are not of the opinion that silica extracted by acid-oxalate (see Jackson et al., 1986) is a reliable indicator of *only* allophane. Acid-oxalate-extractable silica *might* be an indicator of allophane, but it also can be an indicator of imogolite (Dahlgren, 1994; see Chapter 14), which the Hamakua soils contain, and of short-range-ordered aluminosilicate gels that do not qualify as allophane (Jackson et al., 1986). There is a possibility that short-range-ordered phyllosilicates may be dissolved by an acid-oxalate treatment (Fey & LeRoux, 1977; Makumbi & Jackson, 1977). Materials that do conform to the definition of allophane have been found in isolated pockets and almost always in association with imogolite (Patterson, 1971; Wada et al., 1972).

Hudnall (1977) investigated a profile consisting of pyroclastics believed to be deposited during the 1924 (Powers, 1948) phreatic Halemaumau explosions at the Kilauea volcano. The XRD patterns of the Kilauea profile showed that the 10 horizons sampled contained an abundance of x-ray amorphous material. Imogolite and smectite were present (confirmed by TEM), as well as small amounts of an allophanelike material which appeared spherical in TEM photographs. For the most part, TEM showed the amorphous material to be volcanic glass. The point to be made is that the lack of discrete peaks and the appearance of an "amorphous hump" in XRD patterns are not and cannot be criteria for the presence of allophane. Acid-oxalate-extractable SiO_2 and Al_2O_3 were nearly identical in each horizon of the Kilauea profile. Acid-oxalate extracted 15.2 g SiO_2 kg^{-1} and 16.7 g Al_2O_3 kg^{-1} from the horizon that contained imogolite (confirmed by TEM) and 15.4 g SiO_2 kg^{-1} and 15.5 g Al_2O_3 kg^{-1} from the horizon that contained smectite (confirmed by XRD and TEM). Based on this data we cannot say that the silica extracted from the Kilauea profile is an accurate indicator of the percentage of allophane that was present in each horizon. We can only say, based on XRD, TEM, and extraction procedures that the Kilauea profile contains mostly x-ray amorphous materials.

We have found a subsoil collected near a soil-rock boundary at a depth of 180 to 200 cm that appears to be rich in allophane. Figure 10-3 is an XRD pattern of the Maile soil, which is tentatively classified as an hydrous, isomesic, Acrudoxic Hydrudand (H. Ikawa, 1993, personal communication). For curve fitting, the pattern was divided into two sections. The first section (Fig. 10-4) employs a Lorentz-polarization distribution as a background, whereas the second section (Fig. 10-5) employs a Gaussian distribution as a background to approximate the "amorphous hump." In view of the fact that clean glass slides produce "amorphous humps" in XRD patterns, such a feature does not necessarily indicate the presence of allophane.

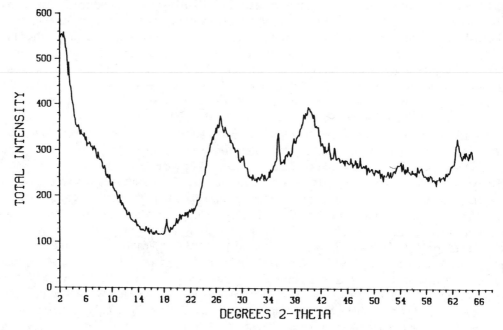

Fig. 10-3. XRD full pattern of the Maile soil collected near a rock-soil interface at a depth of 180 to 200 cm.

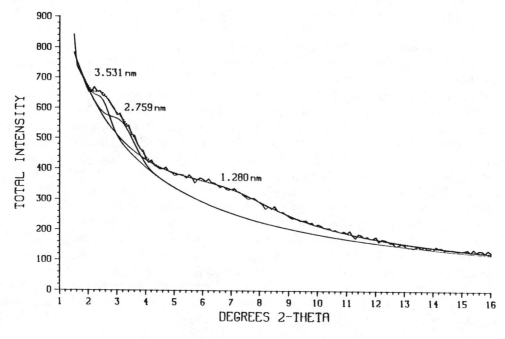

Fig. 10-4. A truncated portion of the Maile soil XRD pattern in Fig. 10-3. Truncation changes the pattern's scaling and provides an improved visual perception of the goodness of curve fit.

In Fig. 10-4, the first maximum was fit with two peaks labeled 3.5 and 2.8 nm. Whether there are actually two discrete peaks in the range of 3.5 to 2.8 nm is debatable because of their location where the Lorentz-polarization factor is changing rapidly. Nonetheless, the positions of the first two peaks in Fig. 10-4 agree with the findings of van der Gaast et al. (1985) who reported very strong peaks in the range of 3.0 to 3.4 nm. The second peak in Fig. 10-4 is very broad, indicating a very short range of order. Although the peak centers at 1.27 nm there is strong likelihood that there was a peak shift due to excessive broadening. Table 10-1 lists d-values that we attribute to allophane compared to d-values from the literature. The d-values that we report in Table 10-1 were found by curve fitting the patterns in Fig. 10-4 and 10-5 (the sharp peaks in Fig. 10-5 were ignored).

Figure 10-5 contains two types of information, the positions and relative integrated intensities of the peaks (areas under the peaks) and the area between an imaginary straight-line background

Table 10-1. Comparison of allophane XRD data obtained from the literature and from the Maile subsoil, island of Hawaii.

Brown (1980)		van der Gaast et al. (1985)		Parfitt (1990)		Present work Fig. 10-4, 10-5	
d	I†	d	I‡	d	I‡	d	I§
		3.0-3.4	VS			3.5	3
						2.8	7
1.5	S	1.1-1.5	w				
				1.2	100	1.27	46 B
0.43	w			0.43	15	0.425	4
0.34	VS			0.34	100	0.337	100
						0.300	6
0.25	S						
				0.22	65	0.225	56
				0.19	5	0.189	8
0.18	vw						
				0.17	10	0.170	6
						0.162	6
0.14	M			0.14	25 B	0.144	30

†VS = very strong, S = strong, M = medium, w = weak, vw = very weak, B = broad.
‡Intensities are our visual estimates from the published XRD patterns.
§Intensities are relative areas of the peaks shown in Fig. 10-4 and 10-5 found by curve fitting and normalized to a maximum of 100.

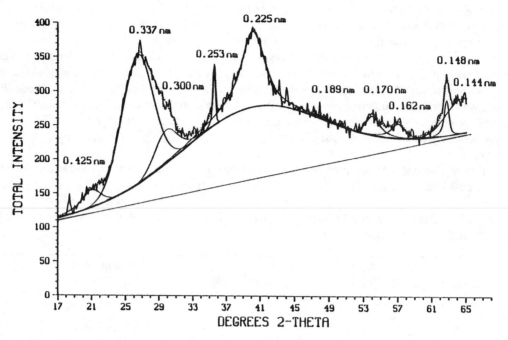

Fig. 10-5. Upper-2Θ portion of the Maile soil XRD pattern in Fig. 10-3. The background comprises a Gaussian distribution and a straight line that are optimized during curve fitting. The area between the Gaussian Distribution and the straight line provides an estimate of the amount of x-ray amorphous materials that are present in addition to the allophane in the sample.

and a Gaussian distribution that is superimposed on the line. When analyzed as a bulk powder or mounted on a zero-background slide, the slope of the imaginary background line and the area between the line and an approximation of the "amorphous hump" can be determined by curve fitting. We intend to make two points from the curve-fit data in Fig. 10-5. First, from Table 10-1 there is good agreement between the peaks believed to be allophane (in the Maile soil) and the literature values. Figure 10-6 is a TEM of the Maile soil sample that produced the XRD patterns in Fig. 10-3, 10-4, and 10-5. The largest spherical objects are approximately 20 nm in diameter. The smallest spheres visible in the micrograph are about 5 nm in diameter. The hollow spheres seen in Fig. 10-6, although larger than most allophanes cited in the literature (for example, Henmi & Wada, 1976; Wada & Wada, 1977), agree with the published morphology and XRD patterns. We must point out, however, that no allophane was found in the top meter of the Maile soil.

Second, more important than the identification of allophane in the Maile soil is that a large proportion of the x-ray amorphous material in the soil does not qualify as allophane. The area enclosed by the Gaussian background and an imaginary straight line beneath the pattern provides an estimate of the *difference* in the amount of x-ray amorphous material (excluding allophane) among samples collected from a soil profile or along a transect. By following the change of the area under the "amorphous hump" and comparing this area with the integrated intensities of the allophane peaks (bands), the proportion of allophane relative to x-ray amorphous material can be estimated. Wada (1989) makes the following statement: "In the last decade, information has been accumulated to indicate that allophane is not amorphous (without form) and some long-range order develops in the structure of allophane." The most important point we wish to convey is that not all x-ray amorphous material is allophane.

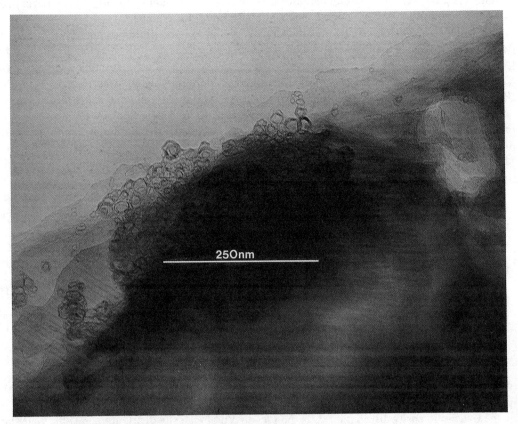

Fig. 10-6. Transmission electron micrograph of the Maile soil.

CURVE-FITTING PROCEDURES

Until about a decade ago, semiquantitative estimates were necessarily made by measuring peak intensities with a ruler or by determining the integrated intensities (areas under the peaks) by use of a planimeter. Strip-chart recorders are still occasionally used for quick survey work and for very rough alignment procedures; however, most XRD data are collected by computer in a step-scan mode. Peak parameters, ie., exact positions (°2θ), intensities (total counts per integration interval), full width at half maximum (FWHM) (°2θ), and shapes (functions that describe the contours of the peaks) can be found for almost any number of isolated peaks as well as peaks in unresolved clusters. Once the closest-fit peak parameters have been found, integrated intensities can be calculated.

Pi'o Pili Pa'a

There are several excellent algorithms available that utilize least-squares minimization. We have had considerable success with the Simplex algorithm (Nelder & Mead, 1965); however, we now routinely use the Levenberg-Marquardt algorithm (Levenberg, 1944; Marquardt, 1963). From our experience, the Simplex method of minimization requires an excessive amount of CPU time to converge if the initial parameters are far from the minimum (the closest fit). On the other hand, Langford et al. (1986) report that there is a minimum of parameters correlation with the Simplex algorithm, meaning that one parameter is not strongly influenced by the others. The Levenberg-Marquardt algorithm converges very quickly, whether the parameters are far from the minimum *or* close to the minimum. A significant fault with this algorithm is its tendency to correlate parameters. We have successfully minimized parameters correlation by modifying the Levenberg-Marquardt algorithm through normalizing and constraining the parameters.

In order to avoid confusion with commercially available curve-fitting computer programs, we named our program Pi'o Pili Pa'a, which is Hawaiian for "curve close fit" (Jones, 1989). The program, written in FORTRAN-77, utilizes a modified version of the subroutine that appears in Bevington (1969). Three major modifications were made to Bevington's subroutine. With Pi'o Pili Pa'a, the parameters are normalized on the basis of 0.0 to 1.0 (all parameters placed into the same order of magnitude). Constraints can be assigned to each parameter by the program user. Parameters can be assigned a status of a constant (not considered in the minimization) or allowed to vary. In addition, Bevington's subroutine was modified to assure that parameters that have "hit" their constraints or that are not contributing to the progression toward a minimum are temporarily removed from the calculation.

Combinations of two or more functions are usually required to fit XRD patterns from oxide-rich soils because of their small diffracting-domain size and lattice strains. The basic functions that are used by most profile refinement programs are:

$$\text{Gaussian} \quad I_g = I_o \exp\{-\ln 2[2(P - P_o)/FWHM]^2\} \qquad [4]$$

$$\text{Cauchy} \quad I_c = I_o \{1 + [2(P - P_o)/FWHM]^2\}^{-1} \qquad [5]$$

$$\text{pseudo-Voigt} \quad I_v = I_g \delta + I_c (1 - \delta) \qquad [6]$$

$$\text{Pearson-VII} \quad I_p = I_o \{1 + (2^{1/m} - 1)[(P - P_o)/FWHM]^2\}^{-m} \qquad [7]$$

where I_g is the Gaussian intensity distribution, I_o is the maximum peak intensity, P_o is the peak position in °2Θ, P is the 2Θ step position, $FWHM$ is the full width at half maximum (°2Θ), I_c is the Cauchy intensity distribution, I_v is the pseudo-Voigt intensity distribution, δ is the Gaussian-Cauchy mixing function (δ can vary between 0.0 for pure Cauchy to 1.0 for pure Gaussian distribution, I_p is the Pearson-VII intensity distribution, and m is the Pearson-VII shape parameter.

The two combinations that are used most often are (i) a split pseudo-Voigt that incorporates a different combination of Gaussian and Cauchy distributions on each side of a peak, and (ii) a split Pearson-VII with different shape parameters, m, on each side of a peak. Pi'o Pili Pa'a also has the facility to use a pseudo-Voigt on one side of a peak and a Pearson-VII on the other side.

Pi'o Pili Pa'a utilizes Pearson-VII shape parameters that are <1. Values of the Pearson-VII shape parameter, m, may vary from 1, a Cauchy distribution, to infinity, a Gaussian distribution. The approximate relationship between the Gaussian-Cauchy mixing function, δ, and the Pearson-VII shape parameter, m, is summarized in Table 10-2.

Table 10-2. Approximate relationship between the Pearson-VII shape
parameter, m, and the Gaussian-Cauchy mixing function, δ.

m	δ	m	δ
1	0.000	10	0.872
2	0.424	20	0.935
3	0.597	50	0.974
4	0.691	100	0.987
5	0.750	∞	1.000

For all values of m between one and infinity, the Pearson-VII distribution is asymptotic to zero. The distribution is asymptotic to values >0 if m is <1. Therefore, if m is <1, as it can be in Pi'o Pili Pa'a, the area under the peak (integrated intensity) cannot be rigorously found by

$$A_{P7} = [(\pi(FWHM)2^{2(1-m)})/(2^{1/m} - 1)^{0.5}][\Gamma(2m - 1)/(\Gamma(m))^2] \qquad [8]$$

where A_{P7} is the area under a Pearson-VII distribution (Langford et al., 1986; Hall et al., 1977; Freiberger, 1960, p. 1173). For m values <1, however, an approximation of the area under a Pearson-VII peak can be determined by the trapezoidal method, which we have found differs by only a maximum of 3% when compared with the area found by the above equation. The comparison was made using an m value of 1 in both instances.

Figure 10-7 illustrates the differences between Gaussian, Cauchy, and Pearson-VII distributions. A pseudo-Voigt distribution with a Gaussian-Cauchy mixing function, δ, of 0.5 describes a distribution midway between the Gaussian and Cauchy distributions. Although, the $K\alpha_1$ and $K\alpha_2$ lines from the x-ray tube are included in the final fit and graphics, only the $K\alpha_1$ line is used by the curve-fitting algorithm. That is, only the intensity, position, FWHM, and shape of the $K\alpha_1$ line is used by the algorithm. For example, if a peak has an intensity of 100 total counts (sum of $K\alpha_1$ and $K\alpha_2$ lines), only the $K\alpha_1$ intensity of 67 counts is considered by the algorithm. Likewise, if the FWHM of a peak is 0.3 °2θ, only the FWHM of the $K\alpha_1$ line, which is <0.3 °2θ, will be included by the algorithm.

Inasmuch as we are often interested in the amount and diffracting-domain size of goethite, the "background" in the vicinity of the goethite *110* peak is important to approximate. By "background" we are referring to the combined intensities of adjacent peak tails and prevailing background. The minerals that produce an interfering tail are halloysite and poorly crystalline kaolinite. In order to approximate the decay of the halloysite *02ℓ* peak, a pseudo-Voigt/modified-

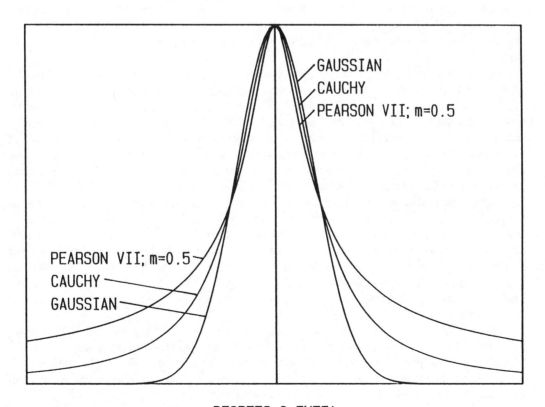

DEGREES 2-THETA

Fig. 10-7. A comparison of peak shapes for Gaussian, Cauchy, and Pearson type-VII distributions.

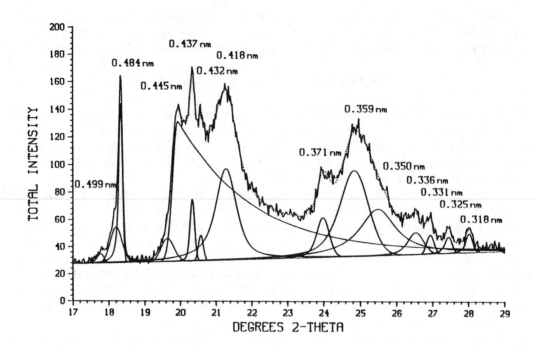

Fig. 10-8. Example of a pseudo-Voigt/modified-exponential distribution (peak labeled 0.445 nm). Peaks labeled 0.437 through 0.318 are riding on the exponential tail of the halloysite *02ℓ* peak. The halloysite tail serves as a background for the peaks riding on it.

exponential option was included in Pi'o Pili Pa'a. Figure 10-8 is a curve-fit XRD pattern that incorporates a pseudo-Voigt/modified-exponential peak. Note that two peaks must be used to approximate a halloysite *02ℓ* peak--a pseudo-Voigt/modified-exponential peak and a pseudo-Voigt peak.

Jones (1989) shows a progression of XRD patterns whereby a "standard" mixture of halloysite, gibbsite, and goethite was curve-fit. Halloysite was curve-fit to establish the profile of the *02ℓ* tail before gibbsite and goethite was added. The XRD pattern from the mixture of the three minerals was then curve-fit. Next, the mixture was heated to 350 °C to destroy the gibbsite and goethite and the pattern was again curve-fit to confirm the profile of the halloysite *02ℓ* tail. The profile of the heated halloysite *02ℓ* tail was remarkably similar to the tail of the unheated sample, thus confirming that the area between the halloysite tail and the goethite profile represented a close approximation of the integrated intensity of goethite in the mixture.

A frequently asked question about curve fitting and peak decomposition is, "are the results unique?" In other words, is there no other possible way the pattern can be curve-fit? The answer is no. The more peak functions that are added to a data-set, the better the fit will be. In fact, since there is always noise in the data, an infinite number of peak functions are required to achieve a perfect fit. Therefore, a user of Pi'o Pili Pa'a, or any other curve-fitting program for that matter, must have a maximum amount of prior knowledge. With the knowledge acquired from the total pattern and from as many other patterns as possible, we know what minerals are present in the sample that produced the pattern in Fig. 10-9.

The pattern in Fig. 10-9 is fit with 16 peaks, all of which are identified. The approximate positions and intensities with a narrow range of constraints were entered as a first approximation. The parameters were then refined to achieve a best fit of the raw data. Within the confines that we placed on the parameters prior to refinement, we can say the results are unique, and we can say with confidence that the results make good mineralogical sense.

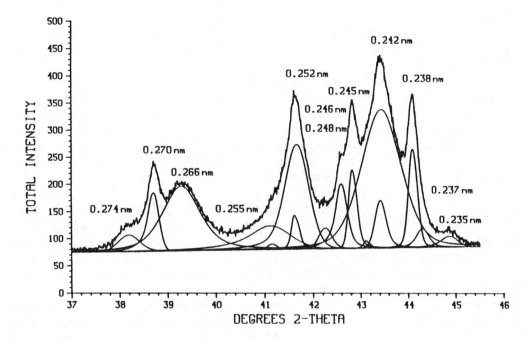

Fig. 10-9. Curve-fit XRD pattern of the Kauai bauxite (Kapaa series) surface soil. Peaks found by curve fitting are from ilmenite, hematite, goethite, maghemite, gibbsite, anatase, and rutile.

Diffracting-Domain Size

Refined peak widths (FWHM of the $K\alpha_1$ line) are useful for estimating diffracting-domain size by use of the Scherrer equation,

$$B = K \lambda(180/\pi)/(b - \beta)\cos\Theta, \qquad\qquad [9]$$

where B is the diffracting-domain size in nanometers, b is the peak FWHM in $^\circ 2\Theta$, β is the instrumental broadening of the diffractometer in $^\circ 2\Theta$, λ is the x-ray wavelength in nanometers, and K is a shape factor that is approximately unity (Klug & Alexander, 1974). Klug and Alexander suggest that K would best be taken as 0.89. Theta (Θ) in the Scherrer equation is the Bragg angle of the peak in radians. Thus, the term $(180/\pi)$ is required to convert degrees to radians (see Jenkins & de Vries, 1978).

There is often confusion when discussing particle size. When separating particle size on the basis of Stoke's law we often refer to aggregates as "particles." In order to avoid confusion, we prefer to use the term "diffracting domain" for a regular repeat of crystal planes between two crystal defects or from a defect to the crystal surface. Ionic substitutions are not considered defects in the sense of our reference to diffracting domains. A crystallite, therefore, consists of a three-dimensional array of diffracting domains. The results obtained by the Scherrer equation represent a *mean* diffracting-domain size.

Peak widths alone should not be used to calculate diffracting-domain sizes; peak shapes must also be considered. The universally accepted assumption is that diffracting-domain size is described by a Cauchy distribution, and strain or crystal-defect broadening is described by a Gaussian distribution (Young & Wiles, 1982; de Keijser et al., 1982; Delhez et al., 1982; Klug & Alexander, 1974; Alexander & Klug, 1950). Therefore, when applying the Scherrer equation, only the Cauchy width should be used. Computer program Pi'o Pili Pa'a finds the Cauchy width independent of the Gaussian width.

Peak asymmetry occurs because of axial divergence of the incident and diffracted beams and in some instances because of crystal defects such as layer-stacking faults in halloysite. In most instances, however, the upper-2Θ side of the peak can be used for diffracting-domain-size estimations inasmuch as axial divergence primarily affects the lower-2Θ side of the peak (Alexander, 1954). We must caution that even when using the Cauchy width of the upper-2Θ side of peaks for diffracting-domain-size estimates, care must be taken not to confuse nonuniform strain with size broadening.

Another point to consider is the widely varying range of mineral shapes (morphologies) that are encountered in soils. Mineral shapes can be tabular, tubular, spherical, cubic, fibrous, acicular (needle-like), etc. Mean diffracting-domain sizes, therefore, apply only to a dimension normal to the $hk\ell$ planes that produce each peak. By curve fitting as many $hk\ell$ peaks as practical, the mean morphology of a mineral's crystallites can be estimated. We have found that morphologies determined by XRD curve fitting do not always agree with the morphologies seen with TEM. In many instances, particles observed by TEM are mosaics of crystallites.

The Scherrer equation applies if two conditions are met. First, the diffracting-domain size must fall in a range where peak broadening is produced. Cullity (1978) states that with extremely exacting experimental techniques, diffracting-domain sizes as large as 200 nm (0.2 μm) can be determined by the Scherrer equation if there is no evidence of strain broadening. Cullity also states that the most accurate size-range to measure is 0 to 50 nm. Jenkins and de Vries (1978) give graphic evidence that the greatest accuracy of the Scherrer equation lies in the range of 2 to 50 nm. Second, peak broadening must not be caused by nonuniform crystal-lattice strains. We have observed that many minerals, halloysite, for example, will produce "peaks" whose position ranges from >1.0 to 0.73 nm (Fig. 10-10). In such an instance, the "peak" in Fig. 10-10 is produced by a continuous variation in the 00ℓ spacings of halloysite. A model of nonuniform strain is given by Cullity, page 287. Although strain, per se, is most likely not involved with the halloysite in Fig. 10-10, the pattern shows that the hydrated phases were present in approximately equal amounts and the diffracting-domain sizes were similar. As a matter of interest, the sample in Fig. 10-10 was a slickenside collected from a cave-in site at the trans-Koolau tunnel presently under construction.

Considerable information can be collected by curve-fitting XRD patterns. We have found that in many instances the absolute amount of a given mineral is not as important for the interpretation

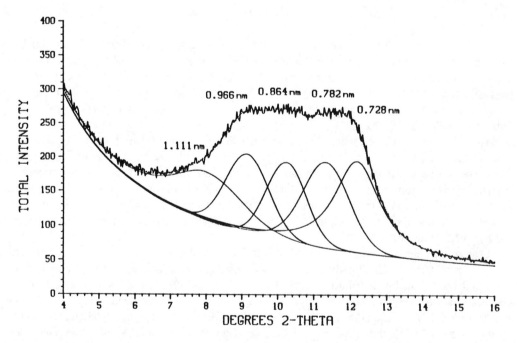

Fig. 10-10. Curve-fit XRD pattern of halloysite that ranges from fully hydrated (1.111 nm) to completely dehydrated (0.728 nm). Although a very good fit was achieved with five peaks, there are more than likely an "n" number of hydration states. This clay mineral is responsible for cave-ins in the trans-Koolau tunnel.

of the chemical or physical behavior of a soil as a specific property of the mineral, for example, diffracting-domain size. Figure 10-11 is a scatter pattern of phosphate sorption vs. goethite diffracting-domain size. The samples were collected from a vertical section at a bauxite deposit on the island of Kauai. The soil for data point number 1 in Fig. 10-11 was collected from a depth of 0 to 10 cm. Sample number 12 was collected from a depth of 260 to 290 cm. Figure 10-11 presents a clear pattern of the decrease in the mean diffracting-domain size of goethite with depth. We also see that if bauxite were to be mined, the phosphate-sorption problem would be about three times greater than if the soil was left intact. Although gibbsite in Kauai bauxite averages about 500 g kg^{-1}, it has very little effect on phosphate sorption because its mean diffracting-domain size is very large. Large enough, in fact, that a portion of it can be seen with a hand lens. Without peak-parameter refinements, the data shown in Fig. 10-11 would have been impossible to obtain.

The methods of peak-parameter refinements described here are most useful for quantitative estimates of *changes* in mineral quantities, mean diffracting-domain sizes, hydration states, and lattice substitutions between suites of samples. We are not prepared to say that curve fitting/peak decomposition provides absolute mineral concentrations; the program was not written for such a purpose. Pi'o Pili Pa'a was written to discern subtle differences between tropical soils rich in poorly crystalline oxides and interstratified phyllosilicates. For profile refinements of samples having little or no x-ray amorphous materials such as, for example, the bauxites found on the island of Kauai, we feel the Rietveld procedures are perhaps the only way of determining absolute mineral concentrations by standardless XRD methods. For Rietveld procedures the reader is referred to Taylor (1991), Post and Bish (1989), and to Bish (1994; see Chapter 9).

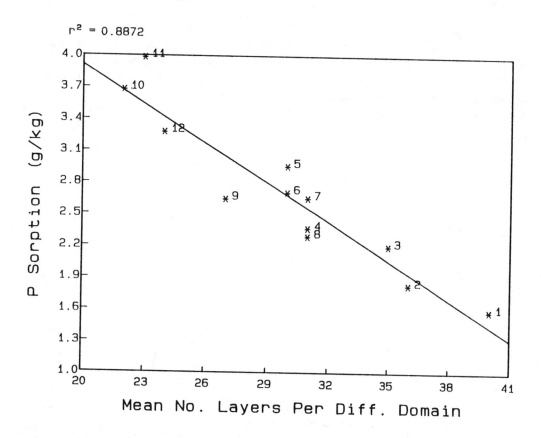

Fig. 10-11. Relationship between phosphate sorption and the diffracting-domain size expressed in terms of the mean number of *110* layers of goethite in the Kauai bauxite (Kapaa series). Data point 1 is from the 0- to 10-cm horizon and data point 12 is from the 260- to 290-cm horizon.

Comparison With the Rietveld Refinement Technique

Whereas the Rietveld refinement technique can be used for quantitative analyses of reasonably crystalline minerals, curve fitting is best used to discern subtle differences between samples. Differences in peak shapes, FWHM, position shifts, and integrated intensity can be very accurately determined by curve-fitting methods. Sample differences such as mean diffracting-domain size of each mineral in a sample, morphology, lattice substitutions for some minerals, and differences in mineral concentrations can be accurately estimated by curve fitting. Unless standards of known composition are available, curve fitting a whole pattern can only provide rough estimates of absolute mineral concentrations.

If standards of known composition are available, then standard curves can be constructed for each sample matrix. The mineral composition of all samples must be approximately similar in terms of percentages and must all contain the same series of minerals. For example, the bauxite soil on the island of Kauai contains gibbsite, goethite, hematite, magnetite, anatase, rutile, and ilmenite. Notice that the soil contains no XRD-detectable phyllosilicates. There is less than 20 g total SiO_2 kg^{-1} soil, whereas the gibbsite content varies from 370 to 600 g kg^{-1} and that of goethite from 190 to 460 g kg^{-1}. By using CoKα radiation and infinitely thick samples, there is a minimum of variation of mass absorption from sample to sample.

Because the bauxite samples from Kauai had a similar matrix, they were used to test the relationship between the information gained from curve fitting and from the Rietveld refinement. First, the clay fraction from each of 10 horizons was analyzed in cavity mounts with a cobalt tube operated at 1 kW. A fixed divergence slit set to one degree was used for all diffraction angles. The samples were step-scanned from 10 to 78 $^\circ 2\theta$ with a step increment of 0.025 $^\circ 2\theta$ for 10 s per step. The patterns were then truncated into two segments, 20 to 22 $^\circ 2\theta$ for the gibbsite *002* peak (Fig. 10-12), and 22 to 27 $^\circ 2\theta$ for the gibbsite *110* and *200* peaks and the goethite *110* peak (Fig. 10-13). The pattern segments were then curve-fit using computer program Pi'o Pili Pa'a to find the integrated intensities of the peaks as well as the other parameters listed above.

The gibbsite *002* peak was fitted with three peaks, the most intense peak being for the mineral and the other two peaks for instrumental broadening. Therefore, to find the integrated intensity for gibbsite *002*, only the area under the third, most intense peak was used. The areas under the

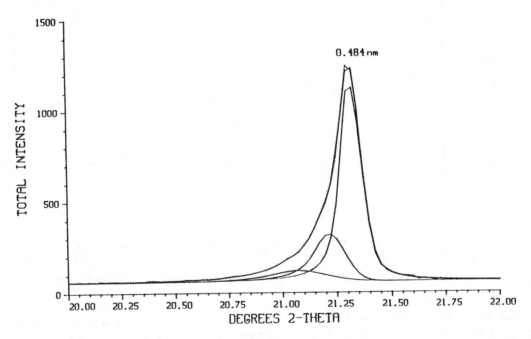

Fig. 10-12. Curve-fit *002* gibbsite peak. Material is the Kauai bauxite, 30- to 50-cm depth, CoKα radiation, 380 g gibbsite per kilogram bauxite.

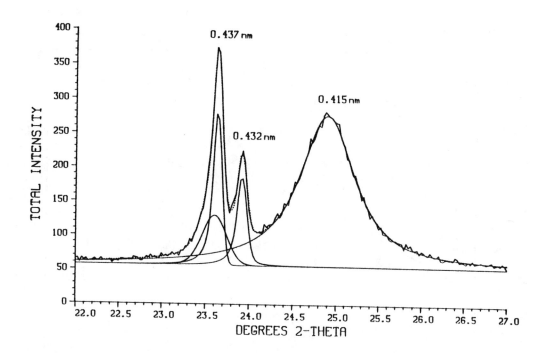

Fig. 10-13. Curve-fit *110*, *200* gibbsite and *110* goethite peaks. Material is Kauai bauxite, 30- to 50-cm depth, CoKα radiation, 460 g goethite per kilogram bauxite.

first two peaks are the instrumental contribution to the overall width of the peak. Figure 10-12 shows that the fit is not exact at the top of the peak. The FWHM of the main peak is only 0.09 °2θ, meaning that the peak is only slightly more than four steps wide at half height. Therefore, the fit at the peak top is at the limit of resolution of the pattern. The misfit is the difference between two steps.

In Fig. 10-13, gibbsite *110* is labeled 0.437 nm and gibbsite *200* is labeled 0.432 nm. The wide peak labeled 0.415 nm is Al-substituted goethite. The broad peak that extends across the two gibbsite peaks accounts for instrumental broadening. Instrumental broadening is involved with the goethite *110* peak also but because the peak is so broad (0.7 °2θ), instrumental width comprises only a small fraction of the overall broadening. The extreme width of the goethite *110* peak is due to small diffracting-domain size and internal lattice strains.

When analyzing plate-shaped minerals such as gibbsite, peak intensities vary as a function of preferential orientation. Figure 10-14 shows the correlation between the concentrations of gibbsite as found by full-pattern Rietveld refinement and the integrated intensities of the *002* gibbsite peaks (areas under the peaks) found by Pi'o Pili Pa'a. Preferential orientation was not corrected in Fig. 10-14 although the correlation coefficient ($r^2 = 0.82$) is reasonably good. In order to estimate a correction for preferential orientation, the integrated intensity of the gibbsite *110* peak relative to the *002* peak was calculated for each horizon. For example, for the 20- to 30-cm horizon (Point 3 in Fig. 10-14 through 10-16), the area of the *110* peak was divided by the area of the *002* peak and then multiplied by 100 to get a relative intensity of 24% for the *110* peak vs. the *002* peak. For the 10 horizons appearing in Fig. 10-14 through 10-16, the most intense *110* peak relative to the *002* peak was 33% (Point 7 on Fig. 10-14 through 10-16). Then on the basis of Point 7, all other relative intensities were normalized with Point 7 = 1.0 and Point 3 = 0.73, etc. The integrated intensities of the *002* peaks from each horizon were then reduced by their respective normalized orientation correction factor and plotted in Fig. 10-15.

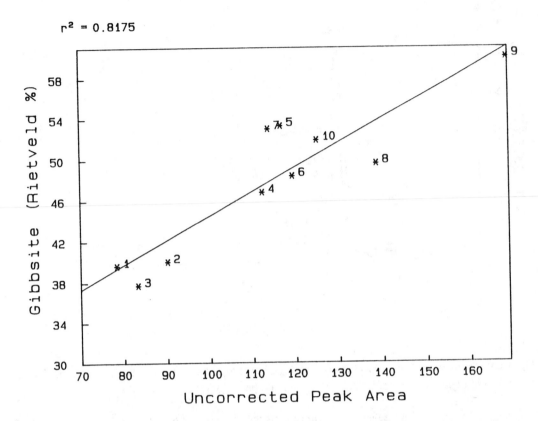

Fig. 10-14. Linear regression of gibbsite concentration (weight %) as found by full-pattern
   Rietveld refinement with the gibbsite *002* peak area that has not been corrected for preferential
   orientation.

For perfectly random orientation, the relative intensity of the *110* peak should be close to 36%
(JCPDS File No. 29-0041, 1986, calculated intensities). Therefore, as the degree of preferential
orientation increases, the intensity of the *002* peak will increase relative to peaks whose index ℓ
= 0. As Fig. 10-15 shows, decreasing the integrated intensities of the *002* peaks improved the
correlation between the two methods ($r^2 = 0.92$). However, additional factors may need to be
considered to improve the correlation between the curve-fit and Rietveld data.

Preferential orientation was apparently not a problem with goethite. Figure 10-16 is a
regression plot of the integrated intensity of the *110* goethite peak vs. the results of the full-pattern
Rietveld refinement. By curve-fitting diffraction patterns from other samples with a similar matrix,
the percentage of a given mineral can be calculated from the slope and intercept of the regression
line for that mineral. The level of confidence is directly reflected by the correlation coefficient for
the appropriate standard curve. For example, the level of confidence will be higher for goethite
on the basis of the standard curve in Fig. 10-16 ($r^2 = 0.97$) than for gibbsite on the basis of the
standard curve in Fig. 10-15 ($r^2 = 0.92$). Even the concentrations of gibbsite found on the basis
of the standard curve in Fig. 10-15 would, for most work, be perfectly acceptable.

One might question, why employ a curve fitting program when Rietveld refinement will
directly give very close estimates of mineral percentages? The amount of difficulty in using the
Rietveld programs relative to the much less complex curve-fitting programs might be cited as the
primary reason for using only curve fitting for the analysis of XRD data. This reason is not
justified, however, because full-pattern Rietveld refinements are required for accurate estimates of
mineral concentrations! The two approaches may be considered complementary rather than
competitive. They differ in the type of information produced and in the time required to obtain
that information. For example, with a 10-s step[-1] integration time, each full pattern of the Kauai

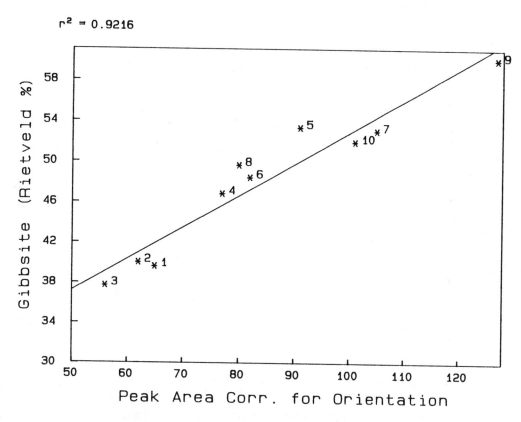

Fig. 10-15. Linear regression of gibbsite concentration (weight %) as found by full-pattern Rietveld refinement with the gibbsite *002* peak area that has been corrected for preferential orientation.

bauxite soil samples required about 10 h of diffractometer time to step from 10 to 78 °2θ. If only curve fitting was to be done for gibbsite and goethite such as in Fig. 10-12 and 10-13, an integration time of 10 s step^{-1} would require about 1 h assuming the same step size in both instances. For a total of 10 samples, diffractometer time for the curve-fit approach would be a little over 10 h to step from 20 to 27 °2θ compared to 100 h for the full-pattern Rietveld refinement approach. One must judge which method to use on the basis of the type of information that is being sought. We use both techniques in our work with oxide-rich soils.

FUTURE TRENDS

Future techniques for the study of oxide-rich soils will involve the increasing use of computer software. There is also a thrust toward the use of synchrotron radiation for the study of all crystalline materials. In particular, oxide-rich soils typically contain minerals with very small diffracting domains that are believed to contain a high degree of lattice strain. X-rays emitted from a tube radiate in a semielliptical pattern from the focal line. To reduce the divergence of the x-rays that impinge upon the sample, collimator plates are placed in the incident beam. There is a considerable loss of intensity at the sample by using thin-plate collimators and there is still a significant horizontal divergence of the beam. Vertical divergence is also large with an x-ray tube. As a result of an inherently divergent x-ray beam, peak shapes are distorted and the resolution of closely spaced peaks is poor, rendering the distinction of diffracting-domain-size broadening from lattice-strain broadening very difficult if not impossible.

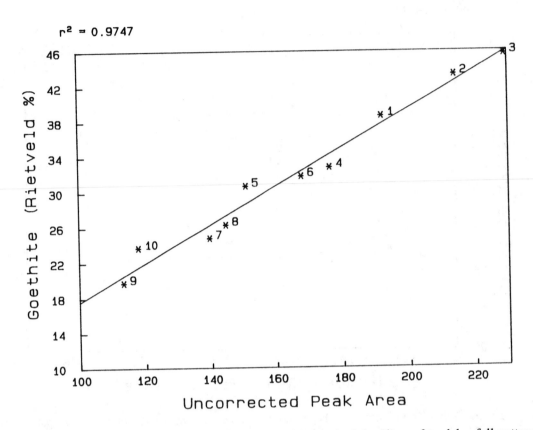

Fig. 10-16. Linear regression of goethite concentration (weight %) as found by full-pattern Rietveld refinement with the goethite *110* peak area. A correction for preferential orientation of goethite was not necessary.

An x-ray beam from a synchrotron source operating at an energy of 2.5 GeV has a vertical divergence of about 0.2 mrad. Such a vertical divergence results in a beam thickness of 0.5 cm at a working distance of 25 m from the source (Finger, 1989). Finger presents a diffraction pattern of the alpha-quartz *212*, *203*, and *301* peaks with a calculated FWHM of $0.02\,°2\theta$ that can be typically achieved with a 2.5-GeV synchrotron source. By comparison, a high-resolution diffractometer with a fine-focus x-ray tube will produce the same pattern with peak widths of $0.1\,°2\theta$ or wider. The x-ray tube sources with both incident and diffracted beam collimators display a certain amount of peak asymmetry. There is almost an order-of-magnitude better resolution and nearly symmetrical peak shapes with synchrotron radiation as opposed to an x-ray tube source. Better resolution and better peak symmetry translate into greatly improved chances of quantifying and separating diffracting-domain-size broadening from lattice-strain broadening.

In the future, there will certainly be an increased use of synchrotron radiation and more sophisticated software to analyze the data from oxide-rich soils. Other than size separations, there will be a minimum of sample pretreatments. Because the beam from a synchrotron source is nearly parallel as opposed to a divergent beam from an x-ray tube, samples must be very fine for synchrotron radiation. Particle sizes that produce acceptable diffraction patterns with an x-ray tube may not produce satisfactory results with synchrotron radiation. Therefore, increasingly stringent requirements for dispersion and clay separation of oxide-rich soils will be needed in order to take full advantage of synchrotron radiation.

SUMMARY AND CONCLUSIONS

Oxide-rich soils are often difficult to disperse for the separation of the clay fraction. These soils are usually strongly aggregated, giving a first impression that their texture is a fine sand or coarser. With persistence, many oxide-rich soils from Hawaii can be reduced to a clay texture by ultrasound or cavitation shaking. Soil pH usually ranges from five to six, sometimes as low as four. These soils can be best dispersed by raising their pH to between 8.0 and 8.5 after the aggregates have been reduced to discrete crystallites.

For XRD, we strongly recommend a minimum of chemical pretreatments for specimen preparation. Oxide-rich soils typically contain high concentrations of iron-bearing minerals that are often more important for some fertility considerations, phosphate sorption for example, than the phyllosilicates. There is usually nothing to be gained by removing "free" Fe. A situation where "free" Fe removal proved to be beneficial was in our study of surface soils containing Asiatic tropospheric dust. The removal of Fe-bearing minerals concentrates the other minerals and allows a higher degree of preferred orientation of the phyllosilicates.

Some "x-ray-amorphous" soils display a series of broad peak-like maxima superimposed on an amorphous scattering band. The scatter band, sometimes referred to as an amorphous hump, can be approximated by one or more Gaussian-shaped curves. By curve fitting the diffraction pattern with computer programs such as Pi'o Pili Pa'a, the integrated intensities and d-values of the peak-like maxima can be estimated independently of the amorphous scatter band. Therefore, the broad XRD maxima that have d-values associated with allophane can be separated from amorphous materials that do not qualify as allophane. The abundance of amorphous materials relative to allophane and crystalline minerals can be estimated by the area between the Gaussian distribution and a straight-line background under the curve.

Curve fitting alone is not appropriate for quantitative estimates of soil mineralogy. If standards with similar matrices are available for comparison, then integrated intensities (areas under the peaks) that are determined by curve fitting can give good estimates of mineral abundances. Curve fitting used with the Rietveld refinement procedure can yield a great deal of information about a sample, such as mean surface-area estimates of single phases in complex mixtures such as soils.

REFERENCES

Alexander, L. 1954. The synthesis of x-ray spectrometer line profiles with applications to crystallite size measurements. J. Appl. Phys. 25:155-161.

Alexander, L. and H. P. Klug. 1950. Determination of crystallite size with the x-ray spectrometer. J. Appl. Phys. 21:137-142.

Atkinson, R. J., R. L. Parfitt, and R. St. C. Smart. 1974. Infra-red study of phosphate adsorption on goethite. J. Chem. Soc. Faraday Trans. I. 70:1472-1479.

Bates, T. F. 1962. Halloysite and gibbsite formation in Hawaii. Clays Clay Miner. 9:315-328.

Betzer, P. R., K. L. Carder, R. A. Duce, J. T. Merrill, N. W. Tindale, M. Uematsu, D. K. Costello, R. W. Young, R. A. Feely, J. A. Breland, R. E. Berstein, and A. M. Greco. 1988. Long-range transport of giant mineral aerosol particles. Nature (London) 336:568-571.

Bevington, P. R. 1969. Data reduction and error analysis for the physical sciences. McGraw-Hill, New York.

Bish, D.L. 1994. Quantitative x-ray diffraction analysis of soils. p. 267-295. In J. Amonette and L.W. Zelazny (ed.) In Quantitative methods in soil mineralogy. SSSA Misc. Publ. SSSA, Madison, WI.

Bish, D. L. and R. C. Reynolds, Jr. 1989. Sample preparation for x-ray diffraction. In D. L. Bish and J. E. Post (ed.) Modern powder diffraction, Reviews in Mineralogy. Vol. 20. Miner. Soc. Am., Washington, DC.

Bracewell, B. L. and W. J. Veigele. 1971. Tables of x-ray mass attenuation coefficients for 87 elements at selected wavelengths. Dev. Appl. Spectr. 9:357-400.

Brown, G. 1980. Associated Minerals. p. 361-410. In G. W. Brindley and G. Brown (ed.) Crystal structures of clay minerals and their x-ray identification. Miner. Soc., London.

Busacca, A. J., J. R. Aniku, and M. J. Singer. 1984. Dispersion of soils by an ultrasonic method that eliminates probe contact. Soil Sci. Soc. Am. J. 48:1125-1129.

Cullity, B. D. 1978. Elements of x-ray diffraction, 2nd ed. Addison-Wesley Publ. Co., Inc., Reading, MA.

Dahlgren, R.A. 1994. Quantification of allophane and imogolite. p. 430-452. *In* J. Amonette and L. W. Zelazny (ed.) Quantitative methods in soil mineralogy. SSSA Misc. Publ. SSSA, Madison, WI.

de Keijser, Th. H., J. I. Langford, E. J. Mittemeijer, and A. B. P. Vogels. 1982. Use of the Voigt function in a single-line method for the analysis of x-ray diffraction line broadening. J. Appl. Cryst. 15:308-314.

Delhez, R., Th. H. de Keijser, and E. J. Mittemeijer. 1982. Determination of crystal size and lattice distortions through x-ray diffraction line profile analysis. Fresenius Z. Anal. Chem. 312:1-16.

Dymond, J., P. E. Biscaye, and R. W. Rex. 1974. Eolian origin of mica in Hawaiian soils. Geol. Soc. Am. Bull. 85:37-40.

El-Swaify, S. A. 1980. Physical and mechanical properties of Oxisols. p. 303-324. *In* B. H. G. Theng (ed.) Soils with variable charge. N. Z. Soc. Soil Sci., Lower Hutt, New Zealand.

Fey, M. V. and J. LeRoux. 1977. Properties and quantitative estimations of poorly crystalline components in sesquioxide soil clays. Clays Clay Miner. 25:285-294.

Finger, L. W. 1989. Synchrotron powder diffraction. p.309-331. *In* D. L. Bish and J. E. Post (ed.) Modern powder diffraction, Reviews in Mineralogy. Vol. 20. Miner. Soc. Am., Washington, DC.

Freiberger, W. F. (ed.). 1960. The international dictionary of applied mathematics. D. Van Nostrand Co., Inc., New York.

Genrich, D. A., and J. M. Bremner. 1972a. A reevaluation of the ultrasonic-vibration method of dispersing soils. Soil Sci. Soc. Am. Proc. 36:944-947.

Genrich, D. A. and J. M. Bremner. 1972b. Effect of probe condition on ultrasonic dispersion of soils by probe-type ultrasonic vibrators. Soil Sci. Soc. Am. Proc. 36:975-976.

Hall Jr., M. M., V. G. Veeraraghavan, H. Rubin, and P. G. Winchell. 1977. The approximation of symmetric x-ray peaks by Pearson type VII distributions. J. Appl. Cryst. 10:66-68.

Henmi, T., and K. Wada. 1976. Morphology and composition of allophane. Am. Miner. 61:379-390.

Hodges, S. C. and L. W. Zelazny. 1980. Determination of noncrystalline soil components by weight difference after selective dissolution. Clays Clay Miner. 28:35-42.

Hudnall, W. H. 1977. Genesis and morphology of secondary products in selected volcanic ash soils from the island of Hawaii. Ph.D. diss. Univ. of Hawaii at Manoa, Honolulu.

Jackson, M. L. 1958. Soil chemical analysis. Prentice-Hall Inc., Englewood Cliffs, NJ.

Jackson, M. L. 1979. Soil chemical analysis: Advanced course. 2nd ed. M.L. Jackson, Madison, WI.

Jackson, M. L., C. H. Lim, and L. W. Zelazny. 1986. Oxides, hydroxides, and aluminosilicates. p. 101-150. *In* A. Klute (ed.) Methods of soil analysis, Part 1. 2nd ed. Agron. Monogr. 9. ASA and SSSA, Madison, WI.

Jenkins, R. and J. L. de Vries. 1978. Worked examples in x-ray analysis, 2nd ed. Philips Technical Library, Springer-Verlag, New York.

Jones, R. C. 1981. X-ray diffraction line profile analysis vs. phosphorus sorption by 11 Puerto Rican soils. Soil Sci. Soc. Am. J. 45:818-825.

Jones, R. C. 1989. A computer technique for x-ray diffraction curve fitting/peak decomposition. p. 52-101. *In* D. R. Pevear and F. A. Mumpton (ed.), Quantitative mineral analysis of clays, CMS Workshop Lectures. Vol. 1. Clay Miner. Soc., Evergreen, CO.

Kanehiro, Y., and L. D. Whittig. 1961. Amorphous mineral colloids of soils of the Pacific region and adjacent areas. Pacif. Sci. 15:477-482.

Klug, H. P., and L. E. Alexander. 1974. X-ray diffraction procedures for polycrystalline and amorphous materials, 2nd ed. Wiley, New York.

Kunze, G. W., and J. B. Dixon. 1986. Pretreatment for mineralogical analysis. p. 91-100. *In* A. Klute (ed.) Methods of soil analysis. Part 1. 2nd ed. Agron. Monogr. 9. ASA and SSSA, Madison, WI.

Lai, S., and L. D. Swindale. 1969. Chemical properties of allophane from Hawaiian and Japanese soils. Soil Sci. Soc. Am. Proc. 33:804-808.

Langford, J. I., D. Louer, E. J. Sonneveld, and J. W. Visser. 1986. Applications of total pattern fitting to a study of crystalline size and strain in zinc oxide powder. Powd. Diffr. 1:211-221.

Levenberg, K. 1944. A method for the solution of certain non-linear problems in least squares. Quart. Appl. Math. 2:164-168.

Liebhafsky, H. A., G. H. Pfeiffer, E. H. Winslow, and P. D. Zemany. 1972. X-rays, electrons, and analytical chemistry. Wiley-Interscience, New York.

Makumbi, M. N., and M. L. Jackson. 1977. Weathering of Karroo argillite under equatorial conditions. Geoderma 19:181-197.

Marquardt, D. W. 1963. An algorithm for least-squares estimation on nonlinear parameters. J. Soc. Indust. Appl. Math. 11:431-441.

Mattson, S. 1932. The laws of soil colloidal behavior: IX. Amphoteric reactions and isoelectric weathering. Soil Sci. 34:209-240.

McClune, W. F. (ed.). 1986. Mineral powder diffraction file data book. International Centre for Diffraction Data, Swarthmore, PA.

Mehra, O. P., and M. L. Jackson. 1960. Iron oxide removal from soils and clays by a dithionite-citrate system buffered with sodium bicarbonate. Clays Clay Miner. 7:317-327.

Moberly, R., Jr. 1963. Amorphous marine muds from tropical weathered basalts. Am. J. Sci. 261:767-772.

Nelder, J. A., and R. Mead. 1965. A simplex method for function minimization. Comput. J. 7:308-318.

Okada, K., S. Morikawa, S. Iwai, Y. Ohira, and J. Ossaka. 1975. A structure model of allophane. Clay Sci. 4:291-303.

Parfitt, R. L. 1990. Allophane in New Zealand - a review. Aust. J. Soil Res. 28:343-360.

Parfitt, R. L., R. J. Atkinson, and R. St. C. Smart. 1975. The mechanism of phosphate fixation by iron oxides. Soil Sci. Soc. Am. J. 39:837-841.

Parfitt, R. L., C. W. Childs, and D. N. Eden. 1988. Ferrihydrite and allophane in four Andepts from Hawaii and implications for their classification. Geoderma 41:223-241.

Parrish, W., and A. J. C. Wilson. 1959. Precision measurements of lattice parameters of polycrystalline specimens. p. 216. In J. S. Kasper and K. Lonsdal (ed.) International tables for x-ray crystallography. Vol. II. Int. Union Crystallogr., Birmingham, England.

Patterson, S. H. 1971. Investigations of ferruginous bauxite and other mineral resources on Kauai and a reconnaissance of ferruginous bauxite deposits on Maui, Hawaii. U.S. Geol. Surv. Prof. Pap. 656:1-70.

Post, J. E., and D. L. Bish. 1989. Rietveld refinement of crystal structures using powder x-ray diffraction data. p. 277-308. In Modern powder diffraction, Reviews in Mineralogy. Vol. 20. Miner. Soc. Am., Washington, DC.

Powers, H. A. 1948. A chronology of the explosive eruptions of Kilauea. Pacif. Sci. 2:278-296.

Reynolds, R. C., Jr. 1985. NEWMOD, a computer program for the calculation of one-dimensional diffraction patterns of mixed-layered clays. Published by the author, 8 Brook Rd., Hanover, NH.

Reynolds, R. C., Jr. 1989. Principles of powder diffraction. p. 1-17. In D. L. Bish and J. E. Post (ed.) Modern powder diffraction. Reviews in Mineralogy. Vol. 20. Miner. Soc. Am., Washington, DC.

Rietveld, H. M. 1969. A profile refinement method for nuclear and magnetic structures. J. Appl. Cryst. 2:65-71.

Sawhney, B. L. 1989. Interstratification in layer silicates. p. 798-799. In J. B. Dixon and S. W. Weed (ed.) Minerals in soil environments. 2nd ed. Agron. Monogr. 9. ASA and SSSA, Madison, WI.

Schulze, D. G. 1981. Identification of soil iron oxide minerals by differential x-ray diffraction. Soil Sci. Soc. Am. J. 45:244-246.

Schulze, D. G. 1984. The influence of aluminum on iron oxides. VIII. Unit-cell dimensions of Al-substituted goethites and estimations of Al from them. Clays Clay Miner. 32:36-44.

Schulze, D.G. 1994. Differential x-ray diffraction analysis of soil minerals. p. 412-429. In J. Amonette and L.W. Zelazny (ed.) Quantitative methods in soil mineralogy. SSSA Misc. Publ. SSSA, Madison, WI.

Schwertmann, U. 1973. Use of oxalate for Fe extraction from soils. Can. J. Soil Sci. 53:244-246.

Sposito, G. 1984. The surface chemistry of soils. Oxford Univ. Press, New York.

Tamura, T., M. L. Jackson, and G. D. Sherman. 1953. Mineral content of low humic, humic and hydrol humic latosols of Hawaii. Soil Sci. Soc. Am. Proc. 17:343-346.

Taylor, J. C. 1991. Computer programs for standardless quantitative analysis of minerals using the full powder diffraction profile. Powd. Diffr. 6:2-9.

Theisen, A. A., and M. E. Harward. 1962. A paste method for preparation of slides for clay mineral identification by x-ray diffraction. Soil Sci. Soc. Am. Proc. 26:90-91.

Uehara, G., and G. Gillman. 1981. The mineralogy, chemistry, and physics of tropical soils with variable charge clays. Westview Press, Boulder, CO.

U.S. Department of Agriculture-Soil Conservation Service. 1976. Soil survey laboratory data and descriptions for some soils of Hawaii. Soil Surv. Invest. Rep. no. 29. Hawaii Agric. Expt. Stn. and the Hawaiian Sugar Planters' Assoc. USDA-SCS. U. S. Gov. Print. Office, Washington, DC.

U.S. Department of Agriculture-Soil Conservation Service. 1984. Soil survey laboratory methods and procedures for collecting soil samples. USDA-SCS Soil Surv. Invest. Rep. no. 1. U. S. Gov. Print. Office, Washington, DC.

van der Gaast, S. J., K. Wada, S.-I. Wada, and Y. Kakuto. 1985. Small-angle x-ray powder diffraction, morphology, and structure of allophane and imogolite. Clays Clay Miner. 33:237-243.

van Olphen, H. 1971. Amorphous clay materials. Science (Washington, DC)171:91-92.

van Olphen, H. 1977. An introduction to clay colloid chemistry, 2nd ed. John Wiley & Sons, New York.

Wada, K. 1989. Allophane and Imogolite. p. 1051-1087. *In* J. B. Dixon and S. W. Weed (ed.) Minerals in soil environments, 2nd ed. SSSA, Madison, WI.

Wada, K., and S.-I. Wada. 1976. Clay mineralogy of the B horizon of two Hydrandepts, a Torrox and a Humitropept in Hawaii. Geoderma 16:139-157.

Wada, S.-I., and K. Wada. 1977. Density and structure of allophane. Clay Miner. 12:289-298.

Wada, K., T. Henmi, N. Yoshinaga, and S. H. Patterson. 1972. Imogolite and allophane formed in saprolite of basalt on Maui, Hawaii. Clays Clay Miner. 20:375-380.

Wada, K., Y. Kakuto, and H. Ikawa. 1986. Clay minerals, humus complexes, and classification of four "Andepts" of Maui, Hawaii. Soil Sci. Soc. Am. J. 50:1007-1012.

Wada, K., Y. Kakuto, and H. Ikawa. 1990. Clay minerals of two Eutrandepts of Hawaii, having isohyperthermic temperature and ustic moisture regimes. Soil Sci. Soc. Am. J. 54:1173-1178.

Whittig, L. D. and W. R. Allardice. 1986. X-ray diffraction techniques. p. 331-362. *In* A. Klute (ed.) Methods of soil analysis: Part 1. 2nd ed. ASA and SSSA, Madison, WI.

Young, R. A., and D. B. Wiles. 1982. Profile shape functions in Rietveld refinements. J. Appl. Cryst. 15:430-438.

11 Qualitative and Quantitative Analysis of Clay Minerals in Soils

R. E. Hughes, D. M. Moore, and H. D. Glass
Illinois State Geological Survey
Champaign, Illinois

The most permanent record of the history of a soil resides in its minerals. The most effective method of identifying and characterizing the minerals in soils is x-ray diffraction (XRD). Questions about the mineral compositions, alteration sequences, effects of parent material, provenances, and type and intensity of weathering in soil can be answered with qualitative and quantitative XRD analysis of clay minerals. Most problems can be solved with carefully taken and prepared samples, and methods that emphasize detailed sampling and relatively rapid analysis over extensive treatments of fewer samples. This paper describes XRD methods that provide useful data about minerals in different types of soils and their parent materials and considers the distinction between qualitative and quantitative data. Soils are multicomponent materials acted upon by pedogenic processes in biogeochemical systems open to exchange of both matter and energy (Sposito, 1989). Soils form in response to pedogenic processes, an aspect of diagenesis (Retallack, 1983). We use the term soil for modern profiles at the surface of the earth and paleosol for a soil that has formed in the geologic past. We include within the definition of soils those that form below standing water and those resulting from entirely abiotic processes at the surface of the earth.

Quantitative phase analysis (QPA), sometimes called quantitative modal analysis, which requires high precision and high accuracy, is difficult under the best of conditions (Brindley, 1980). [We use the term precision here to signify reproducibility. Accuracy is the difference between actual results and the correct answer. Griffiths (1967) has discussed these concepts in detail.] The combination of high precision, high accuracy, and high sample throughput for XRD is not, at present, a possibility for soils and most other clay-rich materials. A fast but precise method of XRD analysis is the method of choice for most soil problems. [We recognize that most of the following tends to fall near one side of the range of approaches common in this field. We are reminded of the 1981 meeting of the Clay Minerals Society and the extended discussion between one of us (HDG) and G. W. Brindley about how the former seldom took time during thousands of analyses to focus on a single sample, whereas the latter could "waste" a month or more on running and rerunning a single sample!] Although single-sample studies drive improvements in the techniques used for larger numbers of samples, the statistical force of large numbers of samples is essential.

The methods we recommend here are particularly suited for screening studies of large numbers of samples. The results of screening studies are an essential first step for identifying the few samples that may require detailed studies. Although time-consuming analyses involving ion exchange, removal of Fe oxides, carbonates, and organic matter, and other treatments may increase accuracy (often with a loss in precision), the reduction in rate of analysis generally is not repaid by the detail gained from these treatments. Although no QPA methods seem currently suited to the analysis of soils, several are presented in abbreviated form because (i) they may suggest better ways to collect data and characterize the minerals in soils; (ii) as computerized collection and analysis of XRD data become more common, more time can be spent on sample preparation; (iii) some problems require accurate mineralogical analysis, and XRD is still the cost-effective method of obtaining those analyses; and (iv) advances in modeling and curve fitting suggest that precision, accuracy, and high sample throughput will be possible soon.

The widths of diffraction peaks for most nonclay minerals are a function of the resolution of the instrument, a limit of resolution called the instrumental signature. Peak areas or peak heights can be used for QPA for these minerals. Conversely, most clay minerals in soils scatter x-rays from such small domains that peaks are broadened beyond the instrumental signature, enough so that programs for peak fitting, area measurements, and peak decomposition are usually required. Almost all methods for QPA are designed for nonclay minerals (Snyder & Bish, 1989) and cannot yet be used for the minerals in soils. Acceptable QPA results, even for nonclay minerals, are obtained only by attention to many details. For enumeration and discussion of such details, for evaluation of the kinds of errors inherent in the process, and for indicating the larger errors involved if details are ignored, see Brindley (1980), Bish and Post (1989), Moore and Reynolds (1989, Chapter 8), and Pevear and Mumpton (1989). For coverage of QPA since Klug and Alexander (1974), in addition to the previous citations, see Hubbard and Snyder (1988) and Snyder and Bish (1989). The physical and geometrical principles that underlie the equations for quantitative analysis can be found in Brindley (1961), James (1965), Klug and Alexander (1974), and Brindley (1980).

Reynolds's (1989) approach to quantitative analysis is specifically for clay minerals with their broad and often asymmetrical peaks. The NEWMOD computer program (Reynolds, 1985) contains options that allow the operator to model the diffraction patterns of clay minerals having very small crystallite sizes and very short structural-defect distances, characteristics that are typical of the clay minerals in soils.

In characterizing the mineral content of soils and paleosols, we make a distinction between quantitative representation (QR) and QPA. By QR, we mean that a set of numbers can be generated that represents the mineral content but does not necessarily reflect an accurate measure of the absolute amounts of minerals present. The QR replaces the awkward and misleading term semiquantitative analysis and more accurately describes data that have been presented under this title. It is an attempt to balance the degree of detail needed, the time available, and limitations imposed by available XRD instruments. When the procedures of QR described here for preparation, measurement, and analysis are rigorously adhered to, data to solve problems encountered in our laboratory, and in many investigations of soils with which we are familiar, are obtainable. For example, reliable estimates of the intensity of weathering, relating parent materials to alteration products, and the relative physical, chemical, and thermal properties of a soil are obtained from even the most rapid comparative methods, as long as the precision of the method is high. In some cases, QR data may be refined to produce a QPA. A distinction also can be made on the basis of purpose for the analysis. If precision is of primary concern and accuracy is of only passing consequence, QR is usually the correct approach.

Because of its high precision, QR can be widely applicable. For example, much of the Pleistocene stratigraphy of Illinois and Iowa, which requires distinguishing among several paleosols and tills, is based on the method of Glass (e.g., Wilman et al., 1963; Wilman et al., 1966; Frye et al., 1968; Hallberg et al., 1978; Glass & Killey, 1987). Using this method, large numbers of samples can be characterized in a short period of time. Clay-mineral analyses are often the best aid for unraveling complicated stratigraphic relationships in soils and rocks because they can be present in similar proportions in most facies of a unit without regard to texture (Glass & Killey, 1987), and unlike fossils, minerals are found at virtually every occurrence of a unit.

The QR requires only that simple procedures be adhered to rigorously. Quantitative phase analysis requires; however, the consideration of so many variables that relatively few samples can be analyzed. Comparisons should be possible from one laboratory or one method to another for QPA; they are often unsatisfactory for QR. There is a continuum between QR and QPA. As methods and instrumentation improve, especially computerized collection and manipulation of diffraction data, more QPA will be possible.

In this chapter, methods of QR are applied to several soils and paleosols from Recent to Ordovician in age. These examples illustrate (i) what the parent material was, (ii) the degree of stability of phases in the parent material, and (iii) what degree of drainage existed. These studies show similarities between patterns of mineral changes during weathering and those of diagenesis during burial and metamorphism.

CLAY MINERALS IN SOILS

The most abundant minerals at the surface of the earth, clay minerals, are hydrous aluminum, iron, and magnesium phyllosilicates, or layer silicates; they are mostly finer than 2 μm in equivalent spherical diameter and vary considerably in chemical and physical properties. Within this family of minerals, most have platy morphology and perfect (001) cleavage, a consequence of their layered atomic structures. A significant amount of what is known about the structural and chemical details of these minerals has been extrapolated from XRD studies of their macroscopic counterparts. This extrapolation is necessary because clay minerals are too small for study by single-crystal x-ray methods.

The compositions and structures of the common clay minerals are described by Grim (1968) and Bailey (1984, 1988). Moore and Reynolds (1989) provide a summary of the characteristics, terms, and conventions used for individual and mixed-layered clay minerals and polytypic stacking sequences. Individual or discrete (as distinct from mixed-layered) clay minerals are classified first as having either a 1:1 or a 2:1 layer structure; within these two classes, they are grouped as either dioctahedral or trioctahedral; and within these groups they are arranged according to layer charge (Table 11-1).

For clay minerals that vary in the c-dimension in response to the activity of water or polar organic solvents, we use the terms expandables ($\sim$1.7 nm after ethylene glycol solvation) and collapsibles ($\sim$1.0 nm after 1 h at $\sim$300 °C). For example, the expandable component of the mixed-layered mineral in soils conventionally called montmorillonite, smectite, or low-charge vermiculite usually is not quite smectite nor vermiculite but rather a mixed-layered mineral with a large proportion of smectite-like layers, the balance nonexpanding layers, and an ordering of R0 (Malla & Douglas, 1987; Suquet & Pézerat, 1988). Therefore we refer to such components simply as expandables. In other instances, it may be useful to describe low-, intermediate-, and high-charge vermiculite phases corresponding to clay mineral peaks at about 1.7, 1.65, and 1.4 nm after ethylene-glycol solvation (Newman et al., 1990). This use of the term vermiculite is particularly appropriate where illite and chlorite are weathering first to high-charge vermiculite and then to low-charge vermiculite.

Table 11–1. Abbreviated classification of discrete clay minerals.

Layer type	Group	Subgroup	Species
1:1	Serpentine-kaolin ($x\sim0$)	Serpentines (Tr)†	Serpentine, berthierine
		Kaolins (Di)	Kaolinite, dickite, nacrite, halloysite
2:1 Platy	Smectite ($x\sim0.2-0.6$)	Tr smectites	Saponite, hectorite
		Di smectites	Montmorillonite, beidellite, nontronite
	Vermiculite ($x\sim0.6-0.9$)	Tr vermiculites	
		Di vermiculites	
	Illite ($x<0.9>0.6$)	Tr illite?	
		Di illite	Illite, glauconite
	Chlorite (x variable)	Tr, Tr chlorites	Chamosite (Fe^{2+}), clinochlore (Mg^{2+}), nimite (Mn^{2+}), and pennanite (Ni^{2+})
		Di,Di,Di,Tr; and Tr,Ci chlorites (uncommon)	
2:1 Fibrous	Sepiolite-palygorskite	Inverted ribbons (with x variable)	

† Tr = trioctahedral and Di = dioctahedral; x = charge per formula unit. Based on Bailey (1980a,b), Brindley (1981), Hower and Mowatt (1966), and Środoń (1984). Modified from Moore and Reynolds (1989).

Our use of the term vermiculite should be distinguished from that of MacEwan and Wilson (1980). Their definition is based upon Mg-exchanged, metamorphic vermiculite solvated with glycerol. It has been our experience that soil and metamorphic vermiculites are as different in their properties as illite and muscovite. MacEwan and Wilson's method (1980) certainly has merit whenever the problem requires these sorts of distinctions, especially for contrast with results from ethylene-glycol-solvated samples. Further exacerbating the problem of distinguishing among species of expandable clay minerals are the problems of variable expansion with ethylene glycol and glycerol and variable results from different workers for ion exchange (Kunze, 1955; Ĉeĉil & Machajdík, 1981; Novich & Martin, 1983). Środoń (1980), in concluding that the two-layer ethylene-glycol complex of dioctahedral smectites varies from 1.65 to 1.73 nm, declared, "Glycerol seems to be less suitable (than ethylene glycol) for identifying illite/smectite because some beidellites and some K-, Rb-, Cs-, or NH_4-montmorillonites form a one-layer, i.e., a vermiculite-type complex." Based partly on the views of W. F. Bradley in the 1940s, we have thought that ion-exchange and similar treatments are often necessary second steps, but these treatments often "create" clay minerals that do not match natural ones. Finally, although we have begun to use these treatments, our ~100 000-sample database shows that XRD analyses after ethylene-glycol solvation and Na exchange from dispersion, faithfully record the alteration sequences in most soils. A large number of analyses with Na, Mg, and glycerol on samples from the existing database will be required before the value of ion exchange and glycerol solvation can be assessed (see our recommendations on procedures in the METHODS section).

As identification techniques improve, more and more samples of apparently discrete clay minerals show evidence of at least small amounts of interstratification of a second component. Mixtures of more than two components have seldom been reported (Weaver, 1956; Ostrom, 1960; MacEwan & Ruiz-Amil, 1975; Klimentidis et al., 1990). However, when the third component is present in small amounts (~ 50 g kg^{-1}) or in a random interlayering sequence, it may escape detection. Therefore, three-component (or more?) mixtures may be more common than current understanding reflects (Reynolds, 1980, p. 250). Mixed-layered clay minerals are either regularly stacked so that they yield a rational series of 00ℓ spacings (these are relatively rare), or they are irregularly stacked and give an irrational series. Irregularly stacked illite/smectite (I/S) is the most common of the mixed-layered clay minerals, perhaps more common than either discrete illite or discrete smectite, especially in soils. Specific names are assigned to mixed-layered clay minerals with regularly stacked components present in a fixed ratio (Table 11-2).

Mineralogists working with the clay-sized minerals in soils seem to have a somewhat different view of mixed-layered clay minerals than those working with geological materials. This difference is one of degree only. Wilson (1987) discussed the differences between soil smectites and those in bentonites. One of the consequences of minerals being weathered in a soil is that domain size, the part of a crystallite that coherently diffracts as a single unit, becomes smaller. Once minerals are removed from the soil-forming processes, domain sizes begin to increase. These changes are

Table 11-2. Regularly stacked, 50/50 mixed-layered clay minerals.

Aliettite	1:1	Talc/trioctahedral smectite
Corrensite	1:1	Two varieties, a low-charge trioctahedral chlorite/trioctahedral smectite and a trioctahedral chlorite/high-charge trioctahedral vermiculite
Hydrobiotite	1:1	Biotite/vermiculite
Kulkeite	1:1	Talc/chlorite
Rectorite	1:1	Dioctahedral mica†/dioctahedral smectite
Tosudite	1:1	Chlorite, which is, on the average, dioctahedral/smectite

† The Nomenclature Committee recognized K-, Na-, and Ca-rectorites corresponding to the different mica components (Bailey, 1982). Modified from Moore and Reynolds (1989).

kinetically controlled. An example of domain size in minerals in soil was provided by Blahoslav Ĉeĉil (Slovak Academy of Sciences, Bratislava, Slovakia, personal communication, 1990). He has studied three smectites that vary in their response to water: one that expands to 1.55 nm, one that expands to 1.25 nm, and one with d_{001} of 1.0 nm that does not expand. Ĉeĉil (1981) suggested that there can be changes from one state to another laterally (Fig. 11-1) where, by convention, they would be called defects, as well as in the c-axis direction. These lateral defects can be detected by comparing the crystallite size determined by electron microscopy with the domain size within the crystallites estimated from the broadening of XRD peaks. In the simplified scheme represented by Fig. 11-1, the x-ray beam would see three domains, each yielding a somewhat different diffracted beam. This would be reflected by broadened peaks with indistinct maxima (Fig. 11-2).

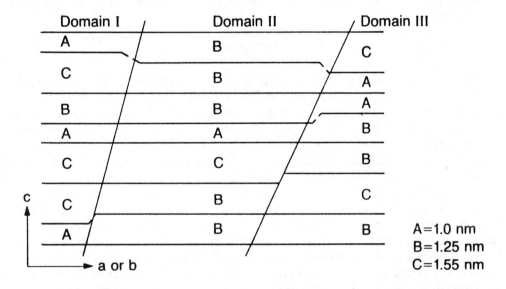

Fig. 11-1. Smaller domains created from fully expandable smectite by wetting and drying, as suggested by Ĉeĉil (1981).

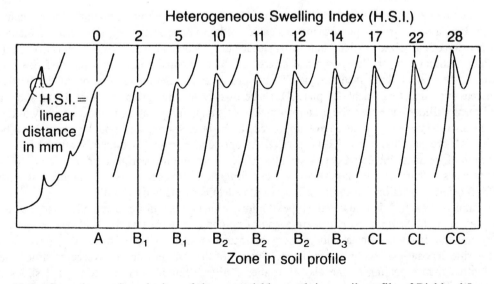

Fig. 11-2. Smearing or broadening of the expandables peak in a soil profile of Richland Loess; inset shows the calculation of the Heterogeneous Swelling Index (HSI) (Frye et al., 1968); CL = leached; CC = calcareous. [Note that the traces run from low 2θ angle (right) to high 2θ angle (left).]

(The three domains illustrated in Fig. 11-1 may diffract as a single domain upon solvation with ethylene glycol, glycerol, or similar agents thus yielding a narrower XRD peak.) In general, as the number of defects increases, materials appear more amorphous to the x-ray beam and standard XRD techniques become less adequate, even for QR. However, it is still possible to use the broadened peaks of a material such as allophane or the amorphous "hump" on XRD patterns to carry out QR.

For discussion of other properties of clay minerals such as the electric double layer, interaction with water, isoelectric points, surface area, surface charge, and zeta potential, we recommend Grim (1968), Gast (1977), Marshall (1977), van Olphen (1977), Newman (1987), and Sposito (1989). For the interaction of clay minerals and organic compounds, refer to Weiss (1969), Mortland (1970), Fripiat and Cruz-Cumplido (1974), Harter (1977), MacEwan and Wilson (1980), and Pinnavaia (1983).

METHODS

Although XRD analysis is the most widely applicable method for studying the mineral content of soils, no one method of preparation satisfies all situations. In this section we describe the sample preparation techniques used in our laboratory and briefly review other analytical methods that may be used directly or modified to quantitatively describe the mineral content of soils.

Preparation techniques depend on (i) the purposes of the analysis, (ii) the supplies and equipment available, and (iii) the material itself. However, there must be a single method of preparation for a given set of samples if XRD data are to be compared within the set. Our discussion is directed toward the analysis of many of the clay minerals in soils, paleosols, and their parent materials.   The methods discussed below vary from the relatively fast sedimented slide-ethylene glycol-diffractogram technique to the laborious analysis of powders and smears of narrow particle-size intervals. We prefer the fastest method that satisfies the needs for precision for the problem at hand. A common misconception is that the newest instrument is always the instrument of choice. We have encountered several problems where soil profiles were more clearly detected in data from our oldest instrument-- a General Electric XRD-5 dating from about 1952.

Sample Preparation and Instrumentation

The three most common methods of preparing a sample for presentation to the x-ray beam are (i) packing a finely ground powder into a sample holder (a bulk pack, BP), (ii) settling a dispersed suspension of clay-sized particles onto a flat surface (an oriented aggregate, OA), and (iii) smearing a finely ground powder, dampened and mixed to a buttery consistency, on a flat surface (smear, SM). The first of these methods yields an approximately random orientation of the particles, whereas the other two orient the platelike shape of clay-mineral grains parallel to the substrate. Sedimented slides have been the principal method of sample preparation in most of our work. As XRD instruments and grinders have improved, more bulk smears and powder packs have been used. Moore and Reynolds (1989) described the options for making sedimented slides of the clay fraction of samples, including drying on glass slides, suction onto porous tiles, suction onto filter membranes with transfer to glass slides, and centrifugation onto porous slides. Hughes and Warren (1989) offered variations on these techniques plus details of methods of QR.

Choice of method depends on the type of information that is required from XRD and the nature of the material. Sedimented slides are a good first choice for soils in the Midwest, because the clay mineral fraction from a wide variety of materials such as soils, tills, silts, gravels, and lacustrine deposits can easily be compared. Smears of the $<2\mu m$ fraction overcome stratification and often reduce peeling of the clay film after drying. For peeling in sedimented slides, we typically dilute the suspension and remake the slides. Rich and Barnhisel (1977) solved peeling problems by adding a small amount of finely ground glass wool to the clay suspension. For clay-rich materials such as underclays, smears of the bulk sample often give the best estimate of degree of weathering. Bulk packs or smears are essential for estimates of amounts of nonclay minerals and the degree to which they have been altered to clay minerals.

Procedures that damage clay minerals or affect the exchangeable ions, other than those necessary for dispersion, should be avoided when possible. If a sample requires treatment to isolate certain clay minerals, the techniques of Jackson (1969) may be necessary. However, the untreated as well as treated samples should be analyzed, and differences between the two should be reported. Techniques of slide preparation have been presented in detail by Hughes and Warren (1989). Sedimented slides require a consistent solids-to-H_2O ratio, a standardized technique of preparation, and consistent drying conditions. Samples with abundant carbonates, organic matter, or Fe oxides are often treated to free and concentrate the clay fraction. A solution of 5% acetic acid is used to remove carbonates. The solution is readjusted to approximately 5% acetic acid periodically until the carbonate reaction ceases. A 10 to 20% solution of laundry bleach (NaOCl) is adequate to remove most organic matter, although the solution may have to be heated to 60 to 100 °C to remove coaly materials in some paleosols. Strong oxidants may affect iron within the structure of clay minerals, and consequently, the apparent layer charge on those clay minerals. Bleach can only be removed by flocculating with a small amount of HCl or salt and pouring off the clear supernate. When the samples must also be treated with acetic acid to remove carbonates, the acid can be added after the bleach reaction is complete. Acid is removed by repeated washing with deionized water. Iron oxides can be removed with most acids and/or dithionite (Jackson, 1969). The proliferation of diffractometers with monochromators or other ways of reducing or eliminating spurious radiation from iron has almost eliminated the need to "clean up" samples before XRD. However, high-Fe soils still lend themselves to Fe removal, especially when the problem requires complete characterization of small amounts of clay minerals.

Samples are normally analyzed by XRD after exposure to ethylene glycol for 2 d at STP or for 24 hours at 60 °C, and sometimes after heating to approximately 300 °C for at least 1 h. Temperatures above 300 °C can be used, although we have observed partial structural breakdown in two types of clay minerals, mixed-layered kaolinite/expandables and berthierine, at 325 to 375 °C (Hughes et al., 1990b; Moore & Hughes, 1991). Unless a controlled-atmosphere chamber is available on the diffractometer, samples must be scanned as soon as possible after removal from the ethylene glycol atmosphere or the heating chamber to avoid loss of glycol or adsorption of water. A small sponge or tissue saturated with ethylene glycol can be placed near the sample holder to maintain a high ethylene-glycol vapor pressure during the XRD scan. Even so, it is good practice to rescan the 001 peak of expandables and collapsibles at regular intervals after collection of the high-angle parts of the diffractogram of glycolated and heated samples to monitor shifts in peak position caused by loss of glycol or adsorption of water.

Samples prepared for QR or QPA should possess certain characteristics:

1. The sample should be longer than the spread of the incident beam at the lowest diffraction angles used; therefore, two quantities need attention, the sample length and the angular divergence of the beam or size of the divergence slit.
2. The sample should be thicker than the penetration of the x-ray beam at the highest diffraction angles used (infinitely thick to the x-ray beam).
3. The sample must be mounted in the diffractometer so that, for all diffraction angles, the angle between the sample surface and the incident beam is equal to the angle between the sample surface and the diffracted beam; i.e., the diffractometer must be properly aligned.
4. A minimum particle-size gradient should exist between the top and bottom of the sample.
5. For random powder mounts, particle size should range ideally from ~5 to 10 μm. For those minerals with particularly good cleavage or high absorption coefficients, the size should be reduced to ~2 μm (Bish & Reynolds, 1989).
6. For oriented aggregates, the deviation from perfect orientation should be determined.

Reynolds (1989) discussed each of these characteristics. For example, he found preferred orientation impossible to control precisely and unmeasurable except as a practice that is too time consuming to be routinely practical. The range of preferred orientation found, even within a suite of similar minerals, is recorded in Table 11-3. It shows the intensity from the 003 peak of five different illites prepared in an identical manner. The parameter, $\sigma*$ is the standard deviation of the angle between real crystallites and perfectly oriented ideal crystallites that are parallel to the sample surface/sample substrate. For angles $> 20°2\theta$ (all diffraction angles used in this paper refer to CuKα radiation), intensity is roughly inversely proportional to $\sigma*$ (Reynolds, 1986; 1989). From Table 11-3 it can be seen that if the Madison illite were selected as an external standard and the amount of illite in the Mancos Shale were being measured, it would be underestimated by a factor of 7.9. The more perfectly oriented samples having low values of $\sigma*$ (i.e., $<10°2\theta$) are less suitable for quantitative analysis than those with moderate to high values of $\sigma*$ (see Table 11-3). Reynolds (1989) found that $\sigma* = 12°2\theta$ is an average value for the filter-membrane peel preparation method. Relative intensity errors caused by sample-to-sample differences in preferred orientation can be minimized by using peaks that are close together and restricted to the high-angle region ($2\theta \geq 10°$). Perhaps the measurement of $\sigma*$ will prove to be amenable to automation.

The six characteristics discussed above can be difficult to control, but their effects on the accuracy and precision of analyses are minimized in general if two important principles are adhered to: (i) choose peaks at $>10°2\theta$ unless prohibited by interference or larger errors introduced because of low peak intensity; and, (ii) select analytical peaks that are as close together (in $°2\theta$) as possible, using the same qualifications given in (i).

Random Powder Mounts

Random powder mounts of bulk and size-fractionated samples are used to improve the detection and quantitative estimate of nonclay minerals, to determine the clay-mineral to nonclay-mineral ratio, and to detect diffraction peaks of the clay minerals from other than the basal spacings. These non-00ℓ peaks can be used (i) to distinguish dioctahedral from trioctahedral clay minerals, (ii) to determine polytypes, and (iii) to identify turbostratic and ordered-disordered stacking. Because the Powder Diffraction File of the Joint Committee on Powder Diffraction Standards (JCPDS) is based on random powder mounts, comparisons can be made directly with their system (see caveat from Snyder & Bish, 1989, p. 106). Random powders also are amenable to the use of internal, external, or additive standards and to the estimation of the crystallite domain size perpendicular to each $hk\ell$-plane of the mineral (Klug & Alexander, 1974).

Perfectly random orientation of all mineral particles in a sample is impossible to achieve. However, the use of side-loading sample holders has greatly improved the degree of randomness possible for most minerals. Wet grinders such as the McCrone Micronizing mill have improved randomness even further with an additional 10 to 20 min preparation time for each sample; however, grinding can damage mineral structures and caution should be exercised. If the closest approximation to randomness is required, a ground slurry of the mineral may be spray dried (Hughes & Bohor, 1970; Bish & Reynolds, 1989).

Table 11-3. Preferred orientations, integrated intensities, and corrected integrated intensities of the *003* peak of five illites.

Source	$\sigma*$	Integrated intensity of *003* in cps	Corrected integrated intensity†
Madison Fm.	4.5	27 790	5.50
Duvernay Fm.	6.0	15 020	5.41
Gunflint Fm.	6.9	12 520	5.96
Sylvan Shale	8.4	7 840	5.53
Mancos Shale	12.5	3 540	5.53

† (*003*,cps) $\times$ ($\sigma*$)$^2 \times 10^{-5}$, modified from Reynolds (1989) with permission.

Smears

Smear slides of the bulk sample and the silt and clay fractions of coarse-grained materials are a method used increasingly in our laboratory. Smear slides overcome the distortions caused by size fractionation of mineral particles with variable densities and sizes. Density and size differences can cause segregation of minerals in sedimented slides. Therefore, some minerals may be missed in the analysis or significantly underestimated or overestimated. Fluorite, sylvite, or corundum can also be added as internal standards to the smear to obtain a complete QPA. We use a vacuum hose to hold the slide for making smear slides (Fig. 11-3).

Wet smears allow detection of the hydrated ($4H_2O$) variety of halloysite. Detecting the 1.0-nm peak of halloysite requires keeping samples at field moisture or wetter, and running them in that condition. The slides can be rerun after drying, mild heating, or both to observe the loss of the 1.0-nm peak and the growth of the 0.7-nm peak. Wet smears also will reveal the exchangeable cation(s) of smectite, for example, by the 1.9-nm peak for Ca-smectite compared to the >4.0-nm peak for Na-smectite. Solutions of pollutants, soil treatments, and polar organic agents in addition to ethylene glycol can be advantageously evaluated by making wet smears of the solution and clay-mineral sample in question.

Instrumentation

The minimum equipment required to carry out QR of soils using XRD is a diffractometer and some routine laboratory equipment such as beakers, glass slides, and glass tubing. Various types of mixers and grinders are often useful, but a simple mortar and pestle and hand stirring are adequate to many tasks. We use diffractometers that range from a Ni-filtered manual unit with a Geiger counter and logarithmic scale chart recorder to a fully computerized instrument with a θ/θ goniometer, and a liquid N-cooled Ge detector. It has software to control the goniometer and shutter, find and measure peak intensities automatically, decompose overlapped peaks, and carry out a search-match for identification. Digitally stored diffraction data that can be manipulated by computer and interactively compared to calculated diffraction tracings will make possible routine QPA of soil samples.

Fig. 11-3. Making a smear slide by holding the slide with a vacuum hose.

Methods of Analysis

We expect a wider range of variation in accuracy than in precision. Replicate analyses should provide a standard deviation of $\pm 5\%$ (or less) of the amounts measured if the constituent minerals are all present in reasonably large quantities, i.e., 200 g kg^{-1} or more. For Quaternary samples, Glass and Killey (1987) have demonstrated precision typically good enough to distinguish strata that differ by only 40 or 50 g kg^{-1} in apparent illite content. Accuracy is another matter. A QPA can be considered acceptable if errors amount to $\pm 10\%$ of the calculated content of major constituents, and $\pm 20\%$ for minerals with concentrations <200 g kg^{-1}. At present, quantitative analysis of minerals by XRD does not approach the accuracy expected from quantitative chemical analysis. We recommend that most projects include ways to measure and report precision, and if possible, accuracy.

Method of External Standards

External standards are derived by analyzing a pure sample of a mineral and recording the intensity of one or more peaks to use as a standard for comparison with unknowns. The standard in this case can be either the best sample of the mineral available or a sample that most closely matches the composition, structure, and crystallinity of unknowns. In some cases the variation within a suite of samples makes it possible to estimate endmember peak intensities for minerals from within the suite. In practice, the intensity of selected peaks of the common clay minerals are measured and a series of factors is generated on the basis of ratios of these measurements. For example, the method used for analysis of soils in our laboratory employs factors for QR of underclays and related Pennsylvanian materials that include ratios of peaks for illite and collapsibles (1.0 nm) to kaolinite + chlorite (0.7 nm) of 1:0.4. This ratio reflects the fact that the appropriate peaks for the best smectite, illite, kaolinite, and chlorite standards available in the late 1960s gave this ratio for their intensities. Until recently, all of our analyses used external standards, and most still do.

The National Institute of Standards and Technology (Gaithersburg, MD) offers LaB_6, Standard Reference Material no. 660, as a standard for peak shape. This material can be used as an external standard and should give XRD peaks with maximum sharpness and symmetry and minimum width. A standard for peak shape and sharpness is important because they often contain a great deal of information.

Method of Additive Standards

In the method of known additions or additive standards, known amounts of a mineral present within the sample are added and the resulting peak intensities are plotted to obtain the original content. The mineral content determined by the additive standard method can then be used as an internal standard. This is a particularly valuable technique when peaks of internal standards would overlap essential peaks in the unknown. A most important problem, over which little control can be exercised, is that of choosing a standard mineral with diffraction characteristics identical to those of the same mineral in the unknowns. For example, all specimens of illite are not identical (Table 11-3). In fact, phases can have wide chemical variation and still be called by a single mineral name; e.g., the analyses of Mg-rich chlorites will be incorrect if an Fe-rich chlorite is used as the standard.

Method of Internal Standards

Internal standards are minerals or chemicals added to, but not present in the sample. Mixtures (usually 50:50) of the standard and the mineral to be measured must be run to determine the reference intensity ratio factors to be used. Pyrophyllite, talc, boehmite, and other minerals have been used for clay mineral internal standards; however, these minerals have XRD spectra that sometimes interfere with essential peaks for common clay and nonclay minerals, and this approach has largely been abandoned (Snyder & Bish, 1989, p. 101ff). Some chemical analyses can be used as internal standards, as long as the analyses accurately measure only a single mineral, e.g., sulfur content with only pyrite or marcasite or CO_2 content with only calcite or dolomite or siderite.

A method designed for larger crystallites than commonly found in the clay minerals in soil is the reference-intensity-ratio method (RIR). It uses the intensity of the phase to be measured (I) compared to the intensity of corundum (I_C). This is convenient because the Powder Diffraction File has I/I_C values for more than 2500 phases; however, uncertainties exist about some of the these values (Snyder & Bish, 1989). Also, because of differences between instruments and methods used in different laboratories, intensity comparisons with corundum must be made under the same conditions as those for the intensity comparisons of unknowns. This method is also referred to as the Chung method (Chung, 1974a,b; 1975). [It is unfortunate that a generic term such as reference intensity ratio (RIR) is used to refer to intensity ratios with respect to corundum. Perhaps this method should be designated RIR_C for corundum or Chung. We use RIR in the general sense.]

Calculated Standards

Intensities for minerals may be calculated to serve as standards and avoid the almost impossible task of having a library of natural samples large enough to meet the needs of all possible sets of samples, especially if mixed-layered clay minerals are encountered. Calculated mineral reference intensities or factors add no new sources of error to quantitative analysis and values unique to a particular set of samples can be created. For example, if a set of samples, or even a single sample, contains an illite with an unusually low *002/001* intensity ratio, suggesting an Fe-rich illite, a computer program such as that of Reynolds (1985) can be used to calculate a tracing with the same *002/001* ratio. The chances of having on hand a standard sample of illite with a similar *002/001* ratio are very low. Excellent agreement has been attained between calculated values for kaolinite, illite, smectite, and mixed-layered illite/smectite and experimental values from pure examples of these clay minerals measured and corrected for deviation from perfect orientation and

Table 11–4. Reference intensity ratios (CuK_α radiation) for the *001* peaks of common clay minerals relative to the intensity of low-Fe illite (left two columns) and selected peaks of clay and nonclay minerals relative to the *10Ī0* peak of quartz (right two columns).†

RIRs based on low Fe illite		RIRs based on quartz *10Ī0*	
Low Fe Illite	1.0	Quartz (0.426 nm)	1.0
Low Fe Smectite *g	0.20	Quartz (0.334 nm)	0.18
High Fe Smectite *g	0.13	K-Feldspar (0.324 nm)	0.39
Low Fe DiVerm *g	0.20	Plagioclase (0.320 nm)	0.39
High Fe DiVerm *g	0.13	Calcite (0.303 nm)	0.21
Low Fe TriVerm *g	0.17	Dolomite (0.290 nm)	0.23
High Fe TriVerm *g	0.084	Pyrite (0.313 nm)	0.45
Illite/Smec *g (0.15I)	0.30	Low Fe Illite	3.4
Illite/Smec *g (0.5I)	0.80	Kaolinite	1.4
Illite/Smec *g (0.85I)	8.4	Low Fe Smectite *g	0.66
High Fe Illite	0.35	Low Fe Collap *ht	1.7
Low Fe Collap *ht	0.49		
High Fe Collap *ht	0.22		
Kaolinite	0.43		
Low Fe Chlorite	0.94		
High Fe Chlorite	0.16		
K/E *ht (0.15K)	1.0		
K/E *ht (0.85K)	0.61		

† Fe = iron; *g = ethylene glycol solvated; Smec = smectite; Verm = vermiculite; Illite/Smec = mixed-layered illite/smectite; (0.15I, 0.1K) = proportion of illite, kaolinite in I/S, K/E; Collap = collapsibles (smec, vermic, I/S); *ht = heated to 300 °C for 1 h; K/E = mixed-layered kaolinite/expandable. Note, collapsibles are assumed to have a 0.98-nm spacing. Chlorite is assumed to be trioctahedral in the silicate layer and brucitelike sheet. The collapsed part of K/E is assumed to be 0.96 nm. RIRs for clay minerals calculated from NEWMOD (Reynolds, 1985).

other instrumental factors (Table 11-4). The NEWMOD program developed by Reynolds (1985) can be used to generate patterns for all of the phyllosilicate clay minerals, the compositional variants of individual clay minerals, and their mixed-layered equivalents. The intensity of the peaks on these patterns is standardized with respect to the principal peak of quartz, and this intensity can be set to match individual instruments.

Full-Pattern Matching

With the availability of digitized diffraction data generated by automated x-ray diffractometers, the complete XRD pattern can be used for QR or QPA rather than only a few of the strongest peaks. Some variations of the use of full patterns minimize the problems of solid solution, preferred orientation, peak overlap, and primary extinction. For this method the full experimental diffraction tracing is fitted to derived synthetic or calculated patterns, or a combination of both. A synthetic pattern is formed by modifying an experimental pattern of a standard mineral. Calculated patterns are just that, calculated from the physical principles of the interaction of matter with radiation given the crystal structure of the mineral in question. Such patterns can be added together to match experimental patterns to determine amounts of minerals in a mixture. The Rietveld method (Rietveld, 1969; Bish, 1994; see Chapter 9) is the most powerful of the methods using calculated patterns, a method originally devised for refining crystal structures using neutron diffraction. The data needed for a Rietveld analysis are similar to those needed for the calculation of a diffraction pattern, i.e., the approximate crystal structure of all of the phases in the sample, easily available for most macroscopic minerals but often difficult to obtain for minerals in soils. The Rietveld method consists of starting with best guesses of the crystal structures of the minerals of interest, then making iterative fittings of the complete experimental diffraction pattern with calculated profiles and backgrounds until a best fit is found. This method uses all classes of $hk\ell$ reflections and calculates absolute intensities for these peaks. Then, quantitative data can be taken from scale factors for each phase in a mixture. A bonus from this method is refinement of the cell parameters of the phases being investigated (Snyder & Bish, 1989).

Reynolds (1989) introduced another method using a full $hk\ell$ pattern. The purpose of this method is not to quantitatively measure phases in a mixture, although it has that potential because several calculated patterns can be added together. Its purpose is to enable the user to quantify the degree of three-dimensional order of clay minerals. The degree to which the phase is turbostratic or the degree of order of polytypic stacking can be measured in experimental tracings by comparing them to calculated ones.

In yet another type of full-pattern method, synthesized patterns are combined for comparison with an unknown experimental pattern. Occasionally, for patterns of minerals that are unavailable, tracings can be calculated or simulated from data in the Powder Diffraction File. For each standard pattern, reference intensity ratios to corundum (see Method of Internal Standards) are calculated on the basis of addition of a known amount of standard. The patterns used to compile the synthetic pattern and the unknown are run using identical conditions. The pattern of the unknown has the background removed and then the combination of synthetic patterns is iteratively matched with the experimental pattern to obtain a quantitative analysis. This method is described by Smith et al. (1987) and Smith (1989).

Methods That Combine X-Ray Diffraction With Supplementary Data

Combining XRD data with information gathered by other methods promises to overcome problems in some analyses. If calcite is the only carbonate in a soil or a rock, measurement of the amount of CO_2 evolved will quantify the amount of calcite in the original sample. Almost any physical or chemical property that is unique to an individual mineral in the sample can be used in a similar way.

Hussey (1972) and Johnson et al. (1985) described a method in which identification of mineral components with XRD was combined with data from chemical analysis, weight losses on heating, and determinations of cation exchange capacity. These data allowed the design of a program of simultaneous linear equations that were used iteratively by adjusting mineral characteristics and mineral quantities until errors were minimized. Davis and Johnson (1989) described a method that uses rigorous sample preparation combined with x-ray transmission (XRT) and XRD results to yield a quantitative mineral analysis. Perhaps of most importance to soil mineralogists is that comparison of calculated with experimental mass absorption coefficients, as measured by XRT with this method, permits the determination of noncrystalline components. "Normative" procedures have been described but we have found these unreliable because, for example, TiO_2 and S are reported respectively as rutile and pyrite, even though our experience suggests that anatase is the most common form of TiO_2 in soils and sedimentary units, and many paleosols contain marcasite as the principal form of FeS_2. Furthermore, TiO_2 is often included within phyllosilicate structures. Of more importance is the fact that normative programs tends to leave the impression that the clay mineral entities, such as the expandable components, are made up of discrete species of clay minerals (montmorillonite, nontronite, saponite) despite the fact that the range of composition of expandables typical of soils and sediments is usually greater and bears little relationship to endmember species.

Our recommendation is to incorporate data from other techniques with XRD results whenever high-quality chemical analyses can be obtained at reasonable cost. However, limited resources may often be better spent first on collecting larger numbers of routine XRD analyses, refining the match between the XRD patterns of natural samples and calculated patterns, carrying out additional treatments such as ion exchange and glycerol solvation, and employing an internal standard for bulk nonclay mineral analysis. The design of an investigation and allocation of analytical resources are discussed in detail in the section Examples of Quantitative Representation.

Interpretation of Results

Several methods can be used to condense the information from the XRD patterns to a form that can be easily communicated. Peak areas give the most accurate results when measuring quantities or chemical variation such as the amount of Fe in chlorite. Peak heights give adequate precision for some problems and are usually easier to obtain. Peak shapes are sensitive indicators when measuring alteration of a mineral by its peak broadening (see HSI in Fig. 11-2 and K/E "shoulder" in Fig. 11-7). XRD patterns from samples representing compositional and structural variations can be sorted or classified into a series of types. This typing method is similar to soil typing and it has similar advantages and shortcomings. On the basis of his interpretation of the relative heights and shapes of the XRD peaks of the clay minerals present, Parham (1964) assigned a letter to each type of underclay (paleosol) of Pennsylvanian-age coals of the eastern and central United States. These types ranged from kaolinite-rich to illite-chlorite-kaolinite-rich to rich in mixed-layered illite/smectite, in what was thought to be a continuous series. Figure 11-4 illustrates the compositions of these three types observed for underclays in the Illinois Basin.

Peak ratios can be used to create indices of mineral content (Glass & Killey, 1987; Gardner et al., 1988; Hughes & Warren, 1989). Calculations of percentages are most useful when the clay minerals are calculated on a 100% basis. Analyses should only be reported in significant figures. Hughes and Warren (1989) should be consulted for further cautions about the reporting of data.

We have found various indices that can be measured precisely to be increasingly useful. The following sections illustrate how they can be used for stratigraphic identification and matching, for distinguishing soils from parent material, and for ranking the degree of alteration of a series of soils. The Type Composition Index of H. D. Glass (as described in Hughes & Warren, 1989) is an example of measuring the degree of alteration with an index. This index adjusts the compositions of weathered zones to an assumed or actual parent composition. Hughes and Warren (1989) define the Diffraction Intensity Index as a ratio of the 0.7- and 1.0-nm peaks, the Vermiculite Index as a positive or negative measure of the relative heights of the 1.4- and 1.0-nm peaks, the Clay Index as a ratio of the corrected peak intensities of all the clay minerals to all the nonclay minerals for the bulk sample, and the Colloidal Index as a calculation of the total

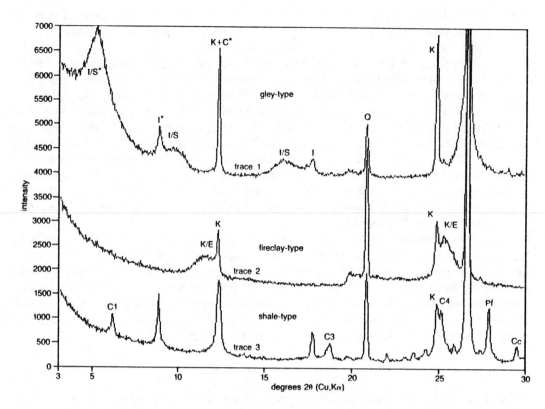

Fig. 11-4. Diffractograms of the three paleosol compositions commonly observed in Pennsylvanian underclays of the Illinois Basin. The shale-type underclay represents preservation of source material with little alteration. The gley- and fireclay-type underclays are developed from the shale-type parent by pedogenesis in two distinct environments (Hughes et al., 1987): I/S* = mixed-layered illite/smectite (* = 1.7 nm); I* = illite (* = 1.0 nm); K+C* = kaolinite + chlorite (* = 0.7 nm); C1, C3, = chlorite *001, 003* peaks; K/E = mixed-layered kaolinite/expandables; Q = quartz; Pf = plagioclase; Cc = calcite.

percentage of "smectite" layers in expandables, in mixed-layered illite/smectite and kaolinite/expandables, and in vermiculite.

EXAMPLES OF QUANTITATIVE REPRESENTATION

A typical problem in our laboratory requires an identification of the age and stratigraphic correlatives of a soil or paleosol and its parent material and a determination of their regional uniformity, their relationships to other units, and their physical and chemical properties. Our purpose may be to evaluate an area for general land use or something more specific, such as a hazardous waste site. Such a project may require several hundred sedimented slide analyses, 50 to 100 random powder packs, 10 to 20 bulk analyses with an internal standard, 10 to 50 analyses of Mg-saturated and glycerol-solvated sedimented slides, 5 to 20 complete chemical analyses of the extremes of composition, and 10 to 50 tests of the interaction of wastes with representative earth materials from the site.

The examples in the following paragraphs are given in the order in which the alterations of parent materials usually occur in soils: (i) oxidation of organic matter and alteration of chlorite, (ii) removal of carbonates, and (iii) alteration of illite to expandables, and illite and expandables to mixed-layered illite/smectite, mixed-layered kaolinite/expandables, vermiculite, kaolinite, palygorskite, and sepiolite. This set of examples is by no means complete, is particularly deficient in the discussion of laterites, in the description of occurrences of Al hydroxyl-interlayered vermiculite, and in illustration of soils formed in climates warmer or colder than those typical of the midcontinent during Quaternary interglacials. Tropical soils are represented by the examples cited for underclays; and an example of relatively arid soils from the High Plains is included for its unique mineralogical features and its stringent sample preparation requirements.

Oxidation and the Loss or Alteration of Chlorite

Oxidation is the first change observed in the parent material of soils and usually produces a distinctive change in color. The lateral and vertical variations in color aid in selecting samples and can be used to estimate the degree of oxidation. The color of XRD slides is often an accurate indicator of the genetic history of a soil; for example, the oxidation associated with well-drained profiles normally produces reds and yellows. If sediments from an oxidized profile are reduced because of a change in drainage (gleyed) (Frye et al., 1960), or deposition in a poorly drained area (an accretion gley), and then reoxidized, colors with green and orange hues are produced. Reds and yellows are usually associated with goethite, whereas shades of green and orange in gleys often occur with lepidocrocite. In other parts of the world, red is often associated with hematite, whereas yellow is only found with goethite. The colors of the gleyed tills and accretion gleys would probably be distinctively different if measured with a color meter. Color associated with pedogenesis appears to us to be an underused parameter.

The first clay mineral to show detectable changes as temperate soils weather is chlorite (Droste, 1956). The extraction of Mg^{2+} by plants and the reaction with the acids produced during hydrolysis of CO_2, oxidation of sulfides and decay of organic matter are responsible for the alteration. Chlorite can either be dissolved and lost from the system during pedogenesis, or be almost entirely altered through mixed-layered chlorite/vermiculite to vermiculite. In Illinois, weathering of chlorite in red tills gives first high-charge and then low-charge vermiculite (calculated as expandables), whereas chlorite in gray tills yields only high-charge vermiculite (expressed as vermiculite). An anomaly in the method of Glass (Hughes & Warren, 1989) causes some of the apparent changes in composition because high-charge vermiculite is not calculated as a weight fraction and is only measured as the Vermiculite Index. Furthermore, because Mg-vermiculites have a less intense 1.4-nm XRD peak than Fe varieties, part of the apparent loss of Mg-rich chlorite may represent a failure to detect the *001* Mg-rich vermiculite peak.

Table 11-5 represents the profile of a gray Wisconsinan till that underlies loess. Pedogenic processes have oxidized the till and altered chlorite to high-charge vermiculite, but have not increased the expandables content. Most of the change in the profile is an apparent increase in illite as chlorite is lost. The large increase in the Vermiculite Index from approximately -35 below a depth of 4.23 m to approximately -15 above this depth indicates the formation of mixed-layered chlorite/vermiculite and vermiculite as described by Droste (1956). In contrast, the profile developed on red Wisconsinan till (Table 11-6), has undergone chlorite weathering that produced both high- and low-charge vermiculite. The increased vermiculite content is reflected in the higher Vermiculite Index (from an average ~ +5 in unaltered till, 3.20 m to 5.18 m, to ~ +20 in the top of the paleosol, 0.91 m to 1.52 m) and the increased expandables content (from ~ 15% in unaltered till, 3.20 m to 5.18 m, to ~30% in the paleosol, 0.91 m to 2.29 m).

The breakdown of chlorite to high-charge and low-charge vermiculites in weathering profiles was used by Newman et al., (1990) to identify and distinguish two superposed tills in the area of Boston, Massachusetts. The Glass method of QR (Hallberg et al., 1978) gave standard deviations of from 0.7 to 2.9% of the measured mineral content. They assigned an Illinoian age to the lower till and a Wisconsinan age to the upper one and based their conclusions on more intense weathering of the lower till than of the upper one. This study also illustrates the advantage of adopting a method of QR that allows the estimation of low- and high-charge vermiculite.

Figure 11-4 shows x-ray diffraction data from the parent material and two pedogenic mineral suites developed in paleosols (underclays) in Pennsylvanian-age coal measures. Chlorite is always lost from fireclay-type underclays but is often preserved in gley-type underclays. The difference

Table 11-5. Example of chlorite loss in a soil profile developed on gray Wisconsinan (Batestown) Till (Wickham, 1979; Champaign County, Illinois). [†]

Material	Depth m (ft)	Cc, cps	D, cps	VI	%E	%I	%KC
R. Loess	1.23 (4)	0	0	0	71	21	8
Mixed	1.83 (6)	0	0	0	27	63	10
W. Bates T.	2.13 (7)	10	16	1	8	82	10
	2.38 (7.8)	14	21	−14	3	88	9
	2.63 (8.6)	15	29	−16	2	88	10
	2.83 (9.3)	?	24	−21	3	88	9
	3.03 (9.9)	16	28	−18	3	87	10
	3.58 (11.7)	9	37	−17	3	84	13
U. Bates T.	4.23 (13.9)	17	32	−30	2	76	22
	4.43 (14.5)	20	27	−34	2	79	19
	4.63 (15.2)	20	26	−36	2	78	20
	4.83 (15.8)	19	36	−36	2	77	21
	5.13 (16.8)	16	26	−25	2	76	22
	5.43 (17.8)	18	30	−34	2	79	19
	5.98 (19.6)	26	29	−21	2	77	21
	6.33 (20.8)	26	36	−34	2	79	19
	6.53 (21.4)	31	36	−18	2	76	22
Piatt T.	6.93 (22.7)	24	30	−6	7	73	20
	7.23 (23.7)	33	46	−8	6	72	22
	7.63 (25)	25	36	−10	6	72	22
	8.03 (26.3)	29	33	−8	5	73	22

[†] R. Loess = Richland Loess; U. and W. Bates T. = unweathered and weathered Batestown Till; Piatt T. = Piatt Till; Cc = calcite; D = dolomite; cps = counts per second; VI = vermiculite index; %E = % expandables; %I = % illite; %KC = % kaolinite + chlorite; ----- = boundary between material types or the top of a soil; ..... = boundary within a material or soil.

Table 11-6. Example of chlorite loss in soil developed on red Wisconsinan (Tiskilwa) Till (Wickham et al., 1988; McHenry County, Illinois). [†]

Material	Depth m (ft)	Color	Cc, cps	D, cps	VI	%E	%I	%KC
R. Loess	0.61 (2)	drab grn yl	0	0	nd	58	30	12
W. Tisk Till	0.91 (3)	peach	12	65	25	34	53	13
	1.22 (4)	dk rd peach	38	61	19	29	58	13
	1.52 (5)	dk peach	36	75	19	28	58	14
	1.83 (6)	dk red peach	48	75	nd	28	58	14
	2.29 (7.5)	peach	34	64	17	31	57	12
	2.74 (9)	peach	35	66	9	25	63	12
	2.90 (9.5)	pale peach	28	81	7	19	67	14
	3.05 (10)	peach	31	70	14	22	64	14
Transition	3.20 (10.5)	copper	39	69	6	19	63	18
	3.66 (12)	rd br	48	76	3	16	65	19
U. Tisk Till	3.96 (13)	rd br	30	68	5	13	64	23
	4.27 (14)	rd br	34	75	7	14	65	21
	4.57 (15)	rd br	27	53	7	15	64	21
	4.72 (15.5)	rd br	35	73	3	13	61	26
	4.97 (16.3)	rd br	38	70	8	15	61	24
	5.18 (17)	rd br	45	63	7	15	61	24

[†] R. Loess = Richland Loess; U. and W. Tisk Till = unweathered and weathered Tiskilwa Till; Cc = calcite; D = dolomite; cps = counts per second; VI = vermiculite index; %E = % expandables; %I = % illite; %KC = % kaolinite + chlorite; ----- = boundary between material types or the top of a soil; ..... = boundary within a material or soil; grn = green; yl = yellow; dk = dark; rd = red; br = brown.

between loss and preservation in these paleosols seems to reflect the fact that weathering in a saturated, gley-like environment leads to significant alteration of illite and orthoclase as a result of K^+ removal by plants, while chlorite is left unaltered. Calcium, Mg, and Fe carbonates are sometimes formed in gley-type underclays as well. In fireclay-type underclays, SiO_2 removal and nutrient extraction by plants, percolation of acid water, or combinations of these two processes, result in a suite of kaolin-group minerals and poorly crystallized expandables (Fig. 11-4) and the rapid loss of chlorite, orthoclase, and plagioclase.

These examples demonstrate two aspects of pedogenesis. First, mineral alteration is dependent on the composition and pH of pore water and the rate of percolation. If certain elements are at low concentrations in connate solutions, the phases containing these elements will be rapidly altered without respect to the average stability of the minerals. Second, the changes associated with plant growth are distinct from those induced by abiotic or "chemical" weathering.

Loss of Carbonates

Calcite and dolomite are common in parent materials of glacial origin. The removal of calcite and then dolomite is the pedogenic process that usually follows oxidation and chlorite alteration. The normal sequence of calcite and then dolomite loss is shown in Tables 11-5 through 11-8. Calcite has been partly removed in the uppermost sample of the profiles in Tables 11-5 and 11-6. Both calcite and dolomite in the upper part of the profile shown in Table 11-7 have been lost by leaching to a depth of about 5 m. The preservation of calcite and dolomite in the zone at 6.1 to 6.3 m (Table 11-7) probably reflects the oxidation of a gleyed till, as indicated by the secondary

Table 11-7. Example of gleyed till formed on Illinoian (Kellerville) Till (Wickham, 1980; Hancock County, Illinois).[†]

Material	Depth m (ft)	Color	Cc, cps	D, cps	HSI	%E	%I	%KC
GP	1.68 (5.5)	nd	0	0	32	82	13	5
	1.83 (6)	nd	0	0	31	80	14	6
SB	3.20 (10.5)	bl org	0	0	7	65	22	13
	3.35 (11)	bl org	0	0	10	75	15	10
BC	4.72 (15.5)	grn yl	0	0	26	83	11	6
	5.03 (16.5)	grn gd	0	0	31	84	10	6
SG	6.10 (20)	or yl br	9	15	21	56	28	16
	6.31 (20.7)	yl or br	14	?	19	55	28	17
GK	6.52 (21.4)	grn br	16	?	26	71	18	11
	7.65 (25.1)	gley gr	14	15	18	67	19	14
	7.86 (25.8)	br gr	17	14	19	63	22	15
	8.05 (26.4)	gley gr	20	10	19	63	22	15
	9.18 (30.1)	ol grn br	14	13	18	58	24	18
	9.39 (30.8)	ol grn br	18	15	20	57	24	19
	9.57 (31.4)	grn br	21	12	16	49	30	21
UK	10.88 (35.7)	till gr	6	17	7	44	34	22
	11.03 (36.2)	dk gr br	13	22	8	42	37	21
	13.87 (45.5)	dk gr br	19	15	8	42	38	20
	14.08 (46.2)	dk gr br	18	21	7	43	38	19
	15.33 (50.3)	brnish gr	10	10	8	44	40	16
	15.64 (51.3)	brnish gr	10	9	7	42	39	19
	16.92 (55.5)	brnish gr	12	18	7	42	36	22

† GP = gleyed Peoria Loess; SB = soil on Berry Clay; BC = Berry Clay; SG = soil on gleyed Kellerville Till; GK = gleyed Kellerville; UK = unweathered Kellerville Till; Cc = calcite; D = dolomite; cps = counts per second; HSI = heterogeneous swelling index; %E = % expandables; %I = % illite; %KC = % kaolinite + chlorite; ----- = boundary between material types or the top of a soil; ..... = boundary within a material or soil; bl org = black (organic); brnish = brownish; gd = gold; gr = gray; ol = olive; or = orange; grn = green; yl = yellow; br = brown.

colors. Table 11-8 illustrates a case where dolomite is preserved only in the least altered till at the base of the profile.

Some Quaternary profiles contain secondary calcite that must be distinguished from primary carbonate (Hughes & Glass, 1984). However, assuming that (i) dolomite was originally present in these materials, and (ii) secondary dolomite does not form in these materials (to our knowledge it has never been reported), we conclude that the absence of dolomite in association with pedogenic calcite is diagnostic of secondary calcite. Unweathered parent materials sometimes vary significantly in calcite:dolomite content (Glass & Killey, 1987). This variation can be used to further separate source materials or to evaluate the type and intensity of pedogenesis.

Weathering of Expandables and Illite

Because expandables and illite are the most abundant clay minerals in the parent materials in many areas, the weathering products of these 2:1 clay minerals dominate pedogenic profiles. The soil profile on a Late Wisconsinan Richland Loess, illustrated by the XRD patterns in Fig. 11-2, shows the "smearing" of the expandables (1.7-nm) peak that is the product of weathering. (This use of "smearing" of XRD peaks should be distinguished from preparation of smear slides.) This figure illustrates the use of percentages as opposed to indices of alteration in the characterization of soils. The heterogeneous swelling index (Fig. 11-2, HSI) measures the peak broadening or "smearing" of the 001 expandables peak, which indicates weathering. The decrease in HSI occurs upward in the profile from unaltered loess to the A horizon of the modern soil (Frye et al., 1968). If the percentage of expandables in this profile is of interest, the area of the 1.7-nm peak would have to be calculated, or preferably, a scan would have to be made of the slide after heating to 300 to 400 °C to measure collapsibles. Although the sequence of alteration observed in this profile matches the general category of those which are oxidized and well drained, the exact nature of the changes in clay-mineral structures that led to the broad XRD peaks in this profile is not clear. However, this type of alteration is similar to that observed in modern and ancient profiles in loess where the expandable clay minerals are weathering to mixed-layered kaolinite/expandables (K/E).

Table 11–8. Example of an accretion gley on Illinoian (Sterling) Till (Frye, et al., 1969; Lee County, Illinois).†

Material	Depth m (ft)	D, cps	HSI	DI	%E	%I	%KC
P. Loess	2.16 (7.1)	10	nd	1.8	63	27	10
	2.41 (7.9)	0	nd	0.73	73	14	13
Rox. L.	2.67 (8.8)	0	nd	0.33	74	9	17
	3.05 (10)	0	nd	0.63	75	12	13
S. Gley	3.56 (11.7)	0	18	nd	82	8	10
	3.68 (12.1)*	0‡	18‡	1.1‡	85‡	9‡	6‡
A. Gley	3.94 (12.9)	0	28	0.80	89	6	5
	4.27 (14)	0	26	0.94	90	6	4
LS Till	4.52 (14.8)	0	22	1.5	82	12	6
	4.95 (16.3)	0	15	3.4	60	33	7
CS Till	5.33 (17.5)	0	nd	4.7	11	78	11
	5.84 (19.2)	27	nd	5.5	10	80	10
	6.10 (20)	37	nd	5.4	7	82	11

† P. Loess = Peoria Loess; Rox. L. = Roxana Loess; S. Gley = soil on accretion gley; A. Gley = accretion gley (Berry Clay); LS Till = leached Sterling Till; CS Till = calcareous Sterling Till; D = dolomite; cps = counts per second; HSI = heterogeneous swelling index; DI = diffraction index; %E = % expandables; %I = % illite; %KC = % kaolinite + chlorite; ----- = boundary between material types or the top of a soil; ..... = boundary within a material or soil.
‡ Indicates lepidocrite.

Because the detection of K/E with few kaolinitic layers is so difficult, this profile may represent an extremely early stage of this type of alteration. A mixed-layered clay-mineral structure that included variable amounts of vermiculite and smectite could give rise to such diffraction patterns. Frye et al. (1968) believed that "smearing" of the expandables peak was due to variations in the amount of expansion in response to solvation with ethylene glycol. Heating at 375 °C shows that, based on the XRD trace of the unaltered loess, the expected amount of peak intensity is present in the A zone of the soil (Fig. 11-2), although the peak of the collapsibles shows significant broadening. This broadening would suggest that a reduction of domain size may be a principal factor in the "smearing" of the expandables peak.

Weathering of Illite to Illite/Smectite, Expandables, and Vermiculite

As is generally true of weathering of other minerals, illite alters to several clay minerals, Al- and Fe-oxide minerals, amorphous oxides, or some combination of these products. We have observed a sequence of reactions of illite to I/S, or rarely high-charge vermiculite or kaolinite. As weathering of illite continues, the initial products are weathered to varieties of vermiculite with a lower charge, to I/S with increasing smectite content, or more generally to expandables, to K/E, to kaolinite, and most often to some combination of these clay minerals. The nature of the illite alteration products depends on the degree of drainage in the profile and the magnitude of nutrient extraction by plants.

Table 11-7 shows data from a profile (6.52-9.57 m) where the alteration of illite (and chlorite) has resulted in the formation of expandables. This profile also illustrates the use of HSI as an indicator of gleying conditions. Pedogenesis is reflected in illite and chlorite loss, and HSI values much higher than would be observed in a well-drained profile. The color change between 8.05 m and 9.18 m indicates the boundary between the oxidized Kellerville Till below and the zone of gleyed Kellerville above. Based on color differences and an apparent loss of expandables, the two samples at 6.10 m and 6.31 m represent a weathering event that followed the gleying of the Kellerville Till. Back-calculating from the composition of these two samples gives reconstructed compositions of 710 g E kg^{-1}, 190 g I kg^{-1}, and 100 g $K+C$ kg^{-1} at a depth of 6.10 m, and 710 g E kg^{-1}, 180 g I kg^{-1}, and 110 g $K+C$ kg^{-1} at the 6.31-m depth. Comparison with the composition of the sample of gleyed Kellerville Till at 6.52 m, which is assumed to be a representative parent material, shows that the assumption of continuity in the gley before late-stage pedogenesis is a good one. [This back-calculation technique is referred to by us as the Type Composition Index (Hughes & Warren, 1989).] An accretion gley (Berry Clay), a deposit of colloidal particles from nearby paleogeographic highs, is present from 4.72 to 5.03 m, as shown by the high HSI and percentage of expandables. Above the unaltered accretion gley at 3.2 and 3.35 m, there appears to be a weathering profile developed on the gley. Furthermore, the higher proportion of expandables than average (700 g kg^{-1}) in the overlying Peoria Loess indicates that the loess has been gleyed.

The data listed in Table 11-8 show the compositional variation from a profile of an accretion gley that overlies Illinoian Sterling Till. The leached zone in the till (4.52-4.95 m) is mixed with material from the accretion gley above and overlies relatively unaltered, high-illite till (5.33 m to 6.10 m). In this instance, the composition changes upward from ~800 g illite kg^{-1} in the unweathered till to 600 to 800 g expandables kg^{-1} in the leached till and mixed gley zone. Above the till, in the accretion gley (3.56- 4.27 m), the expandables content reaches 900 g kg^{-1}. The upper two samples in the accretion gley (3.56-3.68 m) appear to be weathered, as indicated by a decrease in the HSI and percent expandables. Overlying the accretion gley, two loess units can be distinguished by the large increase in the Diffraction Intensity Index and a distinct and unique change in proportions of illite relative to kaolinite + chlorite when Peoria Loess is compared to the Roxana Loess. The Diffraction Intensity Index also changes significantly for samples from the accretion gley, the mixed zone of leached till and accretion gley, and the mostly unaltered till.

Figure 11-4 shows the mineral suite associated with partial alteration of the parent material (shale-type underclay, bottom trace) to a gley-type underclay (top trace). The result is a situation in which illite and orthoclase are altered to I/S and well-crystallized kaolinite, respectively. In many cases chlorite and albite, which are easily weathered in many pedogenic environments, are preserved in gley-type underclays (Hughes et al., 1987). Soil profiles in gley-type underclays are uncommon, and most mineralogical changes associated with weathering are observed at different localities of the same paleosol or in erratic sequence within a single occurrence. A few examples of profiles have been found, however, where the local environment of soil formation seems to have varied systematically through time (Rimmer and Eberl, 1982). The origin of the clay-mineral suite in gley-type underclays can be summarized as progressive alteration to I/S with an increase in smectite content and decrease in ordering along the c-axis (i.e., R3 → R1 → R0) (Rimmer & Eberl, 1982; Hughes et al., 1987). As noted by Hughes et al. (1987) and Moore and Reynolds (1989), this "opening up" of illite due to K^+ removal from illite and the sequence of clay minerals that occurs in a few feet of underclay are virtually identical to the same sequence of clay minerals that occurs over a range of 10 000 ft during the collapse of smectite to illite resulting from burial diagenesis. This parallelism in clay-mineral reaction sequence is similar to the preferential weathering of nonclay minerals in a pattern that is the reverse of Bowen's (1956, p. 60) reaction series of the crystallization of igneous minerals.

Vermiculite, rather than I/S, forms from illite in many cases. Although the cause is unknown, the lower pH and increased percolation rates associated with most vermiculite occurrences suggest that the intensity of illite weathering controls the relative proportions of I/S and vermiculite in weathering profiles. High rates of K^+ extraction by plants in neutral-to-alkaline, poorly drained environments favor the formation of I/S, whereas rapid K^+ extraction in more acidic, well-drained environments tend to favor vermiculite formation. When analyzing soils developed on glacigenic deposits in the Midwest, it is often important to distinguish illite from muscovite and biotite, because the latter two minerals are increasingly abundant in size fractions greater than about 5 μm. It is likely that part of the high-charge vermiculite observed in many soils in the Midwest represents weathered micas. In other soils, such as those developed in loess, vermiculite seems to be developed in association with peak broadening of expandables and with the formation of mixed-layered kaolinite/expandables (K/E) (Fig. 11-5).

Weathering of 2:1 Minerals to Mixed-layered Kaolinite/Expandables and Kaolinite

The extent of K/E formation observed in soils developed on loess, till, and in underclays suggests that this clay mineral may be an excellent indicator of intensity and duration of weathering in temperate to tropical soils where abundant 2:1 clay minerals are present in the parent material (Hughes & Glass, 1984; Hughes et al, 1987; Hughes et al., 1990b). Figures 11-5 and 11-6 illustrate the weathering of expandables to K/E for samples of a sequence of soils developed in loess near Thebes, Illinois. The $001/001$ peak from collapsed K/E with a low proportion of kaolinite to smectite occurs as a shoulder on the high-angle side of the 1.0-nm peak for collapsed expandables and illite (best illustrated by tracing 2 in Fig. 11-7). This shoulder becomes a separate peak and migrates progressively towards the kaolinite peak position as the kaolinite content increases.

The uppermost traces in Fig. 11-5, 11-6, and 11-7 are of the <2-μm fraction of the least-weathered Peoria Loess. This composition is assumed to represent the parent material of the modern soil (Trace 2 in Fig. 11-5, 11-6, and 11-7), and is probably similar to the parent materials for the Sangamonian and Yarmouthian paleosols. Trace 2 (Fig. 5, 6, and 7) is of the <2-μm fraction from the modern soil. Although the expandables peak near 1.7 nm on the glycol-solvated trace of the modern soil appears to have broadened and vermiculite is clearly present, a casual comparison of the upper two traces fails to show the presence of a significant amount of K/E. Exact superposition of the XRD traces from heated samples of parent material and modern soil shows the characteristic elevated intensity of K/E between 1.0 nm and 0.7 nm, and the asymmetry on the high-angle side of the 1.0-nm peak typical of K/E (Fig. 11-7).

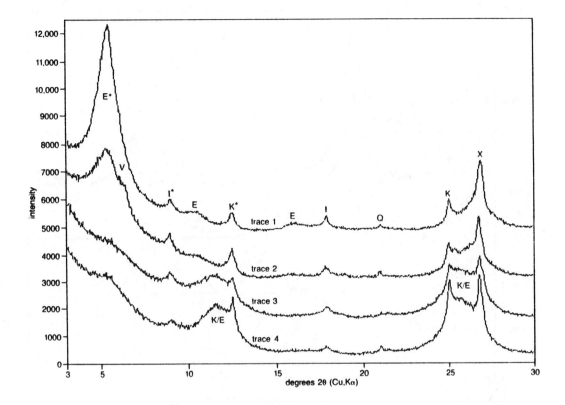

Fig. 11-5. XRD traces of ethylene-glycol-solvated samples from an approximately 20-m (64-ft) boring in loess near Thebes, Illinois. Least-altered Peoria Loess, Trace 1 (Sample 3242CC); Modern soil, Trace 2 (Sample 3242B); Bt of the Sangamonian paleosol, Trace 3 (Sample 3243K); and that of part of the Yarmouthian paleosol, Trace 4 (Sample 3243MM). $E*$ = expandables (* = 1.7 nm); V = vermiculite; $I*$ = illite (* = 1.0 nm); $K*$ = kaolinite (and minor chlorite?) (* = 0.7 nm); K/E = mixed-layered kaolinite/expandables; Q = quartz; X = $E + I + I/S + Q$.

The K/E peaks are easier to identify on the traces of samples from the Sangamonian and Yarmouthian paleosols at Thebes (Fig. 11-5 and 11-6, Traces 3 and 4). Intense smearing of the expandables peak is also apparent in the lower two sets of traces of Fig. 11-5 and 11-6. The K/E peak is distinct enough on both the glycol-solvated and heated traces to estimate the proportion of kaolinitic to expandable layers within the K/E. The proportion of kaolinitic to expandable layers of K/E from the modern soil at Thebes can only be estimated by using a computer program to decompose the three overlapped peaks from about 7 to 14 °2θ. This peak decomposition still requires a good initial guess for the peak position of the K/E. Several lines of evidence suggest that the parent materials for the paleosols were similar to the unaltered Peoria Loess. If that assumption is made, an estimate of a reference intensity ratio for K/E can be made by comparing the loss of intensity at 1.0 nm with the gain in intensity for the K/E peak between 1.0 nm and 0.7 nm.

The K/E is commonly observed in fireclay-type underclays. Figure 11-4 shows an XRD trace for a fireclay-type underclay (middle trace) with a ratio of kaolinitic layers to expandable layers of about 80:20. Most underclays of this type only have diffraction bands for K/E between 1.0 nm and 0.7 nm (Hughes et al., 1987) and lack even the broad peaks observed for K/E in loessal paleosols. Most likely, the lack of peak definition for K/E in underclays is the result of K/E formation from an illite-rich parent material (Fig. 11-4, bottom trace). The higher-charge parent material probably gives rise to K/E with a wider range of expandable layer charges and more variable ratio of kaolinitic to expandable layers.

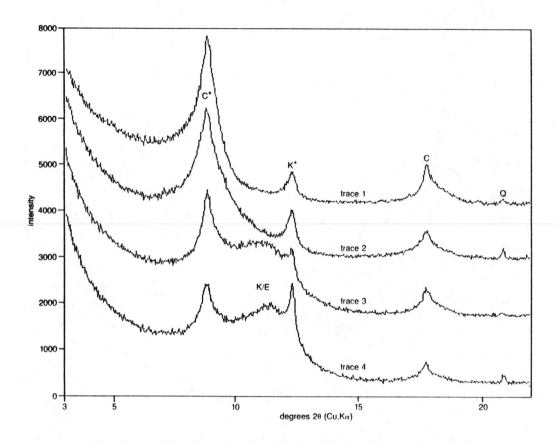

Fig. 11-6. XRD traces of the same set of samples as in Fig. 11-5, but after heating at 325 °C. Symbols as in Fig. 11-5, and $C*$ = collapsibles + illite at 1.0 nm; $K*$ = kaolinite at 0.7 nm. Note the progressive increase in background and ultimately the formation of the peak between 1.0 nm and 0.7 nm.

In contrast to underclays, smectite-rich parent materials in loess give rise to K/E with a relatively narrow proportion of kaolinitic to expandable layers. We typically use the *001* of illite, the *001* of collapsibles, the 0.7-nm kaolinite + chlorite peak, and the peak or diffraction band between 1.0 nm and 0.7 nm for K/E as measurable quantities. This procedure does not resolve the nature of the collapsibles, largely because of the difficulty of distinguishing between the expandable component of K/E, other mixed-layered expandable minerals such as I/S, and discrete expandables such as smectite and vermiculite. Our practice for the present is to separate the intensities to the extent possible and assign all intensity on the traces of heated samples between the 1.0- and 0.7-nm peaks to K/E. Calculated patterns from Reynolds's (1985) NEWMOD show that the phases should be distinct and separable. However, the HSI values in weathered loesses (Frye et al., 1968 and Fig. 11-2) suggest that the structure of expandable clay minerals is being altered in concert with the formation of K/E, and it is perhaps more meaningful to regard the entire system as a single alteration phase and the alteration as a continuum.

The formation of pedogenic kaolinite has been observed in Quaternary soils (Hughes & Glass, 1984; Pye & Johnson, 1988). However, the kaolinite seems to be derived mainly from feldspar in these cases. Kaolinite is definitely formed from 2:1 clays in strata associated with coals (Hughes et al, 1987). An interesting question is why most of the alteration product is K/E in some cases and kaolinite in others. We believe that the ratio of these two phases, or, perhaps more correctly, the extent to which K/E forms as an intermediate during kaolinitization of 2:1 clay minerals, is determined by the same weathering factors that control the relative amounts of I/S and vermiculite or of chlorite and vermiculite, namely, climate, amount and type of plant growth, solution pH, rate and regularity of percolation, etc.

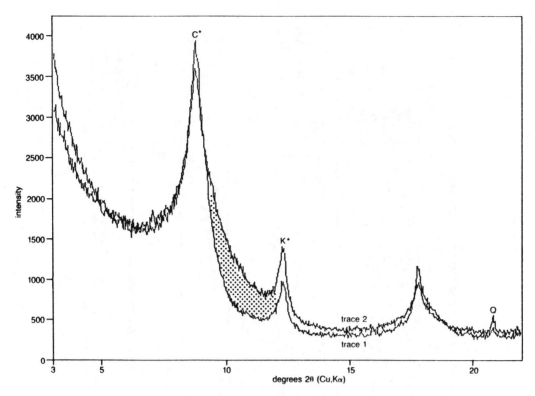

Fig. 11-7. Superimposed XRD traces of heated samples of the Modern soil, Trace 2 (Sample 3242B), and the least-altered Peoria Loess, Trace 1 (Sample 3242CC), showing the elevated background (stippled area between 1 and 2) characteristic of K/E with a small proportion of kaolinitic to expandables layers. The elevated background is between large $C *$ peak and $K *$ peak. Symbols as in Fig. 11-6.

Weathering of 2:1 Clay Minerals to Palygorskite and Sepiolite

Frye et al, (1974) report the alteration of expandable clay minerals in soils of the High Plains to palygorskite and sepiolite (Fig.11-8). Typically, smectite-rich parent material is altered first to palygorskite and then to sepiolite. Because palygorskite and sepiolite contain less SiO_2 than smectite, opaline silica is often observed in the alteration zone. As a result of upward water movement in these soils, SiO_2 is trapped in the top of the profile. The profile in Fig. 11-8 also shows a kaolinite- and illite-rich zone at the top. This unit is thought to represent younger detrital material that is not yet a part of the soil profile and results from increasing erosion during the Pleistocene. It is worth noting that detection of the palygorskite and sepiolite in these samples requires careful extraction of carbonates with a 5% acetic acid solution (see sample preparation and instrumentation above). Stronger acids or too much acid can easily result in the dissolution of the clay fraction and wrongly suggest that only opaline silica and caliche minerals are present in the upper zones of the soil.

Iron and Aluminum Oxides/Hydroxides

Oxide formation in soils often represents a significant increase in phases that are relatively amorphous to the x-ray beam. This loss of intensity is illustrated in Fig. 11-9 and generally should be adjusted for in QR and QPA methods. Hematite, goethite, magnetite, and lepidocrocite all form in temperate soils. The common Fe oxide in Midwest soils is goethite. Lepidocrocite can form during the reoxidation of reduced Fe phases in accretion gleys (Table 11-8). Trace 2 in Fig. 11-9 is for an Fe nodule typical of those from an Ordovician-age paleosol in northwestern Illinois. An

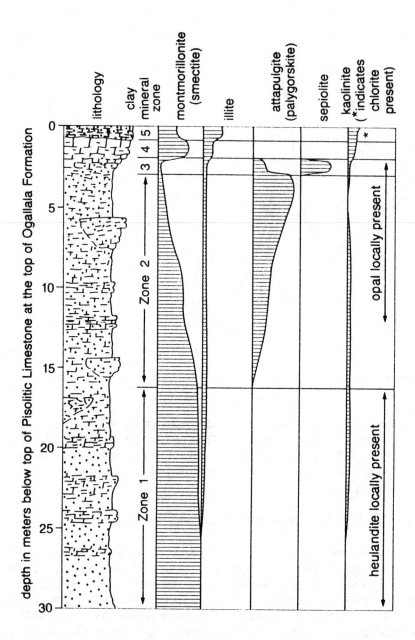

Fig. 11-8. Generalized profile of the mineral trends observed in the High Plains soils of central-eastern New Mexico. Compositions are from acetic acid residues. Reproduced with permission from Frye et al. (1974).

XRD pattern of associated claystones that are from the same paleosol is included in Fig. 11-9 (Trace 1) to illustrate the amount of lost intensity in the ironstone zone (Hughes et al., 1990a). When significant amounts of amorphous material become abundant in a soil, it is often necessary to employ other analytical methods, remove the amorphous constituents with acid or other reagents, or factor them out by ignoring them in the calculation of percentages for the mineral fraction.

Aluminum oxides have only recently been observed in the soils analyzed in our laboratories. This occurrence of boehmite is associated with the paleosol illustrated in Fig. 11-9. However, these phases are common in soils of warmer and wetter regions, and represent an extension of the time and intensity of weathering usually observed in temperate regions. We believe that occurrences of kaolinite (and berthierine) in these Early Paleozoic paleosols may represent pedogenic Al and Fe oxides that have been resilicified during postpedogenic diagenesis.

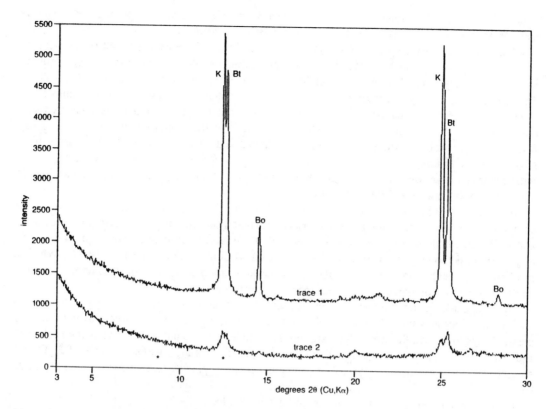

Fig. 11-9. Two XRD patterns showing the Fe-oxide-rich nodules (lower pattern) associated with Native American artifacts from near Sterling, Illinois. Upper trace shows the unaltered composition of a typical artifact. The raw materials for these artifacts are thought to have come from a 400-million-year-old paleosol (Hughes et al., 1990a). *K* = kaolinite; *Bt* = berthierine; *Bo* = boehmite; * near 8.8 and 12.5 °2θ = 1.0 and 0.7 nm, respectively.

FUTURE DEVELOPMENTS

The complexities outlined in this report might suggest that future trends will be away from XRD. Our experience indicates, however, that continued improvements have made this field an area for future growth. New approaches that calculate the variation in width, *d* spacing, or shape of peaks, and that can be carried out rapidly and automatically, add a new level of possible refinements. In preparing this report, we have seen the potential value of studies that compare the results of QR in large databases with the results of special treatments of a few carefully selected samples. The treatments that provide the most useful information are XRD analysis of narrow particle-size intervals, exchange of cations, solvation of samples with glycerol, thermal treatment above 300 °C, measures of the layer charge on expandable layers, and intercalation of kaolinite, halloysite, dickite, and nacrite.

Although we know of no published reports of QPAs that meet the standards for accuracy and precision requirements typical for chemical analyses, the capability to carry out QPA at acceptable levels of cost seems to be at hand. For clay-rich materials, a procedure with bulk-smear samples or powder packs and smears of a size fraction, that contains most of the clay-mineral fraction (usually ~ <5-20 μm), and a reasonable XRD regimen should yield an adequate QPA. If samples contain too much sand or silt for a bulk smear, an XRD trace of the bulk sample and of a smear of a size fraction that contains most of the clay mineral fraction are essential. An internal or chemically determined standard is necessary to calculate the absolute content of the nonclay minerals and to estimate the mass absorption coefficient of the bulk sample. An XRD trace of the bulk, ethylene-glycol-solvated, and ~300 ° C-heated sample is required, although bulk smears make it possible to fulfill these needs in as little as two ~15-min XRD scans.

The final necessary step for a QPA is the detailed fitting of the experimental traces with the computerized results from NEWMOD and the apportioning of the relative clay-mineral contents to the overall clay-mineral weight fraction determined by difference from the amount of nonclays. This method will only give acceptable results when the content of organic matter is minimal, the organic matter is removed prior to analysis, or the organic matter content is measured and corrected for in the analysis. Materials other than organic matter that are amorphous to the x-ray beam will cause problems similar to those of organic material. These materials will require similar steps to reduce or eliminate their effect on the analysis. However, it has been our experience that most materials have low amorphous-silicate or Fe-oxide contents, and therefore, the problem of amorphous inorganics should be insignificant.

Advances will be made in improving the constraints for normative analysis and for using computer systems that draw upon past experience (expert systems). It is less certain that Rietveld and similar methods will become widely used. Partly this is a result of recent improvements in using 00ℓ peaks, and partly it reflects difficulties involved in using $hk\ell$ peaks. The NEWMOD program may also be modified to calculate the mass absorption coefficient of the sample, and this coefficient can be compared to the experimentally determined value.

Although recent improvements in instruments for mineralogical analysis suggest to us that similar improvements will take place in the future, the nature of those new directions is much less clear. Ways to control the atmosphere around the sample during analysis are a valuable first step. For clay-mineral analysis, a chamber that can be adjusted for a high vapor pressure of H_2O and ethylene glycol is essential. Glycerol-clay complexes are apparently stable for extended periods (Novich & Martin, 1983). This chamber can be advantageously made to allow for the maintenance of a dry air condition and for the transfer of heated samples to the sample stage.

SUMMARY

We emphasize XRD methods that use large numbers of samples, meet the constraints of time and resources available, and maintain acceptable levels of precision ($\pm$2-5%). This approach is the necessary first step to define the limits of variation for most problems and provides essential statistical confidence. Many problems require no further analysis. Our choice of methods and use of terms reflect this emphasis; for example, we use the terms expandables and collapsibles, and report our results in those terms, so that the number and duration of analytical determinations can be minimized. We designate those procedures that are concerned primarily with precision and rapid analysis as quantitative representation (QR); those that must maintain acceptable levels of precision and accuracy are commonly referred to as quantitative phase analysis (QPA). Our QPA methods employ random powder mounts with an internal standard, smear mounts of bulk samples and critical size-fractions, and calculated XRD patterns. We feel that in the very near future we will be able to treat samples in the detail required to provide acceptable levels of precision and accuracy and to analyze a quantity of samples large enough to assure true representation. Recent improvements in the understanding of the characteristics of clay minerals and in instrumentation are the basis for our optimism.

The treatments and procedures used for QR, in decreasing order of frequency are: preparation of sedimented and smear slides; ethylene-glycol solvation of slides; exposure of slides to ~300 ° C for 1 h (heating above this temperature causes structural breakdown in several clay minerals); calculation of "model" XRD patterns; preparation of random powder packs of bulk samples; addition of internal standards; removal of organics, Fe oxides, and carbonates; exchange of ions; intercalation of kaolin-group minerals; and separation of magnetic and density fractions.

Studies of soils and paleosols from modern to Paleozoic illustrate the use of QR methods and the types of alteration observed in temperate, tropical, and semiarid environments of pedogenesis. Nutrient extraction by growing plants, pH, and degree of drainage are the principal factors controlling the formation and alteration of clay minerals during weathering. High rates of K^+

extraction by plants, low rates of percolation, and neutral-to-alkaline environments favor the formation of mixed-layered illite/smectite. Higher rates of drainage and lower pH levels, as well as K^+ removal by plants, leads to greater amounts of vermiculite. The nature of the chlorite in gray tills limits its alteration product to high-charge vermiculite, whereas chlorite in red tills alters to high- and then low-charge vermiculite. Mixed-layered kaolinite/expandables form from 2:1 clay minerals and in environments where SiO_2 is removed by plants. Well-crystallized kaolinite forms readily from most parent materials other than 2:1 clay minerals, and from all aluminosilicates where drainage, salinity, and pH are low and SiO_2 removal rates are high.

The XRD methods yield essential information about the composition and genetic history of a soil with the minimum investment of time and money. Questions about past global climates and present land use require detailed understanding of the information stored in the minerals making up soils, much of which has yet to be worked out, and insure the continuing importance of XRD analysis.

ACKNOWLEDGMENTS

The authors gladly cite the assistance of L. R. Follmer, J. M. Masters, W. A. White, C. L. Window, and J. C. Hurley. We also appreciate the helpful reviews of J. E. Amonette, L. R. Follmer, and J. H. Goodwin.

REFERENCES

Bailey, S.W. 1980a. Summary and recommendations of AIPEA nomenclature committee. Clays Clay Miner. 28:73-78.

Bailey, S.W. 1980b. Structures of layer silicates. p. 1-123. *In* G.W. Brindley and G. Brown (ed.) Crystal structures of clay minerals and their x-ray identification. Mineral. Soc., London.

Bailey, S.W. 1982. Nomenclature for regular interstratifications. Am. Miner. 67:394-398.

Bailey, S.W. 1984. Classification and structure of the micas. p. 1-12. *In* S.W. Bailey (ed.) Micas. Reviews in mineralogy 13. Miner. Soc. of Am., Washington, DC.

Bailey, S.W. (ed.). 1988. Hydrous phyllosilicates (exclusive of micas). Reviews in mineralogy 19. Miner. Soc. of Am., Washington, DC.

Bish, D.L., and J.E. Post (ed.). 1989. Modern powder diffraction. Reviews in mineralogy 20. Miner. Soc. of Am., Washington, DC.

Bish, D.L., and R.C. Reynolds, Jr. 1989. Sample preparation for x-ray diffraction. p. 73-99. *In* D.L. Bish and J.E. Post (ed.) Modern powder diffraction. Reviews in mineralogy 20. Miner. Soc. of Am., Washington, DC.

Bowen, N.L. 1956. The evolution of igneous rocks. Dover Publ., Inc., New York.

Brindley, G.W. 1961. Quantitative analysis of clay mixtures. p. 489-516. *In* G. Brown (ed.) The x-ray identification and crystal structures of clay minerals. Miner. Soc., London.

Brindley, G.W. 1980. Quantitative mineral analysis of clay mixtures. p. 411-438. *In* G.W. Brindley and G. Brown (ed.) Crystal structures of clay minerals and their x-ray identification. Miner. Soc., London.

Brindley, G.W. 1981. X-ray identification (with ancillary techniques) of clay minerals. p. 22-38. *In* F.J. Longstaffe (ed.) Short course in clays for the resource geologist. Miner. Assoc. of Canada, Toronto.

Chung, F.H. 1974a. Quantitative interpretation of x-ray diffraction patterns--I. Matrix-flushing method of quantitative multicomponent analysis. J. Appl. Crystallogr. 7:519-525.

Chung, F.H. 1974b. Quantitative interpretation of x-ray diffraction patterns--II. Adiabatic principle of x-ray diffraction analysis of mixtures. J. Appl. Crystallogr. 7:526-531.

Chung, F.H. 1975. Quantitative interpretation of x-ray diffraction patterns--III. Simultaneous determination of a set of reference intensities. J. Appl. Crystallogr. 8:17-19.

Čečil, B., and D. Machajdík. 1981. Potassium- and ammonium-treated montmorillonites--I. Interstratified structures with ethylene glycol and water. Clays Clay Miner. 29:40-46.

Davis, B.L., and L.R. Johnson. 1989. Quantitative mineral analysis by transmission and x-ray diffraction. p. 104-117. *In* D.R. Pevear and F.A. Mumpton (ed.) Quantitative mineral analysis of clays. Clay Miner. Soc., Evergreen, CO.

Droste, J.B. 1956. Alteration of clay minerals by weathering in Wisconsin tills. Geol. Soc. Am. Bull. 67:911-918.

Fripiat, J.J., and M.I. Cruz-Cumplido. 1974. Clays as catalysts for natural processes. p. 239-256. *In* F.A. Donath (ed.) Annual review of earth and planetary sciences, Vol. 2. Annual Rev., Inc., Palo Alto, CA.

Frye, J.C., P.R. Shaffer, H.B. Willman and G.E. Ekblaw. 1960. Accretion-gley and the gumbotil dilemma. Am. J. Sci. 258:185-190.

Frye, J.C., H.D. Glass, and H.B. Willman. 1968. Mineral zonation of Woodfordian loesses of Illinois. Illinois State Geol. Surv. Circ. 427.

Frye, J.C., H.D. Glass, J.P. Kempton, and H.B. Willman. 1969. Glacial tills of Northwestern of Illinois. Illinois State Geol. Surv. Circ. 437.

Frye, J. C., H.D. Glass, A.B. Leonard, and D.D. Coleman. 1974. Caliche and clay mineral zonation of Ogallala Formation, central-eastern New Mexico. New Mexico Bur. Mines Miner. Resour. Circ. 144.

Gardner, T.W., E.G. Williams, and P.W. Holbrook. 1988. Pedogenesis of some Pennsylvanian underclays; ground-water, topographic, and tectonic controls. Geol. Soc. Am. Spec. Pap. 216:81-101.

Gast, R.G. 1977. Surface and colloid chemistry. p. 27-73. *In* J.B. Dixon and S.B. Weed (ed.) Minerals in soil environments. SSSA, Madison, WI.

Glass, H.D., and M.M. Killey. 1987. Principles and applications of clay mineral composition in Quaternary stratigraphy: Examples from Illinois, USA. p.117-125. *In* J. J. M. van der Meer (ed.) Tills and glaciotectonics: Proceedings of an INQUA symposium on the genesis and lithology of glacial deposits. A. A. Balkema, Rotterdam, the Netherlands.

Griffiths, J.C. 1967. Scientific method in analysis of sediments. McGraw-Hill, New York.

Grim, R.E. 1968. Clay mineralogy, 2nd ed. McGraw-Hill, New York.

Hallberg, G.R., J.R. Lucas, and C.M. Goodman. 1978. Semiquantitative analysis of clay mineralogy. Part I. p. 5-21. *In* Standard procedures for evaluation of Quaternary materials in Iowa. Iowa Geol. Survey Tech. Info. Ser. no. 8.

Harter, R.D. 1977. Reactions of minerals with organic compounds in soils. p. 709-739. *In* J.B. Dixon and S.B. Weed (ed.) Minerals in soil environments. SSSA, Madison, WI.

Hower, J., and T.C. Mowatt. 1966. The mineralogy of illite and mixed-layer illite/montmorillonite. Am. Miner. 51:825-854.

Hubbard, C.R., and R.L. Snyder. 1988. Reference intensity ratio measurement and use in quantitative XRD. Powd. Diffract. 3:74-78.

Hughes, R.E., and B.F. Bohor. 1970. Random clay powders prepared by spray drying. Am. Mineral. 55:1780-1786.

Hughes, R.E. and H.D. Glass. 1984. Mixed-layer kaolinite-smectite and heterogeneous swelling smectite as indexes of soil weathering intensity. p. 63. *In* Program with abstracts, Clay Miner. Soc. Annu. Meet,, 21. Baton Rouge, LA.

Hughes, R.E., P.J. DeMaris, W.A. White, and D.K. Cowin. 1987. Origin of clay minerals in Pennsylvanian strata of the Illinois Basin. p. 97-104. *In* L.G. Schultz et al. (ed.) Proc. Int. Clay Conf., Denver. 28 July-2 Aug. 1985. Clay Miner. Soc., Bloomington, IN.

Hughes, R.E. and R. Warren. 1989. Evaluation of the economic usefulness of earth materials by x-ray diffraction. p. 47-57. *In* R.E. Hughes and J.C. Bradbury (ed.) Proc. Forum Geology Industrial Minerals, 23rd. Illinois State Geol. Surv., Mineral Notes 102.

Hughes, R.E., D.M. Moore, T.E. Berres, and K.B. Farnsworth. 1990a. Berthierine pipestones of Native Americans in the mid-continent. p. 64. *In* Program with abstracts, Clay Miner. Soc. 27th Annu. Meet, Columbia, MO.

Hughes, R.E., D.M. Moore, and H.D. Glass. 1990b. The nature, detection, occurrence, and origin of kaolinite/smectite. p. 65. *In* Program with abstracts, Clay Miner. Soc. 27th Annu. Meet., Columbia, MO.

Hussey, G.A. Jr. 1972. Use of a simultaneous linear equations program for quantitative clay analysis and the study of mineral alterations during weathering. Ph.D. diss. Pennsylvania State Univ., University Park.

Jackson, M.L. 1969. Soil chemical analysis-advanced course, 2nd ed. M.L. Jackson, Madison, WI.

James, R.W. 1965. The optical principles of the diffraction of x-rays. Cornell Univ. Press, Ithaca, NY.

Johnson, L.J., C.H. Chu, and G.A. Hussey, Jr. 1985. Quantitative clay mineral analysis using simultaneous linear equations. Clays Clay Miner. 33:107-117.

Klimentidis, R.E., D.R. Pevear, and G.A. Robinson. 1990. Analysis of CMS special clay "CorWa-1," a corrensite pore filling in an Eocene volcaniclastic sandstone from Washington state. p. 72. *In* Program with abstracts, Clay Miner. Soc. Annu. Meet., 27th. Columbia, MO.

Klug, H.P., and L.E. Alexander. 1974. X-Ray diffraction procedures for polycrystalline and amorphous materials. 2nd ed. John Wiley & Sons, New York.

Kunze, G.W. 1955. Anomalies in the ethylene glycol solvation technique used in x-ray diffraction. p. 88-93. *In* W.O. Milligan (ed.) Clays and clay minerals, Proc. 3rd Natl. Conf., Houston, TX. 1954, Publ. no. 395, NAS-NRC, Washington, DC.

MacEwan, D.M.C., and A. Ruiz-Amil. 1975. Interstratified clay minerals. p. 265-334. *In* J. E. Gieseking (ed.) Soil components. Vol. 2. Springer-Verlag, New York.

MacEwan, D.M.C. and M.J. Wilson. 1980. Interlayer and intercalation complexes of clay minerals. p. 197-248. *In* G.W. Brindley and G. Brown (ed.) Crystal structures of clay minerals and their x-ray identification. Miner. Soc., London.

Malla, P.B., and L.A. Douglas. 1987. Identification of expanding layer silicates: layer charge vs. expansion properties. p. 277-283. *In* L. G. Schultz et al. (ed.) Proc. Int. Clay Conf., Denver. 28 July-2 Aug. 1985. Clay Miner. Soc., Bloomington, IN.

Marshall, C.E. 1977. The physical chemistry and mineralogy of soils, Vol. 2. Wiley, New York.

Moore, D.M., and R.C. Reynolds Jr. 1989. X-ray diffraction and the identification and analysis of clay minerals. Oxford Univ. Press, New York.

Moore, D.M. and R.E. Hughes. 1991. Varieties of chlorites and illites and porosity in Mississippian sandstone reservoirs. p. 174. *In* Program with abstracts, 75th Annu. Meet. Am. Assoc. Petrol. Geol., Dallas, TX.

Mortland, M.M. 1970. Clay-organic complexes and interactions. Adv. Agron. 22:75-117.

Newman, A.C.D. (ed.). 1987. Chemistry of clays and clay minerals. Miner. Soc., London.

Newman, W.A., R.C. Berg, P.S. Rosen, and H.D. Glass. 1990. Pleistocene stratigraphy of the Boston Harbor drumlins, Massachusetts. Quat. Res. 34:148-159.

Novich, B.E., and R.T. Martin. 1983. Solvation methods for expandable layers. Clays Clay Miner. 31:235-238.

Ostrom, M.E. 1960. An interlayer mixture of 3 clay mineral types from Hector, California. Am. Miner. 45:886-889.

Parham, W.E. 1964. Lateral clay mineral variations in certain Pennsylvanian underclays. p. 581-602. *In* W.F. Bradley (ed.) Clays and clay minerals, Proc. 12th Natl. Conf., Atlanta, GA. 1963. Pergamon Press, New York.

Pevear, D.R., and F.A. Mumpton (ed.). 1989. Quantitative mineral analysis of clays. Clay Miner. Soc., Evergreen, CO.

Pinnavaia, T.J. 1983. Intercalated clay catalysts. Science (Washington, DC) 220:365-371.

Pye, K., and R. Johnson. 1988. Stratigraphy, geochemistry, and thermoluminescence ages of lower Mississippi Valley loess. Earth Surf. Proc. Landforms 13:103-124.

Retallack, G.J. 1983. A paleopedological approach to the interpretation of terrestrial sedimentary rocks. The mid-Tertiary fossil soils of Badlands National Park, South Dakota. Geol. Soc. Am.. Bull. 94:823-840.

Reynolds, R.C., Jr. 1980. Interstratified clay minerals. p.249-303. *In* G.W. Brindley and G. Brown (ed.) Crystal structures of clay minerals and their x-ray identification. Miner. Soc., London.

Reynolds, R.C., Jr. 1985. NEWMOD, a computer program for the calculation of one-dimensional diffraction patterns of mixed-layered clays. R. C. Reynolds, Jr., Hanover, NH.

Reynolds, R.C., Jr. 1986. The Lorentz-polarization factor and preferred orientation in oriented clay aggregates. Clays Clay Miner. 34:359-367.

Reynolds, R.C., Jr. 1989. Principles and techniques of quantitative analysis of clay minerals by x-ray powder diffraction. p. 4-36. *In* D.R. Pevear and F.A. Mumpton (ed.) Quantitative mineral analysis of clays. Clay Miner. Soc., Evergreen, CO.

Rich, C.I., and R.I. Barnhisel. 1977. Preparation of clay samples for x-ray diffraction analysis. p. 797-808. *In* J.B. Dixon and S.B. Weed (ed.) Minerals in soil environments. SSSA, Madison, WI.

Rietveld, H.M. 1969. A profile refinement method for nuclear and magnetic structures. J. Appl. Crystallogr. 2:65-71.

Rimmer, S.M., and D.D. Eberl. 1982. Origin of an underclay as revealed by vertical variations in mineralogy and chemistry. Clays Clay Miner. 30:422-430.

Smith, D.K., G.G. Johnson, A. Scheible, A.M. Wims, J.L. Johnson, and G. Ullmann. 1987. Quantitative X-ray powder diffraction method using the full diffraction pattern. Powd. Diffract. 2:73-77.

Smith, D.K. 1989. Computer analysis of diffraction data. p. 183-216. *In* D.L. Bish and J.E. Post (ed.) Modern powder diffraction. Reviews in mineralogy 20. Miner. Soc. of Am., Washington, DC.

Snyder, R.L., and D.L. Bish. 1989. Quantitative analysis. p. 101-144. *In* D.L. Bish and J.E. Post (ed.) Modern powder diffraction. Reviews in mineralogy 20. Miner. Soc. of Am., Washington, DC.

Sposito, G. 1989. The chemistry of soils. Oxford Univ. Press, New York.

Środoń, J. 1980. Precise identification of illite/smectite interstratification by x-ray powder diffraction. Clays Clay Miner. 28:401-411.

Środoń, J. 1984. X-ray identification of illitic materials. Clays Clay Miner. 32:337-349.

Suquet, H., and H. Pézerat. 1988. Comments of the classification of trioctahedral 2:1 phyllosilicates. Clays Clay Miner. 36:184-186.

Van Olphen, H. 1977. An introduction to clay colloid chemistry, 2nd ed. Wiley, New York.

Weaver, C.E. 1956. The distribution and identification of mixed-layer clays in sedimentary rocks. Am. Miner. 41:202-221.

Weiss, A. 1969. Organic derivatives of clay minerals, zeolites, and related minerals. p. 737-781. *In* G. Eglinton and M.T.J. Murphy (ed.) Organic geochemistry. Springer-Verlag, New York.

Wickham, J.T. 1979. Glacial geology of north-central and western Champaign County, Illinois. Illinois State Geol. Surv. Circ. 506.

Wickham, J.T. 1980. Status of the Kellerville till member in western Illinois. p. 151-179. *In* G.R. Hallberg (ed.) Illinoian and pre-Illinoian stratigraphy of southeast Iowa and adjacent Illinois. Iowa Geol. Surv. Tech. Info. Ser. no. 11.

Wickham, S.S., W.H. Johnson, and H.D. Glass. 1988. Regional geology of the Tiskilwa till member, Wedron formation, northeastern Illinois. Illinois State Geol. Surv. Circ. 543.

Wilman, H.B., H.D. Glass, and J.C. Frye. 1963. Mineralogy of glacial tills and their weathering profiles in Illinois: Part I. Glacial tills. Ilinois State Geol. Surv. Circ. 347.

Wilman, H.B., H.D. Glass, and J.C. Frye. 1966. Mineralogy of glacial tills and their weathering profiles in Illinois: Part II. Weathering profiles. Ilinois. State Geol. Surv. Circ. 400.

Wilson, M.J. 1987. Soil smectites and related interstratified minerals: Recent developments. p. 167-173. *In* L. G. Schultz et al. (ed.) Proc. Int. Clay Conf., Denver, CO. 28 July-2 Aug. 1985. Clay Miner. Soc., Bloomington, IN.

12   Quantitative Thermal Analysis of Soil Materials

A.D. Karathanasis
University of Kentucky
Lexington, Kentucky

W.G. Harris
University of Florida
Gainesville, Florida

Thermal analysis, as defined by the International Confederation for Thermal Analysis, is "a group of techniques in which a physical property of a substance and/or its reaction products is measured as a function of temperature whilst the substance is subjected to a controlled-temperature program" (Wendlandt, 1986). Much has been published on the topic of thermal analysis, including book chapters devoted specifically to soil mineral applications (Mackenzie & Mitchell, 1970; Mackenzie & Calliere, 1975; Tan et al., 1986). This chapter emphasizes the quantitative applications of thermal analysis to soil minerals. Specifically, the objectives are to: (i) review the principles and instrumentation of thermal-analysis techniques which have potential quantitative applications to soils, (ii) discuss applications and limitations of various techniques in the quantification of specific soil minerals and their properties, and (iii) describe recent advances and project future developments.

Thermal-analysis systems have advanced in recent decades from "home-made" (hybridized) devices to sophisticated computer-controlled systems. Computerization has streamlined calibration, data storage, data analysis, and graphics capabilities. It has even improved the quality of the data by permitting greater precision in the control of the instrument. Recently, thermal analysis has been coupled with spectroscopic techniques to obtain complementary information about the mineral reaction of interest. In short, the traditional thermal-analysis techniques used for soil-mineral analysis have improved in versatility and precision, and several nontraditional techniques are emerging with the potential to further improve soil-mineral quantification.

Several traditional thermal-analysis techniques, measuring different physical properties, have quantitative application to soil minerals. These include thermogravimetry (TG), differential thermal analysis (DTA), and differential scanning calorimetry (DSC), and measure changes in mass, temperature, and enthalpy, respectively (Wendlandt, 1986). Other less common techniques that have the potential for quantitative measurement of soil mineral properties include thermodilatometry (TD) and thermomechanical analysis (TMA), and measure dimensions and mechanical characteristics, respectively. A brief account of the principles and instrumentation employed by these techniques is given next to provide a background for subsequent discussion of quantitative measurements using thermal analysis.

PRINCIPLES AND INSTRUMENTATION

Thermogravimetry

Thermogravimetry (TG) is a technique in which the mass of a substance is monitored as a function of temperature or time as the sample is subjected to a controlled-temperature program (Earnest, 1988). Application of TG is therefore limited to reactions involving a mass change. Thermogravimetry is a refinement of gravimetry, a technique in which the sample is weighed following periods of isothermal heating. Advantages of TG over gravimetry include faster heating

---

rates and shorter analysis times made possible by instrumental sensitivity and smaller sample size. The instrumental requirements of TG, which are dictated by the low mass of the sample and the need for programmable temperature-control (Wendlandt, 1986; Earnest, 1988), consist of a highly sensitive electronic microbalance combined with a furnace and sample enclosure (Fig. 12-1).

Microbalances may be of the deflection or null type (Wendlandt & Gallagher, 1981). The former monitors mass change via direct deflection of the balance beam about a fulcrum. The latter monitors the restoring force (proportional to sample mass) required to continuously maintain the sample at the same position. Modern microbalances are more commonly the null type, since stationary sample-placement minimizes variations attributable to uneven heating within the furnace.

The furnace is operated through a temperature-programming unit, which permits control of heating rates, isothermal settings, or any combination of heating and cooling sequences. Proportional-type furnace-temperature controllers use sophisticated feedback mechanisms to attain heating rate accuracies of within 0.1 °C min^{-1} (Wendlandt & Gallagher, 1981). The sample temperature is monitored by a thermocouple placed close to the sample. The enclosure is commonly purged with an inert gas to suppress oxidation or to minimize possible complications arising from gas evolved during sample reactions. The evolved gas may also be monitored spectroscopically. The data collected in TG are sample mass, sample temperature, and time. Typically, data acquisition and temperature control in modern TG instruments are accomplished by interfacing with a microcomputer.

Commonly, a TG analysis consists of applying a constant heating rate to a sample over the temperature range for which reactions are anticipated. The resulting data would be recorded as a plot of sample mass (or percentage of initial mass) vs. temperature. This technique has been referred to as dynamic thermogravimetry (Wendlandt, 1986). However, TG is not limited to this format (Wendlandt, 1986; Earnest, 1988). For example, mass could be monitored while the sample is cooling (as a function of temperature), or isothermally (as a function of time).

Interpretation and resolution of thermal events are often facilitated by plotting the derivative of the TG curve, which substitutes discrete peaks for subtle inflections. Derivative thermogravimetry (DTG) is particularly useful in resolving reactions with overlapping temperature ranges.

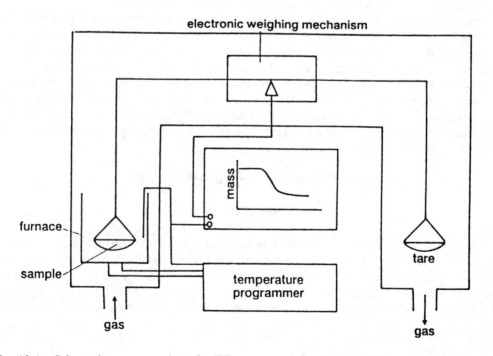

Fig. 12-1. Schematic representation of a TG apparatus (after Brown, 1988, with permission).

Results obtained with TG are subject to the influence of a number of instrumental and sample variables. Some important instrumental factors include furnace-heating rate, furnace atmosphere, and geometry of sample holder and furnace (Wendlandt, 1986). Faster heating rates tend to increase: (i) the onset reaction temperature (T_i), (ii) the temperature of reaction completion (T_f), and (iii) the difference between the onset and completion temperature (T_i-T_f). Gases within the sample chamber (furnace atmosphere) can potentially react with the sample or its reaction products, or in some cases can suppress a reaction of interest. Sample holder and furnace geometry affect, among other things, the exit rate of volatile products which can be important in diffusion-controlled reactions.

Important sample factors affecting TG results include sample mass, degree of packing, particle size, thermal conductivity, and heat of reaction (Wendlandt, 1986; Schilling, 1990). Increasing sample mass and degree of packing can increase DTG peak temperatures due to impedance of the diffusion of volatile reaction products. Larger particles tend to decompose more slowly than smaller ones due to greater intracrystal diffusion distances and lower surface area/mass ratios. Also, coarser materials have more air-filled space which decreases thermal conductivity. The heat of reaction of the sample results in a discrepancy between sample and furnace temperature which for some measurements (i.e., kinetic constants) can be a source of error. Heat generated by oxidation of sample components can be minimized by purging the sample chamber with an inert gas to maintain O_2-free conditions.

Differential Thermal Analysis

Differential thermal analysis (DTA) is a technique of recording the difference in temperature (ΔT) between the sample of interest and a thermally inert reference as the two materials are subjected to identical temperature regimes in an environment heated or cooled at a controlled rate (Mackenzie, 1970). Most applications involve heating in a furnace at a constant rate. The sample and reference are placed in mounts which are (ideally) identical and of low thermal conductivity. The mounts are enclosed symmetrically within a chamber or block, the walls of which serve as a suitable heat sink (Fig. 12-2a).

The ΔT signal is obtained through a pair of differentially-connected thermocouples which link sample and reference. The baseline of a DTA curve is ideally a horizontal line on a ΔT vs. T graph which corresponds to $\Delta T = 0$ (Fig. 12-2c). As the sample is heated, an enthalpy change (chemical reaction or phase transition) within the sample induced by heating results in a deflection from the baseline. The sign of ΔT and direction of baseline deflection are arbitrary, but by convention in DTA, exothermic thermal events are considered "positive" (upward deflection of thermal curve). The complete recording of a thermal event includes the initial deviation from baseline (onset), the maximum deviation (peak), and the return to baseline.

The ΔT signal can be represented by the following equation (Brown, 1988):

$$\Delta T = R \, (dT/dt) \, (C_s - C_r) \qquad [1]$$

where R is thermal resistance, dT/dt is the heating rate, and $(C_s - C_r)$ is the difference in heat capacity between the sample (C_s) and reference (C_r) materials. The value of C_s changes as the sample undergoes a reaction, whereas C_r is constant. Determination of R is difficult since it depends on the thermal properties of the instrument, sample, and reference. Also, derivation of this equation requires the assumption that the heating rate for the sample and reference do not appreciably differ. This assumption is not strictly valid with a DTA apparatus. A differential heating rate can be minimized by diluting the sample with the thermally inert reference, but at the cost of reduced sensitivity (MacKenzie & Calliere, 1975).

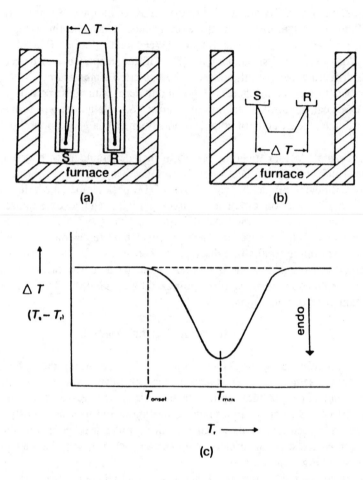

Fig. 12-2. Schematic representations of DTA and heat-flow DSC instrumentation and output. (*a*) configuration for classical DTA, with thermocouples placed within sample (S) and reference (R); (*b*) configuration for heat-flow DSC, showing external placement of thermocouples; (*c*) typical DTA curve, with arrow indicating deflection for an endothermic reaction. The curve would appear the same for heat-flow DSC, except that the signal (ΔT) would be converted to heat flow dq/dt) (after Brown, 1988, with permission).

A theoretical relationship between ΔT obtained from DTA and an enthalpy change (ΔH) in the sample was derived by Kerr and Kulp (1948):

$$m\,\Delta H = gk \int_{t1}^{t2} \Delta T\, dt \qquad\qquad [2]$$

where m is the sample mass, g is a constant dependent on the geometric shape of the sample housing, k is the sample thermal conductivity, and t is time.

The applicability of this equation to DTA is limited by the requirement that the heating rate (dT/dt) be constant, a condition which cannot be strictly met when the sample undergoes an enthalpy change (MacKenzie & Calliere, 1975). The theory also fails to account for changes in the specific heat of the sample as it undergoes a reaction, or for temperature gradients in the sample (Wendlandt, 1986). Nevertheless, it does convey the theoretical proportionality between ΔH and the area under the peak on a ΔT vs. T curve and is valid to the extent that changes in heating rate and other sources of error can be minimized. Subsequent theoretical refinements

(Sewell & Honeyborne, 1957; Cunningham & Wilburn, 1970; Wunderlich, 1981; Wendlandt, 1986) have been made to account more comprehensively for factors which can influence DTA.

As with TG, a number of instrumental and sample factors influence DTA results. Important instrumental factors include heating rate; furnace atmosphere and geometry; sample-holder material and geometry; and thermocouple material and placement (Wendlandt, 1986). Increasing the heating rate generally increases T_i, T_f, peak height, and peak area. Resolution is usually maximized at lower heating rates. The vapor pressure and composition of gases within the furnace enclosure affect the monitored reaction in much the same way as described for TG. Sample-holder and thermocouple effects are varied, complex, and difficult to account for or control in routine analysis other than to hold conditions constant. The latter factors are discussed in detail by Wendlandt (1986).

Differential Scanning Calorimetry

Differential scanning calorimetry (DSC) is a technique in which the energy added to an unknown sample and a reference material is measured as a function of temperature while the samples are subjected to a controlled-temperature program. The term DSC has been applied to instruments which are precise and accurate enough to make calorimetric measurements. In effect, heat flow is measured and heats of reaction can be determined. At least two classes of DSC instruments have evolved that differ in principles of operation: heat-flow and power-compensated DSC (Wendlandt, 1986; Brown, 1988).

Heat-flow DSC is more closely related to classical DTA in principle. Sample and reference are heated in the same furnace, and ΔT is measured as in DTA. However, unlike DTA, the thermocouple junctions are not placed directly in the sample or reference, but are connected directly to two thermally conducting bases (Brown, 1988) (Fig. 12-2b). The sample is placed in a holder, usually a flat pan, which rests on one of the conducting bases. The reference, commonly an empty sample pan, is placed over the other base. The advantage of this configuration over that of DTA is that the signal is essentially independent of the thermal properties of the sample. The resistance term (R) in Eq. [1], which is influenced by the sample and instrument in DTA, is limited primarily to instrumental influence in heat-flow DSC (Brown, 1988). Instrumental effects such as the temperature dependency of instrumental calorimetric sensitivity are more predictable and can be compensated for with calibration.

The objective of power-compensated DSC (Wendlandt, 1986; Brown, 1988) is to maintain the sample and reference at the same temperature during heating. In the strict sense, this can not be accomplished because some temperature deviation ("error signal") is required for the feedback mechanism (O'Neil, 1970). The sample and reference are placed in separate, independently controlled furnaces (Fig. 12-3). The signal, which is the difference in electrical current required to heat the sample and reference at the same rate, can be converted to a heat flow (q) differential (Brown, 1988),

$$\Delta (dq/dt) = (dT/dt) (C_s - C_r) \qquad [3]$$

The resistance term (R) in the signal equations for DTA and heat-flow DSC is not required for power-compensated DSC because for the latter the calorimetric sensitivity is independent of temperature. In effect, no temperature correction of the cell constant is required.

Curves for DSC are indistinguishable from DTA curves, except that by convention power-compensated DSC curves deflect upward for endothermic reactions (positive heat flow). The same instrumental and sample factors that influence DTA results (discussed above) also influence DSC.

Thermomechanical Methods

Thermomechanical methods include thermodilatometry (TD) and thermomechanical analysis (TMA) (Wendlandt & Gallagher, 1981). In TD, expansion and contraction are measured as a

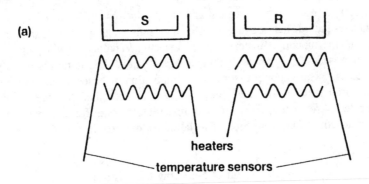

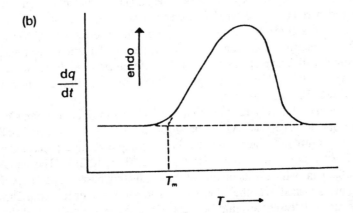

Fig. 12-3. (*a*) Schematic representation of power-compensated DSC, showing separate furnaces and heaters for sample (S) and reference (R). (*b*) Typical curve for this technique, with upward arrow indicating deflection for an endothermic reaction (after Brown, 1988, with permission).

function of temperature under negligible loads (Brown, 1988). The term TMA applies to analyses in which the sample response to tension, compression, flexure, or torsional stress is measured as a function of temperature. The same apparatus can be used for both, and sometimes TD is included under the general heading of TMA (Brown, 1988).

The basis of thermomechanical methods rests upon the fact that the dimensions of most solids expand with heat (as modified by load, if applied). The change in linear dimension with temperature can be represented by:

$$L_2 - L_1 = \Delta L = L_1 \alpha \Delta T \qquad [4]$$

where L_1 is the length at the lower temperature (T), L_2 the length at higher T, and α is the coefficient of linear expansion (Brown, 1988). The validity of this relationship rests upon the near constancy of α, which is reasonable to assume for most materials over small temperature intervals.

Thermomechanical instruments are equipped with a furnace, temperature controller, and a transducer capable of monitoring dimensional changes in the sample under varying applied loads. For TD the dimensions are monitored under minimal load usually by the hydraulic suspension of a vertical rod in contact with the sample. For TMA vertical loads can be applied to the rod, and various probes and accessories can be used depending upon the type of mechanical measurement of interest (i.e., compression, flexure, etc.) (Wendlandt, 1986).

QUANTITATIVE APPLICATIONS OF THERMAL ANALYSIS TO SOIL MATERIALS

Compared to the wide qualitative use of thermal techniques for rapid and inexpensive studies of the mineral constituents of soils, the quantitative applications of thermal analysis have been relatively few. Inherent limitations associated with variability in particle size and/or crystallinity, structural defects, surface chemistry and reactivity, and the presence of amorphous coatings have resulted in considerable differences of opinion concerning the reliability of thermal-analysis methods for quantitative estimation of soil components. In spite of the limitations, more researchers are using TG, DTA, or DSC as primary or complementary tools for quantitative determinatiion of the mineralogical composition of soil materials. In all cases the quantifications are based on the precise measurement of characteristic weight losses or gains (TG) associated with dehydration, dehydroxylation, oxidation, decomposition, or evaporation of a particular mineral, or characteristic energy changes (DTA, DSC) associated with the above or other transformation processes. Tables 12-1 and 12-2 list characteristic TG and DTA/DSC temperature regions commonly used for thermal quantification of soil minerals.

The mass loss associated with a specific thermal reaction at the characteristic temperature region for the mineral is measured between the inflection points of the TG curve, which correspond to the starting and ending point of the reaction. The result is compared to a theoretical mass loss of a pure or a reference mineral (Fig. 12-4a). A distinct problem encountered frequently in this process relates to the difficulty of determining precisely the inflection points for a specific reaction against a background of continuous mass loss (Fig. 12-4b). Another source of error is associated with the assumption that the investigated soil mineral has the same stoichiometry as the ideal pure mineral. Finally, overlapping mass losses from two different minerals, although sometimes resolved by employing isothermal intervals or DTG (Fig. 12-4c), may contribute additional difficulties to the quantification process.

In DTA, the intensities of characteristic endotherms or exotherms are measured and compared to intensities obtained for pure minerals run separately as references. Although quick estimation of intensities may be made using peak heights, better results are obtained when integrated intensities or peak areas are used. The determination of integrated intensities involves the construction of a baseline following either the tangential or the stepped procedure (Fig. 12-5). Direct areal measurements can be made from chart recorder outputs either by planimetry or cutting out the peak and weighing it. However, most modern thermal instruments are equipped with microcomputer processing options that can be used to construct the baseline and calculate the peak area.

Differential scanning calorimetry patterns are very similar to those of DTA, but the area under the peak is calibrated to be proportional to the enthalpy of the particular reaction. Enthalpies (ΔH) are calculated by measuring the area under the peak following a baseline construction as in DTA or directly by the use of a microcomputer processor. The calculated ΔH is compared to that of an ideal or reference mineral to determine the amount of the particular mineral component in the sample. Although starting and ending points are more readily identified in DTA and DSC patterns, certain samples may require special chemical pretreatment(s) to secure interference-free thermal events for accurate peak determinations. Baseline construction may also cause difficulties in the quantification process for some samples.

For reliable quantitative determinations, the experimental conditions (instrument used, heating rate, atmosphere, sample size, packing, etc.) should be identical for reference and unknown samples. Furthermore, the selection of a suitable standard that closely matches the thermal properties of the unknown sample and the preparation of a standard calibration curve that encompasses the concentration range of the unknown samples are prerequisites for reliable quantifications.

Table 12-1. Thermogravimetric (TG) weight loss regions used for quantitative estimation of selected soil minerals.

Mineral	Temperature region °C	Type of reaction	Weight loss† g kg^{-1}	Reference
Oxides/hydroxides				
gibbsite	250-250	$-OH \rightarrow H_2O\uparrow$	312	Karathanasis & Hajek (1982a)
goethite	300-400	$-OH \rightarrow H_2O\uparrow$	101-112	Todor (1976)
Carbonates/sulfates/sulfides/phosphates				
calcite	750-900	$-CO_3 \rightarrow CO_2\uparrow$	440	Todor (1976)
dolomite	500-950	$-CO_3 \rightarrow CO_2\uparrow$	476	Todor (1976)
magnesite	500-750	$-CO_3 \rightarrow CO_2\uparrow$	524	Todor (1976)
siderite	450-600	$-CO_3 \rightarrow CO_2\uparrow$	379	Todor (1976)
Na$_2$CO$_3$	500-1000	$-CO_3 \rightarrow CO_2\uparrow$	415	Asomoza et al. (1978)
gypsum	100-350	$H-OH \rightarrow H_2O\uparrow$	209	Todor (1976)
alunite	160-560	$-OH \rightarrow H_2O\uparrow$	131	Arazi & Krenkel (1970)
pyrite	400-550	$-S \rightarrow SO_2\uparrow$	Variable	Paulik et al. (1982)
strengite	200-400	$-OH \rightarrow H_2O\uparrow$	190	Nathan et al. (1988)
1:1 layer silicates				
kaolinite	400-600	$-OH \rightarrow H_2O\uparrow$	140	Jackson (1975)
halloysite	400-600	$-OH \rightarrow H_2O\uparrow$	122	Jackson (1975)
endellite	400-600	$-OH \rightarrow H_2O\uparrow$	122	Jackson (1975)
antigorite	600-900	$-OH \rightarrow H_2O\uparrow$	129	Todor (1976)
2:1 layer silicates				
Mg-smectite (52% RH)	25-250	$H-OH \rightarrow H_2O\uparrow$	228	Karathanasis & Hajek (1982a)
	600-900	$-OH \rightarrow H_2O\uparrow$	50	Jackson (1975)
Mg-vermiculite (52% RH)	220-250	$H-OH \rightarrow H_2O\uparrow$	24	Karathanasis & Hajek (1982a)
hectorite	500-900	$-OH \rightarrow H_2O\uparrow$	193	Earnest (1988)
mica	500-900	$-OH \rightarrow H_2O\uparrow$	50	Jackson (1975)
talc	850-1050	$-OH \rightarrow H_2O\uparrow$	50	Jackson (1975)
2:2 layer silicates				
chlorite	540-800	$-OH \rightarrow H_2O\uparrow$	123	Jackson (1975)

†Pure mineral on an air-dry weight basis.

Table 12-2. Differential thermal analysis (DTA) or DSC temperature regions and respective heats of reaction (ΔH) used for quantitative estimation of selected soil minerals.

Mineral	Temperature region	Type of reaction	ΔH[†]	Reference
	------ °C ------		--- J g^{-1} ---	
Oxides/hydroxides				
gibbsite	250-250[‡]	-OH → H_2O ↑	1075.3	Karathanasis & Hajek (1982a)
goethite	300-400[‡]	-OH → H_2O ↑	895	Smykatz-Kloss (1974)
quartz	~573[‡]	$\alpha \to \beta$ inversion	5.4	Bartenfelder & Karathanasis (1989)
Carbonates/sulfates/sulfides/phosphates				
calcite	900-990	$-CO_3 \to CO_2$ ↑	1787	MacKenzie (1970)
magnesite	650-750	$-CO_3 \to CO_2$ ↑	1345	Wendlandt (1986)
dolomite	700-950	$-CO_3 \to CO_2$ ↑	1565.5	Dubrawski & Warne (1988)
gypsum	180-220	H-OH → H_2O ↑	657.7	Wendlandt (1986)
alunite	430-590	-OH → H_2O ↑	878.6	Arazi & Krenkel (1970)
1:1 layer silicates				
kaolinite	500-600[‡]	-OH → H_2O ↑	656.9	Karathanasis & Hajek (1982a)
halloysite	500-600[‡]	-OH → H_2O ↑	694.5	Barshad (1952)
2:1 layer silicates				
Mg-smectite/vermiculite	50-250	H-OH → H_2O ↑	592.5	Karathanasis & Hajek (1982a)
Mg-vermiculite	220-250	H-OH → H_2O ↑	159.0	Barshad (1952)

[†]Pure mineral on an air-dry weight basis.
[‡]Tan et al. (1986).

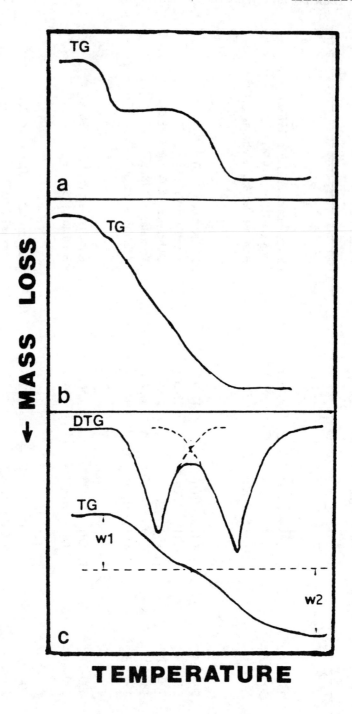

Fig. 12-4. Thermogravimetric curves of soil clay samples showing (*a*) sharp inflection points (*b*) gradual mass loss with no distinct inflection points, and (*c*) the use of DTG to resolve overlapping mass losses.

The most common applications of quantitative thermal analysis to soil materials may be grouped into the following categories:

1. Estimation of the mineralogical composition of soil, clay, or geological samples.
2. Determination of thermal reaction kinetic parameters.
3. Special applications in which certain thermal properties of minerals or materials are measured directly and related to other properties.

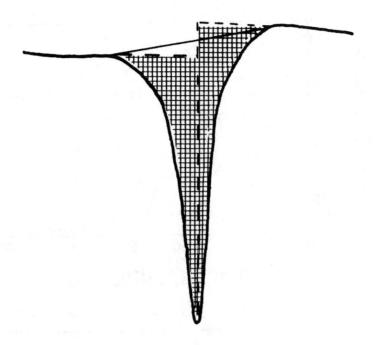

Fig. 12-5. Tangential (———) and stepped (— — —) baseline construction in DTA/DSC peaks.

Quantitative Estimation of Soil Minerals

All four major thermoanalytical techniques (TG/DTG, DTA, and DSC) have been used in quantitative assessments of common soil minerals, with some of them having more specific application to certain minerals than others (Barshad, 1965; Tan & Hajek, 1977; Mackenzie, 1981; Tan et al., 1986). In most cases, however, they have been used in combination or in association with other techniques (XRD, chemical analysis) for the quantitative determination of a particular mineral or for estimation of the total mineralogical composition of a sample. Examples of these applications are given in the following sections.

Hydroxide/Oxide Group of Minerals

Gibbsite, goethite and quartz are three of the most common minerals found in soils, especially in tropical and subtropical regions. Of the three, gibbsite is probably the most successfully quantified by thermal analysis. Indeed, thermal techniques are often considered more effective than XRD for gibbsite quantification because of the interference of the *003* XRD reflection of 1.40-nm minerals with the characteristic 0.482-nm gibbsite peak. Bayerite, which is a polymorph of gibbsite, may also be present in soils, but differentiation of the two polymorphs in natural samples by thermal techniques is impossible. Both minerals lose their hydroxyls in the 250 to 350 °C region according to the reaction:

$$2 \; Al(OH)_3 \; \text{---}> \; Al_2O_3 \; + \; H_2O\uparrow \qquad\qquad [5]$$

The size of the weight-loss (TG) or endotherm (DTA/DSC) associated with this dehydroxylation reaction is used to quantify the amounts of these two minerals (Tables 12-1 and 12-2).

Figure 12-6 shows TG and DSC curves of a citrate-bicarbonate-dithionite (CBD)-treated clay fraction ($<2 \; \mu$m), separated from the Bt2 horizon of a Hiwassee soil from South Carolina. As the TG-weight loss (Fig. 12-6a) and the endotherm (Fig. 12-6b) in the 250 to 350 °C region show,

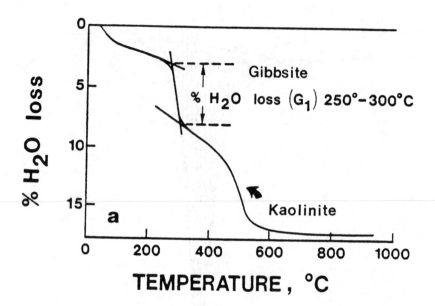

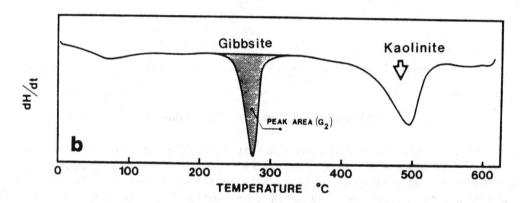

Fig. 12-6. Thermogravimetric and DSC curves of a CBD-treated soil clay fraction from the Bt2
horizon of a Hiwassee soil (South Carolina) containing gibbsite and kaolinite (from
Karathanasis & Hajek, 1982a, with permission).

this sample contains appreciable amounts of gibbsite. Using the extrapolated-onset-and-return
graphical technique a TG weight-loss (G_1) of 54.6 g kg^{-1} is calculated in that region. The amount
of gibbsite present in the sample is calculated from the equation:

$$(G_1/312) \times 1000 = (54.6/312) \times 1000 = 175 \text{ g kg}^{-1} \qquad [6]$$

The value 312 represents the hydroxyl loss (in g kg^{-1}) of a reference pure gibbsite mineral (Table
12-1). A similar quantitative estimation can be made from the DSC curve (Fig. 12-6b) by
measuring the area under the endothermic peak (G2) and employing the equation:

$$(C \times G_2 \times 1000)/(M \times \Delta H_{GIB}) = (125 \times 7.22 \times 1000)/(5.0 \times 1075) = 168 \text{ g kg}^{-1} \qquad [7]$$

where C = instrument constant (mJ cm^{-2}), G_2 = peak area (cm^2), M = sample mass (mg), and
ΔH_{GIB} = heat of reaction (mJ mg^{-1}). In this instance, the two techniques (TG and DSC) yielded
very similar estimates of gibbsite content.

Reasonable agreement ($r = 0.97$, slope 1.23 ± 0.10) between TG and DSC gibbsite estimates
has also been obtained for a set of CBD-treated clay samples fractionated from argillic horizons

of Ultisols in the southern region of the USA, (Karathanasis, unpublished data) and for untreated whole-soil and clay samples of another Hiwassee soil (Tan et al., 1986). Gibbsite quantifications by DTA in clay fractions of four Ultisols of the southeastern USA (Dixon, 1966) produced more precise values than selective-dissolution analysis, apparently because NaOH dissolved Al from minerals other than gibbsite. In another study involving clay fractions from B horizons of 16 Oxisols, Ultisols and Spodosols, DTA estimations of gibbsite compared very favorably ($r = 0.99$, slope $= 1.04$) with values obtained by Na_2CO_3 dissolution (Jorgensen et al., 1970). The combination of DTA with sequential cold and hot Na_2CO_3 dissolution treatments enabled well-ordered gibbsite to be distinguished from poorly ordered aluminous material. Poorly-crystalline $Al(OH)_3$ yields endothermic peaks in the 100 to 250 °C region, which overlaps with endotherms from other minerals such as smectite, vermiculite, halloysite, allophane, etc. The cold Na_2CO_3 treatment presumably removes most of the poorly-ordered $Al(OH)_3$ and allows the quantification of crystalline gibbsite. In a study involving the quantification of crystalline gibbsite present in a mixture with montmorillonite coated with poorly crystalline $Al(OH)_3$, Tullock and Roth (1975) noted that dehydration of noncrystalline Al-hydroxides occurred below 200 °C. Dehydroxylation of crystalline $Al(OH)_3$ occurred mainly in the temperature range of 230 to 300 °C. The TGA weight loss in the latter region was used for crystalline-gibbsite quantification, while poorly crystalline $Al(OH)_3$ was estimated by difference from the total amount of Al retained by the clay.

Interferences in the quantitative estimation of gibbsite are possible in mixtures with hydroxyinterlayered vermiculite or smectite minerals (HIV, HISM). Most naturally occurring Al-hydroxyinterlayers, however, lose their hydroxyls in the 350 to 450 °C region and are expected to interfere less with gibbsite and more with kaolinite quantifications (Barnhisel & Bertsch, 1989). Another more common interference occurs where appreciable amounts of goethite are also present in the sample. One way of resolving this problem is to pretreat the sample with Na-dithionite, which removes goethite and other free Fe oxides and leaves only the gibbsite endotherm (Fig. 12-7). Citrate should be used cautiously in this extraction because in poorly ordered samples it may attack the gibbsite structure (MacKenzie, 1970). Another option is to manually, or through a convolution software progam, resolve the overlapping endotherms (Fig. 12-7a). With TG quantifications, the derivative curve (DTG), could also be very helpful in resolving weight loss contributions from gibbsite and goethite, if the sample is not pretreated with Na-dithionite (Fig. 12-8).

The dehydroxylation of goethite occurs in the 300 to 400 °C region (Tables 12-1 & 12-2) according to the reaction:

$$2FeOOH ----> Fe_2O_3 + H_2O \qquad\qquad [8]$$

This generally coincides with the region in which lepidocrocite and feroxyhite (goethite polymorphs) lose their hydroxyls, and therefore, makes these phases practically indistinguishable. Although the theoretical weight loss for the dehydroxylation reaction is 101 g kg^{-1}, experimentally determined values as high as 112 g kg^{-1} have been reported in the literature. The discrepancy is associated with impurities, poor crystallinity, or Al-substitution in the goethite structure. Similar variability is attached to the heat of the dehydroxylation reaction. Although dissolution methods are preferred for goethite quantification, TG and DSC analyses have yielded accurate estimations in untreated soil or clay samples without gibbsite and reasonable estimations in mixtures with gibbsite, especially when DTG curves were used (Fig. 12-8) (Karathanasis & Hajek, 1982a; Karathanasis et al., 1983). Precise goethite quantification in the presence of gibbsite has also been accomplished by DTA in a H_2 atmosphere (Lodding & Hammel, 1960). In a reducing environment, goethite dehydrates below 300 °C and recrystallizes to magnetite (Fe_3O_4) between 300 and 360 °C. If the H_2 is replaced by N_2 at 400 °C, and then by air, two exotherms are produced at 400 and 775 °C, corresponding to the oxidation of magnetite to maghemite and conversion of maghemite to hematite, respectively. The integrated area under the 775 °C peak (which is usually a doublet) is proportional to the amount of hematite formed, and therefore the amount of goethite present in the sample. Since the endotherm below 300 ° C encompasses the dehydroxylation of

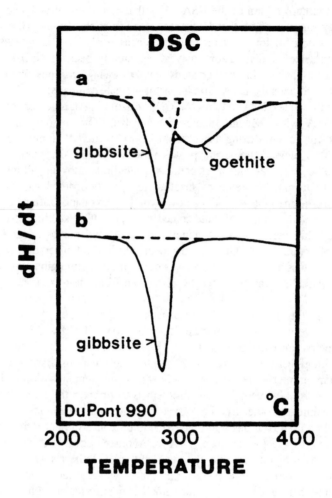

Fig. 12-7. Differential scanning calorimetry curve of a soil clay containing gibbsite and goethite
(a) before and (b) after CBD-treatment.

both gibbsite and goethite, the amount of gibbsite is determined from the difference between the
exotherm and the endotherm peak areas. The double or triple dehydroxylation peak often observed
in natural or synthetic goethites is attributed to either various degrees of Al-substitution, different
types of surface-hydroxyl groups or variable crystallinity (Mackenzie et al., 1981; Schwertmann,
1984) and may also cause difficulties in goethite quantification.

The endotherm at ~573 °C corresponding to the α ----> β phase transition has been used for
quartz quantification by DTA and DSC in soil and geological samples. Sensitivity and precision
of these methods are not favorable for quartz due to the low ΔH of this transition. Also, variations
in the size, shape and temperature of the endotherm reflecting different geological sources, lattice
substitutions, and/or weathering make the quantification process suspect. For 42 geological and
soil samples consisting of nearly 100% quartz, peak-height to peak-width-at-half-maximum ratios
varied by as much as 400% (Fig. 12-9), with sedimentary and soil quartz showing the smallest
ratios (Barwood & Hajek, 1979). In a similar study involving 12 soil and geological samples the
enthalpy of quartz inversion ranged from 1.7 $\pm$ 0.2 to 5.9 $\pm$ 0.1 kJ kg^{-1} (Bartenfelder &
Karathanasis, 1989). Values near 5.0, close to the 5.4 value calculated by the Clausius-Clapeyron
equation, and up to 18.8 kJ kg^{-1} have been reported in the literature. In spite of this variability,
Grimshaw (1953) showed that for four natural quartz samples of widely differing origin the peak
height and area are constant for the same amount of quartz (Fig. 12-10). Encouraged by these

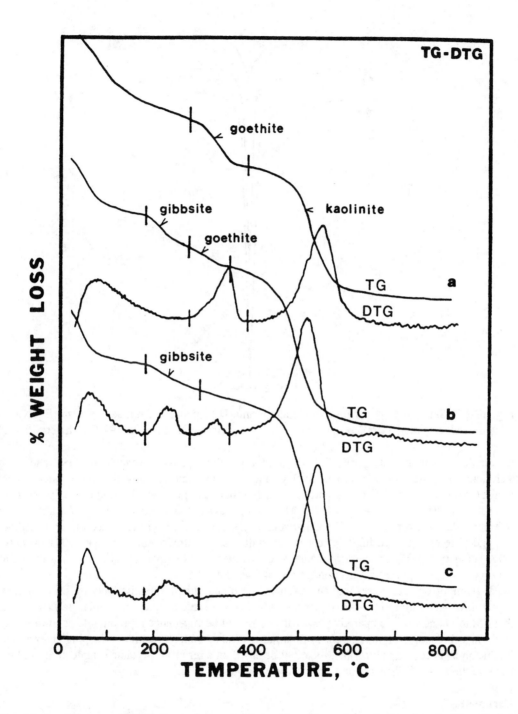

Fig. 12-8. The use of DTG in resolving starting and ending points of mass loss in (a) soil clay containing only goethite, and (b) a different soil clay containing gibbsite and goethite before and (c) after CBD treatment.

results, he suggested that reasonable estimations of quartz can be made in mixtures containing more than 100 g kg^{-1} quartz, within an accuracy of $\pm 1\%$. Other investigators, however, found quartz quantifications by XRD, dissolution, and DTA to vary by as much as 277%, with most of the variation attributed to particle size and packing differences (Mackenzie, 1970).

Care must be taken in the preparation of the samples as extensive grinding apparently produces a noncrystalline layer on the surface of quartz particles, that interferes with the quantification process. For this reason, apparently, DTA-DSC techniques usually under-estimate quartz contents

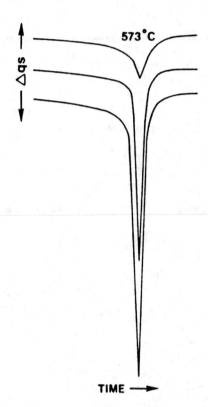

Fig. 12-9.   Single α ----> β inversion endothermic DSC patterns from soil quartz.   Equivalent
mass, all essentially 100% α-quartz by XRD (from Barwood & Hajek, 1979, with permission).

in samples containing microcrystalline or cryptocrystalline quartz.   Analysis by DSC and XRD
techniques of quartz grains occurring in claystone and rock samples showed a significant decrease
in the endotherm area of quartz particles smaller than 20 μm, which resulted in lower DSC
estimates of quartz (Dubrawski, 1987). The use of the XRD-estimated quantity of quartz as the
reference standard for DTA-DSC comparisons is not encouraged, either, since the two methods
respond differently to structural defects and crystallite sizes. Rowse and Jepson (1972) considered
DTA better than XRD or chemical methods for the detection of small quantities of quartz in clay
materials. They reported a detection limit of 4.0 ± 0.4 g kg^{-1}.

It is important, therefore, that any attempt to estimate quartz quantitatively by DTA or DSC
should involve a calibration with a quartz standard of the same source and similar particle size to
that of the unknown. Comparative experiments should be run under reproducible conditions. To
avoid interferences from the presence of kaolinite or other minerals with overlapping endotherms,
it is recommended that quartz quantifications be based on a rerun of the same sample after cooling,
because the α ----> β inversion reaction is reversible.

Carbonates

The TG and DTA analyses have been used extensively for carbonate-mineral quantification in
a variety of geological materials (Paulik et al., 1982; Rowland & Beck, 1952; Todor, 1976; Warne
et al., 1981). Their use for quantification of soil carbonates, however, has been rather limited
(Sobecki & Karathanasis, 1987; Warne & Mitchell, 1979; Webb & Heystek, 1957). Differences
in thermal stability among alkaline-earth and transition-metal carbonates ($Ca^{2+} > Mg^{2+} > Fe^{2+}$)
allows for quantification by thermal techniques, even when multiple carbonate species are present.

D T A

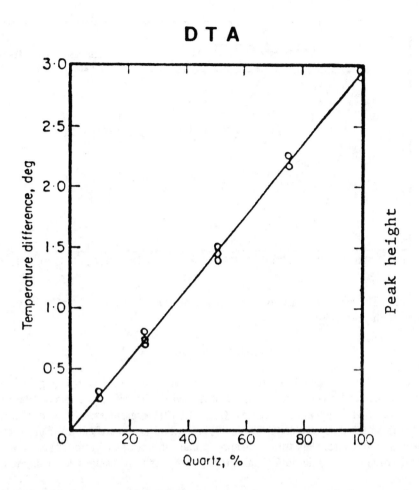

Fig. 12-10. Relationship between peak height and amount of quartz present in four natural samples (from Grimshaw et al., 1953, with permission).

A controlled furnace atmosphere is essential for these quantifications, since the temperature range of decomposition varies dramatically depending on the thermal conductivity of the gas used (Fig. 12-11). The best resolutions of the carbonate-mineral mixtures typically found in soil materials have been accomplished by utilizing a purified CO_2 atmosphere. However, temperature ranges for carbonate thermal decompositions may vary with particle size, sample size, heating rate and amount of soluble salts. Based on literature values and experience obtained in our laboratories, the following ranges are suggested for analysis of the 2- to 50-μm particle size fraction: <580 °C, $FeCO_3$ from siderite; 580 to 690 °C, $MgCO_3$ from magnesite; 690 to 800 °C, $MgCO_3$ from dolomite; and >950 °C, $CaCO_3$ from calcite. It should be noted, however, that soil carbonates and soil minerals in general do not always reproduce the curves of the same minerals in the pure state (Mackenzie, 1970).

Representative TG, DTA and DTG curves of aragonite (calcite polymorph), magnesite, and siderite reference samples are shown in Fig. 12-12. Although aragonite and calcite are usually indistinguishable by thermal analysis, some aragonites may show a small endotherm (~2% weight loss) at about 450 °C, which is associated with the removal of some OH^- groups bonded to Ca^{2+} during the polymorphic transformation of aragonite to calcite (Todor, 1976). Single carbonate species can be quantified by TG analysis based on the amount of CO_2 evolved during their decomposition according to the reactions:

$$CaCO_3 \longrightarrow CaO + CO_2 \uparrow \qquad [9]$$
$$MgCO_3 \longrightarrow MgO + CO_2 \uparrow \qquad [10]$$

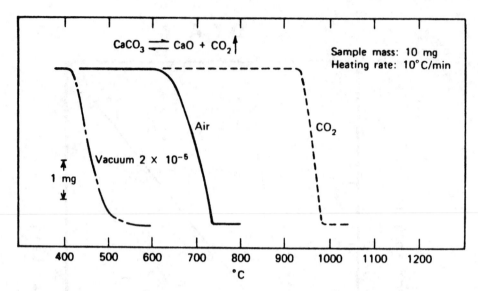

Fig. 12-11. The effect of various atmospheres on the TG curve of $CaCO_3$ (from Wendlandt, 1986, with permission).

$$FeCO_3 ----> FeO + CO_2 \uparrow \qquad\qquad [11]$$

The theoretical weight losses for these reactions are 440, 524, and 380 g kg^{-1}, respectively. Mixtures of these carbonates can also be quantified with relative ease because of their nonoverlapping decomposition temperature regions (Fig. 12-12), regardless of some shifting to lower temperatures in the case of calcite (Todor, 1976). Similar quantifications can be made by DTA from peak area measurements following appropriate calibrations with reference carbonate species.

The decomposition of dolomite [$CaMg(CO_3)_2$] occurs in two stages according to the reactions:

$$CaMg(CO_3)_2 \xrightarrow{600\text{-}800\,°C} MgO + CO_2 \uparrow + CaCO_3 \qquad\qquad [12]$$

$$CaCO_3 \xrightarrow{800\text{-}950\,°C} CaO + CO_2 \uparrow \qquad\qquad [13]$$

The first stage involves exsolution of the dolomite structure into calcite and magnesite, with the latter decomposing immediately. The decomposition starts at about 600 °C and ends at ~800 °C. The second stage involves the decomposition of $CaCO_3$; it begins immediately after the decomposition of $MgCO_3$ has been completed and lasts to about 850 to 950 °C (Fig. 12-13). The mass losses corresponding to the two decomposition stages are determined separately before percentage allocations to the $MgCO_3$ and $CaCO_3$ components are made. If the $MgCO_3/CaCO_3$ ratio is ~1.0, the dolomite is considered proper (Fig. 12-13). If the ratio is higher than 1.0, the dolomite contains $MgCO_3$ in excess and is considered a magnesian dolomite. Ratios <1.0 suggest the presence of a calcitic dolomite.

In the presence of dolomite, magnesite can be resolved because it evolves CO_2 below 600 °C (Fig. 12-14). Thermogravimetric mass losses of carbonate species present in dolomite-calcite, and dolomite-calcite-magnesite mixtures can also be resolved, but with greater difficulty (Fig. 12-14). The availability of combined TG, DTA and DTG curves greatly improves the reliability of such quantitative estimations. In samples where the presence of kaolinite or other aluminosilicate minerals could interfere with carbonate quantifications, two thermal runs are recommended, one before and one after HCl treatment (Fig. 12-15). Still with most soil or clay samples, unless selected or purified samples are used, carbonate quantifications represent total calcium-carbonate-equivalent estimates.

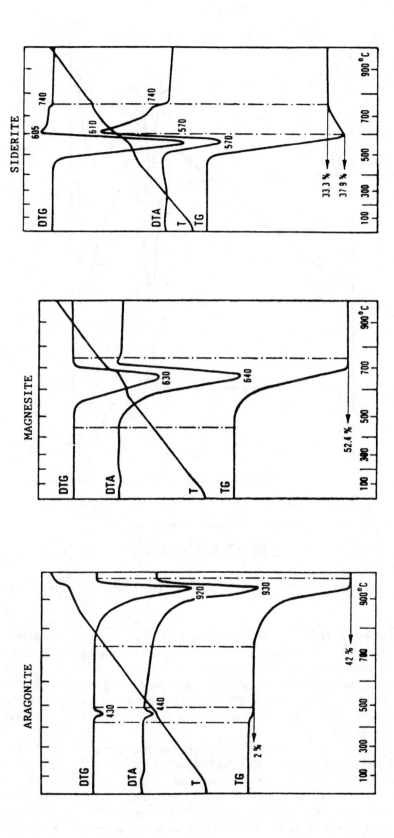

TEMPERATURE

Fig. 12-12. Thermogravimetric, DTA, and DTG curves of aragonite, magnesite and siderite species (from Todor, 1976, with permission).

DOLOMITE

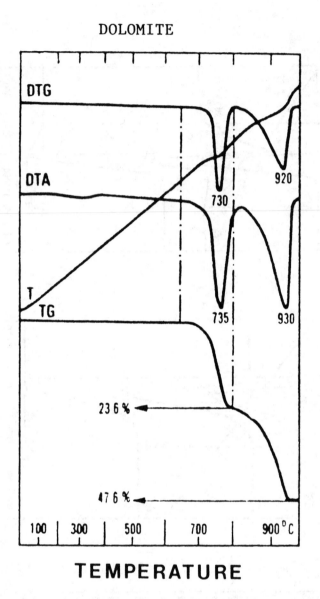

TEMPERATURE

Fig. 12-13. Thermogravimetric, DTA, and DTG curves of dolomite (from Todor, 1976, with permission).

Combining TG with XRD and selective-dissolution analysis, Sobecki and Karathanasis (1987) quantified and calculated the structural formulae of calcite and dolomite phases in silt fractions of calcic horizons in selected Texas Coastal Prairie soils. In spite of the good agreement on total carbonate content among methods, dolomite and calcite estimates differed significantly because of poorly crystalline calcitic dolomite, and low- and high-Mg-calcite carbonate species. The data showed consistent overestimations of calcite at the expense of dolomite by XRD and dissolution methods compared to TG analysis.

Attempts to quantify calcite and dolomite present in concretionary material of calcareous Entisols by DTA under CO_2 atmosphere also produced good peak resolutions and reasonable estimates (Warne & Mitchell, 1979). The carbonate peak definition in most samples was so markedly improved under the CO_2 atmosphere that a detection limit of <2 g kg^{-1} was attainable. Oxidation effects associated with the decomposition of siderite ($FeCO_3$), ferroan dolomite [$(Fe,Mg)Ca(CO_3)_2$] and ankerite [$(Mg,Fe,Mn)Ca(CO_3)_2$], or the presence of organic constituents were also markedly suppressed (Warne et. al., 1981). The presence of montmorillonite in some

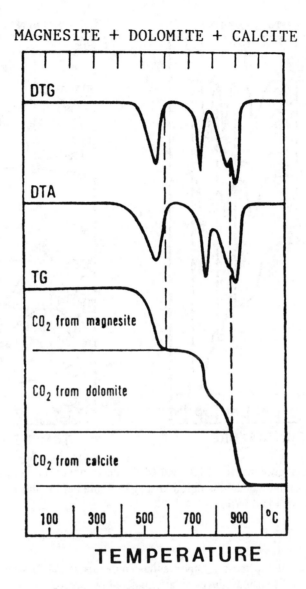

Fig. 12-14. Thermogravimetric, DTA, and DTG curves of a dolomite-calcite-magnesite mixture (from Todor, 1976, with permission).

samples, however, may interfere with the quantification of small amounts of dolomite. This interference could be associated with peak superposition or the formation of new minerals from thermal interactions of carbonate decomposition products with aluminosilicates (Shoval, 1988). Rowland and Beck (1952) were able to detect and quantify dolomite at levels as low as 3 g kg^{-1} in carbonate rocks using DTA equipment with an extremely sensitive recording device and a controlled CO_2 furnace atmosphere. Their quantitative estimations compared favorably with dissolution analysis estimates, whereas XRD failed to yield dolomite peaks in samples containing < 20 g kg^{-1} dolomite.

The applicability of DSC in carbonate mineral studies has not been extensively investigated due to its limited temperature range. Recent studies of dolomite-ankerite solid solutions employing a high-temperature DSC cell (900-1000 °C) under variable atmospheres, however, suggest that enthalpies of decarbonation can be precisely measured, and can also be used to determine the amount of Fe substitution in ankerite structures (Dubrawski & Warne, 1988).

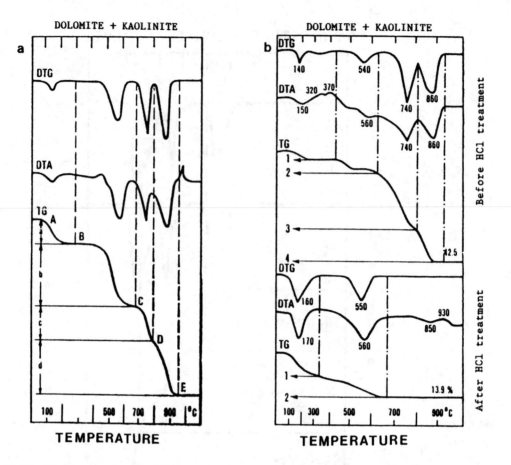

Fig. 12-15. Thermogravimetric, DTA, and DTG curves of (*a*) a dolomite-kaolinite mixture and (*b*) a natural sample containing dolomite and kaolinite before and after HCl treatment (from Todor, 1976, with permission).

Sulfates

Gypsum is one of the most common sulfate minerals found in soils. Quantitative thermal determinations of soil gypsum, however, are hindered by overlapping of the gypsum dehydration endotherms with those associated with adsorbed water from alumino-silicate minerals. In soil fractions without the above interference (sand and silt fractions, concretions, small grains), gypsum can be adequately estimated by thermal analysis. Otherwise, XRD or optical microscopy may be more appropriate techniques.

Gypsum dehydration occurs in two stages (Fig. 12-16). The first stage (~ 140-$150\,°C$ endotherm) corresponds to the dehydration of the first 1.5 moles of water per mole of gypsum and conversion to hemihydrite (bassanite):

$$CaSO_4 \cdot 2H_2O ---> CaSO_4 \cdot 0.5H_2O + 1.5H_2O\uparrow \qquad [14]$$

The second endotherm ($\sim 200\,°C$) is due to the evolution of the rest of the water and conversion to anhydrite:

$$CaSO_4 \cdot 0.5H_2O ---> CaSO_4 + 0.5H_2O\uparrow \qquad [15]$$

Quantitative evaluations of gypsum are based upon the measurement of the heat of reaction associated with the first endotherm, using mainly DTA or DSC techniques. Calibration using a

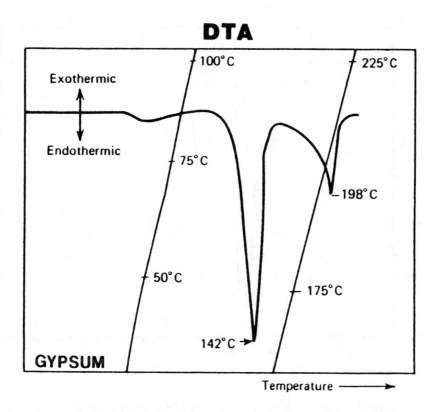

Fig. 12-16. Differential thermal analysis curve of gypsum (from Wendlandt, 1986, with permission).

pure standard of calcium sulfate dihydrate is necessary. Thermogravimetric analysis is not as selective in distinguishing gypsum from hemihydrite.

Gill (1976) determined the gypsum content in a hemihydrite-gypsum mixture by using the area (ΔH) under the first endotherm (~ 142 °C). He obtained a straight line relationship between the amount of gypsum (g kg^{-1}) and ΔH (J g^{-1}). A DSC technique was also described by Dunn et al. (1980) for the quantification of gypsum in lime mixtures. Good agreement was obtained between endotherm peak area and the mass of gypsum present in the analyzed samples. The coefficient of variation for 33 samples of gypsum was 3.65% and, for 25 samples of gypsum in lime, 5.9%, using sample sizes as small as 0.5 mg. Successful estimations of gypsum in mixtures with $Mg(OH)_2$, $Ca(OH)_2$, $CaCO_3$ and MgO have also been accomplished by a combined DSC-DTA technique (Ramachandran & Polomark, 1978). Ideal ΔH values, however, should be carefully applied to soil gypsum as weathering, structural defects, and lattice substitutions may cause considerable deviations. Figure 12-17 shows DSC patterns of two gypsum grains (both essentially 100% gypsum) collected from the C horizon of a Kentucky Alfisol, indicating considerable differences in peak area and intensity. The calculated heats of reactions (ΔH's) were 1216 and 478 J g^{-1}, for the translucent and opaque grain, respectively, compared to 1156 J g^{-1} for the reagent-grade dihydrate.

Quantification of other sulfate minerals has relied basically on XRD, optical microscopy and chemical analytical techniques. One of the few exceptions is the basic aluminum-sulfate mineral alunite [$K Al_3 (SO_4)_2 (OH)_6$], for which a DTA curve is shown in Fig. 12-18 (Todor, 1976). The first endotherm (~ 550 °C) is due to dehydroxylation, the exotherm (~ 745 °C) represents the separation of $Al_2 O_3$ into two phases, and the second endotherm (~ 790 °C) is due to loss of about 66% of SO_3.

Purity characterizations by DTA and DSC (Arazi & Krenkel, 1970) of two kinds of alunites showed that the heat of dehydroxlation was directly proportional to alunite content, as long as excess silica was not present in the sample. The presence of quartz or opal considerably depressed

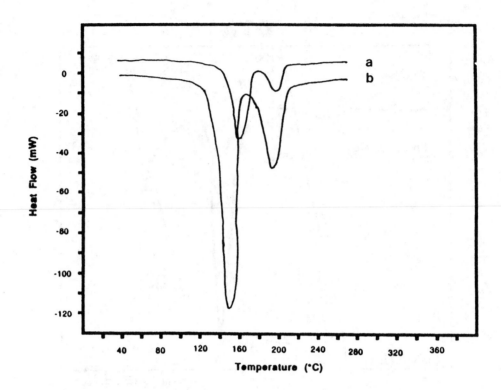

Fig. 12-17. Differential scanning calorimetry curves of (*a*) opaque and (*b*) translucent gypsum grains separated from the C horizon of a Kentucky Alfisol (equivalent mass essentially 100% gypsum by XRD and chemical analysis; Karathanasis, unpublished data).

the enthalpy ($\sim 8\%$) and the temperature (30-50 °C) of dehydroxylation. The presence of kaolinite and alunite in the same mixture makes their quantification problematic by thermal analysis (Pekenc & Sharp, 1974). Similar problems are also encountered in kaolinite-jarosite mixtures because the DTA curve of jarosite [$KFe_3(SO_4)_3(OH)_6$] is similar to that of alunite.

The thermal patterns of pyrite (FeS_2) are notoriously complex and atmosphere dependent. Under an oxidizing atmosphere pyrite produces distinct exothermic peaks around 400 to 450 °C, which have been used to detect small impurities in many types of ores, coal and bauxite samples (Mackenzie, 1970; Dunn et al., 1988). Thermal oxidation of pyrite may cause considerable corrosion on surfaces of the thermal chamber and partly inhibit thermocouple sensitivity.

Phosphates

The weathering products of apatite include a number of phosphate minerals which undergo relatively energetic dehydration and dehydroxylation reactions and a significant weight loss upon heating. Therefore, thermal analysis is potentially useful in detection and quantification of these minerals in soils forming from phosphoritic parent materials.

Some phosphate minerals with potential for thermal quantification include crandallite [$CaAl_3(PO_4)(OH)_5 \cdot H_2O$] (Blanchard, 1971; 1972), wavellite [$Al_3(PO_4)_2(OH)_3 \cdot 5H_2O$] (Blanchard, 1967; Wang et al., 1991), variscite [$(Al,Fe)PO_4 \cdot 2H_2O$] (Blanchard & Denahan, 1966), beraunite [$(Fe^{2+},Fe^{3+})_5(PO_4)_4(OH)_5 \cdot 6H_2O$] (Blanchard & Denahan, 1968), cacoxenite [$Fe_4(PO_4)_3(OH)_3 \cdot 12H_2O$] (Blanchard & Denahan, 1968), strengite [$FePO_4 \cdot 2H_2O$] (Nathan et al., 1988), and brushite [$CaHPO_4$] (Mulley & Cavendish, 1970). Wavellite, for example, loses 260 g kg^{-1} of structural water and hydroxyls in two endothermic reactions between 150 and 320 °C (Blanchard,

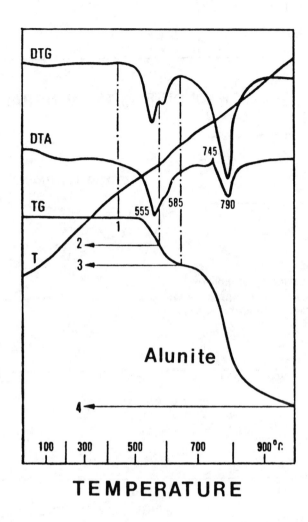

Fig. 12-18. Thermogravimetric, DTG, and DTA curves of an alunite specimen (from Todor, 1976, with permission).

1967). The lower temperature reaction is strong and has been used to quantify wavellite in soils formed from phosphoritic sediments in Florida (Wang et al., 1991). As little as 60 g wavellite per kilogram in a quartz matrix was found to be readily detectable (Blanchard, 1967). As with other minerals, quantification of phosphate minerals can be thwarted by peak overlap and by specimen variations in such properties as particle size, crystallinity, and composition.

Kaolinite and Halloysite

Kaolinite is one of the most commonly occurring minerals in soil and geological materials and one of the most extensively studied by thermal analysis. Pure kaolinite specimens analyzed by TG show a characteristic weight loss of about 140 g kg^{-1} in the 400 to 600 °C temperature region due to dehydroxylation. A small weight loss of 2 to 10 g kg^{-1} occurring between 25 and 200 °C is due to desorbed water and decreases as sample crystallinity increases. Differential thermal analysis curves of pure kaolinite show a symmetric endotherm in the 400 to 600 °C dehydroxylation region, and two exotherms around 1000 °C (spinel crystallization) and 1150 °C (mullite formation). Typical TG, DTG, and DTA curves for a well-crystallized kaolinite sample are shown in Fig. 12-19 (Earnest, 1988). Quantitative estimations of kaolinite are based on either thermogravimetric weight losses (TG), enthalpy of dehydroxylation (DSC), or area under the endothermic peak (DTA), in the 400 to 600 °C temperature region.

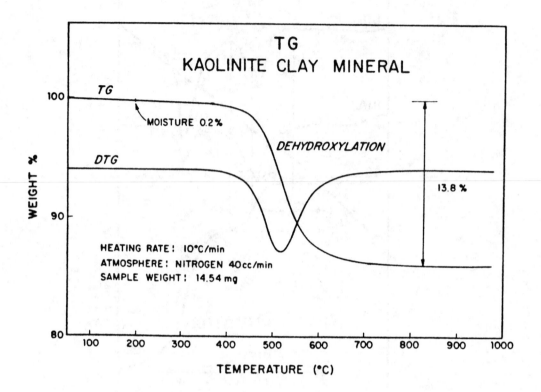

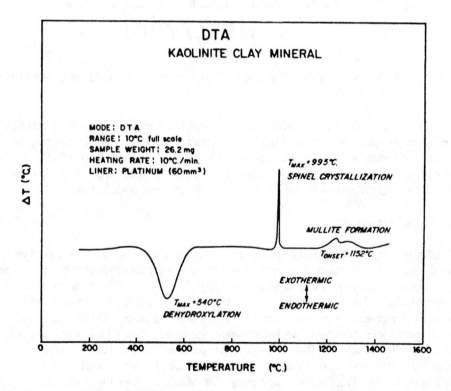

Fig. 12-19. Thermogravimetric, DTG, and DTA curves of a well-crystallized kaolinite specimen
(from Earnest, 1988, with permission).

An example of kaolinite quantification in a CBD-treated clay fraction of the Bt2 horizon of a White Store soil from N. Carolina by TG analysis is shown in Fig. 12-20a. The dehydroxylation weight loss (K_1) in the 400 to 600 °C region is measured first by the onset-and-extrapolated return technique, and from that the kaolinite content is estimated based on a theoretical weight loss of 140 g kg^{-1} for pure kaolinite, according to the equation:

$$(K_1/140) \times 1000 = (55/140) \times 1000 = 393 \text{ g kg}^{-1} \qquad [16]$$

A quantitative estimation of kaolinite in the same sample by DSC analysis (Fig. 12-20b) yields 408 g kaolinite per kilogram. The calculations are based on the equation:

$$C \times (K_2/M) \times (1000/\Delta H_{KLN}) = 125 \times (16.5/7.7) \times (1000/658.9), \qquad [17]$$

with C = instrument calibration constant (mJ cm^{-2}), K_2 = peak area (cm^2), M = sample mass (mg), and ΔH_{KLN} = heat of dehydroxylation (mJ mg^{-1}) of reference Georgia kaolinite used as a standard.

Although the agreement between the TG and DSC kaolinite estimates in this sample is quite good, significant variation may be encountered in some soil samples primarily due to lack of sharp inflections in the TG curve or difficulties in DSC baseline construction. Kaolinite quantities in a set of clay samples from Alfisols, Ultisols and Vertisols of the southern region of the USA, estimated by TG and DSC analysis, varied by as much as 30%, with the DSC values consistently lower than those obtained by TG (Karathanasis, unpublished data). For a similar set of clays,

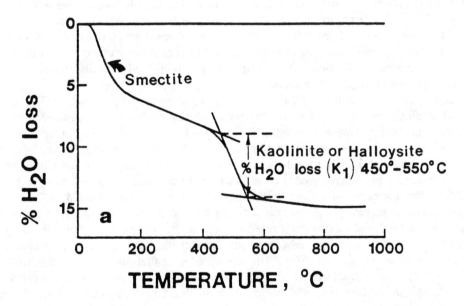

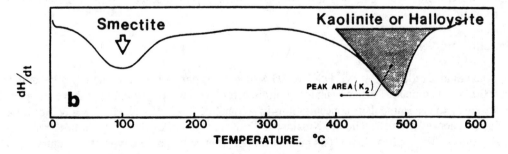

Fig. 12-20. Thermogravimetric (*a*) and DSC (*b*) curves of a CBD-treated clay fraction from the Bt2 horizon of a White Store soil (North Carolina) containing kaolinite (from Karathanasis & Hajek, 1982a, with permission).

however, a very good regression relationship ($r = 0.988$) between the peak area from the DTG curve and kaolinite quantities estimated from TG weight losses was obtained.

Comparisons of kaolinite quantities estimated from DTA and selective-dissolution analysis (SDA) in various clay fractions of four Ultisols (Dixon, 1966) gave satisfactory results for the coarse-clay fractions of the most weathered soils (standard deviation of 0.95%). Relatively large discrepancies between the two methods for estimates of kaolinite were observed, however, in the fine-clay fractions (especially those of less weathered soils). These discrepancies were caused by the presence of large quantities of poorly crystalline materials and expansible minerals which contributed to an overestimation of kaolinite by SDA.

One of the chief deterrents to kaolinite quantification by DTA and DSC is the effect of particle size and degree of crystallinity on the inflection temperature (T_{max}) and on the area and shape of the dehydroxylation peak. According to Carthew (1955), the width and area of the kaolinite endotherm (450-550 °C) decreases as particle size and degree of crystallinity decrease for clay-sized kaolinite. This decrease results from a higher dehydration rate in the smaller particles due to the formation of more dehydroxylation reaction centers. The symmetry of the dehydroxylation peak is usually expressed as the slope ratio of the lower-temperature-side tangent to the high-temperature-side tangent of the endotherm. Well-crystallized and coarse-particle-size kaolinites usually exhibit more symmetrical peaks and yield smaller slope ratios than disordered, fine-particle-size kaolinite varieties (MacKenzie, 1957). This relationship can also be useful in distinguishing halloysite from kaolinite. Halloysite usually exhibits higher slope ratios than kaolinite of the same particle size (Earnest, 1980). Carthew (1955) suggested an empirical relationship between the slope ratio of the DTA peak and the ratio of the area to the width of the peak at half its amplitude to overcome the effect of particle size and degree of crystallinity in quantitative estimations of kaolinite. Gibbs (1965) proposed that other minerals with coincidental thermal reactions be removed from the sample so that a reference with the same thermal properties as the unknown may be used. However, purification of soil kaolinite in most soil samples for use as an internal standard is usually impractical. Therefore, the selection of a kaolinite with similar particle size and crystallinity may be an acceptable alternative.

In soils containing smectite, HIV, mica, or chlorite, corrections for partial overlapping with the kaolinite endotherm may be required. Drawing a line perpendicular to the baseline at T_{max} and a half-peak symmetrical to that of the well-defined temperature side (low-T side for smectite, mica, chlorite; high-T side for HIV) should keep estimation errors to a minimum. In mixtures or soil samples containing both kaolinite and halloysite, the formamide intercalation test (Churchman et al., 1984) appears to provide reasonable quantifications of both minerals (within an error estimate of $\pm 10\%$). The sum of the two minerals (kaolinite and halloysite) is determined by DTA or TG, and their individual percentages are estimated by the integrated intensity (area) ratio of the 0.72- and the 1.04-nm XRD peaks in the intercalated sample. In the absence of kaolinite, halloysite can be quantified similarly from its dehydroxylation endotherm (DTA, DSC) (Mackenzie and Robertson, 1961), or weight loss (TG). The amount of structural OH and the dehydroxylation energy, however, may be more variable than that of kaolinite, depending on degree of crystallinity, mode of formation, surface area, and particle form of the mineral (Minato, 1988). Spherical halloysites usually have higher dehydroxylation energies than tubular halloysites of the same mode of formation.

Smectite and Vermiculite

Quantitative estimates by TG, DSC, or DTA of water vapor sorbed on monoionically-saturated smectite and vermiculite surfaces under controlled relative humidity conditions have been used successfully to determine their concentrations in soil and geological samples (Davis & Holdridge, 1969; Mackenzie, 1970; Smalley & Xidakis, 1979; Karathanasis & Hajek, 1982a,b; Karathanasis et al., 1986). The applicability of this method is highly dependent on the type of the saturating ion and the equilibrium environment, but its sensitivity has been substantially improved with recent advances in thermal technology.

The procedure involves a CBD-pretreatment of the sample, Mg-saturation, washing to remove free Cl⁻, and allowing the samples to air dry. The dried clays are ground to pass a 140-mesh sieve and stored for 48 h in desiccators whose relative humidity is controlled at $52 \pm 2\%$ with $Mg(NO_3)_2$-saturated solutions. Errors due to significant deviations in relative humidity between the controlled chamber and ambient conditions may be minimized by purging the furnace with N_2 gas which has been previously passed through a $Mg(NO_3)_2$-saturated solution. The amount of smectite is calculated from TG weight losses in the 25 to 250 °C temperature region (Fig. 12-21a) or DSC/DTA heat changes/areas of low-temperature endotherms (Fig. 12-21b). When vermiculite is also present in the sample the sum of the two minerals is estimated because both show similar water desorption characteristics.

An example of smectite + vermiculite quantification in an untreated clay fraction from the Bt horizon of an Oktibbeha soil (Alabama) is shown in Fig. 12-21. For TG, the amount of smectite + vermiculite present in the sample is calculated by

$$\text{Smectite} + \text{vermiculite} = (A_1/228) \times 1000 = 50/228 \times 1000 = 220 \text{ g kg}^{-1}, \qquad [18]$$

where A_1 = thermogravimetric weight loss (g kg⁻¹). For DSC,

$$\begin{aligned} \text{Smectite} + \text{vermiculite} &= C \times (A_2/M) \times (1000 \, \Delta H_{sm} + v) \\ &= 125 \times (6.3/6.5) \times (1000/592.5) = 205 \text{ g kg}^{-1}, \qquad [19] \end{aligned}$$

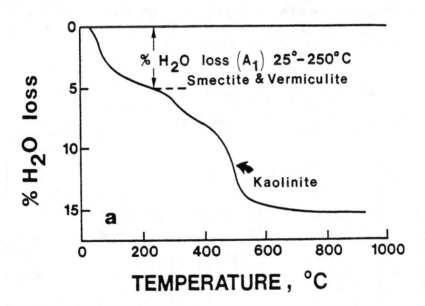

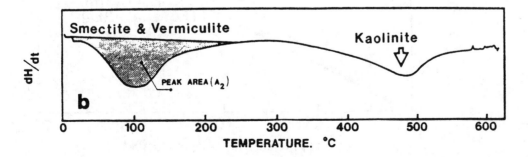

Fig. 12-21. Thermogravimetric and DSC curves of a soil clay fraction separated from the Bt horizon of an Oktibbeha soil (Alabama) containing smectite, vermiculite, and kaolinite (from Karathanasis & Hajek, 1982a, with permission).

where C = instrument calibration constant (mJ cm^{-2}), A_2 = peak area (cm^2), M = sample mass (mg), and $\Delta H_{sm} + v$ = heat of dehydration (mJ mg^{-1}) for pure smectite + vermiculite. The weight loss of 228 g kg^{-1} (Eq. [18]) and ΔH of 592.5 mJ mg^{-1} (Eq. [19]) represent average values obtained by TG and DSC analysis of six reference bentonites containing various amounts of smectite and vermiculite (Karathanasis & Hajek, 1982a). Agreement between TG and DSC estimations of smectite + vermiculite is generally good, unless gibbsite, poorly crystalline or organic materials, illite, HIV, gypsum, zeolites, or phosphates are present in the sample. Because of the variable amount of adsorbed water associated with these materials, the presence of any of them could lead to an overestimation of smectite and vermiculite in the sample. The amounts of hydroxyinterlayered smectite or vermiculite species are usually underestimated by thermal analysis because hydroxyinterlayering blocks a portion of the interlayer space, thus making it inaccessible to water adsorption.

The TG and DSC estimates of the smectite + vermiculite contents of 49 Mg-saturated soil clays from the southern region of the USA. compared quite favorably with values recovered from surface area measurements using BET and Langmuir adsorption isotherms, and CEC determinations (Karathanasis & Hajek, 1982b). Similarly favorable comparisons were obtained between DTA low-temperature-endotherm areas and the amount of montmorillonite present in selected soil clays from Jamaica (Davis & Holdridge, 1969) and Australia (Smalley & Xidakis, 1978). Mackenzie (1970) also used the relationship between sorbed-water content and peak area to quantify montmorillonite and halloysite samples by DTA. He emphasized the importance of controlling the experimental conditions during the montmorillonite determinations because of the curvilinear relationship between moisture content and the DTA peak area when multilayer adsorption occurs (Fig. 12-22). Apparently, the binding energy of the water molecules at higher degrees of hydration is lower, and therefore a smaller supply of energy is required for their release.

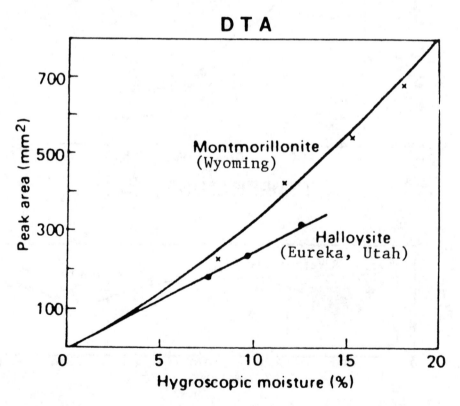

Fig. 12-22. Relationship between sorbed water and DTA peak area for a reference montmorillonite (Wyoming) and halloysite (Eureka, UT) (from Mackenzie, 1970, with permission).

The nature of the exchangeable cation also markedly influences the peak area-water sorption relationship because of differences in cation hydration energies. Table 12-3 demonstrates the effect of three common saturating cations (Mg, Ca, Al) on the amount of water desorbed, the associated enthalpy of dehydration, and the quantities of smectite + vermiculite estimated in a set of soil and reference clays. The data point out the importance of using reference samples saturated with the same ion as the unknown sample if quantitative results are desired.

In soil samples containing both smectite and vermiculite, the amount of vermiculite can be estimated by TG analysis from its characteristic weight loss between 250 and 300 °C, provided that gibbsite or goethite are not present. This weight loss is associated with the dehydration of strongly cation-coordinated H_2O molecules and it is satisfactorily resolved by TG analysis. The value of 27 g kg^{-1} used by Barshad (1952) for pure vermiculite produced consistently lower vermiculite estimates and has been adjusted to 24 g kg^{-1} to improve quantifications in our laboratories (Karathanasis & Hajek, 1982a). A similar approach, utilizing the area associated with the third endotherm (~250-350 °C), was used by Boss (1967) to estimate vermiculite content in biotitic vermiculites employing a sensitive DTA apparatus. Finally, smectite or vermiculite can be quantitatively estimated from TG weight losses or heat changes (peak areas) associated with their 550 to 700 °C dehydroxylation endotherms, provided there are no other minerals with coincidental thermal reactions present. The variable OH content of "normal" (~700 °C endotherm) and "abnormal" (~550-650 °C endotherm) smectites may be a deterrent to TG determinations, but the energy of dehydroxylation (peak area) is the same regardless of normal or abnormal behavior, and it has been used successfully for smectite quantifications by DTA (Mackenzie, 1970).

Other Minerals

Quantitative analysis of mica minerals by DTA or TG is limited by variations in OH content, temperature range of dehydroxylation (300-950 °C), and endotherm area (Jackson, 1975; Fanning et al., 1989). These variations arise from differences in structural and chemical composition. Therefore, most of the thermal studies involving mica minerals are of qualitative importance. Thermogravimetric, DTA, or DSC investigations involving zeolites (Ming & Mumpton, 1989), HIV (Barnhisel & Bertsch, 1989; Karathanasis, 1988; Harris & Hollien, 1988) and palygorskite-sepiolite (Singer, 1989) minerals have also been completely qualitative. Worth mentioning are quantitative studies of hectorite by TG and DTA in hectorite ores (Earnest, 1988), sodium carbonate and sodium chloride salts in saline clays of lacustrine origin in Mexico (Asomoza et al., 1978) and imogolite present in allophane-imogolite mixtures by DTA (Aomine & Mizota, 1973). More recently, Parfitt (1990) was able to estimate imogolite in various soil fractions separated from Andisols in New Zealand using the 430 °C DTA dehydroxylation endotherm. Other minerals present in the Andisols studied (gibbsite, goethite, halloysite), which exhibit DTA endotherms near the 400 °C range did not appear to interfere with the quantifications.

Thermal Reaction Kinetics

Thermogravimetry and DSC/DTA have been extensively used in either isothermal or nonisothermal mode to determine rates, mechanisms, and activation energies of mineral decomposition reactions and other transitions occurring upon heating. Although a detailed treatment of this subject is beyond the scope of this paper, a few examples involving soil minerals will be discussed. In these examples, measured thermal kinetic parameters are used to assess certain qualitative or quantitative aspects of the behavior of these minerals. For more comprehensive treatments of thermal kinetics consult Wendlandt (1986) and Keattch and Dollimore (1975).

Assuming a basic thermal decomposition reaction of the form

$$AB \text{ (solid)} ---> A \text{ (solid)} + B \text{ (gas)},$$ [20]

Table 12-3. Effect of saturating cation on smectite + vermiculite (Sm + V) quantification by TG and DSC (Karathanasis & Evangelou, 1986).

Clay Sample (<2μm)	Mineralogy‡	TG H$_2$O loss†			TG Sm + V			DSC ΔH†			DSC Sm + V		
		Mg	Ca	Al	Mg	Ca	Al	Mg	Ca	Al	Mg	Ca	Al
		----- g kg^{-1} -----			------ g kg^{-1} ------			------ J g^{-1} ------			------ g kg^{-1} ------		
Shrouts Bt	(Sm)$_{30}$,(Mi)$_{55}$,(K)$_{10}$	68	84	76	300	350	340	178	150	134	300	260	260
Sadler Bt	(Sm)$_{25}$,(HIV)$_{25}$,(K)$_{25}$	91	111	100	400	460	450	207	210	196	350	370	380
Memphis Bt	(Sm)$_{25}$,(HIV)$_{15}$,(MI)$_{10}$,(K)$_{25}$	80	87	78	350	360	350	195	202	187	330	350	370
Bledsoe Bt	(Sm)$_{20}$,(HIV)$_{30}$,(K)$_{30}$	80	90	80	350	370	360	177	199	181	300	350	350
Conecuh Bt	(Sm)$_{40}$,(K)$_{35}$	91	116	106	400	480	480	235	249	227	400	430	440
Wilcox Bt	(Sm)$_{50}$,(K)$_{30}$	114	128	115	500	530	520	293	272	245	500	470	480
Camargo Bentonite§	(Sm)$_{80}$,(V)$_{15}$	217	231	210	950	950	950	564	547	487	950	950	950
Wyoming Bentonite	(Sm)$_{75}$,(V)$_{10}$	194	188	173	850	770	780	502	378	317	850	660	620

†For temperature range of 25 to 240 °C.

‡Sm = smectite, Mi = mica, K = mica, K = kaolinite, HIV = hydroxyinterlayered vermiculite, V = vermiculite. Determined by a combination of XRD, TG, XRF, and DSC methods on Mg-saturated clays (Karathanasis & Hajek, 1982a).

§Used as a reference mineral.

the basic TG experimental plot would be one of weight loss against time. To obtain kinetic data in an isothermal mode the sample is heated isothermally at the desired temperature and the information is collected in the form of α against t plots, where α = the fraction of the mineral that has decomposed, and t = the time of heating. Repeating this experiment at different isothermal settings allows determination of the various kinetic parameters.

The nature of the α vs. t plot depends on the geometry of the decomposing sample and particularly the geometry of the reaction interface. In most homogeneous reactions the entire range of decomposition may fit one kinetic expression. In heterogeneous reactions, however, different expressions may be required to fit the various stages of decomposition. Plots of $d\alpha/dt$ against time are used to calculate the different reaction rates at various stages of decomposition and compare them to theoretical expressions in order to deduce the rate controlling processes (mechanisms).

The foundation for the calculation of kinetic data from a TG curve is based on the formal kinetic equation

$$- d\alpha/dt = k\,(1-\alpha)^n \tag{21}$$

where α = amount of sample that has reacted, n = order of the reaction and k = specific rate constant. The temperature dependence of k is expressed by the Arrhenius equation

$$k = A \exp(-E_a/RT) \tag{22}$$

where A = preexponential factor, E_a = activation energy of the reaction (J mol^{-1}), and R = gas constant (8.314 J mol^{-1} K^{-1}). Various mathematical treatments of kinetic equations have been used including differential, integral and approximate formulations (Wendtlandt, 1986). Selected kinetic expressions and corresponding reaction mechanisms commonly used in thermal kinetic studies are shown in Table 12-4.

Following is an example of the isothermal determination of kinetic parameters associated with the dehydration of standard bentonites, and with soil clays having montmorillonitic, mixed, illitic and kaolinitic mineralogy, when saturated with either Al or Ca ions (Karathanasis & Evangelou, 1987). The isothermal temperatures used in the study were 32, 38, 46, and 54 °C. Twenty-mg samples that had been loaded on the TG balance of a 1090 DuPont thermal analyzer and maintained under a constant H_2O vapor pressure of $P/P_0 = 0.995$ were introduced to the heating chamber equilibrated at the preset temperature. The weight losses were then recorded as a function of time until a steady weight was obtained (Fig. 12-23). The kinetic data were analyzed by the approximate reduced-time plot method of Sharp et al., (1966). According to this method, the general kinetic expression of Eq. [21] is written in the following integrated form:

$$F(\alpha) = A(t/t_{0.5}), \tag{23}$$

where $F(\alpha)$ is a function of α, the fraction converted, A is a constant calculated from the actual form of the kinetic expression, t is the reaction time, and $t_{0.5}$ is the reaction's half-life (i.e., time to 50% conversion). In a first-order reaction ($n = 1$), for example,

$$F_1(\alpha) = -\ln(1 - \alpha) = kt = -0.6391(t/t_{0.5}). \tag{24}$$

Nine of the most common kinetic equations (Table 12-4) were tested in this way by calculating values of A and plotting α against $t/t_{0.5}$. To identify the kinetic mechanism involved, the master plot of the theoretical expressions (Sharp et al., 1966) was superimposed over the reduced-time plot obtained experimentally, and the best fit of the experimental data to one of the theoretical expressions was established. Assuming that the reaction is isokinetic (the same kinetic mechanism is followed throughout the reaction) the two plots will lie on a common curve.

Table 12-4. Selected kinetic expressions and reaction mechanisms (Sharp et al., 1966).

Reaction type	Equation†	Rate-Controlling Process
F_1	$-\ln(1-\alpha) = kt = -0.6931(t/t_{0.5})$	Random nucleation, first-order decay law
R_2	$1-(1-\alpha)^{1/2} = kt = 0.2929(t/t_{0.5})$	Phase-boundary reaction, cylindrical symmetry
R_3	$1-(1-\alpha)^{1/3} = kt = 0.2063(t/t_{0.5})$	Phase-boundary reaction, spherical symmetry
D_1	$\alpha^2 = (k/x^2)t = 0.2500(t/t_{0.5})$	One-dimensional diffusion process
D_2	$(1-\alpha)\ln(1-\alpha) + \alpha = (k/r^2)t = 0.1534(t/t_{0.5})$	Two-dimensional diffusion process, cylindrical symmetry
D_3	$[1-(1-\alpha)^{1/3}]^2 = (k/r^2)t = 0.0426(t/t_{0.5})$	Three-dimensional diffusion process, spherical symmetry
D_4	$[1-(2\alpha/3)]-(1-\alpha)^{2/3} = (k/r^2)t = 0.0367(t/t_{0.5})$	Ginstling-Brounshtein equation, three-dimensional diffusion
A_2	$[-\ln(1-\alpha)]^{1/2} = kt = 0.8326(t/t_{0.5})$	Avrami-Erofeev equation, two-dimensional growth
A_3	$[-\ln(1-\alpha)]^{1/3} = kt = 0.8850(t/t_{0.5})$	Avrami-Erofeev equation, three-dimensional growth

†α = fraction reacted, k = rate constant, t = time, x = thickness of layer, r = radius of particle.

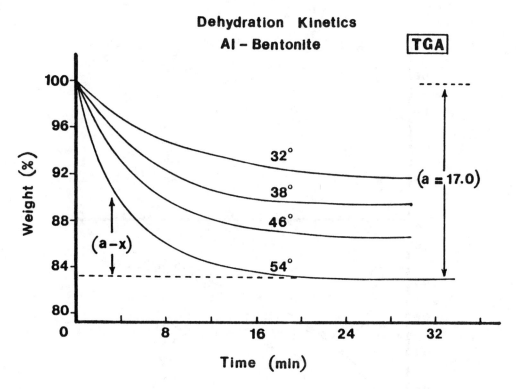

Fig. 12-23   Isothermal TG mass loss (α = a) vs. time (t) plots for the dehydration of an Al-saturated bentonite sample (from Karathanasis & Evangelou, 1987, with permission).

After the reaction mechanism was established, reaction rate constants (k) were calculated for each isothermal temperature. Activation energies of dehydration were calculated by plotting ln k vs. 1000/T (K) and utilizing the Arrhenius equation:

$$\ln k = \ln Z - (E_a/RT)$$ [25]

where k = rate constant (min^{-1}), Z = Arrhenius frequency factor (min^{-1}), E_a = activation energy (J mol^{-1}), R = gas constant (8.314 J mol^{-1} K^{-1}), and T = absolute temperature (K). This equation represents a straight line, the slope of which (-E_a/R) allows calculation of the activation energy (E_a).

Examples of reduced-time plots with the best kinetic expressions and of Arrhenius plots used to obtain E_a values are shown in Fig. 12-24 and 12-25, respectively. This study showed that the mechanism of dehydration of Al-and Ca-saturated clays is mostly independent of the type of the saturating ion, but dehydration rate constants were lower and activation energies of dehydration were generally higher in Al-saturated clays. Hence, increased Al saturation of the exchange phase may slow the dehydration process considerably by reducing dehydration rate constants and by requiring greater energy for the initiation of the desorption reaction.

A comparison of three kinetic expressions for the nonisothermal decomposition of calcium carbonate by TG showed that although the order of the reaction (n) ranged from one-half to two-thirds, the calculated activation energies were reproducible to within $\pm 6\%$ (Sharp & Wentworth, 1969). Another study comparing various nonisothermal techniques at different heating rates and sample masses indicated that E_a values generally decreased with decreasing sample mass at any given heating rate (Gallagher & Johnson, 1973). Isothermal kinetic studies of reference calcite and dolomite samples, and of soil-carbonate samples containing mixtures of calcite, dolomite and Mg-calcite, also obtained kinetic expressions having n = 1/2, with random nucleation or phase-boundary reaction mechanisms (Sobecki & Karathanasis, 1987, unpublished data). The activation energies calculated from this study were approximately 40% lower than E_a values obtained from

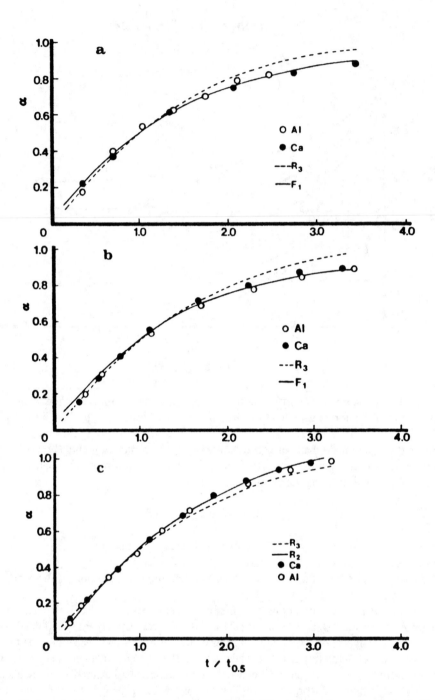

Fig. 12-24. Reduced-time plots for the isothermal (32 °C) dehydration of selected Al- and
     Ca-saturated soil clay fractions and best fit for F1, R2, and R3 kinetics: (a) montmorillonitic;
     (b) mixed; (c) kaolinitic clay (from Karathanasis and Evangelou, 1987, with permission).

nonisothermal techniques, and their magnitude was proportional to the degree of Mg substitution
in the dolomite and Mg-calcite lattice.

   Almost all kinetic studies employing DTA and DSC determinations are based on the general
equation

$$(d\alpha /dt) = A(1 - \alpha)^{n} \; \exp(-E_{a} /RT)$$

[26]

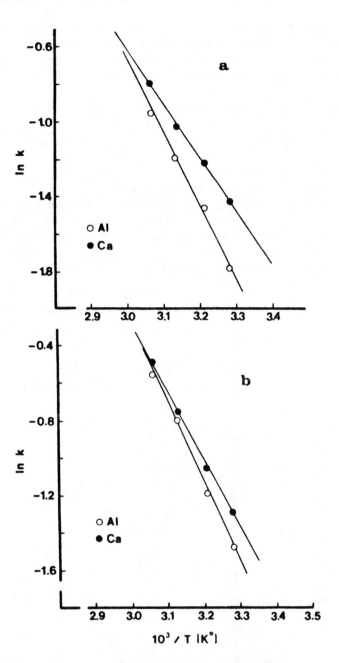

Fig. 12-25. Arrhenius plots for Al- and Ca-saturated clay fractions of (*a*) montmorillonitic, and (*b*) kaolinitic samples from F1 and R2 kinetics, respectively (from Karathanasis & Evangelou, 1987, with permission).

where A is the preexponential factor, α is fraction of sample reacted, t is the reaction time, n is the order of the reaction, E_a is the activation energy (J mol^{-1}), R is the gas constant (8.314 J mol^{-1} K^{-1}), and T is the absolute temperature (K). Various integration and differentiation approaches to Eq. [26] involving different assumptions and approximations have produced a range of mathematical functions that have conditional applicability to kinetic investigations (Wendlandt, 1986).

One of the kinetic expressions most commonly used in DSC/DTA studies is the Kissinger equation:

$$d \, (\ln\beta/T^2_{\,\mathrm{max}})/d \, (1/T) \; = \; -E_a/R, \tag{27}$$

where β is the heating rate, and T_{max} is the temperature at peak maximum. Bartenfelder and Karathanasis (1989) used the Kissinger method to determine the activation energy associated with the $\alpha ---> \beta$ inversion of quartz in sedimentary (loess) and soil samples. The technique employs the relationship of inversion temperature change with different heating rates for first order reactions (Fig. 12-26).

The E_a was calculated from the slope of $\log\beta/T^2_{max}$ vs. $1000/T$ using

$$E_a \text{ (kJ/mole)} = \text{slope x } 2.303 \text{ x } 8.314 \text{ (J K}^{-1} \text{ mol}^{-1}). \qquad [28]$$

The term 2.303 is applied to convert from natural (ln) to base-10 logarithm in the calculations.

Thermal patterns of the Peoria loess sample at heating rates (β) of 2, 5, 10, and 20 °C min^{-1} (Fig. 12-26) illustrate the relationship of heating rate (β) to heat flow and to endotherm inflection maximum. Examples of E_a calculations from $\log(\beta/T^2 \text{max})$ vs. $1000/T$ plots are shown in Fig. 12-27 for Peoria and Loveland loess quartz samples. It is apparent from these plots that quartz from the Peoria loess represents a "high activation energy" ($E_a = 8525$ kJ mol^{-1}) condition, whereas that from the Loveland loess represents a "low activation energy"

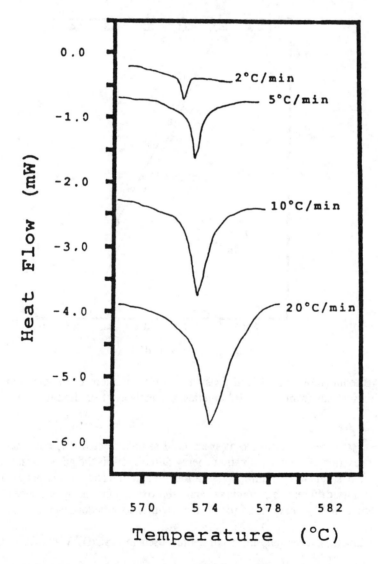

Fig. 12-26. Endothermic peaks and inversion temperatures for the Loveland loess quartz at four heating rates (from Bartenfelder & Karathanasis, 1989, with permission).

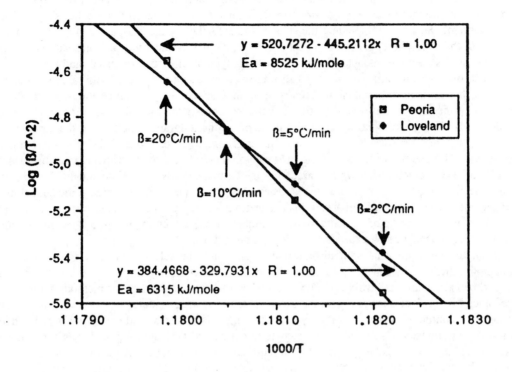

Fig. 12-27. Kinetic plot of Loveland and Peoria loess quartz at four heating rates used for the determination of activation energies (E_a) (from Bartenfelder & Karathanasis, 1989, with permission).

(E_a = 6315 kJ mol^{-1}) condition. Soil-quartz samples generally showed the lowest E_a, apparently reflecting pedogenic deformation effects.

Perhaps the most widely used kinetic method in DTA/DSC studies is that derived by Borchardt and Daniels (1957). The technique was originally described for solutions and was subsequently refined for solids by other researchers. Since sample weight is assumed constant in calculations, the sample should be placed in a hermetically sealed pan to ensure that no significant mass loss occurs during the reaction. To avoid the thermal lag effect, the heating rate should not exceed 20 °C min^{-1}. The other experimental conditions are flexible depending on the sample being analyzed. The method assumes: (i) uniform temperature and identical heat capacity and heat transfer in the sample and the reference material, (ii) the heat is transferred exclusively by conduction, (iii) the reaction follows nth order kinetics, and (iv) the temperature dependence of the reaction rate follows the Arrhenius expression. In its simplified version this equation is expressed as

$$-dN/dt = (-N_0/A)(dH/dt)$$ [29]

where N is the moles of reactant present at time t, N_0 is the initial moles of reactant, A is the total peak area, and H is the rate of heat change. Various versions of this equation have been used successfully for quantitative determinations by DSC and DTA of kinetic parameters for reactions involving inorganic and organic substances and, more recently, polymers (Wendlandt, 1986). Very few applications to soil materials have been reported, however, because of the limited applicability of the Borchardt-Daniels equation to heterogeneous samples (Borchardt, 1960; Piloyan et al., 1966).

Anderson et al. (1977) compared some of the most popular DTA kinetic methods using a homogeneous, irreversible first-order reaction model, from which a theoretical DTA curve was calculated with a mean error in ΔT_{max} of 8%. Five of the methods matched the theoretical E_a value exactly, while the other four predicted it to within ±6%.

The quality of the experimental thermal kinetic data and their subsequent fit into theoretical expressions strongly depends on various experimental factors which influence the shape of the thermogram. Such factors include: the type of the DTA/DSC cell, TG-pan, and sample holder; nature of the atmosphere; particle size, specific heat, thermal conductivity and thermal mass of the sample and its reaction products; density of packing; type of filler material used; and heating rate. The use of a vacuum atmosphere tends to remove gaseous molecule products rapidly from the sample, but an inert or nonflushing atmosphere tends to retain a portion of the products in the vicinity of the sample, thus retarding the forward reaction. Therefore, the use of covered sample holders may have a strong influence on the shape of the thermogram. The smaller the particle size the more active the surface is likely to be. Higher heating rates tend to shift the reaction to higher temperatures. The use of linear heating programs in thermal kinetic experiments modeling DTA, DSC, or TG promotes temperature gradients between the walls of the sample holder and the thermocouple detector. Kinetic analysis is easiest in isothermal experiments, where the temperature of the entire sample is uniform at any instant of time. There are advantages, however, in using the programmed heating of nonisothermal techniques, because they require one experiment (run), with a single sample being exposed to various temperatures.

Whether an isothermal or nonisothermal mode should be selected for a particular thermal-reaction kinetics experiment should depend on the material, the type of reaction, and the experimental conditions involved. The advantages of using nonisothermal methods include: (i) considerably fewer data are required, (ii) kinetic parameters are calculated from a wide temperature range in a continuous manner, (iii) only a single sample is required, and (iv) isothermal methods produce questionable results when the sample undergoes a thermal reaction during the stage at which the temperature is being raised to the required isothermal level. A serious disadvantage of nonisothermal methods is that the reaction mechanism can not always be deduced, and hence, the meaning of the activation energy and order of reaction is questionable. Furthermore, TG curves are strongly influenced by experimental conditions, and hence the kinetic parameters calculated from them may be fictitious or uncertain. Lastly, the Arrhenius equation calculated from homogeneous kinetics cannot be applied to nonisothermal heterogeneous reactions because the conditions of the equation are not fulfilled (Wendlandt 1986).

Special Thermal Applications

This section includes examples of thermal applications for direct quantitative evaluations of certain properties of minerals or soil materials, or use of thermal data to assess compositional changes in mineral structures.

One example is the application of thermomechanical methods (TMA), and especially thermodilatometry (TDA), to determine volume or length changes in clays and ceramic materials. Hajek (1979) and Karathanasis and Hajek (1985) used TMA to measure the coefficient of linear extensibility (COLE) of soil clays. This coefficient has been used to differentiate vertic from typic subgroups in several soil suborders. Field-moist soil peds, approaching field capacity were heated and ped length changes were recorded by a Dupont TMA analyzer (Fig. 12-28). Representative patterns showing soil length change and rate of change (dy) for three soils with high, moderate and low shrink-swell potential are shown in Fig. 12-29. This is a rapid (~ 30 min) and precise ($\pm 5\%$ variation) procedure for soil shrinkage estimation and requires essentially no preliminary preparation. The correlation between COLE determined by TMA and percentage volume change traditionally measured in 32 montmorillonitic soil clays was very good ($r = 0.85**$) for all soil groups considered. Dilatometric curves have also been used as a rapid method for the semiquantitative determination of quartz in ceramic samples, by measuring the volume increase during the $\alpha ----> \beta$ transition of quartz (Yariv, 1989). The peak area under the derivative (dl/dt) curve is proportional to the amount of quartz in the sample. The determination may be somewhat impaired by a decrease in the pore volume of the sample, which partly compensates for the increase in volume of quartz. This problem is more acute with fine or disordered quartz samples that extend the transition range.

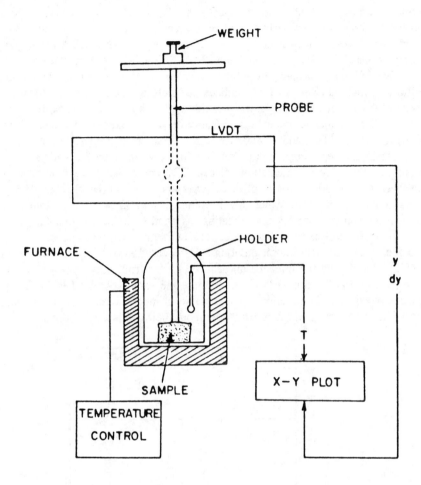

Fig. 12-28. Schematic of a thermomechanical analyzer (from DuPont product bulletin, Wilmington, DE).

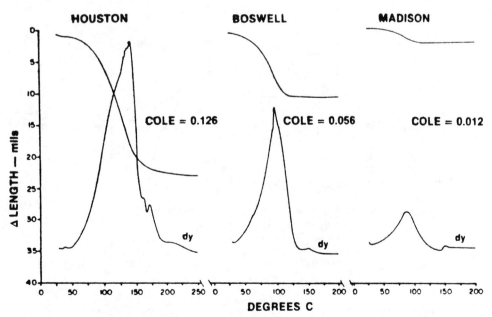

Fig. 12-29. Thermomechanical analysis patterns showing length change, rate of change and TMA-COLE of subsoil samples from Houston, Boswell and Madison soils (from Hajek, 1979, with permission).

Thermodilatometry in combination with TG and DTA reveals many processes occurring during decomposition, transition, or dehydroxylation reactions including crystal modification or recrystallization. These processes are not usually detected by TG or DTA techniques because of overlapping thermal transitions. Simultaneous TDA, TG and DTA curves for kaolinite (Fig. 12-30) reveal that following the dehydroxylation reaction at 400 to 800 °C, kaolinite converts to metakaolinite, which at 950 °C recrystallizes to mullite. The last solid-state transition is accompanied by a 2% increase in volume (Paulik & Paulik, 1978). Thermodilatometry has also been used to identify dickite impurities in kaolinite by an expansion effect at about 650 °C (Schomburg & Störr, 1984), and to measure dimensional changes in zeolite matrices created by water loss or gain, molecule sorption, or ion migration between heteroenergetic sites (Dyer, 1981). Because of its sensitivity to orientation effects, TDA may also have potential application in assessing the degree of orientation of platy particles in clay strata (Mackenzie, 1981).

Thermoluminescence (TL), which involves the measurement of the intensity of radiation released from crystalline samples upon heating, is another technique that has been applied to characterization of marbles, limestones, other crystalline rocks, lunar and meteorite samples, as well as to geological and archaeological dating (McKeever, 1988). Very few of these applications involve soil materials and the information deduced so far from these types of studies is strictly qualitative (David, 1972). Similarly qualitative are the results obtained from the application of thermomagnetometry in the derivation of Curie points of magnetic minerals (hematite, maghemite, ferrihydrite, ilmenite), from which information about their origin can be deduced.

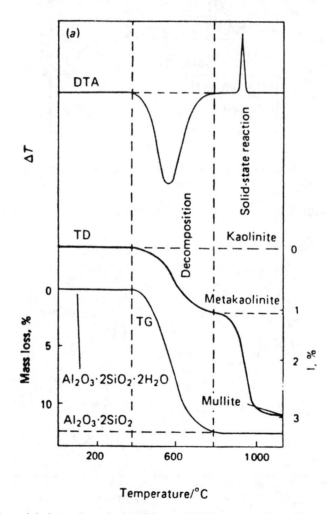

Fig. 12-30.   Differential thermal analysis, TD, and TG curves of kaolinite showing phase transitions and length change (from Wendlandt, 1986, with permission).

Evolved gas analysis (EGA) has enabled determination of the identity and/or amount of volatile materials released during the process of heating soil or geological samples. The most common volatiles evolved from such samples are H_2O, CO_2, SO_2, and, less commonly, H_2 and CO. Evolved gasses can be detected by infrared (IR) spectrometry, gas chromatography (GC), or mass spectrometry (MS) (Morgan, et al., 1988; Charsley et al., 1987). Muller-Vonmoos and Muller (1975) combined DTA with MS to determine organic C, pyrite, and carbonate minerals present in clay samples, while Morgan (1977) used various detectors to estimate the amount of H_2O, SO_2, and CO_2 evolved from clays, shales, and schists. The EGA has also been used for quantitative determinations of mineral mixtures (calcite, magnesite, chrysotile, with greater sensitivity than XRD), decomposition processes of different minerals, pyrolysis of oil shales, and adsorption of organic molecules by clay minerals (Yariv, 1989). Other interesting EGA applications include a combined TG-IR setting for the study of clay interlayer-organic matter complexes in selected fractions of a soil with incipient spodic characteristics (Feldman & Zelazny, 1990), and a pyrolysis-GC-MS series used for the characterization-fractionation of soil organic matter (Bracewell & Robertson, 1977). Special thermoanalytical instrumentation sets combining TG, DTA, TMA, MS, and XRD have also shown promise in archaeological investigations of pottery, clays, pigments, paper, and other materials (Bayer & Wiedemann, 1983).

Some more traditional thermal applications involve the use of DTA peak displacement or deformation for the determination of chemical substitutions or crystal defects in mineral structures. Compared to the relatively sharp decomposition peak of a pure crystal, the DTA curve of an impure or disordered mineral shows a broader, slightly displaced or double peak. The distinction between a decomposition peak influenced only by chemical substitutions and another affected only by crystal defects is difficult and should be supported by other means (XRD, optical analysis). The effect of Fe^{2+} substituting for Mg in the dolomite structure is illustrated in Fig. 12-31 (Smykatz-Kloss, 1984). Similar substitutions for Fe^{3+} by Al^{3+} in goethites (Fig. 12-32), and for Al^{3+} or Fe^{3+} by Mg^{2+} in smectites and chlorites (Fig. 12-33) have been estimated from DTA peak shifting or deformation (Smykatz-Kloss, 1974).

ADVANTAGES AND LIMITATIONS

From the above account, it is evident that thermal analysis provides a set of relatively simple, reliable and rapid techniques that complement other methods in the quantitative mineralogical analysis of soil materials. The quantitative efficiency of these techniques varies with the type of sample and the type of mineral involved, and is generally better in relatively weathered soil samples. Certain mineral components (gibbsite, goethite, kaolinite) can be readily quantified by thermal analysis even though the amounts present are so small as to escape detection by XRD. Thermal analysis is especially useful for the identification and estimation of some highly disordered minerals, which are not detectable or easily resolved by XRD.

Recent improvements in thermal instrumentation have enhanced quantitative applications of thermal analysis, but have not completely removed limitations associated with the nature of the sample (Van der Marel, 1956) or with the analytical technique. Therefore, it is imperative that the experimental conditions during the analysis are maintained constant (instrument, heating rate, atmosphere, sample size, particle size, sample packing, etc.). The selection of a suitable standard that matches the thermal properties of the unknown sample as closely as possible, and the development of a standard calibration curve encompassing the concentration range of the unknown samples are also essential. Certain soil samples, because of their nature, may require complementary treatments (e.g., CBD and HCl extractions) and/or analytical methods, while others may not be suitable at all for thermal quantification.

Some of the problems commonly encountered with the quantitative estimation of soil minerals by thermal analysis include:

1. Difficulties in determination of the reaction onset and end points. Because of low sample crystallinity and impurities contained in many soil minerals, TG inflections are much more gradual than for pure components thus making accurate determinations of weight losses difficult. The use of DTG sometimes is helpful in resolving the inflection points.

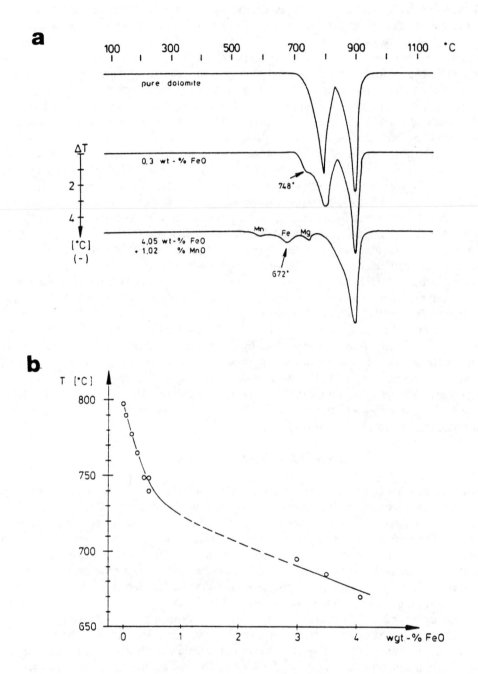

Fig. 12-31. DTA curves of three dolomites with different FeO content (*a*) and the relationship between FeO content in the dolomite structure and peak temperature (from Smykatz-Kloss, 1984, with permission).

2. Overlapping weight-loss temperature regions or DTA/DSC peaks. Common mineral associations having coincidental thermal-reaction regions include gibbsite-goethite, chlorite-carbonates, and chlorite-smectite-mica. The use of various treatments and DTG may improve the resolution of thermal curves and lead to reasonable estimates of mineral quantities.

3. The presence of free or mineral-bound organic complexes. These may yield a continuous weight-loss region in TG patterns extending from 100 to as high as 600 °C, thus interfering with mineral identification and quantification in that region. This situation is particularly common in surface-soil samples. Pretreatment with H_2O_2 may improve the resolution, although the situation

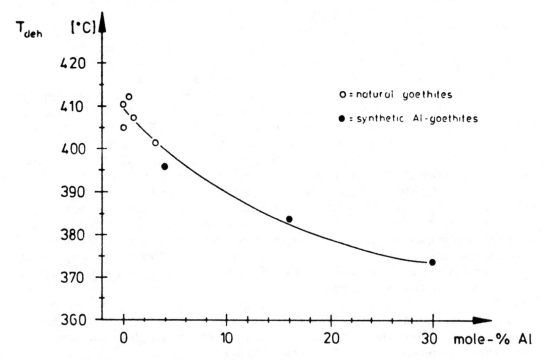

Fig. 12-32. The effect of Al substitution for Fe^{3+} on the DTA dehydroxylation temperature of natural and synthetic goethites (from Smykatz-Kloss, 1984, with permission).

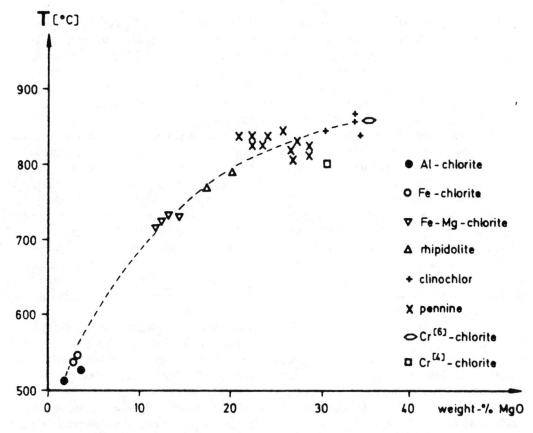

Fig. 12-33. Relationship between DTA dehydroxylation temperature and MgO content of chlorites (from Smykatz-Kloss, 1984, with permission).

with some samples is quite hopeless. The use of combined TG-MS or TG-IR techniques may contribute to the quantification of such samples.

4. The nonideality of the hydroxyl-water content (TG) and heat of dehydroxylation (DSC/DTA) of most soil minerals resulting from structural defects and substitutions. Since quantitative estimations are based on ideal or reference values, there is a certain error associated with them. This error can be minimized by selecting appropriate standards.

5. The variability in adsorbed-water content of 2:1 expanding minerals. Even under controlled moisture and cation-saturation conditions, these minerals exhibit different adsorbed-H_2O contents relative to those of standards as a result of their variable chemical composition. Extensive calibration work using minerals of similar origin could improve the quality of the data.

6. Difficulties in constructing the baselines for overlapping or asymmetric DSC/DTA peaks. Graphical solutions render the curve more quantitatively diagnostic.

Because of the analytical limitations noted above, the accuracy attainable in thermal quantitative estimations will never, even with the most stringent precautions, be expected to be better than 1 to 2%. Normally, an accuracy of 5% might be the best attainable depending on the sample used and the mineral in question. The accuracy depends not only on the sensitivity of the instrument and the reproducibility of the technique used, but also on the nature of the thermal analysis curve itself. Minerals giving inflections where departure from the baseline is gradual are difficult to determine with acceptable accuracy because of the subjective assessments that have to be made in interpolating the baseline. Nevertheless, for certain minerals (gibbsite, gypsum, goethite) detection limits can be lower than 10 g kg^{-1} (TG, DSC) provided steps do not overlap and the material giving each step can be definitely identified. For most practical purposes, however, thermal data are, at best, approximations and should be labelled as such.

RECENT AND FUTURE DEVELOPMENTS IN QUANTITATIVE THERMAL ANALYSIS

The growth of thermal instrumentation over the last 30 yr can be traced from the analog instruments of the early 1960s to today's digital microcomputer-controlled instruments. As this evolution has continued, thermal analyzers have become easier to use by laboratory personnel and at the same time sophisticated enough to perform applications never before possible. This sophistication has been aimed at consistently improving the reproducibility, precision, and sensitivity of thermal-analysis measurements in a wide range of material sciences, and at increasing versatility and productivity while maintaining cost effectiveness.

The introduction of low-mass cooling furnaces and the availability of modern instruments, equipped with sensitive solid-state automatic-recording capabilities has made mineral estimations faster, simpler and more precise. These advances have expanded the quantitative use of thermal analysis to kinetic studies of various dehydration, transition and decomposition reactions and established thermal analysis as a rapid technique for the quantitative evaluation of coal and oil-shale materials as potential fuels. More precise qualitative and quantitative assessments of other properties of minerals and soil materials (chemical composition, structural substitutions, degree of disorder, coefficient of expansion) have also been accomplished with the use of single or multiple thermal-analysis techniques. Current developments focusing on interfacing thermal techniques with other instruments (IR, GC, MS) capable of detecting and measuring thermal decomposition products provide exciting opportunities for quantitative refinement of decomposition- and organomineral reaction products.

Most of the currently available lines of thermal instruments are modular, allowing the tailoring of a system to specific laboratory needs. These needs may range from simple, repetitive testing to complex research and analysis. Capabilities include full data manipulation and storage capability, multitasking and simultaneous multimodule (TG, DSC, DTA, TMA) operation. Extensive libraries of high-quality software are available for all standard thermal analytical techniques as well as special programs for kinetics studies, purity analysis and customization of results (Cassel et al., 1987). The availability of special (pressure-DSC, dual-sample DSC) and combined (TG-DSC, TG-TMA, TG-DTA-DTG) options, the interfacing capabilities with other instruments (IR and mass spectrometers, gas chromatographs) and the upgradability and efficiency of the currently available thermal-analysis systems have made them invaluable tools in material-science research.

Although most of the advances in thermal instrumentation during the last 30 yr came as a result of the flourishing of polymer science research, soil mineralogy has also utilized the fruits of these advances to refine quantitative characterizations of soil and geological materials. As a consequence, the reliability, reproducibility, and detectability of the mineralogy data produced during the last 20 yr have improved dramatically, and new opportunities have been created for some innovative and previously unthinkable research.

Future improvements in automation, sensitivity, versatility, and productivity of traditional thermal techniques will further enhance the quantitative characterization of soil materials, but modern trends indicate that the most promising applications are associated with the hybridization techniques, which combine traditional thermal-analysis units with other single or multiple instruments (Fourier transform IR, GC, MS, titrimetry). Several of these instrument combination technologies (Fig. 12-34) have found valuable applications in polymer studies (Paulik & Paulik, 1978; Wendlandt, 1986; Charsley et al., 1990) and some, involving soil and geological materials, have already been mentioned earlier.

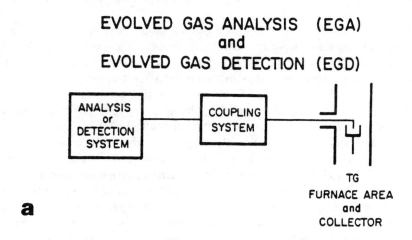

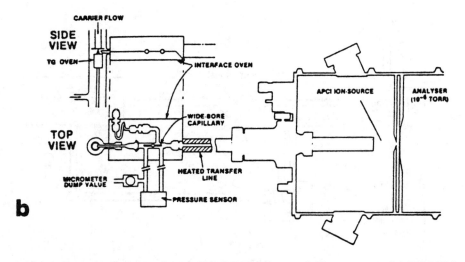

Fig. 12-34. Schematic diagram of (*a*) a TG-EGA system and (*b*) a commercial TG-MS interface for EGA (from Earnest, 1988, with permission).

In combined techniques, two or more specimens, one for each technique, may be involved. Simultaneous techniques utilize the same specimen for both techniques. Sampling modes may be either continuous or intermittent. For the continuous sampling mode, the gaseous products are introduced directly into the detector system via an interface coupling and analyzed. In the intermittent mode, the evolved products are trapped and stored in a low-temperature chamber and later introduced to the detector at certain temperature intervals or after the end of the treatment cycle. The advantages of intermittent sampling are: (i) the detector of the coupling unit does not need to be modified extensively for interfacing purposes thereby freeing the instrument for other analyses, and (ii) detector sensitivity can be optimized for a particular analyte. The disadvantages of this mode are: (i) the gas samples trapped and stored for a certain period of time before analysis may undergo condensation or compositional changes, and (ii) synchronization of thermal events with gas-evolution events may not be achieved. These problems are avoided in the continuous mode, which provides representative analytical data immediately. However, some detectors require very elaborate interfaces due to temperature and pressure gradients between the two systems, and optimum conditions for one technique (i.e. thermal unit) may not necessarily be the same for the other (detector). This problem can be minimized by careful selection of the experimental conditions (small sample, slow heating rate) or by incorporating both modes into the system. A schematic of intermittent and continuous sampling modes for multiple EGA systems is shown in Fig. 12-35.

In many cases a set of detectors specific for one gaseous product can readily and at relatively low cost be attached in series to a suitable thermal instrument. The most effectively utilized, because of its sensitivity and versatility, is the MS system. The recent introduction of relatively inexpensive quadrupole mass spectrometers with simple interfacing options has made thermal analysis and MS a particularly appealing combination. In setting up equipment and assessing results, however, certain important aspects must be kept in mind. For example, (i) a small-volume sample chamber must be utilized to avoid undue dilution of evolved products by the carrier gas; (ii) special consideration should be given to the selected interface so that the concentration of one volatile is not preferentially enriched at the expense of the other; and (iii) the likelihood of the reaction between two volatiles before measurement by the detector should be assessed.

In spite of the many considerations involved, future applications of combined or simultaneous thermal techniques with IR, GC, MS, or titrimetry appear to have great potential for drastically improving resolution and detection limits of minerals with overlapping thermal properties but different evolved products (CO_2, SO_2, H_2O, CO). Another area of research that could benefit substantially from combined thermal analysis applications involves clay-organic interactions, which are especially critical for environmental studies.

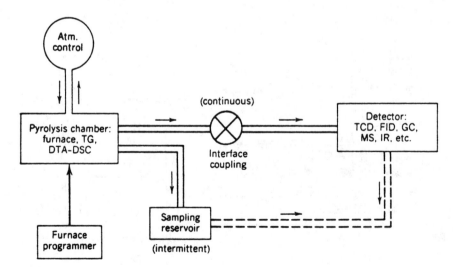

Fig. 12-35. Intermittent and continuous sampling modes for EGA (from Wendlandt, 1986, with permission).

SUMMARY AND CONCLUSIONS

Thermogravimetry (TG), differential thermal analysis (DTA), and differential scanning calorimetry (DSC), have played a key role for many years as complementary tools in qualitative and quantitative characterization of soil materials. Recent advances in thermal instrumentation technology, computerization, and new interfacing capabilities have expanded the range of applications and improved dramatically the sensitivity and efficiency of quantitative interpretations.

When combined with other determinative-mineralogy methods (XRD, chemical, electron microscopy, IR) thermal analysis can provide valuable information which cannot be obtained by any other technique. Because each one of these methods provides complementary information, an accurate quantitative assessment of soil mineralogy can only be made from the application of combined techniques.

Thermoanalytical techniques, especially TG and DSC, have been routinely used for the quantitative mineralogical characterization of soil clays both in the authors' and other colleagues' laboratories for several years. The results have been quite acceptable for many purposes. It should be realized, however, that the complete quantitative characterization of a soil clay is a laborious procedure involving much experimental and interpretative work. Without question, thermal methods have a special place in any such investigation.

ACKNOWLEDGMENTS

Joint contribution from the University of Kentucky and University of Florida Agricultural Experiment Stations (Technical Article no. 91-3-23)

REFERENCES

Anderson, H., W. Besch, and D. Haberland. 1977. Investigation about methods of quantitative evaluation of DTA curves. J. Thermal Anal. 12:59-68.

Aomine, S., and C. Mizota. 1973. Distribution and genesis of imogolite in volcanic ash soils of northern Kanto, Japan. p. 207-313. In J.M. Serratosa (ed.) Proc. Int. Clay Conf., Madrid, Spain. 1972, Div. Ciencias C. S. I. C., Madrid.

Arazi, S.C., and T.G. Krenkel. 1970. Dehydroxylation heat of alunite. Am. Miner. 55:1329-1337.

Asomoza, P.M., M.L. Razo, T.L. Chaidez, and S.R. Cassillas. 1978. Quantitative evaluation of Na_2CO_3 and NaCl content of the clays of the ex-lake of Texcoco (Valley of Mexico) by means of thermogravimetry. J. Thermal Anal. 13:327-339.

Barnhisel, R.I., and P.M. Bertsch. 1989. Chlorites and hydroxyinterlayered vermiculite and smectite. p. 729-788, In J.B. Dixon and S.B. Weed (ed.) Minerals in soil environments, 2nd ed. SSSA, Madison, WI.

Barshad, I. 1952. Temperature and heat of reaction calibration of the differential thermal analysis apparatus. Am. Miner. 37:667-694.

Barshad, I. 1965. Thermal analysis techniques for mineral identification and mineralogical composition. p. 699-742. In C.A. Black (ed.) Methods of soil analysis. Part 1. Agron. Monogr. 9. ASA and SSSA, Madison, WI.

Bartenfelder, D.C., and A.D. Karathanasis. 1989. A differential scanning calorimetry evaluation of quartz status in geogenic and pedogenic environments. Soil Sci. Soc. Am. J. 53:961-967.

Barwood, H.L, and B.F. Hajek. 1979. Differential thermal characteristics of soil and reference quartz. Soil Sci. Soc. Am. J. 443:626-627.

Bayer, G., and H.G. Wiedemann. 1983. Thermoanalytical measurements in archaeometry. Thermochim. Acta 69:167-173.

Blanchard, F.N. 1967. Differential thermal analysis of wavellite. Q. J. Florida Acad. Sci. 30:161-167.

Blanchard, F.N. 1971. Thermal analysis of crandallite. Q. J. Florida Acad. Sci. 34:1-9.

Blanchard, F.N., and S.A. Denahan. 1966. Variscite from the Hawthorn Formation. Q. J. Florida Acad. Sci. 29:163-170.

Blanchard, F.N. 1972. Physical and chemical data for crandallite from Alachua County, Florida. Am. Miner. 57:473-484.

Blanchard, F.N., and S.A. Denahan. 1968. Cacoxenite and beraunite from Florida. Am. Miner. 53:2096-2101.

Borchardt, H.J. 1960. Initial reaction rates from DTA. J. Inorg. Nucl. Chem. 12: 252-254.

Borchardt, H.J., and F.J. Daniels. 1957. The application of differential thermal analysis to the study of reaction kinetics. J. Am. Chem. Soc. 79:41-46.

Borchardt, H.J., and F.J. Daniels. 1957. The application of differential thermal analysis to the study of reaction kinetics. J. Am. Chem. Soc. 79:41-46.

Boss, B.D. 1967. DTA of biotitic vermiculite to determine vermiculite content. Am. Miner. 52: 293-298.

Bracewell, J.M. and G.W. Robertson. 1980. Pyrolysis-mass spectometry studies of humification in a peat and a peaty podzol. J. Anal. Appl. Pyrol. 2:53-62.

Brown, M.E. 1988. Introduction to thermal analysis: techniques and applications. Chapman and Hall, London.

Carthew, A.R. 1955. The quantitative estimation of kaolinite by DTA. Am. Miner. 40:107-117.

Cassel, B., J.S. Mayer and M. Divito. 1987. Thermal analysis: Trends in application software. Am. Labor. January p. 92-96.

Charsley, E.L., N.J. Manning, and S.B. Warrington. 1987. A new integrated system for simultaneous TG-DTA-Mass spectrometry. Thermochim. Acta. 114:47-52.

Churchman, G.J., J.S. Whitton, G.G.C. Claridge, and B.K.G. Theng. 1984. Intercalation method using formamide for differentiating halloysite from kaolinite. Clays Clay Miner. 32:241-248.

Cunningham, A.D. and F.W. Wilburn. 1970. Theory. p. 31-62. *In* R.C. Mackenzie (ed.) Differential thermal analysis. Part I. Acad. Press, London.

David, D.J. 1972. Simultaneous photothermal and differential thermal analysis. Thermochim. Acta. 38:37-45.

Davis, C.E., and D.A. Holdridge. 1969. Quantitative estimation of clay minerals by DTA. Clay Miner. 8:193- 200.

Dixon, J.B. 1966. Quantitative analysis of kaolinite and gibbsite in soils by differential thermal and selective dissolution methods. Clays Clay Miner. 14:83-89.

Dubrawski, J.V. 1987. The effect of particle size on the determination of quartz by differential scanning calorimetry. Thermochim. Acta. 120:257-260.

Dubrawski, J.V. and S.S.J. Warne. 1988. Differential scanning calorimetry of the dolomite-ankerite mineral series in variable atmospheres. Thermochim. Acta. 135:225-230.

Dunn, J.G., B.T. Sturman, and W. Van Bronswijk. 1980. The determination of gypsum and lime in small samples of set plaster by a DSC-Computer method. Thermochim. Acta. 37:337-341.

Dunn, J.G., G.C. De, and P.G. Fernandez. 1988. The effect of experimental variables on the multiple peaking phenomenon observed during the oxidation of pyrite. Thermochim. Acta. 135:267-272.

Dyer, A. 1981. Thermomechanical analysis of zeolites. Anal. Proc. October p. 447450.

Earnest, C.M. 1980. The application of differential thermal analysis and thermogravimetry to the study of kaolinite clay minerals. Appl. Study 30, Perkin-Elmer Corp. Norwalk, CT.

Earnest, C.M. (ed.) 1988. Compositional analysis by thermogravimetry. ASTM, Philadelphia, PA.

Fanning, D.S., V.Z. Keramidas, and M.A. El-Desoky. 1989. Micas. p. 551-634. *In* J.B. Dixon and S.B. Weed (ed.) Minerals in soil environments, 2nd ed. SSSA, Madison, WI.

Feldman, S.B., and L.W. Zelazny. 1990. Characterization of indigenous organo-mineral complexes using TG/FT-IR. p. 349. *In* Agronomy Abstracts, Madison, WI.

Gallagher, P.K., and D.W. Johnson. 1973. The effects of sample size and heating rate on the kinetics of the thermal decomposition of $CaCO_3$. Thermochim. Acta. 6:67-83.

Gibbs, R.J. 1965. Error due to segregation in quantitative clay mineral XRD mounting techniques. Am. Miner. 50:741-751.

Gill, P.S. 1976. Determination of the degree of conversion of gypsum to plaster of paris by DSC. Appl. Brief TA-38. Du Pont Co., Wilmington, DE.

Grimshaw, R.W. 1953. Quantitative estimation of silica minerals. Clay Miner. Bull. 2:2-7.

Hajek, B.F. 1979. COLE determination by thermomechanical analysis. Soil Sci. Soc. Am. J. 43:427-428.

Harris, W.G., and K.A. Hollien. 1988. Reversible and irreversible dehydration of hydroxyinterlayered vermiculite from Coastal Plains soils. Soil Sci. Soc. Am. J. 52:1808-1814.

Harris, W.G., K.A. Hollien, and V.W. Carlisle. 1989. Pedon distribution of minerals in Coastal Plain Paleudulfs. Soil Sci. Soc Am. J. 53:1901-1906.

Jackson, M.L. 1975. Soil chemical analysis-advanced course, 2nd ed. M.L. Jackson, Madison, WI.

Jorgensen, S.S., A.C. Birnie, B.F.L. Smith, and B.D. Mitchell. 1970. Assessment of gibbsitic material in soil clays by differential thermal analysis and alkali dissolution methods. J. Thermal Anal. 2:277-282.

Karathanasis, A.D. 1985. Mineralogical variability within clayey control sections and family mineralogy placement. Soil Sci. Soc. Am. J. 49:691-695.

Karathanasis, A.D. 1988. Compositional and solubility relationships between aluminum-hydroxyinterlayered soil smectites and vermiculites. Soil Sci. Soc. Am. J. 52:1500-1508.

Karathanasis, A.D., and B.F. Hajek. 1982a. Revised methods for rapid quantitative determination of minerals in soil clays. Soil Sci. Soc. Am. J. 46:419-425.

Karathanasis, A.D., and B.F. Hajek. 1982b. Quantitative evaluation of water adsorption of soil clays. Soil Sci. Soc. Am. J. 46:1321-1325.

Karathanasis, A.D., and B.F. Hajek. 1985. Shrink-swell potential of montmorillonitic soils in udic moisture regimes. Soil Sci. Soc. Am. J. 49:159-166.

Karathanasis, A.D., and V.P. Evangelou. 1986. Water sorption characteristics of Al-and Ca-saturated clays. Soil Sci. Soc. Am. J. 50:1063-1068.

Karathanasis, A.D., and V.P. Evangelou. 1987. Low temperature dehydration kinetics of aluminum-and calcium- saturated soil clays. Soil Sci. Soc. Am. J. 51:1072-1078.

Karathanasis, A.D., F. Adams, and B.F. Hajek. 1983. Stability relationships in kaolinite, gibbsite, and Al- hydroxy-interlayered vermiculite soil systems. Soil Sci. Soc. Am. J. 47:1247-1251.

Karathanasis, A.D., G.W. Hurt, and B.F. Hajek. 1986. Properties and classification of montmorillonite-rich Hapludults in the Alabama Coastal Plain. Soil Sci. 142:76-82.

Keattch, C.J., and D. Dollimore. 1975. An introduction to thermogravimetry, 2nd ed. Heyden, London.

Kerr, P.F., and J.L. Kulp. 1948. Multiple differential thermal analysis. Am. Miner. 33:387-419.

Khorami, J., A. Lemieux, H. Menard, and D. Nadeau. 1988. Usefulness of Fourier transform infrared spectroscopy in the analysis of evolved gas from the thermogravimetric technique. p. 147-159. In C.M. Earnest (ed.) Compositional analysis by thermogravimetry.

Lodding, W., and L. Hammell. 1960. Differential thermal analysis of hydroxides in reducing atmosphere. Anal. Chem. 32:657-662.

Mackenzie, R.C. 1957. The differential thermal investigation of clays. Miner. Soc., London.

Mackenzie, R.C. 1970. Differential thermal analysis. Acad. Press, London.

MacKenzie, R.C. 1981. Thermoanalytical methods in clay studies. p. 5-29. In J.J. Fripiat (ed.) Advanced techniques for clay mineral analysis. Elsevier, Amsterdam, the Netherlands.

Mackenzie, R.C., and B.D. Mitchel. 1979. Variable atmosphere DTA in identification and determination of anhydrous carbonate minerals in soils. J. Soil Sci. 30:111-116.

Mackenzie, R.C., and R.H.S. Robertson. 1961. The quantitative determination of halloysite, goethite and gibbsite. Acta. Univ. Carol., Geol. Suppl. 1:139-149.

Mackenzie, R.C., and S. Calliere. 1975. The thermal characteristics of soil minerals and the use of these characteristics in the qualitative and quantitative determination of clay minerals in soils. p. 529-571. In J.E. Gieseking (ed.) Soil components, Vol. II. Elsevier, New York.

Mackenzie, R.C., E. Patterson, and R. Swaffield. 1981. Observation of surface characteristics by DSC and DTA. J. Thermal Anal. 22:269-274.

McKeever, S.W.S. 1988. Thermoluminescence of solids. Cambridge Univ. Press, London.

Minato, H. 1988. Dehydration energy of halloysite by means of DSC methods with the relationships of its mineralogy and modes of ocurrence. Thermochim. Acta. 135:279-283.

Ming, D.W., and F.A. Mumpton. 1989. Zeolites in soils. p. 873-911. In J.B. Dixon and S.B. Weed (ed.) Minerals in soil environments, 2nd ed. SSSA, Madison, WI.

Morgan, D.J. 1977. Simultaneous DTA-EGA of minerals and natural mineral mixtures. J. Thermal Anal. 12:245-263.

Morgan, D.J., S.B. Warrington, and S.J. Warne. 1988. Earth sciences applications of evolved gas analysis: a review. Thermochim. Acta. 135:207-212.

Muller-Vonmoos, M. and R. Müller. 1975. Application of DTA-TG-MS in the investigation of clays. p. 521-530. In I. Buzas (ed.) Thermal Analysis, Proc. 4th Int. Congr. Therm. Anal., Budapest. Vol. 2. Heyden, London.

Mulley, V.J., and C.D. Cavendish. 1970. A thermogravimetric method for the analysis of mixtures of brushite and monetite. Analyst 97:304-207.

Nathan, Y., G. Panzer, and S. Gross. 1988. The thermal analysis of some phosphate minerals: strengite, lipscombite, cyrilorite, and goyazite. Thermochim. Acta. 135:259-266.

O'Neil, M.J. 1970. When is a DSC not a DSC and how does it work anyway? Thermal Analysis Newsl. no. 9. Perkin-Elmer Corp., Norwalk, CT.

Parfitt, R.L. 1990. Estimation of imogolite in soils and clays by DTA. Commun. Soil Sci. Plant Anal. 21:623-628.

Paulik, F., and J. Paulik. 1978. Simultaneous techniques in thermal analysis. Analyst 103:417-437.

Paulik, J., F. Paulik, and M. Arnold. 1982. Simultaneous TG, DTG, DTA, and EG techniques for the determination of carbonate, sulfate, pyrite and organic material in minerals, soils and rocks. J. Thermal Anal. 25:327-340.

Pekenc, E., and J.H. Sharp. 1974. Quantitative mineralogical analysis of alunitic clays. p. 585-591. In I. Buzas (ed.) Thermal Analysis, Proc. 4th Int. Congr. Therm. Anal., Budapest. Vol. 2. Heyden, London.

Piloyan, G.O., I.D. Pryabchikov, and O.S. Novikova. 1966. Determination of activation energies of chemical reactions by differential thermal analysis. Nature (London) 121:1229.

Ramachandran, V.S., and G.M. Polomark. 1978. Application of DSC-DTA technique for estimating various constituents in white coat plasters. Thermochim. Acta. 25:161-169.

Rowland, R.A., and C.W. Beck. 1952. Determination of small quantities of dolomites by DTA. Am. Miner. 37:76-82.

Rowse, J.B., and W.B. Jepson. 1972. The determination of quartz in clay minerals. A critical comparison of methods. J. Thermal Anal. 4:169-175.

Schilling, M.R. 1990. Effects of sample size and packing in the thermogravimetric analysis of calcium montmorillonite STx-1. Clays Clay Min. 38:556-558.

Schomburg, J. and M. Störr. 1984. Dilatometerkurventales der Tonmineralrohstoffe. Academie-Verlag, Berlin.

Schwertmann, U. 1984. The double dehydroxylation peak of goethite. Thermochim. Acta. 78:39-46.

Sewell, E.C., and D.B. Honeyborne. 1957. Theory and quantitative use. p.65-97. In R.C. Mackenzie (ed.) The differential thermal investigation of clays. Miner. Soc., London.

Sharp, J.H., and S.A. Wentworth. 1969. Kinetic analysis of thermogravimetric data. Anal. Chem. 41:2060-2062.

Sharp, J.H., G.W. Grindley, and B.N. Narahari Achar. 1966. Numerical data for some commonly used solid state reaction equations. J. Am. Ceram Soc. 49:379-382.

Shoval, S. 1988. Mineralogical changes upon heating calcitic and dolomitic marl rocks. Thermochim. Acta 135:243-252.

Singer, A. 1989. Palygorskite and sepiolite group of minerals. p. 829-872. In J.B. Dixon and S.B. Weed (ed.) Minerals in soil environments, 2nd ed. SSSA, Madison. WI.

Smalley, I.J., and G.S. Xidakis. 1979. Thermogravimetry of an expansive clay soil from Adelaide: Approximate mineralogical analysis using standard montmorillonites. Clay Sci. 5:189-193.

Smykatz-Kloss, W. 1974. Differential thermal analysis. Application and results in mineralogy. Springer-Verlag, Heidelberg.

Smykatz-Kloss, W. 1984. Determination of impurities in minerals by means of standardized differential thermal analysis. p. 121-137. In R.L. Blaine and C.K. Schoff (ed.) Purity determinations by thermal methods, ASTM Spec. Tech. Publ., ASTM, Philadelphia, PA.

Sobecki, T.M., and A.D. Karathanasis. 1987. Quantification and compositional characterization of pedogenic calcite and dolomite in calcic horizons of selected Aquolls. Soil Sci. Soc. Am. J. 51:683-690.

Tan, K.H., and B.F. Hajek. 1977. Thermal analysis of soils. p. 865-884. In J.B. Dixon and S.B. Weed (ed.) Minerals in soil environments. SSSA, Madison, WI.

Tan, K.H., B.F. Hajek, and I. Barshad. 1986. Thermal analysis techniques. p. 151-183. In A. Klute (ed.) Methods of soil analysis, Part 1. 2nd ed. Agron. Monogr. 9. ASA and SSSA, Madison, WI.

Todor, D.N. 1976. Thermal analysis of minerals. Abacus Press, Kent, England.

Tullock, R.J., and C.B. Roth. 1975. Stability of mixed iron and aluminum hydrous oxides on montmorillonite. Clays Clay Miner. 23:27-32.

Van der Marel, H.W. 1956. Quantitative differential thermal analysis of clays and other minerals. Am. Miner. 41:222-244.

Wang, H.D., W.G. Harris, and T.L. Yuan. 1991. Noncrystalline phosphates in Florida phosphatic soils. Soil Sci. Soc. Am. J. 55:665-669.

Warne, S.S.J., D.J. Morgan, and A. Milodovski. 1981. Thermal analysis studies of dolomite, ferrous dolomite, ankerite series. Iron content recognition and determination by variable atmosphere DTA. Thermochim. Acta. 51:105-111.

Webb, T.L., and H. Heystek. 1957. The carbonate minerals. p. 329-363. In R.C. Mackenzie (ed.) Differential thermal investigation of clays. Miner. Soc., London.

Wendlandt, W.W. 1986. Thermal methods of analysis, 3rd edition. Wiley-Interscience, New York.

Wendlandt, W.W., and R.K. Gallagher. 1981. Instrumentation. p. 3-90. In E. Turi (ed.) Thermal characterization of polymeric materials. Acad. Press, New York.

Wunderlich, B. 1981. Theoretical aspects of thermal analysis. p. 91-234. In E. Turi (ed.) Thermal characterization of polymeric materials. Acad. Press, New York.

Yariv, S. 1989. Thermal analysis of minerals. Thermochim. Acta. 148:421-430.

13   Differential X-Ray Diffraction Analysis of Soil Minerals

D. G. Schulze
Purdue University
West Lafayette, Indiana

Selective-dissolution procedures and x-ray diffraction (XRD) analysis are both common techniques for soil mineralogy research. X-ray diffraction patterns of samples before and after treatment for selective dissolution of one or more components are often used to determine the effect of the dissolution treatment. In the past, these diffraction patterns were simply superimposed to see which parts of the pattern were changed by the dissolution treatment. The ready availability of personal computers and of computer-controlled diffraction equipment over the past decade have made it possible to compare two diffraction patterns by subtracting one digitized pattern from the other (Schulze, 1981). This technique, differential x-ray diffraction (DXRD), makes it easier to observe subtle differences in diffraction patterns, and thus extends the utility of XRD for the study of minerals in complex natural mixtures such as soils and sediments. Differential x-ray diffraction complements data from techniques such as infrared and Mössbauer spectroscopy when the goal is to detect and quantify trace and poorly crystallized phases.

The concept of DXRD is very simple. First, an aliquot of a sample is treated with a chemical agent which selectively dissolves one or more mineral phases. Second, digitized x-ray patterns of both the untreated and the treated samples are obtained. Third, the diffraction pattern of the treated sample is subtracted from the diffraction pattern of the untreated sample to give the DXRD pattern of the material which dissolved (Schulze, 1981). Different chemical extactants dissolve different mineral components, and, although conceptually the two patterns are simply subtracted, in practice some manipulation of the diffraction data is necessary before a useful DXRD pattern is obtained.

The purpose of this chapter is to describe the mechanics of obtaining DXRD data, to give examples of qualitative and quantitative applications of DXRD to soil systems, and to discuss some of the advantages and limitations of the technique. Some of the many potential selective-dissolution procedures will be discussed first, followed by discussions of specimen preparation, data collection, and data reduction.

SELECTIVE DISSOLUTION PROCEDURES

A wide variety of selective-dissolution procedures is available to selectively extract various soil components. Many of the mineral-specific chapters in Dixon and Weed (1989) provide brief reviews of selective-dissolution procedures appropriate for given minerals and generally provide references to the primary literature sources. Only some of the more common procedures are highlighted below.

Iron-oxide minerals have been widely studied using DXRD. Two procedures, the dithionite-citrate-bicarbonate (DCB) procedure (Mehra & Jackson, 1960) and the acid ammonium oxalate procedure (Schwertmann, 1964) have been most used to date. The DCB procedure extracts practically all secondary Fe oxides, without differentiating between mineral phases, while acid ammonium oxalate generally extracts only the poorly crystallized iron oxides, primarily ferrihydrite. Schwertmann and Taylor (1989), Borggaard (1988), and Loveland (1988) provide summaries and additional literature references on these two techniques.

Two additional procedures, pyrophosphate extraction and EDTA extraction, remove Fe, but have not been used with DXRD to date. Pyrophosphate extraction has been widely used for identification of spodic horizons in soils. Pyrophosphate is believed to extract an organically bound Fe fraction, but there is considerable uncertainty as to the exact nature of this fraction (Borggaard, 1988; Loveland, 1988). The EDTA anion has been proposed as an extractant for "amorphous Fe oxides" in soils (Borggaard, 1979), but the mineral phase which is extracted has not yet been identified (Borggaard, 1988). Differential x-ray diffraction may provide useful information regarding the phases extracted by these two techniques.

The DXRD should be very useful for studying poorly crystallized aluminosilicate minerals in soils. Allophane and imogolite can be dissolved using either an acid ammonium oxalate or a NaOH extraction, while "allophane-like" material can be dissolved with Na_2CO_3 (Wada, 1989). An example of a DXRD pattern of allophane is given later in this chapter.

The DXRD may be a particularly useful approach for identifying and characterizing poorly crystallized manganese-oxide minerals in soils, but currently there are few or no studies using DXRD. Manganese-oxide minerals can be dissolved by treatment with H_2O_2 in acid solution (Taylor & McKenzie, 1966; Jackson, 1969). They can also be dissolved with reducing agents such as quinol, dithionite, hydroxylamine hydrochloride, or oxalate (McKenzie, 1977), but some of these, notably dithionite and oxalate, also dissolve Fe-oxide minerals.

Carbonates are another mineral group which could be studied using DXRD. Carbonates are easily dissolved by acidifying the sample. The pH 5, 1 N Na-acetate buffer used to remove carbonates prior to mineral fractionation (Jackson, 1969) is quick and easy to use. An alternative method is to titrate the sample to pH 3.5 to 4.0 with 1 N HCl until a stable pH and lack of effervescence indicate complete destruction of carbonates (Jackson, 1969).

Almost any differential-extraction procedure can be used for DXRD studies, as long as it extracts more than about 10 g kg^{-1} of well-crystallized minerals with sharp diffraction lines or more than about 150 g kg^{-1} of poorly crystallized minerals such as ferrihydrite or allophane.

SAMPLE PREPARATION

Sample preparation is one of the most critical parts of the DXRD procedure. One must attempt to obtain two XRD patterns which are identical to one another, except for the phases which have been removed by the selective-dissolution procedure.

Identical Treatment

Both the treated and untreated samples should be handled the same, insofar as this is possible. For example, if a sonication step is used during the selective-dissolution procedure, then the untreated sample should be sonicated in the same way. Rather than splitting the sample into two aliquots, some prefer to x-ray, treat, and then x-ray the same sample a second time.

Particle Size

If the crystallites within a sample are too large, too few will contribute to each x-ray reflection, and it will be impossible to obtain reproducible x-ray intensities. If relative peak intensities vary from pattern to pattern because of particle-size effects, spurious peaks will result in the DXRD pattern. For samples with mass attenuation coefficients similar to quartz, powders with effective crystallite dimensions of <5-μm should give intensity deviations of <1%, while powders with crystallite sizes <10-μm should give intensity deviations of <2 to 3% (Klug & Alexander, 1974, p. 367). Intensity deviations are proportional to the sample mass attenuation coefficient. Thus, samples with larger attenuation coefficients require smaller particle size to achieve a given intensity deviation. Our experience has shown that the <2-μm size fraction gives reproducible intensities for DXRD. Good results were obtained for the carbonate-containing sample illustrated below by simply grinding the sample to pass a 20-μm sieve. The 20 μm, however, is probably an upper limit, and a somewhat finer material, say <10-μm, would be preferable.

Particle Orientation

Reproducible intensities among patterns can only be obtained when the crystallites are randomly oriented in the samples and when the samples are infinitely thick to x-rays.

Random orientation is best obtained when a cavity mount is used. Good results have been obtained by backfilling the powder into Al sample holders and then gently pressing the powder against unglazed paper to reduce preferred orientation at the sample surface. The procedure is similar to McCreery's procedure described by Klug and Alexander (1974, p. 372-374), except that a piece of unglazed paper is placed between the glass slide and the top surface of the sample holder. The pressing step can introduce some degree of preferred orientation, but if the same amount of pressure is used each time, the degree of orientation should be similar from sample to sample. Some samples give excellent DXRD patterns with no apparent preferred orientation, whereas other samples from different horizons of the same or a closely related soil may show the effects of orientation for some of the phases. The reason for this is not clear. The fact that a specific mineral orients in some samples but not in others could perhaps provide useful information about mineral morphology or mineral associations.

Phyllosilicate minerals have a tendency for increased preferred orientation after removal of the Fe-oxide minerals by DCB treatment. The result is an increase in intensity of the *00l* lines relative to the *hk0* lines and to lines of other minerals such as quartz, making it impossible to completely match all of the peaks common to both patterns. Campbell (A. S. Campbell, 1991, personal communication) has found that some compensation is possible by using a back-filled mount for the untreated sample, and a side-filled mount for the treated sample. Compared to the back-filled mount, the side packing tends to reduce the *00l* lines relative to the other lines in the pattern. The results are somewhat variable, but tend to improve with practice.

Sample Thickness

The samples must also be thick enough that additional increases in sample thickness cause no appreciable increase in diffracted intensity. The thickness of sample needed for maximum diffracted intensity increases as 2Θ increases and as the sample mass attenuation coefficient decreases. Mass attenuation coefficients for common soil minerals are no lower than 30 cm^2 g^{-1} for CuKα, FeKα (Rich & Barnhisel, 1977), and CoKα (Schulze, 1982) x-rays. Sample masses needed to obtain 95, 99, and 99.9% of the theoretical maximum diffraction are tabulated in Rich and Barnhisel (1977). Assuming a worst case of a sample mass attenuation coefficient of 30 cm^2 g^{-1} and a maximum diffraction angle of 80 °2Θ, 0.740 mg mm^{-2} of sample are needed to obtain 99.9% of the observed maximum diffraction. Thus, for a sample holder with 15- by 20 mm of sample area, 222 mg of sample are required. Note that this is a conservative amount. Sample mass attenuation coefficients are generally >30, and the DXRD pattern may not be needed out to 80 °2Θ; thus, about 200 mg of sample are generally sufficient for a sample holder with 15- by 20-mm area.

Oriented Slide Mounts

X-ray patterns obtained from samples sedimented onto glass slides should be used cautiously for DXRD analysis. Preferred orientation is likely to be much greater overall, and differences in preferred orientation before and after selective-dissolution treatment are likely to be accentuated. Only a few tens of milligrams of sample are generally used to prepare glass slide mounts. Such a small mass of sample is easily penetrated by the x-ray beam, resulting in less than maximum diffracted intensity and in some scattering off of the underlying slide. Wang et al. (1993), however, have recently shown that in situ selective dissolution while the sample remains on a nonreflecting quartz plate can be used to good advantage for DXRD analysis when only small amounts of material are available.

DATA COLLECTION

Data collection involves selecting an appropriate step size and counting time per step for a particular sample or group of samples. In theory, long counting times and small 2Θ steps would be most desirable. A $20°$ scan with $0.01°$ steps and a counting time per step of 200 s would, however, take over 4.5 d of continuous x-ray time for one pattern, and a total of 9 d to collect the two patterns needed for DXRD! Realistically, useful data can be collected in much less time. Extremely long data collection times may also result in problems caused by long-term variation in the absolute intensity of the x-ray source, as well as problems associated with temperature and humidity variations.

In practice, the step size need only be small enough so that the sharpest peaks are adequately defined. In general, five to seven points over the sharpest peak are usually sufficient. Quartz peaks are usually the sharpest, and for a peak having a full width at half maximum (FWHM) of $0.1°2\Theta$, a step size of $0.05°2\Theta$ is generally sufficient.

X-ray photons generated by an x-ray tube are emitted randomly with respect to time, and thus enter the detector randomly in time sequence. For events occurring randomly in time sequence, the standard deviation, s_N, of the number of events observed, N, from the true average number, N_o, is given by

$$s_N = N^{1/2},$$ [1]

and the relative standard deviation is given by

$$s_N(\text{rel}) = N^{1/2}/N = N^{-1/2}$$ [2]

(Klug & Alexander, 1974, p. 360-361). As N increases, $s_N(\text{rel})$ decreases rapidly until $N \sim 1000$, and then decreases less rapidly thereafter. For example, for $N = 100$, 1000, and 10,000, $s_N(\text{rel})$ is 0.10, 0.03, and 0.01, respectively. From a practical standpoint, increasing the count time per step from 1 to 10 s will usually have a greater impact on the DXRD pattern than increasing the count time from 10 to 100 s.

In general, the most efficient method for obtaining DXRD data is to maximize both the step size and the counting time per increment. For example, the same total time is needed to collect either a pattern with $0.025°$ steps and 10-s count time, or a pattern with $0.05°$ steps and 20-s count time. The pattern collected with the $0.05°$ steps and 20-s count time, however, is preferable because the longer count time per step significantly decreases the relative standard deviation of each data point.

DATA REDUCTION

Adjusting Intensities

After the selective-dissolution procedure has dissolved part of the sample, the remaining mineral phases are present in higher relative concentrations than before. This increases the XRD intensities of these remaining minerals. In addition, the dissolution of some phases may change the mass attenuation coefficient of the sample. If the mass attenuation coefficient is decreased, the XRD peaks will increase in intensity. If, on the other hand, the mass attenuation coefficient is greater after the dissolution treatment, a decrease in the intensities of the resistant minerals may result. In general, the net result is an increase in intensity after the selective dissolution procedure. Simply subtracting the treated pattern from the untreated pattern would result in "negative" diffraction peaks because of the stronger diffraction peaks in the treated pattern.

To prevent spurious peaks from occurring in the DXRD pattern, the peaks common to both patterns, and presumably unaffected by the selective-dissolution treatment, must match exactly. To do so, the pattern of the treated sample must be multiplied by a scale factor k, which is generally less than unity. If A_i, and B_i are the counts at angle i in patterns of the untreated and treated samples, respectively, then the corresponding intensity in the subtracted pattern, C_i, is defined by the relationship

$$C_i = A_i - k B_i .$$ [3]

The selection of k is obviously quite important to the quality of the DXRD pattern.

Selection of k by Trial-and-Error Subtractions

If k is too small, peaks common to both patterns do not completely cancel. If k is too large, "negative" peaks occur in the DXRD pattern (Fig. 13-1). Schulze (1981) suggested using a trial--and-error approach to finding the proper value of k . The trial-and-error approach works quite well, particularly if interactive computer software is available. By trying different values for k , from a value that is obviously too small, to one that is much too large, it takes only a few minutes to find the k value where peaks common to both patterns just cancel. Figure 13-1 illustrates this approach: $k = 0.80$ gives a DXRD pattern in which the peaks common to both the untreated and treated patterns just cancel, while "negative" peaks occur in the DXRD pattern for $k = 1.00$, and incomplete cancellation occurs for $k = 0.60$. The trial-and-error approach is analogous to focusing a camera; by passing through the focal point several times, the correct focus is found quickly.

The advantages of the trial-and-error approach are: (i) it is quick, (ii) it requires a minimum of sample preparation other than the selective-dissolution procedure, and (iii) it produces the lowest detection limits because the sample is not diluted by the addition of an internal standard. The disadvantages of the procedure are: (i) for some samples it may be very difficult to pick the "best" scale factor because of preferred-orientation effects, and (ii) in some cases, it may be impossible to detect the dissolution of certain phases. For example, if a sample contains only two phases and all of Phase A and 20% of Phase B are dissolved, the trial-and-error approach will lead to a scale factor which will produce a DXRD pattern containing only the pattern of Phase A.

Selection of k Using an Internal Standard

Another approach to determining k is to add an internal standard to the sample before the selective-dissolution treatment and XRD data collection. The scale factor, k , can then be selected so that the internal-standard peaks just cancel out. The internal standard: (i) should not be dissolved by the selective-dissolution procedure, (ii) should not be prone to preferred orientation, (iii) should have relatively few diffraction lines, and (iv) should have diffraction lines that do not coincide with those of the phases of interest. Bryant et al. (1983) used 1-μm α-Al_2O_3 as an internal standard to obtain DXRD patterns of Fe-oxide minerals from Brazilian soils, and α-Al_2O_3 has since been used in a variety of studies with good success.

The advantages of the internal-standard approach are: (i) it reduces the uncertainty in selecting the correct scale factor, and (ii) it decreases the possibility that partial dissolution of a presumably insoluble phase will go unnoticed. The internal standard-approach is not without potential pitfalls, however. It is important that the internal standard be thoroughly mixed with the sample before splitting the sample and doing the selective dissolution. Gentle but thorough grinding in an agate mortar is a good way to mix the standard with the sample. To prevent potential problems caused by inadequate mixing, some workers prefer not to split the sample, but to x-ray, treat, and then x-ray the same sample again. Also, if the standard is somewhat coarser or denser than the sample being studied, the standard may preferentially settle to the bottom of the centrifuge tube during washings to remove the excess salts introduced during the dissolution procedure. If such a

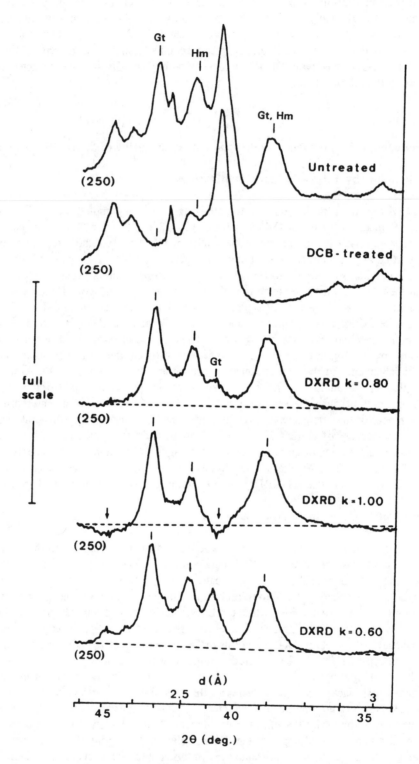

Fig. 13-1. X-ray powder diffraction patterns from untreated and DCB-treated samples, and DXRD
diagrams illustrating the correct ($k = 0.80$) and incorrect ($k = 1.00$, $k = 0.60$) values for
the scale factor. Arrows indicate parts of the pattern which dip below the baseline if the scale
factor is too large, or which rise above the baseline if the scale factor is too small. Samples
are from the <2-μm clay fraction of the Bt horizon of a Typic Paleudalf from Texas.
Numbers in parentheses indicate full-scale counts per second. Gt = goethite, Hm = hematite.
CoKα radiation. Reprinted from Schulze (1981).

segregated sample is dried, material enriched in the standard may remain in the tube when the sample is removed. It is important that all of the sample be recovered from the tube to prevent changes in the ratio of standard to sample.

Aligning 2Θ Scales

A mismatch in the 2Θ position of peaks which are common to both patterns is a common problem in DXRD. The mismatch results in spurious peaks (Fig. 13-2) which often make it difficult to determine the best scale factor, and which are sometimes so prominent that attention is drawn away from the phases of interest. In some cases, the mismatch in the 2Θ scales can give rise to spurious peaks which give the illusion that a particular phase has been dissolved or precipitated. The problem is particularly noticeable for phases like quartz which give very sharp diffraction peaks.

There are three reasons for this mismatch. First, there may be enough freedom of movement within the goniometer gear train so that the starting point for successive patterns is not always the

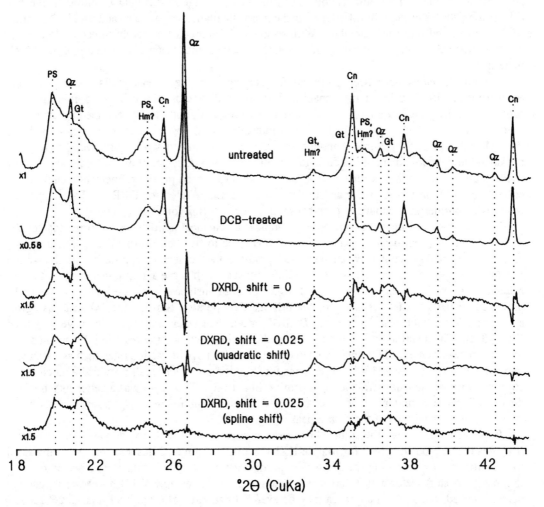

Fig. 13-2. X-ray powder diffraction patterns of an untreated and DCB-treated soil clay, and DXRD patterns obtained with no shift in the 2Θ scale of the untreated pattern, with a 0.025° shift using a quadratic polynomial shifting function, and with a 0.025° shift using a cubic-spline shifting function. x1, x0.58, x1.5 = relative vertical scale expansion; Cn = α-Al_2O_3 (corundum); Gt = goethite; Hm = hematite; Ps = phyllosilicates; Qz = quartz. Hematite identification is uncertain because of strong overlap with phyllosilicate and goethite peaks. Samples contained 100 g kg^{-1} of 1-μm α-Al_2O_3 as an internal standard. Reprinted from Schulze (1986).

same. This can be reduced or eliminated by assuring that the starting 2Θ point is always approached from the same direction. Modern computer-controlled diffractometer drives may do this automatically, but this may not be the case with older systems. Second, small displacements of the sample from the axis of the goniometer can cause significant shifts in the position of the diffraction peaks. The magnitude of this displacement error in radians is given by:

$$\Delta 2\Theta = -2s \, (\cos\Theta)/R, \tag{4}$$

where s is the displacement in mm of the specimen from the axis of the goniometer, and R is the radius of the goniometer circle in millimeters (Jenkins, 1989). Specimen-displacement error is angle dependent, and a 25-μm displacement error can cause a shift of -0.017 to -0.013° 2Θ for peaks occurring between 5 and 80° 2Θ. Specimen transparency is a third cause of mismatch in the 2Θ scale. Specimen-transparency error occurs because the incident x-rays penetrate below the sample surface, causing the average diffracting surface to lie somewhat below the physical surface of the specimen (Jenkins, 1989). This error increases as the absorption of the x-ray beam by the specimen decreases. Thus, diffraction lines can occur at slightly different positions in the two diffraction patterns because of differences in the mass attenuation coefficients and packing densities of the treated and untreated samples. With so many different factors likely to cause slight peak shifts in diffraction patterns, it is not surprising that peak mismatches are quite common in DXRD patterns.

Several strategies have been proposed for dealing with these peak shifts. Campbell and Schwertmann (1985) utilized the capabilities of their diffractometer control to minimize the problem. A sharp peak common to both patterns was selected, and the diffractometer control was used to search the selected peak for maximum intensity. That position was then used to calibrate the 2Θ scale for each pattern. Although this approach seems to work quite well, it requires considerable operator intervention during data collection and does not allow different shifts to be tried unless a new pattern is run. Brown and Wood (1985) used profile refinement techniques to correct for shifts in the 2Θ scale and to determine the scale factor for DXRD. Their approach requires a considerable amount of time-consuming curve fitting for each sample.

Schulze (1986) evaluated two relatively simple mathematical approaches to correcting for 2Θ shifts. In the first approach, a quadratic curve was fitted to five consecutive data points and used to interpolate points lying between measured points. In the second approach, a cubic-spline function was used to connect the measured points with a smooth curve and intermediate points were calculated from the spline curve (Fig. 13-3). In both instances, any desired shift could be produced, regardless of the spacing of the original data points. Shifting the 2Θ scale using the spline function produced a much better DXRD pattern than when the quadratic curve was used (Fig. 13-2). The quadratic curve did not fit the data points well and the resulting distortion of the diffraction peak shape made it impossible to exactly match sharp diffraction lines such as the quartz line at about 26.7° 2Θ (Fig. 13-2). The spline function, on the other hand, provided a much better approximation of the true diffraction-line profile (Fig. 13-3). The result was a better match of the diffraction peaks common to both patterns and virtual elimination of spurious peaks due to mismatch at the positions of the quartz peaks (Fig. 13-2).

Experience has shown the spline shift function to work well with a wide variety of samples. It is thus the method of choice for routine DXRD work. A single shift factor has generally been found sufficient for matching the 2Θ scales, particularly if the angular range scanned is only 20 to 30° 2Θ. A shift factor which varies as a function of angle, using Eq. [4] for example, has not been evaluated, but such an approach may be useful if data are collected over a large 2Θ range.

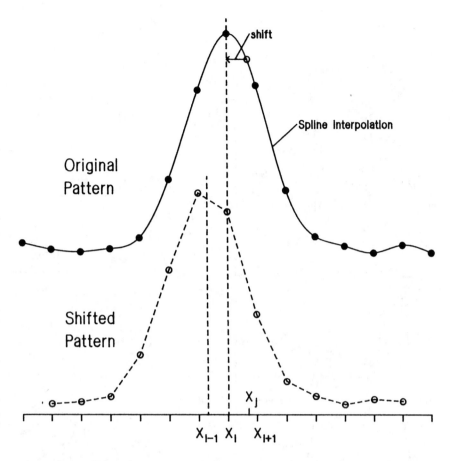

Fig. 13-3. Use of a cubic-spline function to shift an x-ray powder diffraction pattern. Reprinted from Schulze (1986).

IDENTIFICATION OF MINERAL PHASES

The DXRD has been used for a variety of studies to identify and quantify soil minerals. Examples of applications and suggestions for new applications are given below.

Goethite and Hematite

Goethite and hematite are responsible for the brown and red colors of soils. Both minerals are strong pigments and are easily detected visually at concentrations <10 g kg^{-1}. X-ray identification of these minerals at such low concentrations is, however, difficult. Schulze (1981) demonstrated that DXRD could detect hematite in the <2-μm fraction from a smectitic Ustalf containing only 18 g DCB-extractable Fe (Fe$_d$) per kilogram (Fig. 13-4, sample K1154), while goethite was easily detected in the <2-μm fraction from an Ochrept with only 38 g Fe$_d$ per kilogram (Fig. 13-4, Sample 77). Excellent DXRD patterns were obtained from the <2-μm fraction of three Oxisols with Fe$_d$ contents of 77 to 157 g kg^{-1} (Fig. 13-4, samples GO-01, GO-08, 28). Note that in Sample GO-01, a small amount of goethite is detected in the DXRD pattern which is not visible in the untreated pattern. The goethite in samples GO-08 and 28 is highly Al-substituted as indicated by a shift of the goethite diffraction lines to higher angles.

The DXRD method has been used frequently by a variety of researchers for the identification and quantification of goethite and hematite, and occasionally, lepidocrocite and maghemite (Torrent et al., 1980, 1983; Schwertmann et al., 1982; Bronger et al., 1983; Santana, 1984; Curi & Franzmeier, 1984; Palmieri, 1986; Dudas et al., 1988; Boero & Schwertmann, 1989; Yoshinaga et al., 1989; Aniku & Singer, 1990). Torrent (Torrent, 1987; Torrent & Cabedo, 1986; Barrón

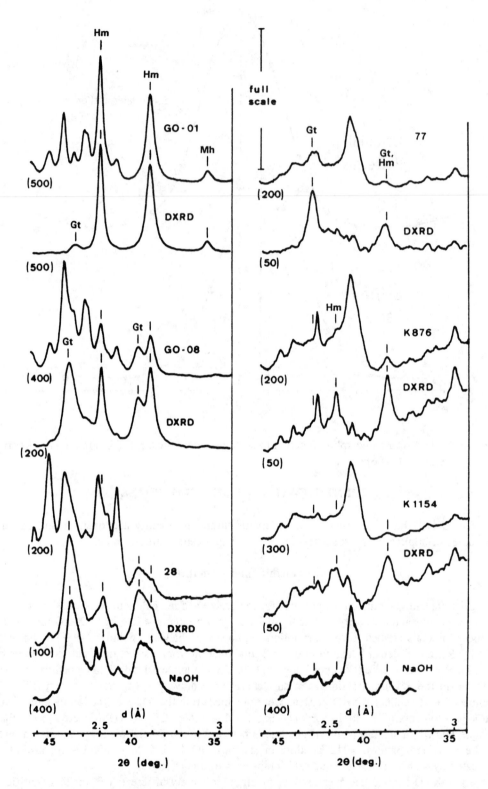

Fig. 13-4. X-ray powder diffraction patterns of untreated soil clays, corresponding DXRD patterns after DCB treatment, and XRD patterns for samples 28 and K1154 after boiling in 5 *N* NaOH for 1 h. DXRD patterns were smoothed using a seven-point cubic-polynomial smoothing function. Numbers in parentheses indicate full-scale counts per second. Gt = goethite, Hm = hematite, Mh = maghemite. CoKα radiation. Reprinted from Schulze (1981).

& Torrent, 1987; Torrent et al., 1980) used boiling 5 M NaOH to concentrate goethite and hematite in Mediterranean soils prior to using DXRD to obtain diffraction patterns for quantification of Fe oxides. Amarasiriwardena et al. (1988) studied mixtures of synthetic goethites and hematites with different Al substitutions. They found that well-crystallized hematite with sharp diffraction lines was readily detected by DXRD at 30 g kg^{-1} in hematite/goethite mixtures, but that detectability worsened as the hematite lines became broader. Nevertheless, DXRD is a very useful technique for directly identifying and quantifying Al-substituted Fe oxide phases.

Ferrihydrite

Poorly crystallized soil minerals like ferrihydrite are difficult or impossible to identify in complex mixtures using standard x-ray powder diffraction patterns. The broad, weak diffraction patterns of these minerals are often lost in the "forest" of sharp diffraction peaks from more crystalline minerals.

The acid ammonium oxalate procedure (Schwertmann, 1964) was known to dissolve almost pure accumulations of ferrihydrite, but there was uncertainty as to whether the Fe dissolved by oxalate in more complex mixtures was really ferrihydrite. The DXRD showed that the oxalate soluble fraction was indeed ferrihydrite (Schulze, 1981; Schwertmann et al., 1982). Examples of DXRD patterns of ferrihydrite are given in Fig. 13-5. Samples with relatively large amounts of oxalate-soluble Fe (Fe_o) such as K2 and FE51f (Fig. 13-5) give very clear DXRD patterns of 6-line ferrihydrite. (Note: these patterns were obtained before the 2Θ shift procedures were developed and the "negative" peaks are due to mismatch in the 2Θ scales of the untreated and treated samples.) Sample K3 is perhaps the most striking demonstration of the capabilities of DXRD. Although oxalate extracted 133 g Fe kg^{-1}, the untreated and oxalate-treated patterns appear essentially the same. Only after subtracting the two and expanding the scale of the DXRD pattern by a factor of 12, does one clearly see the six-line pattern of ferrihydrite. Assuming a bulk chemical composition for ferrihydrite of $5Fe_2O_3 \cdot 9H_2O$ and that no other oxide phases were dissolved by the oxalate treatment, the Fe_o content of 133 g kg^{-1} is equivalent to 229 g kg^{-1} of ferrihydrite. Sample K1, with an Fe_o content of only 29 g kg^{-1} (50 g kg^{-1} ferrihydrite), shows no clear evidence of ferrihydrite in the DXRD pattern. Campbell and Schwertmann (1984) used DXRD to identify ferrihydrite in the <2-μm fraction of placic horizons with Fe_o contents ranging from 232 to 83 g kg^{-1}. Brady et al. (1986) obtained a DXRD pattern of a ferrihydrite-like mineral in an ocherous precipitate containing 296 g $Fe_o kg^{-1}$ that had been collected from a stream receiving acid-sulfate mine drainage. Schwertmann et al. (1982) concluded that the lower limit of detection for ferrihydrite is at an Fe_o content of about 100 g kg^{-1}, or roughly 150 g ferrihydrite per kilogram. This lower limit is also supported by Campbell and Schwertmann (1984).

Poorly Crystallized Aluminosilicates

Both allophane and imogolite are dissolved in acid ammonium oxalate (Wada, 1989) and DXRD should be as useful for studying the distribution of these minerals as it is for studying ferrihydrite. Kodama and Wang (1989) identified allophane, ferrihydrite, and opaline silica in some Canadian soils using DXRD. Instead of using a diffractometer, they used a Guinier-de Wolff focusing camera to obtain simultaneous diffraction patterns of the untreated and treated samples on the same film. The photographs thus obtained were scanned on a microphotodensitometer and registered on a strip chart recorder. DXRD patterns were then calculated manually using 0.1° steps. Campbell and Schwertmann (1985) also report a DXRD pattern of allophane from a sample containing quartz, feldspars, and phyllosilicates.

Figure 13-6 shows an example of the XRD and DXRD diagrams of an allophanic soil clay. Note that the XRD pattern of the oxalate-extracted sample is almost the same as the pattern of the untreated sample, but the DXRD pattern clearly shows the very broad peaks due to allophane at 0.33 and 0.225 nm.

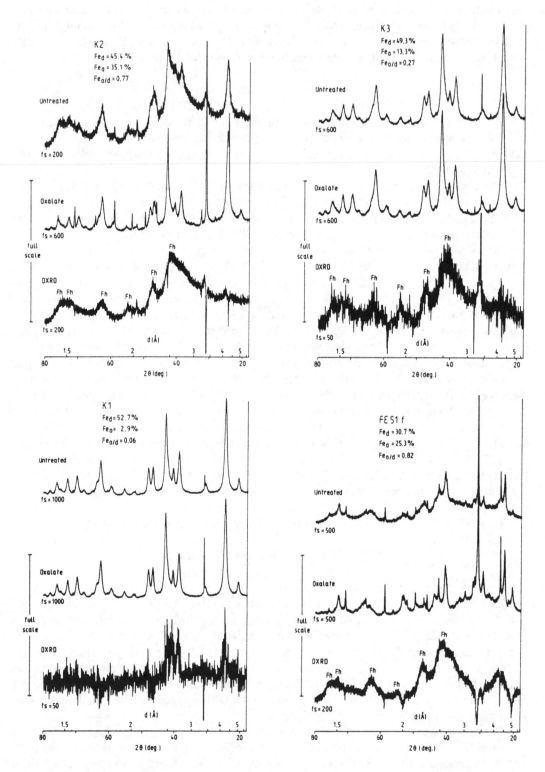

Fig. 13-5. XRD and DXRD patterns of ferrihydrite-containing pipe stems and ferricretes. Extraction procedure: pH 3 NH$_4$-oxalate. fs=xxx indicates full-scale counts per second; Fh = ferrihydrite, Gt = goethite, Qz = quartz. CoKα radiation. Reprinted from Schwertmann et al. (1982).

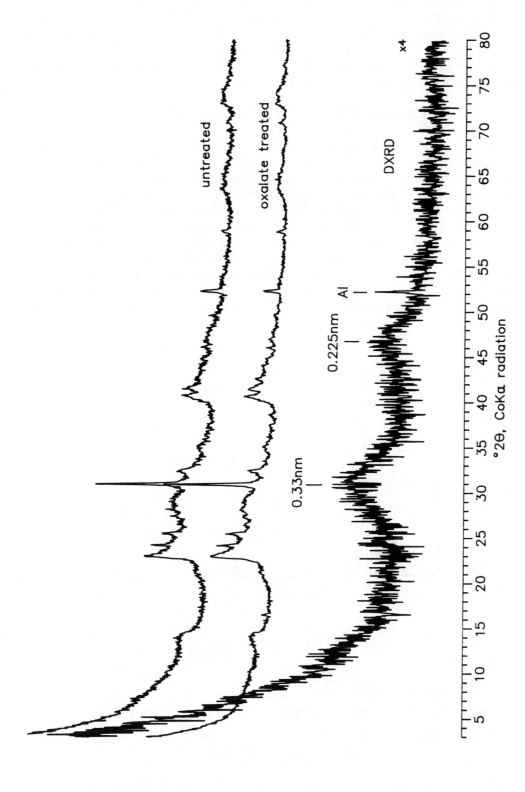

Fig. 13-6. XRD and DXRD patterns of the clay fraction from a Japanese Andisol. Extraction procedure: pH 3 NH_4-oxalate. Al = Al(m) from sample holder.

Carbonates

Carbonate minerals are another group of minerals which are easily dissolved by selective chemical dissolution and which could be easily studied by DXRD. Figure 13-7 shows a DXRD pattern of a calcareous glacial till after dissolving the carbonates in pH 5, 1 M Na-acetate (Jackson, '1969). The DXRD pattern shows only the peaks due to calcite and dolomite. This example is trivial because calcite and dolomite can easily be identified in the untreated XRD pattern. Note, however, that the large quartz peaks were almost completely cancelled using the spline-shift routine to carefully match the 2Θ scales of the untreated and treated patterns.

QUANTITATIVE ANALYSIS OF MINERAL PHASES

The goal in quantitative mineral analysis by XRD is to relate the areas under given diffraction peaks to the concentrations of the minerals in the sample. Bish (Bish, 1994; see Chapter 9) discusses various approaches to quantitative mineral analysis by XRD. Differential x-ray diffraction does not provide any fundamental improvement in these procedures, it only makes it easier to obtain the necessary intensity information when the peaks of interest are weak or convoluted with peaks due to other minerals. In addition, if the dissolved elements are quantitatively removed and analyzed during the selective-dissolution procedure, the chemical data will provide constraints on the absolute quantities of the minerals that dissolved.

In the simplest case, if DXRD shows that only one mineral dissolved during the selective-dissolution procedure, then the chemical data can be used to calculate the amount of that phase present by assuming a given bulk chemical composition. Ferrihydrite, for example, can sometimes be quantified in this way (see above).

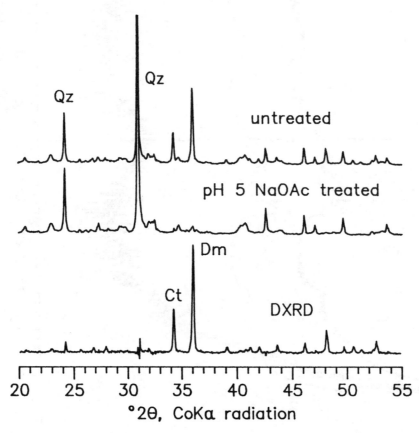

Fig. 13-7. XRD and DXRD patterns from a calcareous loamy glacial till. Sample was ground to pass a 20-μm sieve. Extraction procedure: pH 5, 1 N Na-acetate. Qz = quartz, Ct = calcite, Dm = dolomite.

When two or more minerals are dissolved, the DXRD pattern can be used to obtain the ratios of the dissolved minerals. Torrent et al. (1980) determined the goethite:hematite ratio in the <2-μm fraction of some Xeralfs from Spain using peak-area ratios measured from DXRD patterns. Mixtures of pure synthetic goethite and hematite were used as standards. Absolute quantities of goethite and hematite were then calculated by allocating the difference between dithionite-citrate-bicarbonate extractable Fe (Fe_d) and oxalate-extractable Fe (Fe_o) to goethite and hematite because no other crystalline Fe-oxide phases were present.

The mass attenuation coefficient of the sample must be taken into account in quantitative x-ray analysis when peak areas are used to determine mineral content. One approach is to mix all samples with the same amount of an internal intensity standard, and express the intensity of the peaks of interest relative to the intensity of the internal standard peak. Palmieri (1986) used this approach to quantify goethite, hematite, and maghemite in the <2-μm fraction of 48 A, B, and C horizons of 10 Oxisols from southern Brazil. The clays were dominated by kaolinite, with minor amounts of gibbsite, hydroxy-interlayered vermiculite, quartz, and cristobalite, and occasional traces of mica and smectite. The clay samples were mixed with 100 g kg^{-1} of 1-μm α-Al_2O_3 as an internal intensity standard, the samples were split, and one subsample was extracted using the dithionite-citrate-bicarbonate procedure. The DCB-extractable Fe ranged from 19 to 225 g kg^{-1}, with most of the samples containing about 100 g Fe_d kg^{-1}. The XRD patterns of pressed powder mounts were obtained using CuKα radiation. The areas under the 0.418-nm goethite peak, the 0.366-nm hematite peak, and the 0.295-nm maghemite peak were measured from the DXRD patterns. A single scale factor based on complete cancellation of the α-Al_2O_3 peaks was not satisfactory because preferred orientation sometimes caused artifacts in the areas near the Fe-oxide peaks. The patterns were broken into three sections and different scale factors were determined for each section using the subjective approach. This made it possible to obtain relatively accurate areas for each of the Fe-oxide peaks. The intensities of the Fe-oxide peaks were expressed as the area of the Fe-oxide peaks measured from the DXRD patterns, relative to the area of the 0.206-nm α-Al_2O_3 peak measured on the *untreated* pattern. The Al_2O_3 peak must be measured on the untreated pattern because it is cancelled in the DXRD pattern. There is no inconsistency in measuring the Al_2O_3 peak on the untreated pattern and the Fe-oxide peaks on the corresponding DXRD pattern; the intensity information for the Fe-oxide peaks is already present in the untreated pattern, it is simply easier to extract using the DXRD pattern.

The coefficients needed to convert the intensity ratios to absolute quantities could have been derived from a series of standards measured in the same way as the samples. Matching of samples and standards is often difficult, however, because synthetic specimens or pure natural mineral specimens with sharp diffraction lines are often not suitable standards for natural minerals with broad diffraction lines. With the large number of similar samples being analyzed, Palmieri (1986) used a multiple-regression approach to determine the coefficients needed to convert the intensity ratios to weight fractions. Hematite, goethite, and maghemite were the only crystalline Fe oxides identified. Thus, the following equation could be written:

$$Fe_d - Fe_o = K_{Gt} I_{Gt} F_{Gt} + K_{Hm} I_{Hm} F_{Hm} + K_{Mh} I_{Mh} F_{Mh}$$ [5]

where I_{Gt}, I_{Hm}, and I_{Mh} are the intensities relative to Al_2O_3 of goethite, hematite, and maghemite; F_{Gt}, F_{Hm}, and F_{Mh} are the weight fractions of Fe in goethite, hematite, and maghemite after correction for Al substitution, and K_{Gt}, K_{Hm}, and K_{Mh} are coefficients to be determined. With 48 samples (equations) and only three unknowns, multiple linear regression was used to determine the K coefficients. Multiplication of the I 's by their respective K 's gave the weight fraction of each mineral in each sample.

EVALUATION OF SELECTIVE-DISSOLUTION TECHNIQUES

The DXRD approach has been most widely used for identification and quantification of specific minerals. This approach is also very useful, however, for evaluating selective-dissolution procedures per se. Campbell and Schwertmann (1985) have pointed out that if a selective-dissolution procedure is to be used on soils, it is much more desirable to test such procedures using soil minerals, rather than to use well-crystallized geological or other specimens which may bear little resemblance to soil minerals formed by pedogenic processes. The DXRD and differential infrared absorption spectroscopy are both useful techniques for evaluating new chemical extraction methods. For example, Campbell (A. S. Campbell, 1991, personal communication) has used DXRD to show that citric acid preferentially dissolves carbonate apatite from phosphate rock, leaving a residue with an increased concentration of fluorapatite.

FUTURE DEVELOPMENTS

The emphasis in this paper has been on using selective dissolution to obtain the DXRD pattern of minerals based on their *chemical solubility*. It should be mentioned, however, that it is also possible to use the anomalous x-ray scattering of specific elements to obtain the DXRD pattern of minerals based on their *elemental composition*. In this technique, diffraction patterns are taken using two different x-ray wavelengths, one just below, and one just above the absorption edge of the element of interest. The resulting change in the anomalous-scattering factor of the element whose absorption edge was crossed results in a change in the diffraction pattern of the minerals containing the element. For example, if diffraction patterns are taken just below and just above the Fe absorption edge, the intensities of diffraction peaks from the Fe-containing minerals should change relative to those of the minerals which do not contain Fe. Subtraction of the two patterns, after appropriate normalization to the same 2Θ scale, results in the pattern of the Fe-containing minerals. Wood et al. (1986) demonstrated the feasibility of this technique using a mixture of Co_3O_4 and kaolinite, and diffraction patterns collected using $CoK\alpha$ and $CoK\beta$ radiation. The technique should be applicable to Fe- and Mn-containing minerals if the appropriate wavelengths are used. The advantage of the anomalous-scattering approach is that the same sample is used for both patterns. Although Wood et al. (1986) used a laboratory x-ray source, synchrotron x-ray sources will make this approach much more practical because it is easy to select monochromatic radiation of the desired wavelength.

SUMMARY

Differential x-ray diffraction takes advantage of modern computer-controlled x-ray diffractometers and computer data analysis to make it easier to study trace or poorly crystallized minerals in complex mineral mixtures such as those usually found in soils. Almost any differential-extraction procedure can be used for DXRD studies, as long as it extracts more than about 10 g kg^{-1} of well-crystallized minerals with sharp diffraction lines or more than about 150 g kg^{-1} of poorly crystallized minerals such as ferrihydrite or allophane. Spurious peaks caused by errors in alignment of the 2Θ scale between the untreated and treated patterns can usually be corrected mathematically. Spurious peaks caused by changes in preferred orientation between the untreated and treated pattern can sometimes be corrected by careful sample preparation. Most of the DXRD studies to date have concentrated on Fe-oxide minerals, but Mn oxides, carbonates, and poorly crystallized aluminosilicate minerals such as allophane and imogolite, are major soil-mineral groups which could be studied using DXRD.

ACKNOWLEDGMENTS

I thank A. S. Campbell, U. Schwertmann, and J. E. Amonette for their suggestions for improving the manuscript. I also thank L. P. van Reeuwijk, ISRIC, Wageningen, The Netherlands, for providing the Andisol sample. Purdue Agricultural Experimental Station Journal no. 12 810.

APPENDIX

A copy of a DXRD computer program incorporating the spline-shift routine can be obtained by sending a blank diskette to Darrell G. Schulze, Agronomy Department, Purdue University, West Lafayette, IN 47907. The program is written in BASIC and runs under the DOS operating system.

REFERENCES

Amarasiriwardena, D. D., L. H. Bowen, and S. B. Weed. 1988. Characterization and quantification of aluminum-substituted hematite-goethite mixtures by x-ray diffraction, and infrared and Mössbauer spectroscopy. Soil Sci. Soc. Am. J. 52:1179-1186.

Aniku, J. R. F., and M. J. Singer. 1990. Pedogenic iron oxide trends in a marine terrace chronosequence. Soil Sci. Soc. Am. J. 54:147-152.

Barrón V., and J. Torrent. 1987. Origin of red-yellow mottling in a Ferric Acrisol of southern Spain. Z. Pflanzenernähr. Bodenkd. 150:308-313.

Bish, D.L. 1994. Quantitative x-ray diffractrion analysis of soils. p. 267-295. In J. Amonette and LW. Zelazny (ed.) Quantitative methods in soil mineralogy. SSSA Misc. Publ. SSSA, Madison, WI.

Boero, V., and U. Schwertmann. 1989. Iron oxide mineralogy of Terra Rossa and its genetic implications. Geoderma 44:319-327.

Borggaard, O. K. 1979. Selective extraction of amorphous iron oxides by EDTA from a Danish sandy loam. J. Soil Sci. 30:727-734.

Borggaard, O. K. 1988. Phase identification by selective dissolution techniques. p. 83-98. In J. W. Stucki et al. (ed.) Iron in soils and clay minerals. D. Reidel, Dordrecht, the Netherlands.

Brady, K. S., J. M. Bigham, W. F. Jaynes, and T. J. Logan. 1986. Influence of sulfate on Fe-oxide formation: Comparisons with a stream receiving acid mine drainage. Clays Clay Miner. 34:266-274.

Bronger, A., J. Ensling, P. Gütlich, and H. Spiering. 1983. Rubification of Terra Rossae in Slovakia: Mössbauer effect study. Clays Clay Miner. 31:269-276.

Brown, G., and I. G. Wood. 1985. Estimation of iron oxides in soil clays by profile refinement combined with differential x-ray diffraction. Clay Miner. 20:15-27.

Bryant, R. B., N. Curi, C. B. Roth, and D. P. Franzmeier. 1983. Use of an internal standard with differential x-ray diffraction analysis for iron oxides. Soil Sci. Soc. Am. J. 47:168-173.

Campbell, A. S., and U. Schwertmann. 1984. Iron oxide mineralogy of placic horizons. J. Soil Sci. 35:569-582.

Campbell, A. S., and U. Schwertmann. 1985. Evaluation of selective dissolution extractants in soil chemistry and mineralogy by differential x-ray diffraction. Clay Miner. 20:515-519.

Curi, N., and D. P. Franzmeier. 1984. Toposequence of Oxisols from the Central Plateau of Brazil. Soil Sci. Soc. Am. J. 48:341-346.

Dixon, J. B., and S. B. Weed. 1989. Minerals in soil environments. 2nd ed. SSSA, Madison, WI.

Dudas, M. J., C. J. Warren, and G. A. Spiers. 1988. Chemistry of arsenic in acid sulfate soils of northern Alberta. Commun. Soil Sci. Plant Anal. 19:887-895.

Jackson, M. L. 1973. Soil chemical analysis--Advanced course. 2nd ed. M.L. Jackson, Univ. of Wisconsin, Madison, WI.

Jenkins, R. 1989. Instrumentation. p. 19-45. In D. L. Bish and J. E. Post (ed.) Modern powder diffraction. Reviews in mineralogy. Vol. 20. Miner. Soc. Am., Washington, DC.

Klug, H. P., and L. E. Alexander. 1974. X-ray diffraction procedures for polycrystalline and amorphous materials. 2nd ed. John Wiley & Sons, New York.

Kodama, H. and C. Wang. 1989. Distribution and characterization of noncrystalline inorganic components in Spodosols and Spodosol-like soils. Soil Sci. Soc. Am. J. 53:526-534.

Loveland, P. J. 1988. The assay for iron in soils and clay minerals. p. 99-140. In J. W. Stucki et al. (ed.) Iron in soils and clay minerals. D. Reidel, Dordrecht, the Netherlands.

McKenzie, R. M. 1977. Manganese oxides and hydroxides. p. 181-193. *In* J. B. Dixon and S. B. Weed (ed.) Minerals in soil environments. SSSA, Madison, WI.

Mehra, O. P., and M. L. Jackson. 1960. Iron oxide removal from soils and clays by a dithionite-citrate system buffered with sodium bicarbonate. p. 317-327. *In* A. Swineford (ed.) Clays and clay minerals. Proc. 7th Natl. Conf., Washington, DC.. 1958. Pergamon Press, New York.

Palmieri, F. 1986. A study of a climosequence of soils derived from volcanic rock parent material in Santa Catarina and Rio Grande do Sul states, Brazil. Ph.D. diss. Purdue Univ., West Lafayette, IN (Diss. Abstr. DA8709846).

Rich, C. I, and R. I. Barnhisel. 1977. Preparation of clay samples for x-ray diffraction analysis. p. 797-808. *In* J. B. Dixon and S. B. Weed (ed.) Minerals in soil environments. SSSA, Madison, WI.

Santana, D. P. 1984. Soil formation in a toposequence of Oxisols from Patos de Minas region, Minas Gerais state, Brazil. Ph.D. diss. Purdue Univ., West Lafayette, IN (Diss. Abstr. DA8423422).

Schulze, D. G. 1981. Identification of soil iron oxide minerals by differential x-ray diffraction. Soil Sci. Soc. Am. J. 45:437-440.

Schulze, D. G. 1982. The identification of iron oxides by differential x-ray diffraction and the influence of aluminum substitution on the structure of goethite. Ph.D. diss. Institut für Bodenkunde, Technische Universität München, 8050 Freising-Weihenstephan, German Federal Republic (Diss. Abstr. DA8406070).

Schulze, D. G. 1986. Correction of mismatches in 2Θ scales during differential x-ray diffraction. Clays Clay Miner. 34:681-685.

Schwertmann, U. 1964. Differenzierung der Eisenoxide des Bodens durch Extraktion mit Ammoniumoxalat-Lösung. Z. Pflanzenernähr. Bodenkd. 105:194-202.

Schwertmann, U., E. Murad, and D. G. Schulze. 1982. Is there Holocene reddening (hematite formation) in soils of axeric temperate areas? Geoderma 27:209-223.

Schwertmann, U. and R. M. Taylor. 1989. Iron oxides. p. 379-438. *In* J. B. Dixon and S. B. Weed (ed.), Minerals in soil environments. 2nd ed. SSSA, Madison, WI.

Taylor, R. M., and R. M. McKenzie. 1966. The association of trace elements with manganese minerals in Australian soils. Aust. J. Soil Res. 1:79-90.

Torrent, J. 1987. Rapid and slow phosphate sorption by Mediterranean soils: Effect of iron oxides. Soil Sci. Soc. Am. J. 50:78-82.

Torrent, J. and A. Cabedo. 1986. Sources of iron oxides in reddish brown soil profiles from calcarenites in southern Spain. Geoderma 37:57-66.

Torrent, J., U. Schwertmann, H. Fechter, and F. Alferez. 1983. Quantitative relationships between soil color and hematite content. Soil Sci. 136:354-358.

Torrent, J., U. Schwertmann, and D. G. Schulze. 1980. Iron oxide mineralogy of some soils of two river terrace sequences in Spain. Geoderma 23:191-208.

Wada, K. 1989. Allophane and imogolite. p. 1051-1087. *In* J. B. Dixon and S. B. Weed (ed.) Minerals in soil environments. 2nd ed. SSSA, Madison, WI.

Wang, H. D., G. N. White, F. T. Turner, and J. B. Dixon. 1993. Ferrihydrite, lepidocrocite and goethite in coatings from east Texas vertic soils. Soil Sci. Soc. Am. J. (In press.)

Wood, I. G., L. Nicholls, and G. Brown. 1986. X-ray anomalous scattering difference patterns in qualitative and quantitative powder diffraction analysis. J. Appl. Cryst. 19:364-371.

Yoshinaga, N., Y. Kato, and M. Nakai. 1989. Mineralogy of red- and yellow-colored soils from Thailand. Soil Sci. Plant Nutr. 35:181-205.

14 Quantification of Allophane and Imogolite

R. A. Dahlgren
University of California
Davis, California

The identification of short-range-order aluminosilicate materials in a wide variety of soils has increased the need for techniques to quantify their abundance. The most common short-range-order aluminosilicate materials, allophane and imogolite, have been detected in a wide range of soil environments, but are most commonly associated with Andisols and Spodosols. These short-range-order materials are metastable with respect to crystalline minerals such as kaolinite or gibbsite (Percival, 1985; Dahlgren & Ugolini, 1989), but they have been observed to persist for long periods of time (Steven & Vucetich, 1985; Parfitt, 1990a). Short-range-order materials are known to play an important role in the surface chemistry of the soil because of their large specific-surface areas and high proportion of reactive sites (Theng et al., 1982; Wada, 1989) and their significance has been recognized in soil classification. Accumulations of these materials in soils have been proposed as diagnostic properties for the Andisol and Spodosol soil orders (Soil Survey Staff, 1990).

To better understand the significance of allophane and imogolite in soils, quantification techniques must be employed to determine their abundance. While no individual method currently in use is ideal for quantifying short-range-order aluminosilicates, semiquantitative estimates can be made using selective-dissolution techniques, infrared (IR) spectroscopy, and differential thermal analysis (DTA), which is restricted to imogolite. Appropriate standard materials must be employed for the latter two methods. Other methods, such as differential x-ray diffraction (DXRD) and nuclear magnetic resonance (NMR) spectroscopy, have been able to provide semiquantitative estimates of imogolite concentrations when present in significant amounts (Barron et al., 1982; Campbell & Schwertmann, 1985). All current techniques are prone to interferences from other soil materials; however, care can be taken to minimize these interferences.

Three categories of short-range-order aluminosilicates (Parfitt, 1990a) will be emphasized in this chapter: imogolite, the Al-rich allophanes, and Si-rich allophanes. Imogolite consists of bundles of well-defined fine tubes with inner and outer diameters of 1.0 and 2.0 nm, respectively, and has an Al/Si molar ratio of 2:1. The external surface of the tube is composed of a gibbsite-like structure whereas the interior of the tube consists of isolated SiO_4 groups each coordinated with three Al atoms in the gibbsite-like sheet (Cradwick et al., 1972). The atomic arrangement in the imogolite tubes is regular along the axis although the diameters of the tubes can vary. Some randomness is also involved in the arrangement of tube units to form thread-like bundles. The imogolite structure has longer-range order than allophane.

Imogolite is commonly found in association with allophane. The Al-rich allophanes (also termed protoimogolite allophanes or imogolite-like allophanes) are related to imogolite by having the same local atomic arrangement and chemical composition (Al/Si = 2:1); however, morphologically they consist of many discrete particles having the shape of hollow spherules with outside diameters of 3.5 to 5 nm. The Si-rich allophanes with an Al/Si ratio close to 1:1 differ chemically and structurally from imogolite. They retain the hollow-spherule morphology of Al-rich allophanes while the atomic arrangement indicates a portion of the Si is polymerized (Parfitt et al., 1980). The NMR spectra indicate that Si-rich allophane contains Si-O tetrahedra bonded on the inside and outside surfaces of the Al-O, OH octahedral sheet, and that the proportion of the outside Si-O tetrahedra increases with decreasing Al/Si ratios.

Allophanes with Al/Si ratios between one and two have been proposed based on the results of chemical dissolution. The IR spectra of some of these allophanes show the presence of Al-rich allophane, together with a polymerized silicate structure characteristic of Si-rich allophane. This may indicate that allophane phases with Al/Si ratios between 1:1 and 2:1 consist of mixtures containing various proportions of Al- and Si-rich allophanes rather than a single phase of allophane (Parfitt et al., 1980; Parfitt, 1990a). In addition, allophanes with Al/Si ratios >2 and <1 may be present but have not been isolated (Wada, 1989). The incorporation of significant concentrations of Fe in natural allophanes and imogolite has not been reported.

This chapter outlines available quantification methods for short-range-order aluminosilicate materials, describes the basic principles of instrumentation and methodology, and discusses their value and potential limitations in various aspects of mineralogical investigation. Quantification techniques examined in detail in this chapter are: (i) selective dissolution, (ii) IR spectroscopy, and (iii) DTA. Other potentially useful techniques such as DXRD, NMR, and transmission electron microscopy (TEM) are briefly discussed.

PRINCIPLES AND TECHNIQUES

Selective Dissolution

Most of our current data on concentrations of allophane and imogolite in soils are based on selective chemical dissolution. Allophane and imogolite are differentiated from other poorly ordered inorganic phases and from organically bound Al based on their relative resistance to dissolution by various chemical reagents. These selective-chemical methods depend on the high degree of structural disorder and large specific-surface area of short-range-order materials, which results in a higher dissolution rate than for crystalline minerals. A high ratio of dissolution reagent to sample weight ensures more efficient dissolution and avoids saturation of the solution with dissolved products. These extractions can be performed on the untreated <2-mm soil fraction or on the clay-size fraction. Following dissolution, the extracted Al and Si (optionally Fe) may be determined by ultraviolet/visible spectroscopy, atomic absorption spectroscopy, or inductively coupled plasma optical emission spectrometry (ICP).

A combination of selective-chemical methods is generally necessary to distinguish short-range-order aluminosilicates from other phases dissolved by a given reagent. The most widely used combination is acid-oxalate dissolution to assess the total amounts of Al and Si in noncrystalline weathering products, and pyrophosphate dissolution to determine the amount of Al in organically complexed forms. The difference between these two extracts is then used to estimate the concentrations of allophane and imogolite, along with the Al/Si molar ratio of the materials.

Pyrophosphate and Similar Extractants

Theory. Pyrophosphate solutions (pH 10, 12-16 h) have been used for a number of years to extract organic compounds and associated Al and Fe from soils (Alexsandrova, 1960). This method has been further developed and tested by McKeague (1967) and Bascomb (1968) to determine its effectiveness in numerous types of soil. The mechanism responsible for release of metals from humic complexes involves competitive binding of metals by the pyrophosphate ligand. The high pH of the extracting solution and the monovalent cation associated with pyrophosphate (usually Na) help promote the dispersion of organometal complexes from the solid phase. Once the organometal complexes are in solution, the metals are extracted by complexation with pyrophosphate ligands in a reaction driven by the high concentration of pyrophosphate (0.1 M).

Both the monomeric and hydroxy-polymeric forms of Al react with organic matter to yield stable Al organocomplexes (McLean, 1976). Higashi (1983) showed that pyrophosphate extracted Al quantitatively from synthetic organocomplexes; however, the effectiveness of pyrophosphate in extracting Fe from synthetic complexes was much lower. In contrast to Al-humus complexes, Fe has a greater stability in oxides and oxyhydroxide structures than in humic complexes (Wada & Higashi, 1976; Goodman, 1987). These oxyhydroxides (typically ferrihydrite and goethite) are commonly associated with organic matter and become peptized during the pyrophosphate extraction procedure, adding to Fe concentrations in the extracted solution (Kassim et al., 1984; Madeira & Jeanroy, 1984).

In addition to pyrophosphate, EDTA (pH 7, 1 h) and alkaline tetraborate (pH 9.5, 16 h) have been used to estimate the concentrations of Al complexed by humus (Farmer et al., 1980; Higashi & Shinagawa, 1981; Higashi et al., 1981). Some soil components are also peptized by alkaline tetraborate and caution is required in interpreting results (Madeira & Jeanroy, 1984). Comparison of the quantity of Al extracted by the three reagents generally follows the order: pyrophosphate > EDTA > alkaline tetraborate (Madeira & Jeanroy, 1984). Based on an extensive evaluation of these reagents, McKeague and Schuppli (1985) recommended the use of pyrophosphate for estimating the organically bound Al. None of the methods appear effective at specifically extracting Fe from organometal complexes.

Recommended Method. Procedures for the common selective-dissolution techniques vary somewhat with regard to soil/solution ratios and shaking time employed. The methods outlined are either obtained from standard procedure manuals or reported from the results of studies which have determined the optimum conditions for a particular procedure.

Pyrophosphate (0.1 M, pH 10) is typically used at a soil/solution ratio of 1:100. Samples are shaken for 16 h and then centrifuged at 20,000 rpm for 30 min after adding 2 mL of a 0.1% flocculating agent solution (e.g., Superfloc, American Cyanamid Co., Wayne, NJ) per 100 mL of extractant (McKeague, 1967; McKeague & Schuppli, 1982).

Specific methods for determining Al, Fe, and Si concentrations in all selective-dissolution extracts are given in Searle and Daly (1977) and Jackson et al. (1986). The ICP atomic emission spectrometry is an alternative method that provides excellent results with a minimum of sample pretreatment (Soltanpour et al., 1982).

Acid-Oxalate and Similar Extractants

Theory. Acid-oxalate solutions were first used by Tamm (1922) and later modified by Schwertmann (1959, 1964) to extract amorphous oxides and hydrous oxides from soils. Acid-oxalate has been shown to completely dissolve allophane and imogolite from a variety of soils. These results have been verified by IR spectroscopy and electron microscopy. In addition to allophane and imogolite, acid-oxalate is known to dissolve ferrihydrite, poorly crystalline lepidocrocite, maghemite, magnetite, and Al associated with humus, as well as some of the Al from the hydroxy-Al interlayer of 2:1 layer silicates (Fordham & Norrish, 1983; Shoji & Fujiwara, 1984; Schwertmann, 1985; Parfitt & Childs, 1988). The method relies on low pH and on the complexing ability of the oxalate ligand for Al and Fe to dissolve short-range-order materials.

Acid-oxalate extractable Al (Al_o) and Si (Si_o), together with pyrophosphate extractable Al (Al_p), are used to estimate the contents of allophane and imogolite and the Al/Si molar ratio of the allophanic material (Childs et al., 1983; Farmer et al., 1983; Parfitt & Wilson, 1985). The Al/Si molar ratio is estimated using the formula $(Al_o - Al_p)/Si_o$, which corrects for the Al dissolved from humic complexes by the acid-oxalate treatment. Aluminum-rich allophane and imogolite typically have an Al/Si ratio close to 2.0 and an ideal chemical composition of $SiO_2 \cdot Al_2O_3 \cdot 2H_2O$. Based on this formula, pure Al-rich allophane and imogolite contain approximately 140 g Si per kilogram. Therefore, an estimate of the allophane and imogolite concentrations can be made by dividing the measured concentration of Si_o by 0.14, or equivalently, by multiplying by a factor of seven. This provides the best estimate of allophane and imogolite from dissolution data (when the Al/Si ratio is close to 2.0) because the Si_o originates almost exclusively from allophanic materials. Parfitt

(1990a) provides a listing of the appropriate factors for converting Si_o values to soil allophane and imogolite percentage for a range of Al/Si ratios from 1.0 to 3.5 (Table 14-1). For Al/Si ratios other than one or two, care must be taken when extrapolating to the allophane and imogolite concentration. The material dissolved during the treatment may consist of mixtures of Al- and Si-rich phases of allophane as well as Al- and Si-bearing mineral phases other than allophane and imogolite.

Acid-hydroxylamine solutions have also been shown to be effective for dissolving allophane and imogolite from soils (Lee et al., 1989). A comparison study by Lee et al. (1989) showed that, in general, the amounts of Al and Si extracted by acid-hydroxylamine were not significantly different from those extracted by acid-oxalate; however, hydroxylamine extracted significantly more Si than acid-oxalate in some samples with low extractable Si levels. A similar study by Wang et al. (1987) found approximately twice as much Si dissolved by hydroxylamine as by oxalate in the B horizons of Spodosol and Spodosol-like soils. The major advantage of the hydroxylamine method is that it does not dissolve magnetite. However, this is not a critical factor for estimating allophane and imogolite concentrations because dissolution of magnetite does not affect extractable Al concentrations. The acid-oxalate extraction is currently the methodology of choice for estimating allophane and imogolite concentrations.

Recommended Method. The acid-oxalate method that follows is adapted from Schwertmann (1964). Optimum conditions for extraction of allophane and imogolite with acid-oxalate solutions have been evaluated by Higashi and Ikeda (1974) and Parfitt (1989). The acid-oxalate solution is prepared by adding 700 mL of 0.2 M ammonium oxalate to 535 mL of 0.2 M oxalic acid. The pH is adjusted to 3.0 by addition of either the oxalate or oxalic acid solution. Similar results are obtained whether ammonium oxalate or sodium oxalate is used. Since ammonium oxalate is more soluble, it is easier to prepare than sodium oxalate. A soil/solution ratio of 1:100 should be used and the samples shaken for 4 h in the dark. Parfitt (1989) showed that extractable Al and Si concentrations increased with shaking time up to 2 h and remained constant as extraction times were increased beyond 4 h. Therefore, it may be possible to reduce the time of shaking on soils that contain low concentrations of allophane and imogolite. To ensure complete dissolution of allophane and imogolite, the soil/solution ratio should be increased to 1:200 if extractable Al exceeds 50 g kg^{-1} soil. After shaking, 2 mL of a 0.1% flocculating agent solution (e.g., Superfloc) is added per 100 mL of extractant and centrifuged for 10 min at 2000 rpm. To ascertain that the acid-oxalate dissolution was complete, a subsample of soils should be examined by electron microscopy to verify the complete removal of allophane and imogolite.

Dithionite-Citrate

Theory. A solution of sodium dithionite (also known as sodium hydrosulfite or sodium hyposulfite), sodium citrate, and sodium bicarbonate (DCB) is a common extractant for removal of both noncrystalline and crystalline Fe oxides (Mehra & Jackson, 1960). The mechanism of dissolution for Fe involves reduction of Fe oxides by the dithionite and chelation of the solubilized

Table 14-1. Conversion factors used with Si_o to estimate the concentrations of allophane/imogolite as a function of Al/Si atomic ratio (from Parfitt, 1990a).

Al/Si ratio	Factor	Al/Si ratio	Factor
1.0	5	2.5	10
1.5	6	3.0	12
2.0	7	3.5	16

Fe and Al by the citrate. Sodium bicarbonate is added to buffer the solution pH near 7.3 to obtain the optimum extraction effectiveness. The DCB treatment may dissolve a small fraction of allophane and imogolite, especially if these materials have very poor structural order (Shoji & Ono, 1978; Farmer et al., 1983). Wada and Greenland (1970) were able to dissolve a noncrystalline aluminosilicate using DCB and termed this material "allophane-like constituents." No distinct material corresponding to allophane-like constituents has been isolated to determine whether it represents a phase separate from, or a part of, allophane (Henmi & Wada, 1976; Parfitt et al., 1980; Wada & Wada, 1980).

The DCB-extractable Al (Al_d) may be combined with Al_o and Si_o to provide an estimate of the Al/Si ratio for allophanic materials. Like pyrophosphate, the DCB treatment dissolves organically complexed Al by forming Al-citrate complexes. Similar to acid-oxalate, DCB removes some of the Al from the hydroxy-Al interlayer of 2:1 layer silicates and any Al contained in Fe oxyhydroxides such as ferrihydrite, lepidocrocite, and goethite. The inability of pyrophosphate extractions to account for these latter two sources of Al may lead to high estimates of the Al:Si ratio of allophanic materials when using the $(Al_o - Al_p)/Si_o$ formula. For example, Dahlgren and Ugolini (1991) obtained more realistic estimates of the Al/Si ratio of imogolite using $(Al_o - Al_d)/Si_o$ in Spodosols formed in tephra. The Al/Si ratio estimates using Al_p ranged from 2.3 to 3.2 while estimates made substituting Al_d for Al_p ranged from 2.0 to 2.2. These Spodosol B horizons contained relatively large concentrations of ferrihydrite and an abundance of hydroxy-Al interlayered 2:1 layer silicates, which are the likely sources of Al contributing to the high Al/Si ratio of allophane and imogolite estimated using Al_p rather than Al_d. When significant quantities of Al-substituted crystalline Fe-oxyhydroxides (e.g., lepidocrocite and goethite) are present, however, the use of Al_d in place of Al_p may provide low estimates of the Al/Si atomic ratio because these components are dissolved by DCB but not by acid-oxalate. In addition, Iyengar et al. (1981) found DCB to be generally more effective in removal of hydroxy-Al interlayer material than acid-oxalate; however the differences observed would not significantly affect the calculated Al/Si atomic ratio of the allophanic material. Regardless of the formula used, caution must be taken in estimating the Al/Si atomic ratio of allophane and imogolite. This is especially true with low levels of extractable Al and Si, where division by a small number can result in large errors.

Recommended Method.   The DCB method is that reported by Jackson et al. (1986). To 5 g of soil, 40 mL of 0.3 M sodium citrate and 5 mL of 1 M $NaHCO_3$ are added. The suspension is heated in a water bath to 75 to 80 °C. When the suspension reaches the desired temperature, 1 g of $Na_2S_2O_4$ is added, and the suspension is immediately stirred for 1 min, followed by intermittent stirring for 5 min. A second 1-g portion of $Na_2S_2O_4$ is added and occasional stirring continued for another 10 min. The suspension is allowed to cool and then centrifuged for 10 min at 2000 rpm. If dispersion persists, 10 mL of saturated NaCl solution should be added to the suspension prior to centrifugation. The treatment is repeated until the residue is light grey, but two treatments are often sufficient.

Sequential Selective Dissolution

Sequential selective-dissolution techniques have also been used, primarily on the clay-size fraction, for quantification of allophane and imogolite (Shoji & Saigusa, 1977; Shoji & Ono, 1978; Yamada et al., 1978; Dahlgren & Ugolini, 1991). For example, the clay fraction is first treated with DCB to dissolve any "allophane-like constituents" (if they exist), Al associated with Fe-oxides and humic substances, and a portion of the hydroxy-Al interlayer of 2:1 minerals. The residue from this treatment is then treated with acid-oxalate to dissolve the allophane and imogolite fraction. The residue from acid-oxalate can be further treated with boiling NaOH to estimate the amount of opaline Si and poorly crystalline halloysite (Shoji & Saigusa, 1977). Sequential extractions are typically coupled with analysis by IR spectroscopy, electron microscopy, and x-ray diffraction (XRD) to help verify the source of the extractables. It is also possible to perform sequential extractions (pyrophosphate, DCB, and acid-oxalate) on the <2-mm soil fraction and eliminate the soil pretreatments necessary to isolate the clay fraction. However, verification of the selective-dissolution treatments by IR spectroscopy, electron microscopy, and XRD would still require the isolation of the clay fraction for analysis.

Selective Dissolution Summary

Table 14-2 summarizes the effectiveness of pyrophosphate, DCB, and acid-oxalate for dissolving various soil components. Pyrophosphate dissolves only small amounts of Al and Si from allophane and imogolite, whereas from 5 to 20% of the allophane and imogolite present are dissolved by the DCB treatment (Shoji & Ono, 1978; Parfitt & Henmi, 1980; Farmer et al., 1983). Acid-oxalate treatment, on the other hand, results in virtually complete removal of allophane and imogolite from most samples, as confirmed by electron microscopy.

Advantages and Limitations of Method

Due to the potential interferences involved with all of the proposed quantification methods, the use of more than one method is recommended. Selective-dissolution techniques are routinely used in soil characterization studies and provide a good first estimate of the quantities of allophane and imogolite present. Based on the results of selective-dissolution techniques, a decision can be made to independently verify allophane and imogolite estimates using IR spectroscopy or DTA.

Selective-dissolution techniques are simple to apply and a large number of samples can be processed rapidly. Dissolution techniques provide an estimate of the chemical composition (Al/Si ratio) of the allophane and imogolite that is necessary not only for quantification, but also for distinguishing Si-rich allophane (Al/Si, 1:1) from Al-rich allophane or imogolite (Al/Si, 2:1). The Al/Si ratio can be estimated using either $(Al_o - Al_p)/Si_o$ or $(Al_o - Al_d)/Si_o$, but caution must be employed. Ratios below one or considerably greater than two should be used with caution because allophane and imogolite structures with these compositions have not been isolated from soils. The amount of allophane and imogolite in the sample is estimated by multiplying Si_o by a factor that accounts for the Al/Si ratio of the material (Table 14-1). For allophane and imogolite with an Al/Si ratio of 2:1, the factor is seven.

Table 14-2. Effectiveness of pyrophosphate, oxalate, and dithionite reagents for dissolving various soil components: none, <1%; poor, <10%; moderate, 10-80%; good, >80% (adapted from Parfitt & Childs, 1988).

Soil Component	Pyrophosphate pH 10, 0.1 M	Oxalate pH 3, 0.2 M	Dithionite-citrate
Al humus	Good[A,B]	Good[C]	Good[A]
Ferrihydrite	Some dispersion[C]	Good[D,E]	Good[D]
Allophane	Poor[F,G]	Good[F,G]	Poor-mod.[F,H]
Imogolite	Poor[H]	Good[F,G]	Poor-mod.[F,H]
Gibbsite	None[C]	Poor[C]	None[C]
Halloysite	None[C]	None[I]	None[I]
Opaline Si	None[K]	None[K]	None[K]
Layer silicates	None[K]	None[K,L]	Poor[K]
Hydroxy-interlayer materials	None	Moderate[M]	Moderate[M]

[A] Wada and Higashi, 1976.
[B] McKeague and Sheldrick, 1977.
[C] Parfitt and Childs, 1988.
[D] Schwertmann and Taylor, 1977.
[E] Schwertmann et al., 1982.
[F] Parfitt and Henmi, 1982.
[G] Farmer et al., 1983.
[H] Parfitt and Henmi, 1980.
[I] Theng et al., 1982.
[K] Wada, 1989.
[L] Bhattacharyya and Ghosh, 1986.
[M] Iyengar et al., 1981.

Acid-oxalate-extractable Si concentrations less than 1 to 2 g kg^{-1} (i.e., 7-14 g allophane and imogolite per kilogram soil) should not be considered significant due to the lack of specificity by acid-oxalate, which may attack the surfaces of crystalline minerals. This problem is compounded if the samples have been ground prior to analysis, resulting in the creation of fresh mineral surfaces. High concentrations of ferrihydrite may also contribute to Si$_o$ and lead to an overestimation of allophane and imogolite concentrations. Ferrihydrite is dissolved by acid-oxalate and may contain 20 to 60 g Si per kilogram strongly adsorbed to its surface (Carlson & Schwertmann, 1981). For example, consider a soil containing 50 g ferrihydrite per kilogram that has 50 g Si per kilogram adsorbed to the surface of the ferrihydrite. The oxalate extraction would yield 2.5 g Si$_o$ per kilogram soil (0.5 x 0.5 = 2.5) and, given a conversion factor of seven corresponding to an Al/Si atomic ratio of two (see Table 14-1), would overestimate the allophane and imogolite concentration by 17.5 g kg^{-1} soil (2.5 x 7 = 17.5). This potential source of error is of minor concern for soils that contain more than 100 g allophane and imogolite per kilogram.

Infrared Spectroscopy

Theory

The IR spectroscopy is a rapid and economical method for identification of allophane and imogolite and can be made quantitative with suitable calibration. Absorption of IR radiation by minerals is a function of the atomic masses and the lengths, strengths, and force constants of interatomic bonds in the structure of the material. The IR absorption is also controlled by the overall symmetry of the unit cell and the local site symmetry of each atom within the unit cell. Since an IR spectrum is dependent on the interatomic forces in a particular sample, it is a sensitive probe of the microscopic structure and bonding within minerals. Detailed theoretical treatments of vibrational spectroscopy applied to inorganic solids may be found in Hadni (1967) and Turrell (1972). A number of excellent reviews on the application of IR methods to the study of clay minerals include Farmer and Russell (1967), Farmer et al. (1968), Farmer (1968, 1974), White (1971, 1977), White and Roth (1986), and Russell (1987).

The IR spectrometers are classified into two groups: (i) sequential dispersive spectrometers and (ii) multiplex nondispersive interferometer spectrometers, more commonly referred to as Fourier-transform IR (FT-IR) spectrometers. In dispersive units, the detector monitors the intensity of narrow band-widths of radiation, dispersed by diffraction gratings and/or prisms in a monochromator. In double-beam instruments, the beam produced by the source is chopped and passed through two compartments, one containing the sample, and the other a reference. The radiation passes through the sample and is dispersed by a system of gratings and filters that are a part of the monochromator. The slits of the monochromator select a narrow resolution element (frequency range) and the energy of each resolution element is measured by a detector. The spectrum is collected sequentially in time with each resolution element scanned successively. The scan time is determined by the time required for the instrument to view each resolution element, multiplied by the number of these elements in the spectrum. A medium resolution scan from 4000 to 200 cm^{-1} can be completed in about 12 to 30 min, although high resolution scans may require several hours. The power and versatility of a dispersive instrument can be greatly enhanced by interfacing it with a microcomputer to provide datalogging capabilities.

For adequate characterization of allophane and imogolite, the IR spectrum should be recorded over the range 4000 to 250 cm^{-1}, with the 348 cm^{-1} absorption band being the primary band for allophane and imogolite quantification. A major limitation of many commercially available spectrometers is they are not designed for analysis below 400 cm^{-1}. Without special features, the relatively low energy of sources generally used for far-IR spectrometry severely limits the performance.

In nondispersive FT-IR spectrometers, the detector continuously monitors the full wavenumber range of radiation emitted by the IR source. The FT-IR spectrometers are interferometers and use a microcomputer to transform their output, an interferogram, into an absorption spectrum. A single scan usually takes 0.25 to 1 s and contains all of the spectral information. Desired signal-to-noise ratios are then obtained by averaging multiple scans until a satisfactory spectrum is obtained. Therefore, FT-IR spectrometers have a definite advantage for quantitative work due to the speed and precision with which they can acquire data and the computing capabilities to perform difference spectra.

The IR spectroscopy has been used to estimate the content of allophane and imogolite in the clay-size fraction based on the IR-absorption band at 348 cm^{-1} (Farmer et al., 1977). However, a number of conditions must be met in order to use IR spectroscopy as a quantitative technique (Russell, 1987): (i) the particle size of the sample should be <2 μm in order to ensure maximum absorption and minimum scattering by the sample; (ii) the samples must be weighed accurately and sample transfer must be precise during preparation of the sample for analysis; (iii) the sample must be uniformly dispersed in the dispersion media and the pressed disc must be of uniform quality; (iv) either one or more IR absorption bands must be sufficiently isolated from bands of other components to prevent interferences, or corrective methods for dealing with interferences must be possible; and (v) an appropriate standard material must be available to produce a calibration curve.

The most difficult problems encountered in quantification of allophane and imogolite are spectral interferences from other minerals and difficulty in selecting an appropriate standard. Allowances must be made for interference by aluminous dioctahedral layer silicates which also absorb in the 348 cm^{-1} region. These corrections may be accomplished through the use of differential IR spectroscopy, in which spectra of the sample taken before and after acid-oxalate extraction or thermal dehydroxylation are subtracted to obtain a difference spectrum. An appropriate standard material must also be used to prepare a calibration curve. This requirement is problematic due to the range in order and chemical composition displayed by allophane and imogolite. For example, 1 mg of allophane (Al/Si of 2:1) has a relative absorbance at 348 cm^{-1} of approximately 0.17, while 1 mg of imogolite with the same chemical composition has a relative absorbance of approximately 0.27 (Farmer et al., 1977; Parfitt & Henmi, 1982). Thus, a knowledge of the chemical composition and morphology of the allophanic material is required before selection of an appropriate standard material is possible. The chemical composition can be estimated using selective-dissolution techniques while differentiation of allophane from imogolite requires the use of TEM or DTA analysis. Nevertheless, with proper precautions, estimates of allophane and imogolite concentrations obtained by IR methods show good agreement with results obtained by selective-dissolution techniques (Parfitt & Henmi, 1982).

Recommended Methods

Clay-size materials used for IR analysis should be free of organic matter to maximize the dispersion of the clay fraction and to minimize complications in interpreting the IR spectrum. However, complete removal of organic matter, from the interference perspective, is not required because the IR absorption bands from organic matter are in a different spectral range and do not interfere with the allophane and imogolite spectrum. For routine analysis, the clay material may be mixed thoroughly with KBr as the dispersing media. Other dispersing media, such as petroleum jelly or nujol (a liquid paraffin) may also be used in the far-IR region where the absorption peak for allophane and imogolite quantification occurs. Grinding of the clay-size material should be minimized, and unknown samples should be prepared identically to standard samples to minimize the effects of grinding on the analysis. Henmi and Yoshinaga (1981) showed a complete loss of the 348 cm^{-1} IR absorption band after only 4 min of dry grinding for imogolite. Their data indicate that grinding caused a breakdown of Si-O-Al linkages and a subsequent condensation of the SiO_4 tetrahedra that were released. A 1:100 dilution of clay in oven-dried KBr should be thoroughly mixed in a stainless-steel capsule containing ball bearings using a vibrating mixer (e.g., Wig-L-Bug mixer, Crescent Dental Manufacturing, Lyons, IL) or gently in a conventional agate mortar and pestle (usually for 1-5 min). One-hundred milligrams of the material is then pressed in an evacuated die to give a 13-mm diam. disc. The disc should be heated overnight at 150 °C and allowed to cool in a desiccator to remove adsorbed water, which absorbs radiation in the IR region. The spectrum is then recorded and the absorption peak at 348 cm^{-1} used for quantification (Farmer et al., 1977; Farmer et al., 1979).

To correct for interferences from aluminous dioctahedral layer-silicate clays, which also absorb IR radiation near 348 cm^{-1}, differential IR spectroscopy can be used. Allophane and imogolite can be dissolved from the sample using the acid-oxalate extraction (Fig. 14-1) or dehydroxylated by heating at 350 °C overnight (Fig. 14-2). When using the acid-oxalate treatment to removed allophane and imogolite, the weight loss associated with the extraction must be carefully measured and the amount of clay added to the KBr correspondingly adjusted to account for the weight loss. An internal standardization procedure, such as deuterated brucite, may also be used with the KBr disc technique to compare the ratio of band absorbances of the unknown vs. that of the internal standard with the calibration curve. With a dual beam spectrophotometer, a difference spectrum can be obtained by placing a second sample, from which the allophane and imogolite have been removed by acid-oxalate extraction or thermal dehydroxylation, in the reference beam. The spectrum obtained should reflect only the contribution of the allophane and imogolite to the 348-cm^1 absorption band.

A more convenient method is to record the spectrum of the sample, heat the KBr disc to 350 °C to dehydroxylate the allophane and imogolite, and then record the spectrum of the dehydroxylated sample. With computer datalogging capabilities, the two spectra can be computer subtracted to give a difference spectrum. To manually subtract the two spectra, draw the baseline as nearly as possible where the pen-tracing would go if the absorption band were not present. Integration of the absorption band can be accurately performed by cutting out the area of the absorption band (or a copy of the original spectrum) and weighing the mass of the cut-out area on an analytical

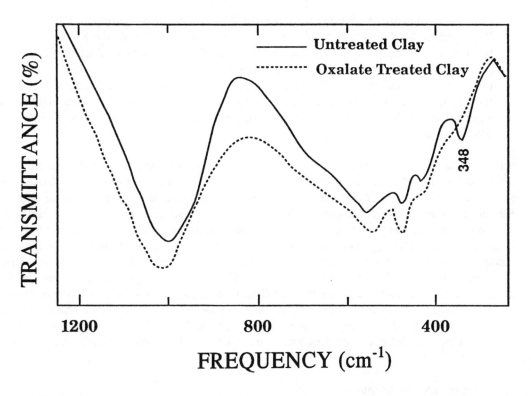

Fig. 14-1. IR spectrum of the untreated clay fraction and of the residual fraction following acid-oxalate treatment for the clay-size fraction from the Bs horizon of a tephritic Spodosol (Dahlgren & Ugolini, 1991).

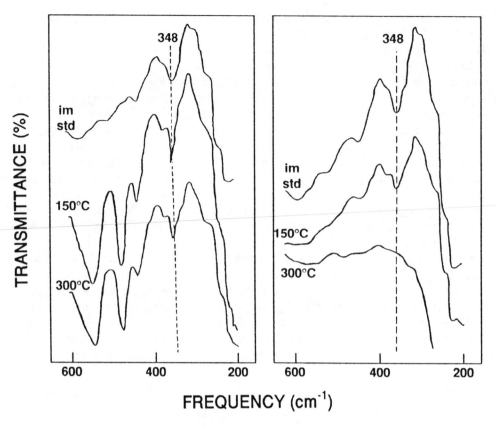

Fig. 14-2. Infrared spectrum showing the effects of dehydroxylation by heating to 350 ° C on the IR absorption band at 348 cm^{-1}. An imogolite standard material from the Kitakami pumice bed is shown for reference. A clay isolated from an Alfisol (left) was dominated by halloysite giving an absorption band at 348 cm^{-1} and gibbsite giving the shoulder at 370 cm^{-1}. Selective dissolution indicated traces of allophane and imogolite; however, heating did not significantly reduce the absorption band at 348 cm^{-1}. An Andisol clay (right) contained appreciable amounts of imogolite as shown by the disappearance of the 348 cm^{-1} peak following dehydroxylation.

balance. The area of the absorption peak is proportional to the concentration of allophane and imogolite present and can be compared to a calibration curve for quantification. Dehydroxylation has an advantage because it eliminates differences that may occur in the preparation of separate KBr discs. When analyzing samples on different days, dispersive IR spectrometers require both wavelength and intensity calibrations to verify that mechanical settings and power levels are identical to the preceding analysis. Instrument calibration can be performed by obtaining a spectrum of a standard material, such as a thin polystyrene film, and comparing the observed peak positions to the accepted standard values. A set of allophane and imogolite standards should also be analyzed on both days to assure reproducibility of the standard curve in day-to-day analysis.

A standard curve can be determined by adding varying amounts of the pure component to the dispersing medium and recording the absorption intensity at 348 cm^{-1}. Concentrations should be selected which yield band intensities within the range of 20 to 60% transmittance (0.2-0.7 absorbance).

Advantages and Limitations of Method

The IR spectroscopy provides a rapid method for semiquantitative analysis of allophane and imogolite in the clay-size fraction. The method requires only a small amount of material for analysis (1-5 mg). The degree of silicate-anion polymerization can also be estimated from IR absorption in the 900 to 1000 cm^{-1} region. A high degree of polymerization, such as in Si-rich allophanes, results in a higher Si-O stretching frequency (~ 1000 cm^{-1}) than for imogolite or Al-rich allophanes (~ 960 cm^{-1}) (Wada, 1989).

The major limitation of IR spectroscopy in allophane and imogolite quantification is obtaining an appropriate standard. Relative IR absorption values at 348 cm^{-1} for 1 mg of allophane range from 0.05 to 0.20 for various types of allophanes, while imogolite has an absorbance of approximately 0.27 (Farmer et al., 1977; Parfitt & Henmi, 1982). Therefore, a calibration curve must be prepared with an allophane or imogolite reference material that has similar IR absorption characteristics to that being quantified. Relatively pure samples of allophane and imogolite have been isolated from pumice beds in New Zealand and Japan and may serve as appropriate standards for quantification methods. Synthetically produced allophane and imogolite samples were shown by Farmer et al. (1979) to display similar IR absorption characteristics to natural samples and may also be used for standardization. Use of synthetic standards should be validated by selective dissolution procedures to determine the suitability of comparing natural samples to synthetic standards. The chemical composition of samples and standards can be compared using selective-dissolution analysis, while differentiation of Al-rich allophane from imogolite (both having Al/Si ratios of 2:1) requires use of TEM or DTA analysis. Concentrations of allophane and imogolite in samples that contain a mixture of the two materials cannot be accurately determined by IR spectroscopy due to their different absorptivities.

A second limitation of IR spectroscopy methods is the spectral interferences at 348 cm^{-1} arising from aluminous dioctahedral layer silicates. Table 14-3 shows the common soil minerals and their absorbance in the IR absorption band near 348 cm^{-1}. Allowances for small quantities of these minerals can be made using differential-IR methods, which involve the removal of allophane and imogolite by dehydroxylation or acid-oxalate treatment. Spectra showing the result of acid-oxalate treatment and dehydroxylation are shown in Fig. 14-1 and 14-2, respectively.

Thermal Analysis

Theory

Thermal analysis is actually a group of techniques (Karathanasis & Harris, 1994; see Chapter 12), in which physical properties of a substance and/or its reaction products are measured as a function of temperature, while the substance is subjected to a controlled temperature increase (Lombardi, 1980; Patterson & Swaffield, 1987). Imogolite concentrations have been determined using differential thermal analysis (DTA) (Aomine & Mizota, 1973; Parfitt, 1990b). The parameter measured in DTA is the difference in temperature between a sample and reference material, while both are subjected to a controlled temperature increase. The resulting DTA curve

Table 14-3. Absorbance of IR radiation near 350 cm^{-1} for 1 mg of clay in a 13-mm KBr disk, heated to 150 °C overnight (adapted from Farmer et al., 1977).

Mineral	Wavenumber	Absorbance
Imogolite	348	0.27
Allophane	348	0.05-0.17
Kaolinite	348	0.67
Halloysite	348	0.52
Montmorillonite	345	0.18
Illite	342	0.11
Gibbsite	370	1.20

indicates differences in temperature, between sample and reference material, occurring over a given temperature range. Temperature increases reflect exothermic reactions while temperature decreases indicate endothermic reactions occurring within the sample.

A quantitative determination of imogolite concentrations can be made by DTA; however, DTA analysis is not diagnostic for allophane. On heating, imogolite and allophane both lose physically adsorbed water at <200 °C, but the structural (hydroxyl) water in imogolite is more thermally stable than in allophane (Russell et al., 1969), being lost in a discrete endothermic reaction near 400 °C (Wada & Yoshinaga, 1968). Presumably, the rigidity of the tubular structure in imogolite results in its greater thermal stability compared with allophane.

Allophane and imogolite have similar DTA peaks showing a large endothermic peak between 50 and 300 °C which arises from removal of large amounts of adsorbed H_2O, and an exothermic peak at 900 to 1000 °C corresponding to the exothermic recrystallization to mullite and/or gamma alumina (Wada, 1989). In addition, imogolite gives a characteristic endothermic peak at 390 to 420 °C, which is due to dehydroxylation and results in about a 13% loss in weight (Yoshinaga & Aomine, 1962; MacKenzie et al., 1989). This peak is absent in the DTA patterns for most other minerals occurring in association with imogolite and thus can be used for quantification. The height and area of the endothermic peak show a linear relationship with imogolite concentrations and detection limits as low as 20 g kg^{-1} have been reported (Aomine & Mizota, 1973; Parfitt, 1990b).

Recommended Method

The DTA is not diagnostic for allophane and therefore provides an estimate of imogolite concentrations only. For quantification of imogolite, DTA is carried out using 50-mg samples placed in a platinum liner seated in the sample cup (Aomine & Mizota, 1973; Parfitt, 1990b). The temperature of the clay sample is compared to that of a reference sample (e.g., 50 mg of Al_2O_3) while being heated at a rate of 10 to 20 °C min^{-1}. Samples may be heated in either an air or N_2 atmosphere. The height or area of the 400 to 430 °C endothermic peak is measured from an extrapolated baseline and compared to a calibration curve. Grinding of the clay sample should be minimized to prevent the breakdown of the imogolite structure. Dry grinding of imogolite for more than 4 min resulted a complete loss of the 400 °C endotherm used to quantify imogolite (Henmi & Yoshinaga, 1981).

Advantages and Limitations of Method

Because DTA is appropriate for quantification of imogolite, but not for allophane, it may be used as a tool to quantitatively distinguish between these two minerals. Concentrations of imogolite can be obtained directly by DTA analysis with detection limits as low as 20 g kg^{-1}. Allophane concentrations and chemical composition can then be estimated by subtracting the imogolite content from concentrations obtained from selective dissolution.

Quantification is based on the height or area of the endothermic reaction peak near 400 °C (Fig. 14-3) (Parfitt, 1990b). The other minerals that commonly occur in association with imogolite and have DTA endotherms close to 400 °C are halloysite (520-570 °C), gibbsite (300-320 °C), and goethite (270 °C). Ferrihydrite has no endotherm above 200 °C (Henmi et al., 1980), while soil organic matter gives an exotherm close to 300 °C (Mackenzie, 1970). All of these endotherms are distinct from the imogolite endotherm and do not present any interference problems. The only potential interference appears to arise from an endotherm near 470 °C which is believed to be associated with the dehydroxylation of hydrated glass (Parfitt, 1990b). Interferences for a particular sample can be determined by removing imogolite from the sample using acid-oxalate and performing DTA on the residue.

Appropriate imogolite standards must be used to obtain a calibration curve for DTA. Because the chemical composition of imogolite is relatively consistent, it may be possible to use naturally occurring imogolite isolated from gel films which occur in a number of pumice beds in Japan

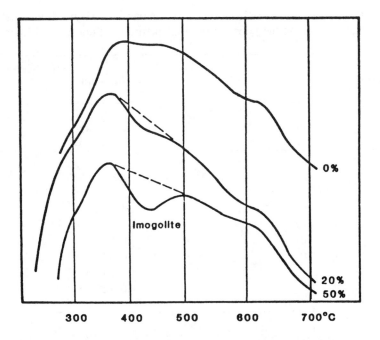

Fig. 14-3.   Differential thermal analysis curves for allophane and allophane mixed with 20%
imogolite and 50% imogolite (from Parfitt, 1990b; used by permission).

(Yoshinaga & Yamaguchi, 1970). Use of synthetically produced imogolite standards may also be
appropriate; however, a series of tests must first be performed to compare the thermal properties
of natural and synthetic forms of imogolite.

Other Techniques

Differential X-Ray Diffraction

Recent developments in differential x-ray diffraction (DXRD) techniques (Schulze, 1994; see
Chapter 13) make possible semiquantitative estimates of allophane and imogolite concentrations
when present in substantial quantities. In DXRD, an x-ray diffraction pattern of a sample from
which allophane and imogolite have been removed is subtracted from the pattern obtained before
their removal. Treatment of the clay material by acid-oxalate is the best choice for removal of
allophane and imogolite. A resistant mineral such as quartz may be used as an internal standard
to calculate the scale factor used to adjust the pattern being subtracted, in order to compensate for
any increase in intensity of the residual components. If the sample does not contain any suitable
internal standard, alumina may be added (Bryant et al., 1983). Use of an internal standard
minimizes spurious peaks through the application of an incorrect scale factor.

Diffraction data for allophane and imogolite are summarized in Table 14-4. The diffraction
peaks are very broad with individual peaks ranging from 0.2 to 0.8 nm in width. Even poorly
ordered materials like allophane and imogolite may vary in their degree of order, and this
variability may produce corresponding shifts in diffraction maxima and intensity. Therefore, the
difficulty in matching calibration standards to the allophane and imogolite in samples is a serious
limitation. Grinding of clay samples prior to x-ray analysis has also been shown to severely affect
the intensity of diffraction maxima. Due to the poorly ordered nature of allophane and imogolite,
detection limits for these components by DXRD are on the order of 200 g kg^{-1} clay (Campbell &
Schwertmann, 1985). Accuracy and precision are also limited by the insensitive nature of this

Table 14-4. Diffraction data for Al-rich allophane and imogolite. Many of the most intense peaks are very broad (vb) due to the poorly crystalline nature of allophane and imogolite.

Al-rich allophane		Imogolite	
d-spacing	intensity	d-spacing	intensity
nm	relative %	nm	relative %
1.2 vb	70	1.6 vb	100
0.34 vb	100	0.79	70
0.22 vb	50	0.56	35
0.14	20	0.37	20
		0.33 vb	65
		0.225 vb	25

method. At present, DXRD appears most useful as a qualitative technique for evaluating the effectiveness of various selective-dissolution reagents.

Nuclear Magnetic Resonance Spectroscopy

High-resolution, solid-state ^{29}Si and ^{27}Al NMR spectroscopy has been used to determine the coordination of Si and Al in allophane and imogolite (Barron et al., 1982; Goodman et al., 1985; Shimizu et al., 1988; MacKenzie et al., 1989). It has been found that isotropic ^{29}Si shifts in silicates depend primarily on the degree of condensation of Si-O tetrahedra. The observed chemical shift of imogolite (-78 ppm) is consistent with the proposal that the Si tetrahedra are isolated by coordination through O with three Al atoms and one proton. Due to the varying degree of Si tetrahedra condensation in allophanes, they show complicated ^{29}Si-NMR spectra with several peaks between -79 and -97 ppm; these spectra indicate that allophanes contain Si atoms in the orthosilicate structure as well as several different states of polymerization.

Quantification of imogolite may be possible based on the -78 ppm chemical shift for imogolite. Barron et al. (1982) were able to quantify imogolite in clay fractions when it was present in significant amounts using dipolar decoupling, magic-angle spinning and cross-polarization ^{29}Si-NMR techniques, although they did not report detection limits for the method. Possible interferences from allophanes, and from amorphous silicates and aluminosilicates, must be considered in the development of quantitative techniques. Use of NMR spectroscopy for imogolite quantification may become more practical as technological advances are made in the field.

Transmission Electron Microscopy

Transmission electron microscopy (TEM) is an excellent tool for examining the morphology of allophane and imogolite (Wada et al., 1978; Gilkes, 1994; see Chapter 6). It is also useful for verifying the effectiveness of various chemical reagents for selective dissolution of allophane and imogolite. For observation, C-coated collodion films supported on Cu grids are placed in a dilute clay suspension. The grids are air dried and observed at a magnification of 10 000× to 20 000× with a beam current of 75 to 100 μA and an accelerating voltage of approximately 80 kV.

Routine quantification of allophane and imogolite by TEM is not practical. However, TEM is well suited to the detection of trace quantities of allophane and imogolite which cannot be measured by other methods. The technique relies on its ability to distinguish the thread-like imogolite structure from the hollow-spherule morphology characteristic of the allophanes. The use of TEM for verifying the effectiveness of selective-dissolution studies is another important application (Fig. 14-4).

APPLICATION TO SOIL MINERALOGY

Sample Handling and Pretreatments

Selective-dissolution techniques can be applied to the <2-mm fraction or to the clay-size fraction (<2-μm) of soils, whereas IR spectroscopy and DTA are generally applied only to the clay-size fraction. To minimize the potential detrimental effects of sample pretreatment, it is perhaps best to perform selective-dissolution techniques on untreated, field-moist samples. Comfort et al. (1991) showed that air drying significantly decreases the amount of Al extracted by a number of commonly used reagents when compared to fresh field-moist soils and to field-moist soils that had been refrigerated at 4 °C or frozen at -5 °C before analysis. This study along with previous studies confirmed that air drying alters the surface chemistry of short-range-order materials; therefore, extractions are best performed on field-moist soils that have been refrigerated (Bartlett & James, 1980; Haynes & Swift, 1985). Whether to grind soils to a finer particle size prior to selective-dissolution procedures is a debatable question. Grinding soils has been shown to destroy the structures of imogolite, kaolinite, and halloysite (Henmi & Yoshinaga, 1981) and to expose new surfaces which are more prone to dissolution by chemical treatments (Wang et al., 1987). However, a compromise must be struck between the risk of altering the mineralogical composition of the sample by grinding and the need to use material with a small enough particle size to ensure a reasonably representative sample.

If the clay-size fraction is to be isolated, the soil samples should be refrigerated at 3 °C at field moisture content to prevent irreversible surface alterations which may prevent effective dispersion of the clays. It is very important to keep the amount of chemical pretreatment to a minimum. For quantification, complete dispersion and isolation of the clay fraction is critical. It is often necessary to remove organic matter to aid the dispersion process. Treatment with H_2O_2, NaOBr, or NaOCl may be used to remove organic matter, although H_2O_2 has been shown to generate oxalic acid that may dissolve allophane and imogolite (Anderson et al., 1982). Samples are then washed with deionized water using a centrifuge to remove soluble salts. To assist in dispersion, the samples can be sonicated at approximately 100 W for 5 min. If the samples do not completely disperse, addition of HCl or NaOH to bring the suspensions to pH 4 or pH 10 may promote the dispersion of variable-charge minerals. Size fractions may be obtained by sedimentation or centrifugation. Allophane and imogolite are concentrated in the <0.5-μm size fraction, but may be present in appreciable quantities in the 0.5- to 2-μm size fraction (Farmer et al., 1980). Following isolation of the clay fraction, the clay should be desalted using dialysis or ethanol-acetone rinsing. Freeze drying the clay suspension produces a fine powder that is easy to work with and requires a minimum of grinding prior to analysis.

Standard Reference Materials

The limited availability of appropriate standard reference materials poses a major barrier to the advancement of quantification techniques for allophane and imogolite. The problem is less severe for imogolite because the range in chemical composition and structural order is relatively narrow regardless of source. Most of the research to date has used imogolite isolated as gel films from pumice beds (e.g., the Kitakami Pumice Bed, Japan). Synthetic imogolite may also be appropriate as a standard for IR spectroscopy and DTA. While synthetic and natural imogolite are not identical, they do produce very similar IR spectra (Farmer & Fraser, 1979; Farmer et al.,

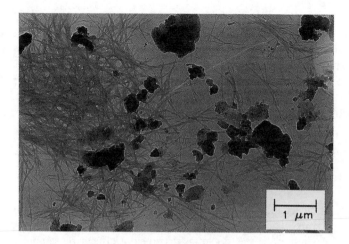

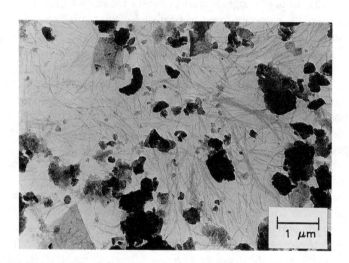

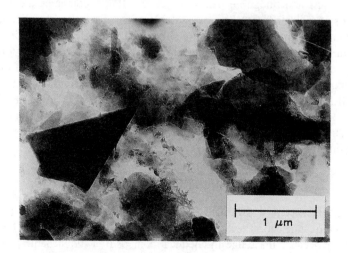

Fig. 14-4.   Transmission electron micrographs of (*a*) untreated sample, (*b*) DCB-treated sample, and (*c*) acid-oxalate treated sample.   Microscopy can be used to verify the effectiveness of selective-dissolution techniques for dissolving allophane and imogolite from soil material.

1979). Naturally occurring allophanes have a much wider range in chemical composition and structure, resulting in variable IR-absorption properties. Synthetic allophanes with variable compositions and structure may serve as appropriate standards; however, little research has addressed this issue (Farmer et al., 1979).

Example from Comparative Studies

Data from four studies comparing allophane and imogolite concentrations in the clay-size fraction determined by both IR spectroscopy and acid-oxalate extractable Si are shown in Fig. 14-5 (Russell et al., 1981; Parfitt & Henmi, 1982; Childs et al., 1983; Dahlgren & Ugolini, 1991, unpublished IR data). In general, the two methods are very consistent (slope = 0.98) with a maximum scatter of approximately 10% between methods. Both quantification methods worked well at low and high concentrations. However, the detection limit for the IR method is approximately 50 g kg^{-1} whereas that of the acid-oxalate method is on the order of 10 g kg^{-1} or better.

Additional Topics

Crystalline Water

Allophane and imogolite have a high total water content due to their large surface areas and high proportion of reactive sites. Thermal analysis shows an endothermic weight loss of about 25% associated with waters of hydration (<200 °C) (MacKenzie et al., 1989). Henmi and Yoshinaga (1983) have shown that the $H_2O(+)$ content of imogolite is correctly estimated as the difference in its weight at 140 and 900 °C. For imogolite with a composition $SiO_2 \cdot Al_2O_3 \cdot 2H_2O(+)$, this corresponds to a $H_2O(+)$ content of 180 g kg^{-1}. Results of other

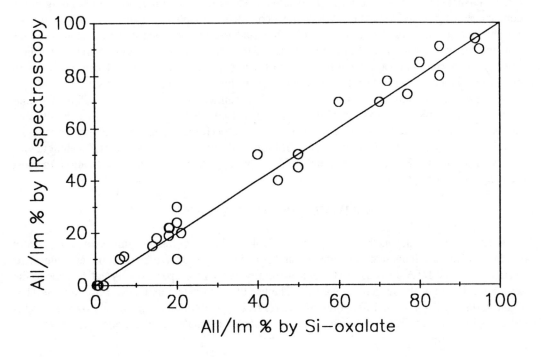

Fig. 14-5. Comparison of allophane and imogolite concentrations quantified by oxalate-extractable Si and by IR spectroscopy using absorption at 348 cm^{-1}.

studies show similar values for imogolite (Parfitt & Hemni, 1980; Parfitt et al., 1980). In contrast, the $H_2O(+)$ concentrations reported for allophane show a much wider range, i.e., from 140 to 290 g kg^{-1} (Parfitt Hemni, 1980; Parfitt et al., 1980; Childs et al., 1990).

Fluoride Reaction

Specific adsorption of fluoride occurs rapidly on allophane and more slowly on imogolite because it is more ordered than allophane. Because ligand-exchange reactions result in a release of OH$^-$, an increase in solution pH to values greater than 9.0 is observed when NaF reacts with allophanic materials. Fieldes and Perrott (1966) proposed this reaction as the basis for a rapid field test for allophane. The test is acceptable as a qualitative tool in most cases, but it is important to keep in mind that the reagent will react with any available Al associated with hydroxyls and is not specific to allophane and imogolite. For example, F$^-$ will also react with poorly crystalline Fe compounds to release OH$^-$ and thus yield a high pH. In addition, Shoji and Ono (1978) have examined several soils that did not contain allophane and imogolite but gave high NaF pH values due mainly to Al associated with humus. Therefore, the usefulness of NaF pH as an indicator for the presence of allophane and imogolite is limited, even as a qualitative tool.

FUTURE DEVELOPMENTS

Future improvements in allophane/imogolite quantification require the establishment of standard reference materials for these phases having various chemical compositions and degrees of structural order. Characterization of the standard reference samples will provide data necessary to determine the quality of existing techniques. Following establishment of natural reference materials, a critical evaluation of synthetic allophane and imogolite materials should be undertaken.

Advances in surface-spectroscopy techniques will serve to better characterize the possible continuum between Al-rich and Si-rich allophanes. Elemental analysis and elemental profiling of individual allophane spherules would provide valuable information concerning the microvariability among allophanes. Continuing technological advances in the field of NMR spectroscopy may lead to its wide-spread use for imogolite quantification. Due to the wide range of Si-O bonding in the allophane series, it is unlikely that NMR spectroscopy will be applied to allophane. Evaluation of new chemical extractions for selectively dissolving allophane and imogolite may lead to better estimates for these materials. At the present time, the acid-oxalate extraction provides the best results with which to compare the effectiveness of new chemical extractants. Recent advances in image analysis may provide a practical means of quantifying electron-microscopy data.

As new techniques and improvements in existing technologies materialize, it will be essential to perform comparisons between methods. If simple techniques, such as selective dissolution, provide comparable results, it will not be necessary to perform expensive and time-consuming analysis for allophane and imogolite quantification. It appears that the existing techniques of selective dissolution, IR spectroscopy, and DTA analysis provide adequate quantification for most research purposes at present.

SUMMARY AND CONCLUSIONS

While no individual method currently available is ideal for quantifying short-range-order aluminosilicates, semiquantitative estimates can be made using selective-dissolution techniques, IR spectroscopy, and DTA analysis. Selective-dissolution results are operationally defined due to the lack of specificity of the reagents involved. The use of acid-oxalate-extractable Si with the Al/Si atomic ratio of allophane and imogolite as determined by the combination of acid-oxalate results and pyrophosphate or DCB results do yield in reasonable estimates of allophane and imogolite concentrations. These methods have a low detection limit (approximately 10 g kg^{-1}) and the results are in good agreement with those obtained by IR spectroscopy.

The IR spectroscopy is an effective method of quantification when appropriate standards are available and corrections can be made for interferences from aluminous dioctahedral layer silicates. The IR methods have a detection limit of approximately 50 g kg^{-1} that varies somewhat depending on the specific absorbance of the allophane and imogolite phase found in the samples. Differences in absorbance between allophane and imogolite, and within the range of known allophanes, makes accurate quantification of mixtures of these phases difficult. Interferences from other mineral constituents can be eliminated by differential IR, which compares spectra taken before and after treatment of the sample by acid-oxalate or heating to destroy allophane and imogolite. An appropriate standard material must be employed to generate a calibration curve for the method.

The DTA provides a good estimate of imogolite concentrations, but is not directly applicable to allophane. Availability of an appropriate imogolite standard is necessary for development of a calibration curve. Detection limits of approximately 20 g kg^{-1} have been obtained. In samples that contain both imogolite and allophane, allophane may be estimated by subtracting the imogolite concentration obtained by DTA from the total concentration determined by selective dissolution.

Although present quantification techniques are not perfect, they do provide adequate accuracy for most soil research purposes. It must also be stressed that a subsample of the soil should be analyzed by an independent technique (e.g., TEM) to ensure the applicability of the primary technique for allophane and imogolite quantification. The lack of commercially available standard reference materials is a major limitation to current quantification techniques. Establishment of reference standards will provide materials of known composition and structure which can be used for calibration standards and development of new techniques.

ACKNOWLEDGMENTS

I am very grateful to J.L. Boettinger and anonymous reviewers for their many helpful comments. I am also indebted to all those who have contributed to the study of allophane and imogolite.

REFERENCES

Alexsandrova, L.N. 1960. The use of pyrophosphate for isolating free humic substances and their organic-mineral compounds from the soil. Sov. Soil Sci. 1960:190-197.

Anderson, H.A., M.L. Berrow, V.C. Farmer, A. Hepburn, J.D. Russell, and A.D. Walker. 1982. A reassessment of podzol formation processes. J. Soil Sci. 33:125-136.

Aomine, S. and C. Mizota. 1973. Distribution and genesis of imogolite in volcanic ash soils of northern Kanto, Japan. p. 207-231. In J.M. Serratosa (ed.) Proc. Int. Clay Conf., Madrid, Spain, 1972, Div. Ciencias C. S. I. C., Madrid.

Bhattacharyya, T. and S.K. Ghosh. 1986. Comparison of some extractants for quantitative estimation of allophane content in soils. J. Indian Soc. Soil Sci. 34:387-392.

Barron, P.F., M.A. Wilson, A.S. Campbell and R.L. Frost. 1982. Detection of imogolite in soils using solid state ^{29} Si NMR. Nature (London) 299:616-618.

Bartlett, R. and B. James. 1980. Studying dried, stored soil samples--some pitfalls. Soil Sci. Soc. Am. J. 44:721-724.

Bascomb, C.L. 1968. Distribution of pyrophosphate-extractable iron and organic carbon in soils of various groups. J. Soil Sci. 19:251-268.

Bryant, R.B., N. Curi, C.B. Roth, and D.P. Franzmeier. 1983. Use of an internal standard with differential x-ray diffraction analysis of iron oxides. Soil Sci. Soc. Am. J. 47:168-173.

Campbell, A.S. and U. Schwertmann. 1985. Evaluation of selective dissolution extractants in soil chemistry and mineralogy by differential x-ray diffraction. Clay Miner. 20:515-519.

Carlson, L. and U. Schwertmann. 1981. Natural ferrihydrites in surface deposits from Finland and their association with silica. Geochim. Cosmochim. Acta 45:421-429.

Childs, C.W., R.L. Parfitt and R. Lee. 1983. Movement of aluminum as an inorganic complex in some podzolised soils, New Zealand. Geoderma 29:39-155.

Childs, C.W., R.L. Parfitt and R.H. Newman. 1990. Structural studies of silica springs allophane. Clay Miner. 25:329-341.

Comfort, S.D., R.P. Dick, and J. Baham. 1991. Air-drying and pretreatment effects on soil sulfate sorption. Soil Sci. Soc. Am. J. 55:968-973.

Cradwick, P.D.G., V.C. Farmer, J.D. Russell, C.R. Masson, K. Wada, and N. Yoshinaga. 1972. Imogolite, a hydrated aluminium silicate of tubular structure. Nature (London) 240:187-189.

Dahlgren, R.A. and F.C. Ugolini. 1989. Formation and stability of imogolite in a tephritic Spodosol, Cascade Range, Washington, U.S.A. Geochim. Cosmochim. Acta 53:1897-1904.

Dahlgren, R.A. and F.C. Ugolini. 1991. Distribution and characterization of short-range-order minerals in Spodosols from the Washington Cascades. Geoderma 48:391-413.

Farmer, V.C. 1968. Infrared spectroscopy in clay mineral studies. Clay Miner. 7:373-387.

Farmer, V.C. (ed.). 1974. The infrared spectra of minerals. Miner. Soc., London.

Farmer, V.C. and J.D. Russell. 1967. Infrared absorption spectrometry in clay studies. Clays Clay Miner. 15:121-142.

Farmer, V.C. and A.R. Fraser. 1979. Synthetic imogolite, a tubular hydroxyaluminium silicate. p. 547-553. In M.M. Mortland and V.C. Farmer (ed.) Proc. Int. Clay Conf., Oxford, England, 1978. Elsevier, Amsterdam.

Farmer, V.C., J.D. Russell, and J.L. Ahlrichs. 1968. Characterization of clay minerals by infrared spectroscopy. p. 101-110. In Trans. Int. Congr. Soil Sci., 9, 3, Adelaide, Australia..

Farmer, V.C., A.R. Fraser and J.M. Tait. 1979. Characterization of chemical structures of natural and synthetic aluminosilicate gels and sols by infrared spectroscopy. Geochim. Cosmochim. Acta 43:1417-1420.

Farmer, V.C., J.D. Russell and M.L. Berrow. 1980. Imogolite and proto-imogolite allophane in spodic horizons: Evidence for a mobile aluminium silicate complex in podzol formation. J. Soil Sci. 31:673-684.

Farmer, V.C., J.D. Russell and B.F.L. Smith. 1983. Extraction of inorganic forms of translocated Al, Fe and Si in a podzol Bs horizon. J. Soil Sci. 34:571-576.

Farmer, V.C., A.R. Fraser, J.D. Russell and N. Yoshinaga. 1977. Recognition of imogolite structures in allophanic clays by infrared spectroscopy. Clay Miner. 12:55-57.

Fieldes, M. and K.W. Perrott. 1966. The nature of allophane in soils. III. Rapid field and laboratory test for allophane. N. Z. J. Sci. 9:623-629.

Fordham, A.W. and K. Norrish. 1983. The nature of soil particles particularly those reacting with arsenate in a series of chemically treated samples. Aust. J. Soil Res. 21:455-477.

Gilkes, R.J. 1994. Transmission electron microscope analysis of soil materials. p. 177-204. In J. Amonette and L. W. Zelazny (ed.) Quantitative methods in soil mineralogy. SSSA Misc. Publ. SSSA, Madison, WI.

Goodman, B.A. 1987. The characterisation of iron complexes with soil organic matter. p. 677-687. In J.W. Stucki et al. (ed.) Iron in soils and clay minerals. D. Reidel Publ., Dordrecht, the Netherlands.

Goodman, B.A., J.D. Russell, B. Montez, E. Oldfield, and R.J. Kirkpatrick. 1985. Structural studies of imogolite and allophanes by aluminium-27 and silicon-29 nuclear magnetic resonance spectroscopy. Phys. Chem. Miner. 12:342-346.

Hadni, A. 1967. Essentials of modern physics applied to the study of the infrared. Pergamon, Oxford, England.

Haynes, R.J. and R.S. Swift. 1985. Effects of air-drying on the adsorption and desorption of phosphate and levels of extractable phosphate in a group of acid soils, New Zealand. Geoderma 35:145-157.

Henmi, T. and K. Wada. 1976. Morphology and composition of allophane. Am. Miner. 61:379-390.

Henmi, T. and N. Yoshinaga. 1981. Alteration of imogolite by dry grinding. Clay Miner. 16:139-149.

Henmi, T. and N. Yoshinaga. 1983. The content of structural OH in imogolite and allophane. p. 29. In 1983 Meet. Clay Sci. Soc. Jpn., abstracts.

Henmi, T., N. Wells, C.W. Childs and R.L. Parfitt. 1980. Poorly-ordered iron-rich precipitates from springs and streams on andesitic volcanoes. Geochim. Cosmochim. Acta 44:365-372.

Higashi, T. 1983. Characterisation of Al/Fe-humus complexes in Dystrandepts through comparison with synthetic forms. Geoderma 31:277-288.

Higashi, T. and H. Ikeda. 1974. Dissolution of allophane by acid oxalate solution. Clay Sci. 4:205-211.

Higashi, T. and A. Shinagawa. 1981. Comparison of sodium hydroxide-tetraborate and sodium pyrophosphate as extractants of Al/Fe-humus "complexes" in Dystrandepts, Japan. Geoderma 25:285-292.

Higashi, T., F. De Coninck and F. Gelaude. 1981. Characterization of some spodic horizons of the Campine (Belgium) with dithionite-citrate, pyrophosphate and sodium hydroxide-tetraborate. Geoderma 25:131-142.

Iyengar, S.S., L.W. Zelazny, and D.C. Martens. 1981. Effect of photolytic oxalate treatment on soil hydroxy-interlayered vermiculites. Clays Clay Miner. 29:429-434.

Jackson, M.L., C.H. Lim and L.W. Zelazny. 1986. Oxides, hydroxides, and aluminosilicates. p. 101-150. In A. Klute (ed.) Methods of soil analysis. Part 1. 2nd ed. Agron. Monogr. 9. ASA and SSSA, Madison, WI.

Karathanasis, A.D., and Harris, W. G. 1994. Quantitative thermal analysis of soil materials. p. 360-411. *In* J. Amonette and L.W. Zelazny (ed.) Quantitative methods in soil mineralogy. SSSA Misc. Publ. SSSA, Madison, WI.

Kassim, J.K., S.N. Gafoor, and W.A. Adams. 1984. Ferrihydrite in pyrophosphate extracts of podzol B horizons. Clay Miner. 19:99-106.

Lee, R., M.D. Taylor, B.K. Daly and J. Reynolds. 1989. The extraction of Al, Fe and Si from a range of New Zealand soils by hydroxylamine and ammonium oxalate solutions. Aust. J. Soil Res. 27:377-388.

Lombardi, G. 1980. For better thermal analysis. Int. Conf. Thermal Analysis and Univ. of Rome, Italy.

Mackenzie, R.C. 1970. Differential thermal analysis. Acad. Press, London.

MacKenzie, K.J.D., M.E. Bowden, I.W.M. Brown, and R.H. Meinhold. 1989. Structure and thermal transformations of imogolite studied by ^{29}Si and ^{27}Al high-resolution solid-state nuclear magnetic resonance. Clays Clay Miner. 37:317-324.

Madeira, M.A.V. and E. Jeanroy. 1984. Mise en evidence de goethite en suspension dans les extraits pyrophosphate et tetraborate de certain sols greseau du Portugal. Can. J. Soil Sci. 64:505-514.

McKeague, J.A. 1967. An evaluation of $0.1M$ pyrophosphate and pyrophosphate-dithionite in comparison with oxalate as extractants of the accumulation products in podzols and some other soils. Can. J. Soil Sci. 47:95-99.

McKeague, J.A. and B.H. Sheldrick. 1977. Sodium hydroxide-tetraborate in comparison with sodium pyrophosphate as an extractant of 'complexes' characteristic of spodic horizons. Geoderma 19:97-104.

McKeague, J.A. and P.A. Schuppli. 1982. Changes in concentration of iron and aluminum in pyrophosphate extracts of soil and composition of sediment resulting from ultracentrifugation in relation to spodic horizon criteria. Soil Sci. 134:265-270.

McKeague, J.A. and P.A. Schuppli. 1985. An assessment of EDTA as an extractant of organic-complexed and amorphous forms of Fe and Al in soils. Geoderma 35:109-118.

McLean, E.O. 1976. Chemistry of soil aluminum. Commun. Soil Sci. Plant Anal. 7:619-636.

Mehra, O.P. and M.L. Jackson. 1960. Iron oxide removal from soils and clays by a dithionite-citrate system buffered with sodium bicarbonate. Clays Clay Miner. 7:713-727.

Parfitt, R.L. 1989. Optimum conditions for extraction of Al, Fe, and Si from soils with acid oxalate. Commun. Soil Sci. Plant Anal. 20:801-816.

Parfitt, R.L. 1990a. Allophane in New Zealand - a review. Aust. J. Soil Res. 28:343-360.

Parfitt, R.L. 1990b. Estimation of imogolite in soils and clays by DTA. Commun. Soil Sci. Plant Anal. 21:623-628.

Parfitt, R.L. and T. Henmi. 1980. Structure of some allophanes from New Zealand. Clays Clay Miner. 28:285-294.

Parfitt, R.L. and T. Henmi. 1982. Comparison of an oxalate-extraction method and an infrared spectroscopic method for determining allophane in soil clays. Soil Sci. Plant Nutr. 28:183-190.

Parfitt, R.L. and A.D. Wilson. 1985. Estimation of allophane and halloysite in three sequences of volcanic soils, New Zealand. Catena Suppl. 7:1-8.

Parfitt, R.L. and C.W. Childs. 1988. Estimation of forms of Fe and Al: A review, and analysis of contrasting soils by dissolution and Moessbauer methods. Aust. J. Soil Res. 26:121-144.

Parfitt, R.L., R.J. Furkert and T. Henmi. 1980. Identification and structure of two types of allophane from volcanic ash soils and tephra. Clays Clay Miner. 28:328-334.

Patterson, E. and R. Swaffield. 1987. Thermal analysis. p. 99-132. *In* M.J. Wilson (ed.) A handbook of determinative methods in clay mineralogy. Blackie, Glasgow and London.

Percival, H.J. 1985. Soil solutions, minerals and equilibria. N.Z. Soil Bur. Sci. Rep. 69.

Russell, J.D. 1987. Infrared methods. p. 133-173. *In* M.J. Wilson (ed.) A handbook of determinative methods in clay mineralogy. Blackie, Glasgow and London.

Russell, J.D., W.J. McHardy, and A.R. Fraser. 1969. Imogolite: A unique aluminosilicate. Clay Miner. 8:87-99.

Russell, M., R.L. Parfitt, and G.G.C. Claridge. 1981. Estimation of the amounts of allophane and other materials in the clay fraction of an Egmont loam profile and other volcanic ash soils, New Zealand. Aust. J. Soil Res. 19:185-195.

Schulze, D.G. 1994. Differential x-ray diffraction analysis of soil minerals. p. 412-429. *In* J. Amonette and L.W. Zelazny (ed.) Quantitative methods in soil mineralogy. SSSA Misc. Publ. SSSA, Madison, WI.

Schwertmann, U. 1959. Experimentelle Untersuchungen uber die sedimentare Bildung von Goethit und Hämatit. Chem. Erde. 20:104-135.

Schwertmann, U. 1964. Differenzierung der Eisenoxide des Bodens durch Extraktion mit saurer Ammoniumoxalat-Lösung. Z. Pflanzenernähr. Bodenkd. 105:194-202.

Schwertmann, U. 1985. The effect of pedogenic environments on iron oxide minerals. Adv. Soil Sci. 1:171-200.

Schwertmann, U., and R. M. Taylor. 1977. Iron oxides. p. 145-180. *In* J.B. Dixon and S.B. Weed (ed.) Minerals in soil environments. SSSA, Madison, WI.

Schwertmann, R, D.G. Schulze and E. Murad. 1982. Identification of ferrihydrite in soils by dissolution kinetics, differential x-ray diffraction and Moessbauer spectroscopy. Soil Sci. Soc. Am. J. 46:869-875.

Searle, P.L. and B.K. Daly. 1977. The determination of aluminum, iron, manganese and silicon in acid oxalate soil extracts by flame emission and atomic absorption spectrometry. Geoderma 19:1-10.

Shimizu, H., T. Watanabe, T. Henmi, A. Masuda, and H. Saito. 1988. Studies on allophane and imogolite by high-resolution solid-state ^{29} Si- and ^{27} Al-NMR and ESR. Geochem. J. 22:23-31.

Shoji, S. and M. Saigusa. 1977. Amorphous clay materials of Towada Ando soils. Soil Sci. Plant Nutr. 23:437-455.

Shoji, S. and T. Ono. 1978. Physical and chemical properties and clay mineralogy of Andosols from Kitakami, Japan. Soil Sci. 126:297-312.

Shoji, S. and Y. Fujiwara. 1984. Active aluminum and iron in the humus horizons of Andosols from northeastern Japan: Their forms, properties, and significance in clay weathering. Soil Sci. 137:216-226.

Soil Survey Staff. 1990. Keys to soil taxonomy. 4th ed. SMSS Tech. Monogr. no. 19. Blacksburg, VA.

Soltanpour, P.N., J. Benton Jones, Jr., and S.M. Workman. 1982. Optical emission spectrometry. p. 29-65. *In* A.L. Page et al. (ed.) Methods of soil analysis, Part 2. 2nd ed. Agron. Monogr. 9. ASA and SSSA, Madison, WI.

Steven, K.F. and G.C. Vucetich. 1985. Weathering of Upper Quaternary tephras in New Zealand. Part II. Clay minerals. Chem. Geol. 53:237-247.

Tamm, O. 1922. Eine Methode zur Bestimmung de anorganischen Komponente des Gelkomplexes in Boden. Madd. Fran Stat. Skogs. Stockholm. 19:387-404.

Theng, B.K.G., M. Russell, G.J. Churchman and R.L. Parfitt. 1982. Surface properties of allophane, halloysite, and imogolite. Clays Clay Miner. 30:143-149.

Turrell, G. 1972. Infrared and raman spectra of crystals. Acad. Press, London.

Wada, K. 1989. Allophane and imogolite. p. 1051-1087. *In* J.B. Dixon and S.B. Weed (ed.) Minerals in soil environments, 2nd ed. SSSA, Madison, WI.

Wada, K. and N. Yoshinaga. 1968. The structure of imogolite. Am. Miner. 54:50-71.

Wada, K. and D.J. Greenland. 1970. Selective dissolution and differential infrared spectroscopy for characterization of "amorphous" constituents in soil clays. Clay Miner. 8:241-254.

Wada, K. and T. Higashi. 1976. The categories of aluminum- and iron-humus complexes in Ando soils determined by selective dissolution. J. Soil Sci. 27:357-368.

Wada, K., N. Yoshinaga, H. Yotsumoto, K. Ibe, and S. Aida. 1970. High resolution electron micrographs of imogolite. Clay Miner. 8:487-489.

Wada, S.-I., and K. Wada. 1980. Formation, composition and structure of hydroxy-aluminosilicate ions. J. Soil Sci. 31:457-467.

Wang, C., P.A. Schuppli and G.J. Ross. 1987. A comparison of hydroxylamine and ammonium oxalate solutions as extractants for Al, Fe and Si from Spodosols and Spodosol-like soils in Canada. Geoderma 40:345-355.

White, J.L. 1971. Interpretation of infrared spectra of soil minerals. Soil Sci. 112:22-31.

White, J.L. 1977. Preparation of specimens for infrared analysis. p. 847-863. *In* J.B. Dixon and S.B. Weed (ed.) Minerals in soil environments. SSSA, Madison, WI.

White, J.L. and C.B. Roth. 1986. Infrared spectrometry. p. 291-330. *In* A. Klute (ed.) Methods of soil analysis, Part 1. 2nd ed. Agron. Monogr. 9. ASA and SSSA, Madison, WI.

Yamada, I., M. Saigusa, and S. Shoji. 1978. Clay mineralogy of Hijiori and Numazawa Ando soils. Soil Sci. Plant Nutr. 24:75-89.

Yoshinaga, N. and S. Aomine. 1962. Imogolite in some Ando soils. Soil Sci. Plant Nutr. 8:22-29.

Yoshinaga, N. and M. Yamaguchi. 1970. Occurrence of imogolite as gel film in the pumice and scoria beds of western and central Honshu and Hokkaido. Soil Sci. Plant Nutr. (Tokyo) 16:215-223.

SUBJECT INDEX